I0891039

Histories of Maize In Mesoamerica

This book is dedicated to
Richard S. MacNeish and Kent V. Flannery
for their pioneering research on the origins of maize
and early agriculture in Mesoamerica

Histories of Maize in Mesoamerica

Multidisciplinary Approaches

John E. Staller
Robert H. Tykot
Bruce F. Benz

EDITORS

Routledge
Taylor & Francis Group
LONDON AND NEW YORK

First published 2010 by Left Coast Press, Inc.

Published 2016 by Routledge
2 Park Square, Milton Park, Abingdon, Oxon OX14 4RN
711 Third Avenue, New York, NY 10017, USA

Routledge is an imprint of the Taylor & Francis Group, an informa business

This volume is an abridgement of Staller, Tykot, and Benz Histories of Maize, ISBN 978-1-59874-462-0, originally published by Academic Press under ISBN 978-0-12-369364-8 in 2006.

Library of Congress Cataloging-in-Publication Data available from the publisher

ISBN 978-1-59874-496-5 paperback

Contents

PART IV

Histories of Maize: The Language of Maize in Mesoamerica

Introduction to the Histories of Maize in Mesoamerica

JOHN E. STALLER
The Field Museum

The goal of the editors of this volume on maize science was to bring together contributions that would individually incorporate and collectively assemble a comprehensive multidisciplinary set of data that developed particular lines or types of evidence from specific time periods (and regions) throughout the pre-Columbian geographic range of maize cultivation in Mesoamerica[1]. Another primary goal in organizing this volume was to be holistic, and include research from the social, biological and earth sciences. The abridged volume on Mesoamerica is organized into four sections dealing with different aspects and regions of research on maize. The scope and breadth of the research takes into account recent methodological and technological innovations from the physical, biological and social sciences. These recently developed technical and methodological approaches provide ever-increasing detail and direct evidence on the antiquity, evolution, cultural importance and role of maize in ancient Mesoamerica. We believe that more recent methodologies and approaches such as isotope analysis regarding paleodiet and DNA analysis have essentially transformed our understanding of the origins, economic roles, and importance of maize as well as other domesticates to sociocultural developments in prehistory. My colleagues Robert H. Tykot, Bruce F. Benz and I hope that the readers of this volume agree that the research presented herein have established this to be the case.

One of our two European contributors observed that such a book could never have been realized had it been organized and published outside of North America (see Chapter 1). Rather it would have been broken up into several books specialized on the respective scientific discipline and specialization concerned. These volumes would have presumably included research that was specifically geared to the interested specialists in those fields. Archaeological research on the domestication of grains in the Old World have generally developed within competing models that consider the spread of agriculture through 'acculturation' or 'waves of advance,' while in the Americas they have been predominantly couched within foraging/farming dichotomies that are specific and distinct to different regions of the hemisphere and their associated time periods [see 4, 9, versus 5, 6, 12, 19, 23].

Although the Old World approaches on the spread of agriculture and domestication lend themselves well to models used or tested by human geneticists and linguists, they have not worked as well for archaeologists and ethnobotanists in the Americas. In the Old World, emphasis has been placed upon initial causes or events, while in the Americas, archaeologists and ethnobotanists have mainly focused upon earliest presence, as well as the developmental and/or evolutionary processes associated with plant domestication and maize agriculture [3. 10, 13]. The Old World emphasis on migration and diffusion of plant domestication also takes away from the general focus upon the distinctions that important cultigens had to different regions and time-periods, while in the Americas such concerns have been clearly evident in methodological approaches to understanding the archaeological record (1, 3, see e.g., Chapters 1, 7, and 10). Despite differences in theoretical and methodological approaches to plant domestication in general and economic plants (mainly grains) in particular, the assumption that maize, like wheat and barley in the Old World, provided the economic basis for the development of civilization has been a central thesis among archaeologists in both hemispheres as well as Latin America [8, 13]. Remarkably, many of the contributions in this volume challenge those basic assumptions.

Although the chapters in this volume appear to support the contention that maize was a major economic staple, some contributions indicate that when and why this occurred is dramatically different than had been previously suggested in the literature [20]. Other contributors present evidence to suggest that the way maize affected sociocultural processes is in fact far more complex and varied than had been originally assumed. One of the primary themes that run through many of

[1] The creation of the original 2006 volume *Histories of Maize: Multidisciplinary Approaches to the Prehistory, Linguistics, Biogeography, Domestication, and Evolution of Maize* was inspired in part through collaboration with several colleagues, T. Michael Blake, John P. Hart, and Robert G. Thompson, while putting together a four-part symposium on maize titled, *The Stories of Maize* that was presented at the 69th Annual Meeting of the Society for American Archaeology, March 31st to April 4th, Montreal, Quebec, Canada.

these contributions, particularly the paleodietary evidence in Part III is that maize was not initially the important economic food source that many archaeologists assumed (see e.g., Chapters 2, 4, 7, 12, and 14). In some cases, it never achieved economic importance in certain regions, although it did seem to play an important role in other aspects of sociocultural development, this is particularly the case for some regions of South America. Another important divergence from previous assumptions brought out by this volume, is that maize was only domesticated once. Its ancestor, teosinte (*Zea mays* ssp. *parviglumis*), was domesticated in the Balsas River drainage of central Mexico (14; Chapter 2, Benz and Staller, this volume). This differs from previous hypotheses regarding a tripartite origin of maize promoted by Paul C. Mangelsdorf [13] and others and by extension the possibility of multiple domestication events in different regions of the Americas [see also 17, 8, 11, 15]. These data will have a profound impact on our understanding of the phylogeny of maize landraces and their spread into different regions of the Americas [1, 7, 14].

Since the publication of the DNA microsatellite data on extant populations of maize and teosinte by Matsuoka et al. in 2002 [see 14], research on ancient maize, and more specifically on its origins has been at a historical crossroads. These important data suggest that like most of the Old World staples, maize was only domesticated once, but rather than focus upon the migrations of farming populations, or the acceptance of maize agriculture in diverse regions, what the contributions in this volume suggest is that there will be an ever greater appreciation for research on maize from the social, and particularly the biological sciences. The botanical evidence has historically influenced archaeological interpretation but the recent evidence from molecular biology suggests that such data may now sets the limits for what is possible regarding the ancient origins and early spread of maize in the Americas [7, 14, 18, 19].

The holistic approach we have inherited from the founders of American Anthropology is largely responsible for the multidisciplinary organization of this volume [see e.g., 12, 19, 21, 22, 23, 16]. These luminaries provided American archaeology with the possibility that such multiple lines of evidence could ever have been brought together as a single reference source on maize.

Many recent advances to our knowledge come from new scientific techniques and approaches to the direct study of archaeological maize collections and the physical remains of the human populations who consumed it (see e.g., Chapters 12 and 13). The development of Accelerator Mass Spectrometry (AMS) radiocarbon dating has had a profound effect in our understanding of the chronological spread of maize in the Americas, and greatly revised our previous assumptions of its antiquity based upon indirect dating techniques [2, 3, 20, 18]. The recent techniques involving isotope analysis, including research on phytoliths, have provided detailed information on the antiquity and role of maize to ancient cultures throughout the Americas and how such evidence is identified when organic remains and other traces of such evidence are not preserved archaeologically (see e.g., Chapter 5). These state of the art scientific approaches and their associated methodologies stand in contrast to the more traditional forms of analyses such as historical linguistics, and archaeological analysis of stratigraphy and the classification and detailed study of artifacts.

The first part of this volume deals with the molecular, biological and microbotanical research, which has so greatly affected recent research on maize. This section also includes a detailed analysis of the chronology of its spread in the Americas (3; Chapter 4). Recently developed techniques in maize DNA research have also revised our earlier perceptions of the antiquity and spread of maize to different regions of the Americas and provided evidence for the previously unknown presence of undomesticated teosinte genes in some early landraces (Chapter 3).

Part II is organized by geographic region going from evidence for maize domestication in Chihuahua northern Mexico to its spread into other areas of the Mesoamerica. Considerations of the scientific, theoretical and methodological approach also influenced the organization of this volume. The topical divisions are in two parts: Part II: deals with the domestication and production of maize in Mexico and Central America, while Part III with the paleodietary results from the Maya regions and Soconusco. The chapter by Benz (Chapter 2) on maize in the Americas addresses some of these biological and chronological data and the underlying biases in previous research methodologies when maize was still believed to have multiple origins [see e.g., 13]. Some chapters present archaeological and ethnobotanic data derived from the latest advances in the study of maize origins, morphology and microfossil analysis — asking the question: what can such research on ancient maize tell us about the origin, history, and spread of this important cultigen? Contributors trace the early movements of maize into the American SW and provide innovative and original insights into its role in sociocultural development and adaptation (Chapter 6). Chapters on the physical characteristics of archaeological maize remains and microfossils also include an assessment of methodological approaches on microfossils in carbon residues that appear to provide greater detailed information on the identification and spread of ancient maize lineages (Chapter 2, see also, Chapters 5, 8, and 9). These chapters suggest that the future of maize research will be more heavily influenced by molecular biology, particularly the maize genome project, and botanical research on plant morphology, as scholars will attempt to quantify, identify and trace those genes, traits, and morphological characteristics related to human as opposed to natural selection (Chapter 7). Another important aspect of such future research indicated by these chapters is an increased focus upon maize landraces and their phylogenetic relationship(s) to one another and their cultural roles among indigenous and latino cultures in Mexico and Central America.

An economic staple throughout the Western Hemisphere at the time of European contact, the evolution and spread of maize (*Zea mays* L.) are topics of major archaeological research in the Americas for well over a century. Part III of the volume deals with the stable carbon isotope analysis and paleodiet and directly addresses these previous concerns with direct quantitative evidence of its economic importance. Researchers working in the Maya regions as well as the southern coast in Soconusco discuss the dietary, social, and economic implications of stable carbon isotope analyses from human skeletal remains (Chapters 12, 13, and 14). Research using strontium isotopes and elemental analyses of biochemistry involving human skeletons as well as plants and animals can now be used to determine whether people, plants, or animals were displaced or brought in from other areas or regions than where they were identified archaeologically. Strontium isotopic research is generating data that has facilitated our understanding of how maize was manipulated and used by ancient societies and challenges our previous assumptions of how maize was dispersed and its role in ancient economies.

In recent years, research using a variety of new methods and techniques in stable carbon isotope analysis has clarified and provided detailed data on the dietary importance of maize in distinct cultural settings and time periods (Chapter 11). Isotopic research on ancient human skeletons, particularly in the past decade, has greatly expanded our understanding of human adaptation, and in some cases, required maize specialists from the natural and social sciences to revise long held theories on the spread and impacts of maize to the development of sociocultural complexity. The summary by Henry Schwarcz (Chapter 15) represents one of the most comprehensive syntheses of these data in the published literature, as well as integrates the theoretical and methodological basis for such biochemical research (see also Chapter 11). Results from various contributions indicate that maize became a primary staple in this culture area much later than had been previously thought, and that its role in sociocultural development is much more complex and varied in some regions of Mesoamerica, particularly in the areas adjacent to where it is originally believed to have been domesticated (see e.g., Chapter 13). In some regions of Mesoamerica, maize became a food staple later than previously suggested in the prehistoric sequence, and in some regions its adoption and role in the ancient economy was highly varied although by the period of the conquest it represented the primary staple over most of Mesoamerica — in contrast to some other regions of the Americas. These data provide refreshing and informative insights into the spread and economic importance of maize, and in many ways challenge our previous assumptions of its importance and role in sociocultural development.

Many recent scientific advances in our knowledge surrounding the increasing dependence upon plant domestication and particularly the role of maize in ancient economies are explored in these chapters. Most of the research is archaeological and many contributions incorporate the most recent multidisciplinary evidence in order to build consensus on primary issues surrounding and origins and spread of maize that are based on internally consistent lines of evidence. The innovative and original approaches presented in this volume will provide a basis for future of multi and interdisciplinary research on this important New World cultigen.

Part IV represent a natural extension of the previous parts of the volume in its interdisciplinary breadth and scope, and is distinguished to some extent in that the research primarily concerns social sciences — the ethnohistory, archaeology, and contextual associations of ancient maize. Numerous ethnohistoric documents, language dictionaries and codices are examined to explore the social and symbolic significance of maize to sociocultural development. Ethnohistoric accounts generally emphasized maize as the preeminent grain of the pre-Hispanic New World, a plant that was critical to sociocultural developments in Mexico, Central and South America at the time of contact (Chapter 10). These accounts were largely biased by the importance of cereal grains in the Old World and have had lasting impacts archaeological assumptions regarding the economic role of maize in Native American economies (Chapter 1). Conquistadores, explorers, clerics, and historians attempted to categorize and classify the exotic plant and animal species of this hemisphere, into something, which was comprehensible and intelligible to their Old World patrons. Consequently, some species were misinterpreted or associated with species in Asia and the Old World. For example, maize was commonly referred to in several early pre-Linnaean botanicals as 'Turkish wheat' or 'Corne of Asia'.

The linguistic section, Part IV, reflects the holistic approach taken by the founders of early American anthropology and the four-field approach to anthropological science [23, 24; Chapter 1]. The chapters presented here are multidimensional in scope and comprehensive in the regions covered. Recent multidisciplinary lines of evidence have recorded the changing role of maize to sociocultural development in different chronological, geographic, and cultural settings. The ethnographic, linguistic, paleobotanic and archaeological evidence have generated ever more detailed evidence of complex sets of data regarding the phylogeny, chronology, evolution, and the sociocultural, and socioeconomic significance of this important New World cultigen. The different social and symbolic roles maize played are explored in diverse chronological and cultural settings (Chapters 2, 16, 18, and 19, Benz and Staller, this volume). Other chapters emphasize the significance of Native American terminology and classification of maize among different regions of Mesoamerica (Chapters 17, 19, and 20). Some contributors use historical linguistics such as glottochronology, or explore folk systems of classification, others use myth, song, iconography and ancient texts to explore the biogeography of this plant among the multi-branched language families of Mesoamerica: Mayan, Mixe-Zoquean, Oto-Manguean, and Uto-Aztecan (Chapters 16, 17, 18, 19, and 20). Cecil Brown (Chapter 19) historical linguistic analysis traces the spread of the terminology surrounding maize by various indigenous linguistic groups throughout the hemisphere. Results from this ambitious contribution indicate that such linguistic data are largely consistent with the most current archaeological and radiocarbon evidence of its spread. These contributions emphasize the importance of historical linguistics and language to our understanding of the antiquity, meaning and the roles of maize in widely dispersed and economically diverse cultures.

Hopkins (Chapter 18) examines indigenous folk taxonomies in order to explore the meaning and uses of maize among ethnographic cultures and by extension their ancient ancestors. Stross (Chapter 17) analyzes images of maize, depicted in iconography, discussed in narratives and stylized in glyphs, to gain an understanding of the ideological and mythological significance that this plant had to Maya civilization. Janis Alcorn and her coauthors (Chapter 16) explore the mythological origins and cultural significance of maize as it is expressed in language and revealed in legend and song of the Teenek (Huastec) of San Luis Potosi and Veracruz currently residing in the moist tropical forests and dry forest zones on the eastern side of the Sierra Madre Oriental. In prehistoric times they also lived along the Gulf Coast up through Tamaulipas and into the dry areas west of the mountains – including the area of the cave where Richard MacNeish discovered teosinte and small maize ears decades ago.

The ethnographic and linguistic evidence presented in this part of the volume represent an affirmation of the holistic American anthropological approach, and reaffirm what Philip Phillips [16] once said, "archaeology is anthropology or it is nothing at all." The linguistic chapters are in this part of the volume to emphasize the anthropological roots of American archaeology and reaffirm what was stated at the beginning of this introduction. Much has changed and been redefined regarding the spread and significance of maize in the past decade [10].

In the volume summary, Benz and Staller explore the multidisciplinary research on maize in Mesoamerica and surrounding regions to show how the data presented in this volume are in some cases a natural extension of the previous results, and in other ways a dramatic departure with conclusions and data that directly challenges the conventional wisdom and provide compelling evidence to suggest that many of our current assumptions and preconceptions are no longer tenable. This summary on the volume and the research on maize presented by the various contributors reaffirms the power of integrating multiple lines of internally consistent data in light of the previous claims and assumptions that have been made in the important and often controversial history of science surrounding the origins, evolution and economic importance of maize to the rise of Mesoamerican civilization.

References Cited

1. B. Benz (1999). On the origin, evolution and dispersal of maize. In M. Blake (ed.) *Development of Agriculture and Emergence of Formative Civilizations in Pacific Central and South America. The Prehistory of the Pacific Basin.* WSU Press, Seattle, pp. 25-38.
2. B. Benz, A. Long (2000). Early evolution of maize in the Tehuacán Valley, Mexico. *Current Anthropology* 41(3): 459-465.
3. M. Blake, J.E. Clark, B. Voorhies, G. Michaels, M.W. Love, M.E. Pye, A.A. Demarest, B. Arroyo. (1995). Radiocarbon Chronology for the Late Archaic and Formative Periods on the Pacific Coast of Southeastern Mesoamerica. *Ancient Mesoamerica* 6: 161-183.
4. V. G. Childe (1939). *Man Makes Himself.* New York: Oxford.
5. K.V. Flannery (1973). The origins of agriculture. *Annual Review of Anthropology* 2: 271-310.
6. K.V. Flannery, (Ed.) (1986). *Guilá Naquitz. Archaic Foraging and Early Agriculture in Oaxaca, Mexico.* Academic Press. Orlando, Florida.
7. F.O. Freitas, G. Bandel, R.G. Allaby, T.A. Brown, (2003). DNA from primitive maize landraces and archaeological remains: implications for the domestication of maize and its expansion into South America. *Journal of Archaeological Science* 30: 901–908.
8. W.C. Galinat (1988). The origin of corn. In G.F. Sprague and J.W. Dudley (Eds.) *Corn and Corn Improvement*, 3rd edition, American Society of Agronomy, Crop Science Society of America, and Soil Science Society of America, Madison, Wisconsin, pp. 1–31.
9. I. Hodder, (1990). *The Domestication of Europe*, Blackwell, Oxford, UK.
10. S. Johannessen, C.A. Hastorf (Eds.) (1994). *Corn and Culture in the Prehistoric New World.* Westview Press, Boulder.
11. Y.T.A. Kato (1984). Chromosome morphology and the origin of maize and its races. *Evolutionary Biology* 17:219–253.
12. A.L. Kroeber (1944). *Configurations of Culture*. University of California Press, Berkeley and Los Angeles.
13. P.C. Mangelsdorf (1974). *Corn: Its Origin, Evolution and Improvement*, Harvard University Press. Cambridge, MA.
14. Y. Matsuoka, Y. Vigouroux, M.M. Goodman, J. Sanchez, E. Buckler, J. Doebley (2002). A single domestication for maize shown by multilocus microsatellite genotyping, *Proceedings of the National Academy of Sciences USA* 90: 6080–6084.
15. B. McClintock, Y.T.A. Kato, A. Blumenschein (1981). *Chromosome Constitution of Races of Maize (Constitución Chromosoma de Razas de Maíz)*. Colegio de Postgraduados, Chapingo, Mexico.

16. P. Phillips, G.R. Willey (1953). Method and theory in American Archaeology: An operational basis for cultural-historical integration. *American Anthropologist* 55: 615-633.
17. D.R. Piperno, K.H. Clarey, R.G. Cooke, A.J. Ranere, D. Weiland (1985). Preceramic maize in central Panama: Phytolith, Pollen evidence. *American Anthropologist* 87: 871-878.
18. B.D. Smith (2005). Reassessing Coxcatlan Cave and the early history of domesticated plants in Mesoamerica. *Proceedings of the National Academy of Sciences USA* 102: 9438-944.
19. B.D. Smith (2001). Documenting plant domestication: the consilience of biological and archaeological approaches. *Proceedings of the National Academy of Sciences USA* 98: 1324-1326.
20. J.E. Staller (2003). An examination of the palaeobotanical and chronological evidence for an early introduction of maize (*Zea mays* L.) into South America: A Response to Pearsall. *Journal of Archaeological Science* 33(1): 373-380.
21. G.R. Willey (1964). *An Introduction to North American Archaeology.* Volume 1: North America. Prentice-Hall Inc. New York.
22. G.R. Willey (1971). *An Introduction to South American Archaeology.* Volume 2: South America. Prentice-Hall Inc. New York.
23. G.R. Willey, P. Phillips, (1958). *Method and Theory in American Archaeology*. The University of Chicago Press, Chicago.
24. G.R. Willey, J.A. Sabloff. (1980). *A History of American Archaeology*. W. H. Freeman & Co., San Francisco.

IMPORTANT: This book is an updated and abridged version of the 2006 volume *Histories of Maize: Multidisciplinary Approaches to the Prehistory, Linguistics, Biogeography, Domestication, and Evolution of Maize.* You will find that some cross references in this book apply to chapters from the previous edition that have not been reprinted here. You will find those references in the original book. .

PART I

Histories of Maize:

Genetic, Morphological, and Microbotanical Evidence

CHAPTER

1

Differing Approaches and Perceptions in the Study of New and Old World Crops

TERENCE A. BROWN

Faculty of Life Sciences, University of Manchester, Manchester, United Kingdom

Glossary

Amplified fragment length polymorphism (AFLP) A type of DNA sequence variation revealed by the polymerase chain reaction.

Ancient DNA DNA preserved in ancient biological material.

Mesolithic-Neolithic Transition A term used here to describe the period of Old World prehistory associated with a broad package of social and cultural changes that included adoption of agriculture in Southwest Asia.

Mitochondrial DNA haplogroups Groups of related DNA sequences that indicate the genetic structure of human populations.

Old World agriculture Used here to mean the agricultural system based on wheat, barley, and other crops that originated in the Fertile Crescent of Southwest Asia approximately 10,000 years ago.

Postprocessualism A form of interpretative archaeology, which, among other things, rejects a positivist view of science.

Predomestication cultivation A period during which the progenitors of a crop are managed as a distinct population that is not completely isolated, in a reproductive sense, from wild members of the species.

Wave of advance A hypothesis that suggests agricultural spread into Europe was driven by large-scale migration of farmers in a southwest to northwest direction across the continent.

The great advances that have occurred in recent years in understanding the origins, spread, evolution, and uses of maize have paralleled similar advances with the principal crops of **Old World agriculture,** namely wheat and barley. Although there has been extensive transfer of ideas and results, the two research communities have developed their work with some independence, leading to an interesting, and possibly informative, divergence in scientific approaches and perceptions. For example, work on the spread of maize cultivation has always been informed by a clear perception that the factors responsible for adoption of agriculture at a particular site are complex and multifaceted, and hence the factors operating at one site may be different to those at another site. In contrast, ideas regarding wheat and barley cultivation have developed within the context of the competing **wave of advance** and acculturation models for agricultural spread, as described by human geneticists. These models, with their emphasis on cross-continental events, tended to draw attention away from the different effects of local factors, which subsequently have received relatively little attention in Europe. This chapter explores the differing approaches and perceptions in the study of New and Old World crops and attempts to identify areas where greater cross-fertilization of ideas may be beneficial to the two research communities.

INTRODUCTION

To an archaeologist, one of the most intriguing features of the development of agriculture is the way in which this process occurred independently and concurrently in different parts of the New and Old Worlds. To the casual observer, an equally intriguing example of independence and concurrence is provided by the way in which the archaeological study of agriculture has developed on the two sides of the

Atlantic. The point is reflected by this book: of the 48 chapters presented on the prehistory, biogeography, domestication, and evolution of maize, only 2 come from groups located in countries outside of the Americas. One might suggest that a volume initially based on a symposium held at the Annual Meeting of the Society for American Archeology (SAA) might inevitably contain contributions that come predominantly from American archaeologists, but the geographical spread of the contributors to this volume does in fact give an accurate indication of where the vital research into the history of maize is being carried out. It is difficult to provide comparable data on the geographical spread of archaeologists working on the Old World crops, such as wheat and barley, because no equivalent symposium on the prehistory of these crops has been held in recent years. If such a symposium were held, then there would probably be a greater mixing of archaeologists from the New and Old Worlds, but a substantial majority of the contributors would be from Europe. And, although there would be notable exceptions, it is perhaps doubtful if many of the speakers or members of the audience from the SAA symposium on maize would speak at or attend a symposium on wheat and barley. Of course, it is not necessary for archaeologists to meet at conferences to exchange ideas, and there has always been an extensive intellectual interchange among groups working on the New and Old World crops, both through direct contacts and indirectly via the electronic and print media. Nonetheless, the degree of independence with which work on these two agricultural systems has developed has led to interesting, and possibly informative, divergences in approaches and perceptions. In this chapter, I attempt to draw out these divergences and (from a European perspective) to identify areas where cross-fertilization of ideas may be beneficial to the two research communities.

DIFFERENT EMPHASES IN NEW AND OLD WORLD AGRICULTURE

The SAA symposium on "The Stories of Maize," on which this book is based, is an interesting point from which to begin a consideration of the contrasts between the American and European approaches to prehistoric agriculture. The symposium organizers attempted, with considerable success, to draw together researchers addressing different aspects of maize, including its origins, spread, and evolution, the different roles and uses of the crop, and the plant's social and symbolic significance. In combination, the presentations covered a huge range of research strategies, encompassing scientific techniques such as DNA, phytolith, and isotope analyses, archaeobotanical and biogeographical approaches, and ethnohistoric, ethnographic, and linguistic studies. The symposium ran for 4 sessions held over 2 full days of the SAA meeting, with papers presented by 46 regular speakers and 8 discussants. The audience—which numbered more than 200 at times—was mixed, and this was certainly not one of those symposia where the speakers speak mainly to themselves. In short, the symposium formed a central and important part of the SAA meeting as a whole.

If a large symposium devoted to the archaeology of maize can be so successful and so well received at the major American archaeology conference, then why has there been no recent symposium devoted to the archaeology of wheat or barley at a major European conference? Or, to phrase the question differently, why should maize be considered a central research topic in the archaeology of the New World whereas wheat and barley are looked on as a specialist's subject within European archaeology? I believe that the answer lies in a fundamental difference in emphasis in the way in which agriculture is studied in the New and Old Worlds. In the Americas, it is clear from the nature and success of the SAA symposium that the approach taken to the study of agricultural origins and development places major emphasis on the crops themselves. In Europe there is equal recognition of the central role agriculture played by the Southwest Asian crop assemblage—principally wheat and barley, but also comprising lentils, peas, flax, bitter vetch, and chickpea [34]—but the difference is that the origin and spread of agriculture is rarely looked on as an event in itself [33]. Instead, the adoption of agriculture is placed within a broader social context, originally called the "Neolithic Revolution" by V. Gordon Childe [7], but now more generally referred to as the **"Mesolithic-Neolithic Transition."** Agriculture, therefore, becomes just one of a package of changes, including pottery and ground stone tools, that together constitute a significant step in human cultural evolution [32]. This explains why a proposal for a symposium on "The Histories of Wheat" would probably receive short shrift from the organizers of a major European archaeology conference and would not attract a particularly large audience. On the other hand, symposia on The Mesolithic-Neolithic Transition are frequently held, and although these adopt a broad approach, they inevitably include contributions from plant biologists in general and cereal geneticists in particular.

Reasons for the Difference in Emphasis

Whether the notion of the Neolithic Revolution is reasonable as an interpretation of and means of studying the origins of agriculture in the Old World, and whether or not agriculture in the New World was part of a similarly broad and far-reaching set of social and cultural changes, are interesting questions, but not the ones I am raising here. The point I wish to explore is why there should be such a striking difference in emphasis in the New and Old Worlds.

There are almost certainly a number of underlying causes to the differences in emphasis apparent today. One particularly fascinating question is the extent that notions regarding the singularity, or otherwise, of the invention of agriculture have played a role in the development of research into maize and wheat cultivation. In the New World, the natural distribution of teosinte in Mexico and Central America places a geographical limitation on the region where maize agriculture might have originated [23], but the initial interpretation, based on the substantial morphological and genetic variation displayed by modern maize, was that the crop was domesticated on multiple occasions [10, 18]. The alternative possibility, that maize has a single origin, has only really had strong supporting evidence since the publication in 2002 of analyses of microsatellite variation in maize cultivars and in wild teosinte populations; phylogeographical analysis of these data indicate that maize is monophyletic and arose from a single domestication in southern Mexico about 9000 years ago [20]. The study of maize archaeology has, therefore, developed with an open picture of the origins of the crop and hence without any strong basis for placing those origins within the social and cultural context of communities living in the region at the appropriate time. In contrast, for more than a decade, the prevailing view, based on various forms of genetic evidence, has been that the Old World crops were each domesticated once [34], a view that appeared to be confirmed in the later 1990s by examination of large **amplified fragment length polymorphism (AFLP)** datasets for wheat and barley [3, 14, 22]. Further, the distributions of the wild progenitors of the Old World crops indicated that their site(s) of domestication must fall within Southwest Asia, one of the regions of the world that has been subjected to the most intense archaeological survey. It is not surprising that the presumed singularity of agricultural origins was, therefore, linked with prevailing archaeological thought on the development of culture and technology in Southwest Asia. Once this connection had been made it was, perhaps, inevitable that Old World agriculture would become looked on, conceptually, as part of a broader social transition, and that the inherent importance of this transition would become linked, at least in the mind of the layman, to the guns, germs, and steel of the modern day [8]. In the New World, studies of agriculture have developed without such distractions.

Interestingly, the notion of a single origin for maize has become prevalent just at the time that Old World researchers are beginning to reject the emphasis previously placed on single versus multiple origins as overly simplistic. This leads on from recognition that a single, geographically localized domestication for any Old World crop is inconsistent with the gradual transition from gathering to cultivation to domestication that is apparent in the archaeological record. In this regard, it has been suggested [28] that genetic monophyly might not reflect a single, localized domestication for a particular crop, but instead be caused by the emergence, following a more complex origin, of a superior landrace from which all modern varieties are descended. Similarly, I have argued [17] that the genetic diversity of modern wheat supports the notion that this crop underwent a substantial period of **predomestication cultivation,** during which the crop was partially, but not completely, isolated in a reproductive sense from its wild cousins. Partial isolation entails that the crop be managed as a distinct population, but one into which there is some gene flow from neighboring wild plants, possibly from different parts of the natural range if cultivation of the predomesticated crop spread. Eventually, a part of the predomesticated crop might acquire full reproductive isolation, resulting in persistence of the morphological traits that are associated with domestication, simply as a result of cultivation spreading outside of the natural range of the wild plants [17]. Hypotheses such as these give a greater complexity to agricultural origins than is evident from the simple interpretation of genetic monophyly as an indicator of a single domestication. It could be argued that persistence of the spurious link between the genetic monophyly of a crop and its assumed domestication by a single community at a single point in time has held up progress in the study of Old World agriculture for a decade or more.

Outcomes of the Difference in Emphasis

How have the different perceptions of agriculture influenced research in the New and Old Worlds? One outcome that appears to be having a substantial impact on the development of research is the differing levels of importance applied to regionality in studies of the spread and subsequent secondary development of maize and wheat cultivation.

In the Old World, particularly in Europe, the trajectories by which the cultivation of wheat and its associate crops spread have been traced by identifying the earliest appearance in different regions of the trappings of agriculture (e.g., quern stones for grinding grain, preserved remains of the crops themselves) and other artifacts looked on as indicative of the cultural package of which agriculture formed a part. This work has defined two paths for the spread of agriculture into Europe: one follows the Danube and Rhine valleys through the center of the continent and into the northern European plain and is associated with Linearbandkeramik pottery, and the second takes a coastal route through Greece and Italy to Iberia and eventually Britain [4, 5, 25]. Since the initial description of these trajectories in broad-brush strokes, more detailed investigation, based largely on new excavation, has defined in greater detail the time of arrival and the local development of agriculture at different points within Europe [13]. This work has identified clear and important regional differences in the way in which agriculture became established and was exploited throughout Europe, but the implications and importance of this regionality have struggled to assume the high position that they warrant on the European archaeological agenda. This is because research in the Old World has been dominated by discussions about the human dynamics responsible for the spread of agriculture [26]. Largely but not exclusively based on human genetics, this discussion was initiated by the publication of Luca Cavalli-Sforza's research into the geographical distribution of nuclear allele frequencies in human populations [6]. Although this work was global in coverage, particular interest was stimulated by the distribution patterns in Europe because these revealed the existence of a genetic gradient across the continent, from southeast to northwest (or vice versa). The date at which this gradient was set up cannot be determined from the genetic data, but the coincidence between the human migration implied by this gradient and the movement of humans that could have accompanied the spread of agriculture, led to the genetic data being used as convincing evidence that agriculture arrived in and was carried through Europe by a human wave of advance, echoing ideas that had been current in European archaeology since the mid-1960s [1, 2]. The wave of advance model has clear and important implications for issues such as demographic change in prehistoric Europe [33], and perhaps explains why the term "Neolithic Revolution" gained such popularity. The wave of advance model also implies uniformity in the manner in which agriculture arrived in different parts of Europe, and even suggests that the subsequent development of agriculture may have followed a similar pattern throughout the continent, in that the same cultural group, regardless of location, oversaw this development. Against this background, regional variations apparent in the archaeological record become of limited importance, on the assumption that they are variations on a theme.

The balance is becoming redressed, but only as the result of a debate among different camps of human geneticists that has itself drawn attention away from the more specific issues regarding agricultural spread. The debate was engendered first by studies of **mitochondrial DNA haplogroups,** which, being inherited solely through the female line and not being subject to recombination, are more amenable than nuclear alleles to the type of population and phylogenetic analyses needed to infer the patterns of past migrations [31]. This work has thrown the wave of advance model into doubt by showing that less than 10% of modern Europeans are descended from individuals likely to have moved into Europe at the time that agriculture arrived [27]. A similar downgrading of any wave of advance is also the outcome of studies of genetic markers on the Y chromosome [29]. Only in the past few years has it been possible to reconcile the human genetic data with the notion that that there are many different ways by which agriculture can diffuse from one point to another, ranging from total displacement of the nonagricultural population at one extreme to adoption of agriculture solely by transfer of technology with no population movement at the other [35]. Once this point has been accepted it is a small step to the realization that across a landmass the size of Europe, with the environmental and cultural variability found across the continent, the spread of agriculture, and its subsequent local development, could have been a highly variable and regional process.

Research into Old World agriculture has, therefore, reached a stage where regionality, after being buried under broader debates for more than a decade, is now reemerging as a central issue. The contrast with the situation in the New World is striking, as this volume amply illustrates. The specific information presented in many of these chapters on maize cultivation in regions that, on occasion, are just a few hundred square kilometers in size provides a picture of New World agriculture substantially richer than anything that is available for much of Europe. Of course, a converse to the European situation, in which an emphasis on regionality is allowed to obscure the bigger picture regarding the spread of agriculture, would be equally debilitating for the development of research, but this has also been avoided in the New World. The time of arrival of maize cultivation in most parts of North and South America is well understood [12, 19, 24], and although genetic data have so far been relatively underexploited as a means of studying crop movement, the original work by Barbara McClintock and colleagues based on chromosome morphologies [21] is gradually being supplemented by more modern studies of DNA markers and sequences [9, 20]. In this regard, the successful recovery and analysis of **ancient DNA** from preserved maize cobs [9, 11, 16] points a promising way forward, the use of archaeological remains enabling studies of the prehistoric spread of maize to be based on specimens whose biogeography has not become complicated by the movement of genotypes that has occurred during the post-Columbian period.

DIFFERENT PERCEPTIONS OF THE ROLE OF SCIENCE

I have argued that the interpretation of Old World agriculture as part of a cultural package is one of the underlying reasons for the differences in the development of agricultural studies in the New and Old Worlds. I have also suggested that human genetics have diverted thought in Europe from the regional issues that have received greater attention in the New World. One final question is the extent to which the prevailing academic climate has influenced agricultural research in the Old and New Worlds.

Over the past 25 years, the most significant event in academic archaeology has been the rise of **postprocessualism.** Adoption of the set of ideas that make up postprocessualism has reinvigorated archaeological theory, particularly with regards to the establishment of agriculture in Europe [15], but this reinvigoration has involved a move away from a scientific approach to archaeology. Postprocessualism does not reject science itself, it merely rejects the way science is practiced [30]; but the result is the same: the scientific contribution to archaeology is deemphasized. One outcome is that academic archaeology, particularly in the U.K., is currently divided into two camps, the members of which, often being faculty colleagues, may talk to one another, but who rarely establish intellectual contact. Science-based archaeology follows its own agenda and is often uninformed by the work of theoreticians, and vice versa. In fact, agriculture is one area where science and theory have maintained a reasonably productive relationship, as evinced by the manner in which science has adapted to the notion of agriculture as part of a cultural package. But the division of archaeology into different cultures is clearly detrimental to the development of any area of the subject, and although these dangers are recognized, especially among those who devise undergraduate syllabuses, it will be several years before archaeology in the U.K. will be able move forward again as a unified discipline. This is perhaps the main reason why a symposium equivalent to "The Stories of Maize" will probably not be held on Old World agriculture in the near future. A clear feature of research into New World agriculture is a mutual acceptance, understanding, and appreciation of the different forms of archaeological research, from the physical, biological, and social sciences through to linguistics, ethnohistory, and ethnography. It is debatable whether, at present, the same degree of intellectual fusion could be achieved among the various approaches to the study of Old World agriculture. Until this problem is solved, the crops themselves will not assume the central role in archaeological studies of Old World agriculture that maize plays in the New World, and the emphasis in Europe will continue to be on agriculture as part of a broader cultural package. But this is not necessarily a reason for discontent. A nonuniform approach on the two sides of the Atlantic becomes beneficial when looked on as two parallel but different experiments in the study of agriculture. Comparisons of our successes and failures can then help both groups make progress toward their goals.

Acknowledgments

I thank Keri Brown for his comments on this paper. The opinions expressed are solely my own.

References Cited

1. A. J. Ammerman, L. L. Cavalli-Sforza. (1971). Measuring the rate of spread of early farming in Europe. *Man, 6,* 674–688.
2. A. J. Ammerman, L. L. Cavalli-Sforza. (1973). A population model for the diffusion of farming into Europe. In: A. C. Renfrew, (Ed.), *The explanation of culture change: Models in prehistory.* London: Duckworth. pp. 343–358.
3. A. Badr, K. Müller, R. Schäfer-Pregl, H. El Rabey, S. Effgen, H. H. Ibrahim, C. Pozzi, W. Rohde, F. Salamini. (2000). On the origin and domestication history of barley (Hordeum vulgare). *Molecular Biology and Evolution, 17,* 499–510.
4. G. Barker. (1985). *Prehistoric farming in Europe.* Cambridge: Cambridge University Press.
5. P. Bogucki. (1996). The spread of early farming in Europe. *American Scientist, 84,* 242–253.
6. L. L. Cavalli-Sforza, P. Menozzi, A. Piazza. (1994). *The history and geography of human genes.* Princeton: Princeton University Press.
7. V. G. Childe. (1935). Changing methods and aims in prehistory. *Proceedings of the Prehistoric Society, 1,* 1–15.
8. J. Diamond. (1997). Location, location, location: The first farmers. *Science, 278,* 1243–1244.
9. F. O. Freitas, G. Bandel, R. G. Allaby, T. A. Brown. (2003). DNA from primitive maize landraces and archaeological remains: Implications for the domestication of maize and its expansion into South America. *Journal of Archaeological Science, 30,* 901–908.
10. W. C. Galinat. (1988). The origin of corn. In: G. F. Sprague, J. W. Dudley (Eds.), *Corn and corn improvement,* 3rd ed. Madison, WI: American Society of Agronomy, Crop Science Society of America, and Soil Science Society of America. pp. 1–31.
11. P. Goloubinoff, S. Pääbo, A. C. Wilson. (1993). Evolution of maize inferred from sequence diversity of an Adh2 gene segment from archaeological specimens. *Proceedings of the National Academy of the Sciences of the United States of America, 90,* 1997–2001.
12. M. M. Goodman. (1978). Historia e origem do milho. In: E. Paterniani (Ed.), *Melhhoramento e Producao do Milho no Brasil.* São Paulo, Brazil: Fundação Cargil. pp. 30–65.
13. D. R. Harris (Ed.). (1996). *The origins and spread of agriculture and pastoralism in Eurasia.* London: UCL Press.
14. M. Heun, R. Schäfer-Pregl, D. Klawan, R. Castagna, M. Accerbi, B. Borghi, F. Salamini. (1997). Site of einkorn wheat domestication identified by DNA fingerprinting. *Science, 278,* 1312–1314.
15. I. Hodder. (1990). *The domestication of Europe.* Oxford: Blackwell.
16. V. Jaenicke-Després, E. S. Buckler, B. D. Smith, T. Gilbert, A. Cooper, J. Doebley, S. Pääbo. (2003). Early allelic selection in maize as revealed by ancient DNA. *Science, 302,* 1206–1208.
17. M. K. Jones, T. A. Brown. (In press). Selection, cultivation, and reproductive isolation: A reconsideration of the morphological and molecular signals of domestication. In: T. Denham, J. Iriarte, L. Vrydaghs, (Eds.), *Rethinking agriculture: Archaeological and ethnoarchaeological perspectives.* London: UCL Press.
18. T. A. Kato. (1984). Chromosome morphology and the origin of maize and its races. *Journal of Evolutionary Biology, 17,* 219–253.
19. P. C. Mangelsdorf. (1974). *Corn: Its origin, evolution and improvement.* New York: Harvard University Press.
20. Y. Matsuoka, Y. Vigouroux, M. M. Goodman, J. Sanchez, E. Buckler, J. Doebley. (2002). A single domestication for maize shown by multilocus microsatellite genotyping. *Proceedings of the National Academy of the Sciences of the United States of America, 99,* 6080–6084.
21. B. McClintock, T. A. Kato, A. Blumenschein. (1981). *Chromosome constitution of the races of maize: Its significance in the interpretation of relationships between races and varieties in the Americas.* Chapingo, Mexico: Colegio de Postgraduados.
22. H. Özkan, A. Brandolini, R. Schäfer-Pregl, F. Salamini. (2002). AFLP analysis of a collection of tetraploid wheats indicates the origin of emmer and hard wheat domestication in southeast Turkey. *Molecular Biology and Evolution, 19,* 1797–1801.
23. D. R. Piperno, K. V. Flannery. (2001). The earliest archaeological maize *(Zea mays L.)* from highland Mexico: New accelerator mass spectrometry dates and their implications. *Proceedings of the National Academy of the Sciences of the United States of America, 98,* 2101–2103.

24. K. O. Pope, M. E. D. Pohl, J. G. Jones, D. L. Lentz, C. von Nagy, F. J. Vega, I. R. Quitmyer. (2001). Origin and environmental setting of ancient agriculture in the lowlands of Mesoamerica. *Science, 292,* 1370–1373.
25. T. D. Price (Ed.). (2000). *Europe's first farmers.* Cambridge: Cambridge University Press.
26. C. Renfrew. (2000). Towards a population prehistory of Europe. In: C. Renfew, K. Boyle (Eds.), *Archaeogenetics: DNA and the population history of Europe.* Cambridge: McDonald Institute for Archaeological Research. pp. 3–11.
27. M. Richards. (2003). The Neolithic invasion of Europe. *Annual Review of Anthropology, 32,* 135–162.
28. F. Salamini, H. Özkan, A. Brandolini, R. Schäfer-Pregl, W. Martin. (2002). Genetics and geography of wild cereal domestication in the Near East. *Nature Reviews. Genetics, 3,* 429–441.
29. O. Semino, G. Passarino, P. J. Oefner, A. A. Lin, S. Arbuzova, L. E. Beckman, G. De Benedictis, et al. (2000). The genetic legacy of Paleolithic Homo sapiens sapiens in extant Europeans: A Y chromosome perspective. *Science, 290,* 1155–1159.
30. M. Shanks, C. Tilley. (1992). *Reconstructing archaeology: Theory and practice,* 2nd ed. London: Routledge.
31. B. Sykes. (2000). Human diversity in Europe and beyond: From blood groups to genes. In: C. Renfew, K. Boyle, (Eds.), *Archaeogenetics: DNA and the population history of Europe.* Cambridge: McDonald Institute for Archaeological Research, Cambridge, 2000. pp. 23–28.
32. J. Thomas. (1993). Discourse, totalisation and the "Neolithic." In: C. Tilley (Ed.), *Interpretative archaeology.* London: Berg. pp. 357–394.
33. J. Thomas. (1996). The cultural context of the first use of domesticates in continental Central and Northwest Europe. In: D. R. Harris (Ed.), *The origins and spread of agriculture and pastoralism in Eurasia.* London: UCL Press. pp. 310–322.
34. D. Zohary. (1996). The mode of domestication of the founder crops of Southwest Asian agriculture. In: D. R. Harris, (Ed.), *The origins and spread of agriculture and pastoralism in Eurasia.* London: UCL Press. pp. 142–158.
35. M. Zvelebil. (2000). The social context of the agricultural transition in Europe. In: C. Renfew, K. Boyle (Eds.), *Archaeogenetics: DNA and the population history of Europe.* Cambridge: McDonald Institute for Archaeological Research. pp. 57–79.

CHAPTER

2

Maize in the Americas

BRUCE F. BENZ

Biology Department, Texas Wesleyan University, Ft. Worth, Texas

Glossary

Agriculture The dependence of human populations on domesticated plants or animals, or both.

Bottleneck Demographic process in which the number of reproductive individuals within a population is reduced provoking a reduction in the population's genetic variation. Founder population.

Domestication The process of taming plants and animals to facilitate their usefulness to humans. The evolution of a mutualistic relationship of the plant or animal on human behaviors that ensure their continued survival.

Domestication syndrome A syndrome is a suite of characters that together reflect a response to some action. In the case of human behavior, a **domestication syndrome** refers to morphogenetic characters such as rigid, nondisarticulating rachis; soft, and sometimes disarticulating chaff (glumes, lemma, and palea) surrounding the grain; reduced defensive traits such as the reduction of silica in the epidermal tissues or defensive secondary metabolites; monopodial branching, or lack of branching, resulting in production of one or only a few inflorescences (fruits); uniform germination of the grain or seeds on planting and uniform ripening on harvesting; uniformity of morphological characteristics of the plant; inflorescence; and fruit or seeds that facilitate harvesting.

Epistasis Multiple genes affecting a single trait.

Evolutionary rate Amount of morphological or genetic change over time.

Landraces Distinct populations of traditional varieties with limited geographic range cultivated in North and South America. Hypotheses abound, and evidence indicates these landraces are the product of culturally prescribed selection for desired agronomic and culinary attributes.

Radiocarbon contemporaneity Radiocarbon dates whose uncertainty (sigma or standard deviation) overlaps such that their probability of being different is not expected by chance. Radiocarbon dates that estimate the same age.

Speciation Macroevolution; the evolution of new species.

Teosinte An English term adapted from the Nahuat "teocintli" (good or evil grain) used widely to refer to the seven taxa of wild grasses that are closely related to maize.

Biological and archaeological evidence concerned with the early evolution of maize is reviewed in this chapter in an attempt to resolve conflicting interpretations about the process and its antiquity. Estimates of **bottleneck** size derived from molecular biology and archeology are internally consistent; domesticated populations were small, perhaps for a few hundred generations during which maize's genetic diversity was significantly, perhaps dramatically reduced. Estimates of the antiquity of the domestication process from molecular biology are internally consistent yet antedate some of the archaeological evidence.

PURPOSE AND SCOPE OF REVIEW

Archaeological evidence for plant domestication provides critical documentation of the human manipulation of plant population genetics: absolute dating and cultural context. The focus of archaeological study of maize's origin and evolution has been the temporal documentation of the domesticates' initial appearance. There has been little archaeological concern for documenting sustained agricultural selection—continued domestication—after recording the initial steps humans took to select for traits that distinguish the domesticated species from its wild ancestor. Archaeological evidence for maize domestication throughout the Americas is abundant. Yet, there is little evidence on domestication process [2]. This stems in part from the historical misidentification of maize's ancestor [26] and from archaeological focus on "firsts" and less interest in the processes of maize production and population manipulation.

Three lines of evidence have been exploited by archaeologists to document the domestication process: study of inflorescence morphology, pollen size, and phytolith shape and abundance. The evidence each provides is discussed in the attempt to winnow salient facts about maize's domestication. However, before discussing archaeological study of maize domestication and **agriculture** in the Americas, I review the biological efforts to understand maize domestication. My choice in focusing on biological evidence first is biased by my belief that biological evidence has been instrumental in directing our attention at the appropriate archaeological questions but recognizes that most biological research on maize produces hypotheses that can lead only to inferences about the historical cultural processes that produced maize. Archaeological evidence, in contrast, is uniquely suited to testing these biological hypotheses.

INTRODUCTION

As mentioned previously, much of the archaeological evidence has documented only when a particular morphological or allelic manifestation (phenotype) of domestication first appears in the archaeological record. The biological community has pursued other evidence that contemplates documenting processes associated with domestication. Molecular biology has focused not only on the genes attributed to domestication but has used these and numerous other genetic markers to discern how humans affected **teosinte** populations in their search for more productive or more readily harvestable phenotypes. The biological focus on domestication genes has touched on two areas of biological importance for understanding human behaviors toward plant harvesting and production: 1) the amount of time in human generations spent manipulating teosinte to manage the harvesting process that eventually domesticated teosinte and ensured a more productive and reliably harvested phenotype; and 2) the size of teosinte and maize populations that humans focused on to achieve these goals. The results of the latter have been more cogent and consistent than the former; although, the results of each are worthy of review not only for the trends they document but also for pointing out human behaviors that can be inferred from them.

The appearance of domestication syndrome traits has long been the focus of archeological inquiry [23]. The first, or earliest appearance of a glume or rachis fragments in the archeological record is taken as a horizon marker, a harbinger of change in human behavior. With a few exceptions, study has rarely focused on the quantitative changes in maize ear morphology to detect and document the rate of change in political, social, or technological adaptations of the cultivators. Instead, research has sought to document difference in assemblages that attribute such changes to environmental or cultural factors. Fortunately, this is changing. Attempts are being made (this volume contains numerous examples, e.g., Chapters 7 and 31) to document the rate of evolutionary change in maize that is inextricably linked to human behavioral change and that will permit one to identify the critical periods during which human effort and focus changed. The possibility of asking these questions is practical and possible using morphological traits of the corn cob and genetic markers and should be possible using phytoliths and pollen.

The chronological placement of domesticated plants, maize in particular, has been fraught with challenges. Archaeological context provided the bias for a long held misconception about the age of domesticated maize in the Tehuacan Valley because its age was inferred from associated radiocarbon determinations whose reliability for estimating corn cob age was low. Attempts to reconcile differences between the age of maize specimens on which accelerator mass spectrometry (AMS) dates have been successful and their contextually associated radiocarbon dates have produced mixed results [44]. Fortunately, the morphology of the remains of maize argues that its initial domestication occurred sometime during the Middle to Late Archaic Periods in Mesoamerica before maize's early appearances at Guila Naquitz.

Chronological determination of maize's origin has long been a contentious issue. The issue stems not so much from the vagaries of relative dating as from the different types of archaeological evidence brought to bear on the subject. Pollen was long the chink in the armor of George Beadle's teosinte hypothesis—the pollen from the Bellas Artes pollen core was clearly maize and believed to be more than 80,000 years old. Mangelsdorf [26, p. 183] believed that wild maize existed long before domesticated maize. Much has been written since then to counter this argument.

Pollen and phytoliths have kept the antiquity of maize controversy alive [38, 40, 41]. Phytoliths are more readily extracted from archaeological sediments than pollen, even though identification is more challenging than pollen simply because the shapes of these epidermal silica inclusions are more varied and less distinctive than maize pollen. Simultaneous study of these two types of botanical evidence has produced intriguing results that will continue to drive the search for pre-2000 BC maize in Central and South America. Pollen and phytoliths are eminently suited for documenting early agriculture in and around lakes and in rock shelters and open sites inhabited by humans during this critical period.

The motivation for writing this chapter should be made clear. Knowing that the evidence for maize agriculture in Tehuacan is unequivocal and knowing that the domesticated teosinte inflorescences from Guila Naquitz were of domesticated plants, the time depth of the human manipulation of teosinte begs for resolution. Juxtaposing the genetic evidence with the archaeological, geochronological, and paleoenvironmental evidence and literature should help resolve the controversy surrounding initial domestication and dispersal.

Genetic Evidence of Teosinte Domestication

The polemic surrounding the ancestor of maize was resolved using molecular techniques first deployed to understand taxonomic relationships in the genus *Zea*. Isozymes [13] clearly indicate that maize and teosinte are closely related, as crossing/hybridization relationships demonstrated long before [26, 28]. Significant breakthroughs followed the development of molecular techniques, which enabled the identification of genes involved in the domestication process [15, 16, 17]. Early attempts to compare genetic relationships between maize and teosinte [18, 24] focused on genes that had little obvious involvement in the domestication process; yet, even they demonstrated evidence of human manipulation of teosinte populations. Simultaneous efforts focused on mapping the genes that distinguish maize and teosinte [15, 16]. Doebley identified at least five regions of the *Zea* genome that were involved in maize's **speciation** [14]. Each of these genes produces a phenotype that is a part of the domestication syndrome. And in some cases, the gene's postulated actions appear to be **epistatic** on other characters of the domestication syndrome (except for *Tga1* [50]).

One of these genes affects plant architecture in the way typical of other domesticated cereals (the so-called domestication syndrome [23]) that affects apical dominance thereby reducing lateral branch lengths and ear number per plant. Teosinte is a highly branched plant (Figure 2-1) with numerous inflorescences protected in the axil of leaves on the main stem and branches. Maize typically produces one or at most two branches, each with only one or, infrequently, two or more inflorescences. Molecular mapping and expression studies identified the teosinte branched 1 *(Tb1)* gene as responsible for the branching difference between maize and teosinte and is epistatic on other domestication syndrome traits such as paired spikelets.

Differences in seed dispersal between wild *Zea* and maize are complex. They encompass not only the development of "free-threshing" grain but the development of a tough, rigid rachis that will tenaciously hold the grain until it ripens and the development of protective husks surrounding the rigid

FIGURE 2-1 Author with two teosinte plants (*Zea mays* ssp. *parviglumis* Iltis and Doebley) along roadside in Balsas region of southern Mexico. Photo taken March 2003. (Courtesy of Bruce Benz)

rachis with attached spikelets harboring ripening grain as well. So far, the gene(s) responsible for rachis rigidity in maize remain(s) unidentified. However, a gene for free-threshing grain has been identified. In teosinte the grain is tightly enclosed within a cupulate fruitcase. This fruitcase is composed of a segment of the stiff, lignified, silica-impregnated rachis (the cupule) and lower glume of the spikelet. The grain of teosinte is neither visible nor easy to liberate from the highly dispersible fruitcase. Maize grains are exposed within the spikelets but are protected by the whorls of leaf sheaths developing from the stem beneath the ear. The grains are otherwise unprotected though as the spikelets have emerged from the cupule and borne at a right angle to the axis of the inflorescence, they have become more readily harvested. Wang and colleagues [50] have demonstrated that this morphological difference is the result of a single nucleotide difference between maize and teosinte.

Teosinte *(Zea luxurians)* grain has high nutritional value [22, 24, 26]. With higher total nitrogen and methionine content than maize, teosinte is purported to be preferred over maize by livestock (Hernandez Xolocotzi, personal communication). There is no difference in lysine, tryptophan, or niacin, however. Maize, like many cereal grains, is deficient in two critical amino acids, lysine and tryptophan, which leads to nutritional deficiencies among populations with little dietary diversity and high dependence on maize. Maize appears to have a higher fat content than teosinte. Maize, like other cereals, possesses a wealth of genetic variation for carbohydrate storage: dent, flint, flour, and sweet are but four phenotypes produced by a complex metabolic pathway affecting the way glucose is stored in the grain's endosperm. The nutritional value of the grain has been subject to human selection [24, 26, 46, 49] over the millennia since domestication. Jaenicke-Deprés and colleagues and Whitt and coworkers documented genetic variation in starch and protein [25, 49] storage enzymes that demonstrate that human selection has significantly reduced DNA sequence polymorphism in maize over teosinte.

Traits associated with the domestication syndrome that currently lack genetic documentation of human manipulation include rachis rigidity (the disappearance of natural rachis disarticulation), paired spikelets, increased grain size, and polystichy, all traits that are present in the earliest archaeological material of domesticated teosinte [3–5]. Numerous reports identify a plethora of genetic loci in maize that were affected by humans imposing stringent selection pressures on the descendants of wild *Zea mays* ssp. *parviglumis* [25, 53].

Genetic Evidence of Population Manipulation

The search for the genetic basis of maize's difference from teosinte yielded understanding about the process of maize domestication. Molecular biologists document the genetic changes responsible for the domestication syndrome traits, but more critically, they have also posed and answered questions about the length of time humans have intervened in teosinte populations and the intensity with which subsequent manipulation of maize populations affected morphological and agronomic change to satisfy human needs. Gene sequence comparisons document population bottlenecks. Such bottlenecks occur among extant traditional agriculturalists [39] and should be expected to have occurred among pre-Hispanic and historical agriculturalists as well.

The model of neutral evolution applied to such comparisons posits mutation as the raw material and genetic drift as the prime mover for genetic change. This model, applied to gene sequences, serves as a means of calculating the amount of genetic change expected to occur by mutation that serves as the baseline for determining how much time was necessary for such genetic differences to have occurred. Comparing genetic sequences between wild and domesticated populations permits statistical testing of this neutral model. Such models also permit statistical testing of hypotheses [18, 24, 46–48, 52] about the length of time humans have experimented with teosinte populations, the intensity of human selection, and of effective population sizes [7].

Numerous studies [18, 22, 24, 46–48, 50–53] have estimated the effective population size and bottleneck duration from gene sequence data. Their estimates assume that random genetic change can mimic a molecular clock that, in turn, describes the structure and behavior of ancestral populations before speciation. Some genes are more suited to the task than others. First, the gene must be neutral, that is, it can show no clear evidence within its DNA sequence that shows significant evidence of systematic fixed differences between taxa. Second, the sequences must exhibit systematic variation demonstrating that ancestor and domesticate are in separate lineages. Third, the number of mutations or segregating sites should be roughly similar among populations.

The Alcohol Dehydrogenase 1 *(adh1)* locus studied by Eyer-Walker and colleagues [18] fits these assumptions. To calculate the size of the bottleneck, the authors assumed that the first successful attempt to domesticate teosinte took place roughly 7500 years ago. Based on this assumption Eyer-Walker provided three estimates of the size of the bottleneck: 20 individuals if the duration of the bottleneck lasted 10 years, 1157 individuals if the bottleneck lasted 500 years, or 16,588 individuals if the bottleneck lasted for 7500 years. Assuming that the domestication bottleneck lasted for 2800 years (from 7500 to 4700 years ago—incorrectly using radiocarbon years for sideral time), the time from estimated domestication to maize's appearance in Tehuacan, the bottleneck involved 5600 plants, a number roughly equivalent to 2 plants per year providing seed for subsequent plantings. If the prehistoric population size of teosinte *(Zea mays* ssp. *parviglumis)* was roughly 940,000 plants, as estimated by similar methods (estimate of gene sequence polymorphism in *Zea mays* ssp. *parviglumis*), then the bottleneck represented only 6% of the long-term population of teosinte.

Hilton and Gaut [24] derived additional estimates of the population bottleneck on the basis of a nonessential seed storage protein—globulin 1 *(glb1)*. The major difference from the previous work is that glb1 shows no fixed nucleotide differences between maize and *Z. mays* ssp. *parviglumis* suggesting the two taxa are closely related. Coalescent simulations of the genetic bottleneck corroborate the estimates by Eyer-Walker and colleagues [18] of the size of the maize founder population. If the bottleneck lasted 10 generations, the founder population may have consisted of as few as 7 individuals. If duration of the bottleneck lasted 100 generations, the population might comprise less than 70 individuals; if it lasted 1000 years, the bottleneck was 659 individuals; and if it lasted 2800 years, the total size of the founder population was only 1830 individuals, roughly 1 individual per generation (Table 2-1).

Inclusion of *Tripsacum* in the studies of *adh1* and *glb1* sequence diversity also permitted the authors to make the

TABLE 2-1 Estimated Population Sizes of Genetic Bottleneck Occurring with the Domestication of Maize. Effective population size and the size of the bottleneck are correlated. The values in the table are of two types: the duration of the bottleneck is assumed, whereas the size of the bottleneck is predicted for *adh1, glb1,* the 12 genes on chromosome 1, and the single nucleotide polymorphisms, whereas the size of the bottleneck is assumed and the duration of the bottleneck predicted for tb1. In either case, the size of the bottleneck and its duration are strongly correlated.

Gene/loci	Duration of bottleneck in years	Size of bottleneck ne	Reference
adh1	10	20	[18]
	500	1157	
	7500	16,588	
	2800[a]	5600	
glb1	100	70	[24]
12 genes	100	892	[46, 47, 48]
Chromosome 1	500	2500	
	2800	12,500	
SNPs	2800	3500	[53]
tb1	313	1000	[51]
	1023	10,000	
Morphology	185	200	[7]

[a]The estimated time difference between the first domesticating genetic change and the last efforts of domesticating selection (7500–4700 BP), see text.

following observation: *Tripsacum* and *Z. mays* ssp. *mays* possess no shared sequence polymorphisms indicating that maize did not evolve as the result of a hybridization between a member of the genus *Zea* and *Tripsacum.* Although the study of gene sequence divergence provides useful clues to the processes involved in maize domestication hypothesized by others, the authors conclude that "there are no good independent estimates for the true duration of the original domestication event" [24, p. 870].

White and Doebley [52] examined sequence diversity of a gene associated with one of the genetic marker loci (quantitative trait loci [QTL]) implicated as one of the domestication syndrome traits [15, 16]. In this, and the two aforementioned studies of sequence divergence, the amount of sequence polymorphism in maize is significantly less than that of its ancestor. This reduction in polymorphism should be expected of most speciation events resulting from domestication. The speciation process involves one or, at most, a few individuals diverging from the ancestral population because they exhibit some beneficial trait that selection can focus on taking with them in the process only a fraction of the total genetic variation present in the ancestral population. The first study of ancient DNA from maize performed by Goloubinoff and colleagues [20] found quite different results. Sequences analyzed from three pre-Columbian maize samples, five modern maize samples, and one sample each from four different *Zea* species and *Tripsacum pilosum,* showed an equivalent amount of sequence diversity between pre-Columbian and modern maize. These results suggest the latter had not acquired different sequence polymorphism through drift or recombination since evolving from their last common ancestor. It is possible that small number of samples might explain the lack of difference.

Wang and colleagues [51] also examined the relationships between domesticated maize and two of maize's closest wild relatives, the Mexican annual teosintes (*Z. mays* ssp. *parviglumis* and *Z. mays* ssp. *mexicana*), using the *tb1* gene that Doebley's [51] laboratory had documented as a crucial locus in the transformation of teosinte into maize. Wang and colleagues were able to sequence the entire *tb1* gene sequence and ascertain that the main differences between maize and *Z. mays* ssp. *parviglumis* are located in the nontranscribed region—the regulatory region—of the gene, not in the transcribed region. Additional insights were gained from this. The estimated time of the domestication bottleneck is estimated to have occurred over a 300- to 1000-year time span. This estimate is based on selection coefficients between 0.04 and 0.08. (These are equivalent to relative annual differences in fitness of 100 maize-like alleles to 94 or 92 teosinte alleles, respectively).

Matsuoka and colleagues corroborated *Zea mays* ssp. *parviglumis'* identity as maize's closest wild relative using microsatellite coverage of the entire genome instead of focusing on a single gene region [29]. Microsatellites document a single origin for maize from *Zea mays* ssp. *parviglumis* and date the time of divergence between these two species somewhere between 5689 and 13,093 generations before the present (which the authors equate to years). This 95% confidence interval surrounds the estimated average time of separation of 9188 years BP or ca. 7200 BCE. This places the highest probability of the first human domesticating selection activities on teosinte as having occurred more than 3500 years before maize was incorporated into the Tehuacan cave deposits.

The phylogenetic results of Matusoka and colleagues [29] also indicate a clear diversification of maize in central Mexico and offer a crucial evolutionary hypothesis of maize's spread throughout the Americas from the highlands of central Mexico. "[T]he phylogenies and PCA suggest two lineages or paths of dispersal. One path traces through western and northern Mexico into the southwestern U.S. and then into the eastern U.S. and Canada. A second path [presumably later] leads out of the highlands to the western and southern lowlands of Mexico into Guatemala, the Caribbean Islands, the lowlands of South America, and finally the Andes Mountains" [29, p. 6084].

Even though the number of collections used to generate this phylogeny is a magnificent effort, its sample of maize is noticeably depauperate: 264 plants represent more than 300 **landraces** cataloged for the Americas, in which only 75% of the Mexican races are included. Yet, the proposed routes of dispersal lend credence to previous hypotheses concerning dispersal into North America north of Mexico, Central America, and the Central–South American connection [28], although it contradicted previous hypotheses concerning the phylogeny of Mexican maize [3]. It also created a polemic surrounding the altitudinal location of maize's domestication instead of resolving such a debate. Though the races identified as basal in the cladogram and cluster diagram by Matsouka and colleagues [29] are all collections occurring today above 1900 meters, the populations of annual teosinte consistently showing greatest genetic similarity to Mexican maize currently grow at elevations between 950 and 1400 meters. San Marcos, Coxcatlan, and El Riego caves in the Tehuacan Valley are all located at elevations between 1100 and 1300 meters. Only Guila Naquitz, with only three corn cobs found dating on or before 4250 BCE, is located at 1926 meters above sea level [36], whereas the Tamaulipas caves are located at 1500 meters. It seems likely that the sample taken by Matsouka and colleagues missed some of the extant basal lineages, because samples representing these lineages were not included in the study or these lineages are extinct. The biogeography of the so-called basal lineages identifies only high altitude populations some of which are traditional varieties still cultivated in the uplands (most are recent INIFAP collections) where isolation and extreme habitats have limited introductions, whereas the lowlands have experienced more extensive and rapid turnover of landraces due to the ease of invasion by higher yielding varieties [1]. If this hypothesis is accurate, one would also expect that all the recently made collections included in the study by Matsuoka and colleagues would tend to show up toward the base of the phylogeny because they have retained the greatest number of shared derived alleles with ancient varieties, whereas those collected during the National Research Council efforts in the 1940s and 1950s would have lost some variation in the process of gene bank storage and repeated increases of gene bank collections. Supporting data available online suggest that most of the collections identified as basal were made recently by members of the Instituto Nacional Forestal y Agropecuario or other institutions.

Work on three domestication-related genes and single nucleotide repeats has shed even greater light on the phylogenetic relationships within *Zea mays* [49, 51, 53]. Whitt and colleagues [49] identified vital genetic differences between maize and teosinte responsible for the changes in the structure of starch storage in maize grains from study of a variety of inbreds, a few landraces, teosinte populations, and a few varieties of sweet corn. Their efforts documented genetic changes in main genes in the amylose storage pathway that reduces the amount stored as amylopectin and increases the amounts of amylose stored. These genetic changes give maize endosperm greater ability to absorb protein and fat and the ability to paste and resist sheer stress. In effect, the documented genetic changes made the starchy endosperm more malleable for making tamales and tortillas. Gene sequence comparisons between maize and *Zea mays ssp parviglumis,* like those with *c1* [22], *tb1* [51], and *glb1* [24], indicate a significant bottleneck is associated with this difference—a threefold to sevenfold reduction in nucleotide diversity in the maize sequences can be attributed to stringent selection. No estimates of the size of the bottleneck or its duration were made.

Doebley [50] succeeded in identifying and characterizing the gene that freed the grain from the teosinte fruitcase. This gene is also responsible for the loss of lignification and reduced silica deposition in the glume and cupule of the maize cob. Wang and colleagues [50] identify the difference between maize and teosinte as a single nucleotide polymorphism resulting in a single amino acid difference in the first exon of *Tga1*. In contrast to other known genetic differences between maize and teosinte, this mutation appears to be a novelty, with no similar variation in *Z. mays* ssp. *parviglumis*. Statistical comparisons of nucleotide diversity between maize and *Z. mays* ssp. *parviglumis* indicate the maize allele has significantly less diversity than the teosinte allele, which again documents (in the promoter and first exon) a genetic bottleneck. The size and duration of the bottleneck is shown to vary with selection intensity. The intensity of selection appears to average about 0.04, with the estimated time since fixation ranging from about 7000 to 12,000 generations (Table 2-2).

Wright and colleague [53] also reaffirmed this domestication bottleneck using 774 single nucleotide polymorphisms in 14 U.S. inbred lines and 16 inbred populations of *Z. mays* ssp. *parviglumis.* Wright and colleagues document significant effect of human selection on at least 30 of the 774 assayed sites implicating that a large number and variety of genes were affected by human selection during the bottleneck that created maize. Coalescent simulations, such as those performed in previous work [18], depend on the assumption that the domestication bottleneck lasted for a maximum of 2800 years (see Table 2-1). The resulting estimate of the genetic bottleneck is about 3500 plants or about

TABLE 2-2 Estimated Divergence Times between *Zea mays* and *Z. m.* ssp. *parviglumis* or Time to Fixation of Maize Allele in Domesticated Populations

Gene	Divergence time (years BP)	Reference
c1	12,650	[13, 22]
Tga1	5000–15,000	[50]
Microsatellites	5689–13,093	[29]

10% of the teosinte population estimated to be in existence today.

Estimates of genetic bottlenecks and their duration vary slightly depending on the assumptions used. Wright and colleagues [53] have depended on an estimate provided by Hugh Iltis in 1983 for maize's domestication. His estimate was based on limited knowledge about the maize from Tehuacan at the time. Those few estimates that have attempted to independently verify when domestication was initiated push the date back even further.

Archaeological-Macrobotanical-Evidence of Teosinte Domestication and Maize Agriculture

Early evidence for human domestication of teosinte was recovered from a small ash lens between zones A and B1 at Guila Naquitz in the Mexican state of Oaxaca. A date on charcoal fragments from Zone B1 yielded a date of about 5800 BC [36]. Two noncarbonized cobs yielded AMS dates that have an average sample age of 4386 BC. The rachis of all three specimens is strongly indurate, rigid, and has a glossy, mottled sheen that closely resembles that of teosinte fruitcases. Two of the specimens (Figure 2-2) have two rows of grain like the teosinte female inflorescence; the third had four rows of grain. This subtle difference—two versus four rows of grain—is one of few vital characteristics that distinguish maize and teosinte. The size of the grain is not significantly different from that of the earliest maize from San Marcos Cave in the Tehuacan Valley [4, 5], although the inflorescences from San Marcos do not have the same cupule shape or mottled rachis sheen seen in the Guila Naquitz specimens. The San Marcos specimens are also significantly later—approximately 700 years, the earliest dating to 3443 + 38 BC. The earliest cobs from San Marcos are **contempo-**

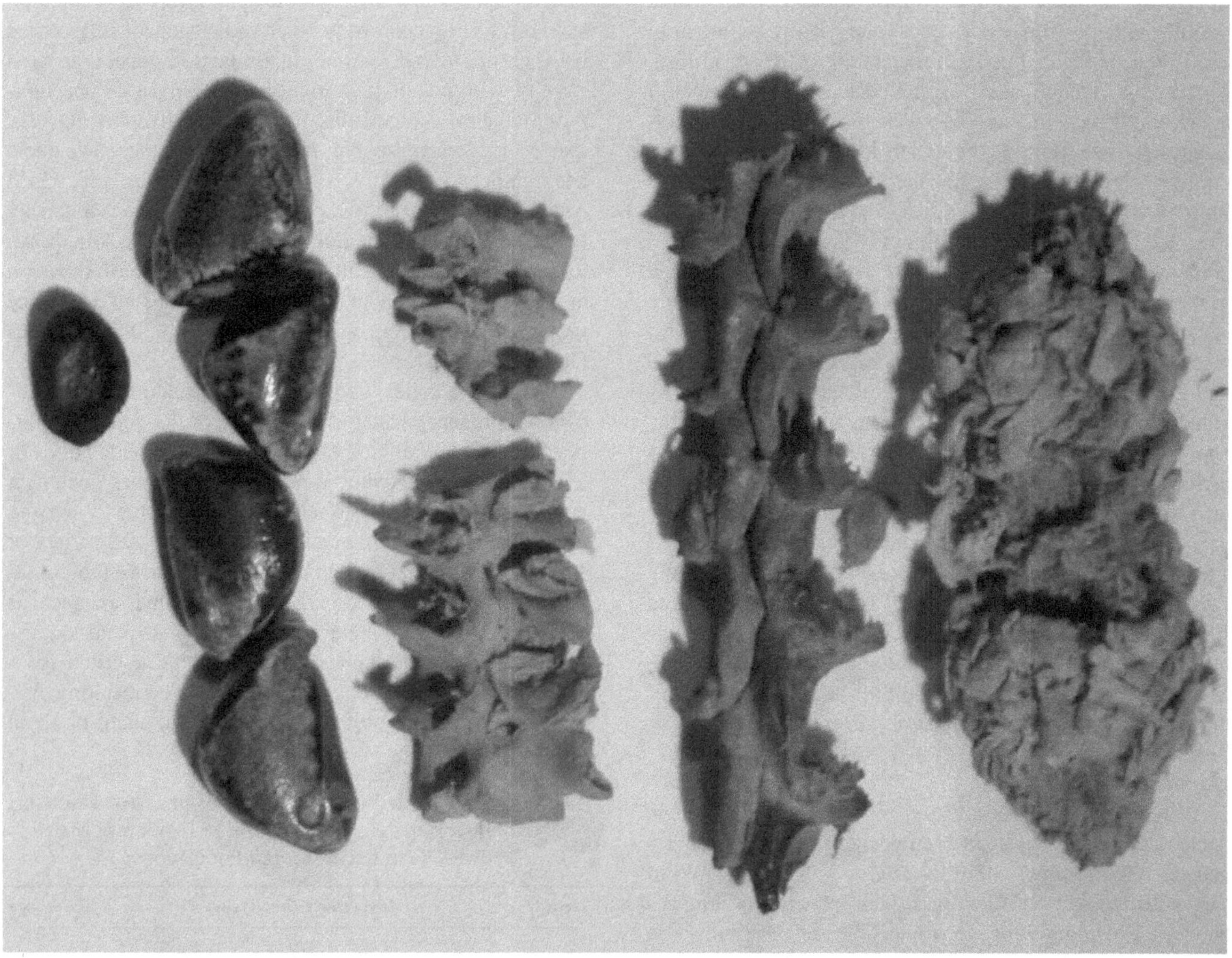

FIGURE 2-2 Archaeological evidence of early maize evolution from Mesoamerican sites. From left to right: Carbonized teosinte fruitcase fragment from Zohapilco. Modern fruitcases from *Zea mays* ssp. *parviglumis*. Teosinte cob fragments from Guila Naquitz C9. Distichous four-rowed cob fragment from D10. Cob from San Marcos Cave 3–5 Zone F. Scale in millimeters. (Courtesy of Bruce Benz) See color insert.

raneous with the earliest at Coxcatlan (Benz, unpublished data, [45]). Early specimens from San Marcos are typical of maize with one significant exception: A significant number of specimens (ca. >5%) are distichous and four-rowed like the specimen from unit D10 [3] at Guila Naquitz.

Intense selection documented by high **evolutionary rates** of morphological change in the maize rachis indicates increasing human dependence on maize in Tehuacan during the latter half of the fourth millennium BC [6, 7]. Estimates of genetic bottlenecks for the earliest transition in Tehuacan are based on the calculation of evolutionary rates. The earliest estimates of bottleneck size vary from 29 to 270 individuals. These estimates are based on a well dated transition that measured 185 years and coincide with the highest overall rates of change. Yet, if we assume that abundance of these remains of early maize from San Marcos accurately reflects their importance, one is led to believe that even though fully domesticated and subject to intense selection, maize was of limited importance in the human diet at this time.

If maize was of limited value, the rapid rates of morphological change might seem counterintuitive. That humans would have had such an evolutionary effect on a species so completely dependent on them for survival—requiring protection from marauding macaws and other seed eating birds, deer, coati, and multiple species of rodents—as well as reproduction, is suggested by the aforementioned genetic evidence for either narrow and short-lived or larger and longer-lived bottlenecks. Seed stores set aside for the next years' plantings were likely marked by occasional seed loss due to seed predators while stored or when they were first planted. Cultivators take their greatest risk when replanting gathered stores of seed because this seed, once domesticated, becomes relatively defenseless; and hence it might be better used by feeding the family or stored until later when protection can be afforded the germinating grains. Loss of stored seed is still relatively common among dry land farmers today. The less effort expended in protecting planted seed, the greater the probability the domesticate population will experience narrow and perhaps repeated short-term bottlenecks. Absent human protection, the more likely morphological variation will appear to be nondirectional because seed predation would be expected to focus and eliminate the traits that are advantageous to humans. On the other hand, evidence for strong directional selection would suggest that cultivators are making the effort to foster the development of certain desirable characteristics among the range of possible phenotypes that achieve reproductive maturity. Directional selection documented at San Marcos and Coxcatlan would be likely to achieve the rapid rate of change at the estimated small population sizes [5, 7].

Once maize appears in the rock shelters of Tehuacan, its cobs tend to increase in abundance and size until Spanish contact (see Chapter 5). Shortly after maize first appears in Tehuacan, it also appears on the Pacific coast of Chiapas (see Chapter 4).

Pollen Evidence for Use of *Zea* and Climate Change and Phytoliths Document Neotropical Plant Domestication

Evidence that climate changed significantly in Mesoamerica during the Paleoindian and Archaic periods is varied. Flannery [19] documents a faunal change during the Early Ajuereado phase in Tehuacan (<6500 BC). The faunal evidence for climate change at Guila Naquitz is less certain. Statistical tests failed to demonstrate differences between modern and Preceramic assemblages of rodents in all but three species suggests some environmental changes were occurring. James Shoenwetter's [41] and Shoenwetter and Smith's [42] analyses of the pollen from Guila Naquitz also shows consistently similar assemblages of native non-weedy species in the Preceramic deposits at Guila Naquitz. Shoenwetter and Smith's analysis does, however, show significant amounts of pollen that can be attributed to environmental disturbance of local vegetation over the course of occupation at Guila Naquitz, some associated with the earliest macrobotanical evidence for teosinte domestication. The evidence is ambiguous, however, and is possibly related to local or global environmental change.

Despite the meager evidence for change detected in archeological sediments, palynological, isotopic, and sedimentary evidence highlight the mid-Holocene (7000 to 5000 BP) as a period of profound change [54] even while there is a paucity of evidence for such change in Mexico. According to Steig, although land air temperatures appear to have declined across the polar regions, lower latitude change appears less dramatic and less obvious. Paleobotanical evidence suggests tropical and subtropical areas either cooled or became more arid or both. In some cases, climatic shifts appear to have been abrupt. Atmospheric and oceanic circulation patterns also appear to have changed considerably during the mid-Holocene. Surface temperature in the North Atlantic warmed between 8000 and 5000 years ago. The tropical Pacific and Antarctic surface waters cooled as well between 7000 and 5000 years ago. Atmospheric CO_2 increased by 10 ppm between 7000 and 5000 years ago at a rate of about 0.5 ppm per century. These changes potentially accompanied a decrease in terrestrial biomass at low latitudes because of increasing aridity and decreasing temperatures that accompany a decrease in the net fixation of carbon [54]. Steig's evidence suggests global change.

Others have documented similar environmental change in Mesoamerica. For example, changes in forest cover appear at a time when *Zea* pollen appears in the coastal Laguna Catrina near the port of Veracruz on the Gulf Coast [43]. Pollen grains larger than 65 μm appear first around

4100 BC, persist until 3300, and then reappear around 2400 BC to persist for more than 3000 years. Two short-lived, discrete stratigraphic occurrences of maize pollen appear again around AD 1100–1200. Goman and Byrne [21] corroborate the presence of *Zea* in the Gulf Coast lowlands between 2900 and 2800 BC at Lake Pompal in Los Tuxtlas slightly above sea level (700 m).

Archaeological work near the Olmec site of La Venta further east along the Gulf Coast has yielded even earlier evidence of *Zea* [40]. Pollen grains from estuary sediments dating to around 5100 BC are smaller than is typically acceptable for maize pollen. Larger *Zea* pollen grains appear shortly thereafter and persist for at least 1000 years. Their appearance and persistence is associated with disturbance in much the same way as at Laguna Catarina, suggesting human manipulation of suitably productive environments. These works document changing local conditions that are associated with the appearance of *Zea* pollen.

Few palynological studies of midaltitude and high-altitude sites describe local vegetation change associated with the entrance of *Zea* during the late to middle Holocene in Mesoamerica. Rue's work at Yojoa Lake and Petapilla Bog [55] indicates maize is present in southern Honduras by 3800 BCE Shoenwetter's analysis of Guila Naquitz sediments yielded a few grains of *Zea* pollen that were deemed too small to represent maize dating as early as 7300 BC and persist until 6700 BC [41, 42]. Work in central Mexico [30, 35] document a much later—only 2000 BC—presence of *Zea* associated with vegetation change. These sites document maize's earliest cultivation in central Mexico locales above 1700 meters. Buckler and colleagues [9] also argue for climatic shifts in Mesoamerica during the Holocene. Buckler and colleagues posit changes that contradict Steig's [54] summary suggesting that the Late Pleistocene was characterized by a cold and dry regime that was supplanted around 9000 BP by warm and moist conditions until about 6000 BP. This warm and moist period was followed by essentially modern conditions, effectively supporting the palynological evidence from Guila Naquitz by Shoenwetter and Smith [42]. But climatic changes were probably not uniformly felt across the Mesoamerican landscape. It is possible that Mexico's rugged terrain would have had local influences on the larger-scale global effects noted earlier.

Evidence of semipermanent settlement in the context of local environmental change is provided by the late Christine Niederberger [33, 34]. Playa Phase (I and II) occupations (ca. 8000–6500 BP, ca. 6100 BC) at Zohapilco on the Cerro de Tlapacoya near Chalco, Mexico, produced a variety of plant and animal remains in association with hearth areas. Although no well-defined living floors were unearthed, the hearths, rock alignments, and wide diversity of associated plant and animal remains document occupation during at least two distinct seasons that suggests long-term residence. This apparent multiseason occupation is made more intriguing by the finding of a single teosinte fruitcase fragment (see Figure 2-2) and more than a few pollen grains. This archeological evidence from the central highlands suggests semipermanent human residence in an environment long inhabited by teosinte *(Z. mays* ssp. *mexicana).*

Small *Zea* pollen grains in San Andres mentioned previously dated around 5100 BC are small enough to be teosinte. Maize pollen—identified on the basis of size of pore/axis diameter—appears by 4000 BC. Shoenwetter identified pollen grains from strata that predate the first maize ears at Guila Naquitz by 4290 BC. *Zea* pollen from San Andres, located in the Gulf Coastal Lowland Forests, associated with forest clearance suggests that such considerable effort on the part of the human inhabitants to plant maize could be construed as evidence for semipermanent habitation. Taken together with the pollen and fruitcase fragment from Zohapilco, this evidence suggests the human–*Zea* association is earlier than the macrobotanical evidence from Guila Naquitz and Tehuacan attests.

Plants, and some protozoans, incorporate silica into their cells to reduce grazing pressures by herbivores. The incorporation of silica into plants' epidermal cells whose shapes and sizes vary leads to a diversity of shapes that are taxonomically useful. Archaeologists have found them useful, oftentimes, as diagnostic indicators of the human–plant species associations [38]. Phytoliths have been the cornerstone of contentious arguments concerning the earliest domesticated plants in South and Central America; in particular they form the data base for identifying the antiquity of maize's dispersal southeastward from its Mesoamerica heartland. Piperno and Pearsall [38] identify early evidence for maize's cultivation in South America during the first half of the fifth millennium BC associated with the Vegas culture on the coast of Ecuador. Phytoliths associated with agricultural sites of Valdivia are also identified by Piperno and Pearsall between 4200 BC and 2200 BC, though the former remain controversial. Macrobotanical evidence of maize is present only in the latest or Terminal Valdivia site of Anllulla before 1900 BC [38]. The macrobotanical evidence of coastal maize cultivation in South America is later and coincides with maize cultivation in Mesoamerica at midelevations in Tehuacan and Tamaulipas but predates the earliest macrobotanical evidence for maize use in Pacific Coastal Mesoamerica. The phytolith evidence for maize, in contrast, antedates macrobotanical evidence at midelevations Mesoamerican locations and is coincident or later than pollen evidence of *Zea*'s presence along the Mesoamerican Gulf Coast depending on whether one accepts size as indicative of *Zea* or maize.

Further inland in lowland Ecuador, Lake Ayauch produced maize pollen and phytoliths dating around 3800 BC. The microremains are associated with vegetation disturbance indicated by *Cecropia* and charcoal. Similarly, pollen of *Zea* occurs in lake core sediments at Hacienda Lusitania before 4000 BC. At the nearby Hacienda Lusitania, maize pollen first appears around 5600 BC [8, 31, 32, 38].

At Abeja in the Colombian Amazon, maize pollen appears and persists in association with vegetation disturbances from 3450 to 3000 BC. Other pollen cores cited by Piperno and Pearsall [37, 38] provide evidence that vegetation disturbance accompanies human occupation of the eastern Amazon around 1600 BC, almost two millennia after such disturbance occurs in association with pollen evidence of *Zea* in the South American lowlands. In the Roosevelt's Parmana cultural sequence in the Orinoco Basin, maize pollen occurs around 800 BC.

If one accepts the aforementioned pollen and phytoliths evidence, then Coastal Peru obtained maize relatively late in comparison with coastal Ecuador and Colombia. Not until 2500 BC was maize present in 50% of sites. This is at least 1500 years later than South American sites located no more than 600 miles distant and 1000 years later than the earliest Mesoamerican macrobotanical evidence of maize [11, 12, 38].

Panamanian sites also document vegetation change, and in some cases this is associated with maize presence in Central America. The Aguadulce Rockshelter located on the Pacific Coastal Plain yielded phytoliths that Piperno attributes to *Zea* that resemble teosinte (67% decorated phytoliths) more than maize (>40% decorated phytoliths) somewhere in the 5800 to 4200 BC time frame. *Zea* phytoliths continue up through the stratigraphic sequence to 2920 BC [38].

In summary, the microfossil evidence suggests that *Zea* occurred well outside the current geographical range of its two closest wild relatives *Z. mays* ssp. *parviglumis* and *Z. mays* ssp. *mexicana* before evidence documenting maize's domestication. The pollen grains attributed to teosinte identified in at least three sites in present-day Mexico—Guila Naquitz, San Andres, and Zohapilco—before the first domesticated teosinte cobs appear in Guila Naquitz is compelling. It is intriguing to imagine how and why teosinte might have been dispersed and manipulated to have been so widely distributed so early and yet leave so little evidence of use. Skeptics will argue that perhaps movement of pollen or phytoliths through the profile might displace the evidence by as much as 10 centimeters and that such displacement explains the early geographically disjunct sites [43]. But can such an argument explain all the early occurrence of pollen in Mexico and South America?

JUXTAPOSING THE ARCHAEOLOGICAL AND GENETIC EVIDENCE FOR EARLY MAIZE

I have reviewed the genetic evidence that specifically addresses mechanisms and timing of teosinte domestication. Genetic evidence argues that humans domesticated teosinte, through a short and perhaps repeated, narrow genetic bottleneck or a long protracted bottleneck. Some of this genetic evidence suggests the duration of the bottleneck could have been as short as 300 years. Results of evolutionary rate studies based on macrobotanical evidence indicate focused selection did occur at least once over a short period of time (~200 years) that could have occurred with a population of ca. 200 plants. Thus there appears to be a certain degree of agreement that a small short-lived bottleneck may be responsible for the genetic polymorphisms in neutral gene sequences, in gene sequences that reflect strong selection and in quantitative variation of archaeological maize cobs from Tehuacan.

But what about timing? When did humans first manipulate teosinte to the extent that a founder population gave rise to speciation? The genetic estimates have large confidence intervals (see Table 2-2), though most coincide with an average date around 10,000 to 11,000 years ago.

Abundant archaeological evidence on this point is still lacking because all of the pollen and phytoliths data describe few grains or silica bodies when they first appear. *Zea* pollen and phytoliths suggest that teosinte may have been widely dispersed, perhaps as early as 6500 to 7000 years ago. This is a significant discrepancy from estimates based on genetic evidence, but the discrepancy is that the pollen and phytolith evidence is too young.

Returning to a point made previously and elsewhere [2], the antiquity of teosinte domestication will be most accurately and precisely determined by archaeologists. Other methods of estimating when human behaviors touched off the cascade of events leading to maize can never be accurately calculated simply because they are rife with assumptions about demographics, selection intensities, and the number of appropriate genes or nucleotides. But if archaeology is the most appropriate method of making this determination, then why is there a discrepancy? The genetic estimates, which are vague and dependent on assumptions that compromise their accuracy, are nevertheless internally consistent. The estimates of bottleneck size and duration with a variety of neutral gene sequences and estimates of generations separating two gene sequences with fixed differences are corroborated with quantitative estimates based on archaeological material. Each lends credence to the other. If the genetic estimates are consistent, then the problem resides in the archaeological camp, the discipline most capable of making the most precise and accurate determination. Archaeologists have focused their efforts on identifying where and when maize was domesticated in only a few areas of Mesoamerica to answer this question. Many proposals have been tendered to reexamine these areas but have been rejected on the basis that Flannery [19], MacNeish [27], or others could not possibly have missed something. More importantly, the areas of Mesoamerica where the

Archaic or Paleoindian periods have been well studied are not within the current geographic range of maize's closest wild relative. We need new enthusiasm and tenacity to conduct the systematic archaeological work in this region. And this work needs to be carried out before grazing and deforestation take their toll on the few populations that remain so that appropriate estimates of yield and harvesting efficiency can be carried out. The time to address the problem of when teosinte fell within the sights of early Mesoamericans is now. It is a great time to be interested in the origin and evolution of maize.

Acknowledgments

This work was supported by NSF 91490. J. Staller critiqued an earlier version though it should not be mistaken for an endorsement of its contents, I am nevertheless grateful to him for calling my attention to my uncritical reading of some of the published literature cited herein.

References Cited

1. M. Bellon, S. Brush. (1994). Keepers of the maize in Chiapas. *Economic Botany, 48,* 196–200.
2. B. Benz. (1999). Reconstructing the racial phylogeny of Mexican maize: where do we stand? In: S. Johnannssen, C. A. Hastort (Eds.), *Corn and culture in the prehistoric New World.* Boulder, CO: Westview Press. pp. 157–180.
3. B. Benz. (1999). On the origin, evolution and dispersal of maize. In M. Blake, *Development of agriculture and emergence of formative civilizations in Pacific Central and South America. The prehistory of the Pacific Basin.* Seattle: WSU Press. pp. 25–38.
4. B. Benz. (2001). Archaeological evidence of teosinte domestication from Guilá Naquitz, Oaxaca. *Proceedings of the National Academy of the Sciences of the United States of America, 98,* 2104–2106.
5. B. Benz. (In press). Evidencia arqueológica sobre la evolución del maíz. In: A. Casas, B. Rendon, (coords.), *Procesos de evolución de plantas bajo domesticación en Mesoamérica.*
6. B. Benz, H. H. Iltis. (1990). Studies in archaeological maize I: The "wild" maize from San Marcos cave reexamined. *American Antiquity, 55,* 500–511.
7. B. Benz, A. Long. (2000). Early evolution of maize in the Tehuacan Valley, Mexico. *Current Anthropology, 41,* 459–465.
8. W. Bray, B. Herrera, M. Schrimpff, P. Botero, J. Monslave. (1987). The ancient agricultural landscape of Calima, Colombia. In: W. Denevan, K. Mathewson, G. Knapp, (Eds.), *Pre-hispanic agricultural fields in the Andean region.* Oxford, UK: British Archaeological Reports International Series. pp. 443–481.
9. E. Buckler, D. Pearsall, T. Holtsford. (1997). The impact of early Holocene climate and plant ecology on Central Mexican Archaic subsistence. *Current Anthropology, 39,* 152–163.
10. M. Bush, D. Piperno, P. Colinvaux. (1989). A 6000 year history of Amazonian maize cultivation. *Nature, 349,* 303–305.
11. T. Dillehay, P. Netherly, J. Rossen. (1989). Middle Preceramic pubic and residential sites on the forested slope of the western Andes, Northern Peru. *American Antiquity, 54,* 73–79.
12. T. Dillehay, J. Rossen, T. Neatherly. (1997). The Nanchoc tradition: The beginning of Andean civilization. *American Scientist, 85,* 46–55.
13. J. Doebley. (1990). Molecular evidence and the evolution of maize. *Economic Botany, 44*(3), 6–27.
14. J. Doebley. (2004). The genetics of maize evolution. *Annu Rev Genet, 38,* 37–59.
15. J. Doebley, A. Stec. (1991). Genetic analysis of the morphological differences between maize and teosinte. *Genetics, 129,* 285–295.
16. J. Doebley, A. Stec. (1993). Inheritance of the morphological differences between maize and teosinte: Comparison of results for two F_2 populations. *Genetics, 134,* 559–570.
17. J. Doebley, A. Stec, L. Hubbard. (1997). The evolution of apical dominance in maize. *Nature, 386,* 485–488.
18. R. L. Eyre-Walker, J. Gaut, H. Hilton, D. L. Feldman, B. S. Gaut. (1998). Investigating the bottleneck leading to the domestication of maize. *Proceedings of the National Academy of the Sciences of the United States of America, 95,* 4441–4446.
19. K. Flannery. (1967). The vertebrate fauna and hunting patterns. In: D. Byers (Ed.), *The prehistory of the Tehuacan Valley* Vol. 1. Austin: The University of Texas Press.
20. P. Goloubinoff, S. Paabo, A. Wilson. (1993). Evolution of maize inferred from sequence diversity of an adh2 gene segment from archaeological maize. *Proceedings of the National Academy of the Sciences of the United States of America, 90,* 1997–2001.
21. M. Goman, R. Byrne. (1998). A 5000 year record of agriculture and tropical forest clearance in the Tuxtlas Veracruz, Mexico. *The Holocene, 8,* 83–89.
22. M. Hanson, B. Gaut, A. Stec, S. Fuerstenberg, M. Goodman, E. Coe, J. Doebley. (1996). Evolution of anthocyanin biosynthesis in maize kernels: the role of regulatory and enzymatic loci. *Genetics, 143,* 1395–1407.
23. J. Harlan, J. DeWet, E. Price. (1973). Comparative evolution of cereals. *Evolution, 22,* 311–325.
24. H. Hilton, B. Gaut. (1998). Speciation and domestication in maize and its wild relatives: Evidence from the Globulin-1 gene. *Genetics, 150,* 863–872.
25. V. Jaenicke-Després, E. Buckler, B. Smith, M. Gilbert, A. Cooper, J. Doebley, S. Pääbo. (2003).Early allelic selection in maize as revealed by ancient DNA. *Science, 302,* 1206–1208.
26. P. C. Mangelsdorf. (1974). *Corn: Its origin, evolution and improvement.* Cambridge, MA: Belknap Press of the Harvard University Press.
27. R. MacNeish. (1967). A summary of subsistence. In: D. Byers (Ed.), *The prehistory of the Tehuacan Valley.* Vol 1. Austin: University of Texas Press. pp. 290–309
28. B. McClintock, T. A. Kato Y., A. Blumenschein. (1981). *Chromosome constitution of races of maize (Constitución chromosoma de razas de maíz).* Chapingo, Mexico: Colegio de Postgraduados.
29. Y. Matsuoka, Y. Vigouroux, M. Goodman, J. Sanchez G., E. Buckler, J. Doebley, A single domestication for maize shown by multilocus microsatellite genotyping. *Proceedings of the National Academy of the Sciences of the United States of America, 99,* 8060–8064.
30. S. Metcalfe, F. Street-Peirott, R. Brown, P. Hales, R. Perrott, F. Steininger. (1989). Late Holocene human impact on lake basins in central Mexico. *Geoarchaeology, 4,* 119–141.
31. S. Monsalve. (1985). A pollen core from the Hacienda Lusitania. *ProCalima, 4,* 40–44.
32. S. Mora, L. Herrera, I. Cavelier, C. Rodriguez. (1991). Cultivars, anthropic soils and stability. *University of Pittsburgh Latin American Report No. 2.* Pittsburgh, PA: Department of Anthropology, University of Pittsburgh.
33. C. Niederberger. (1976). *Zohapilco: Cinco millenios de ocupacion humana en un* sitio lace de la cuenca de Mexico. Coleccion Cientifica 30. Mexico, D.F.: INAH, 1976.
34. C. Niederberger. (1979). Early sedentary economy in the basin of Mexico. *Science, 203,* 131–142.
35. S. O'Hara, A. Street-Perrott, T. Burt. (1993). Accelerated soil erosion around a Mexican highland lake caused by prehispanic agriculture. *Nature, 362,* 48–51.

36. D. Piperno, K. Flannery. (2001). The earliest archaeological maize *(Zea mays L.)* from Mesoamerica: New accelerator mass spectrometer dates and their implications. *Proceedings of the National Academy of the Sciences of the United States of America, 98,* 2101–2103.
37. D. Piperno, M. Bush, P. Colinvaux. (1990). Paleoenvironments and human occupation in late Glacail Panama. *Quaternary Research, 33,* 108–116.
38. D. Piperno, D. Pearsall. (1998). *The origins of agriculture in the Lowland Neotropics.* San Diego: Academic Press.
39. H. Perales, B. Benz, S. Brush. (2005). Maize diversity and ethnolinguistic diversity in Chiapas, Mexico. *Proceedings of the National Academy of the Sciences of the United States of America, 102,* 949–954.
40. K. Pope, M. Pohl, J. Jones, D. Lentz, C. von Nagy, F.Vega, I. Quitmyer. (2001). Origin and environmental setting of ancient agriculture in the lowlands of Mesoamerica. *Science, 292,* 1370–1373.
41. J. Schoenwetter. (1974). Pollen records of Guila Naquitz Cave. *American Antiquity, 39,* 292–303.
42. J. Shoenwetter, L. Smith. (1986). Pollen analysis of the Oaxacan Preceramic. In: K. Flannery (Ed.), *Guila Naquitz.* Orlando, FL: Academic Press.
43. A. Sluyter. (1997). Regional, Holocene records of the human dimension of global change: sea level and land use change in prehistoric Mexico. *Global and Planetary Change, 14,* 127–146.
44. B. Smith. (2005). Reassessing Coxcatlan Cave and the early history of domesticated plants in Mesoamerica. *Proceedings of the National Academy of the Sciences of the United States of America, 102,* 9438–9444.
45. M. Stuiver, P. Reimer. (1993). Extended 14C database and revised CALIB radiocarbon calibration program. *Radiocarbon, 35,* 215–230.
46. M. Tenaillon, M. Sawkins, A. Long, R. Gaut, J. Doebley, B. Gaut. (2001). Patterns of DNA sequence polymorphism along chromosome 1 of maize (*Zea mays* ssp. *mays*). *Proceedings of the National Academy of the Sciences of the United States of America, 98,* 9161–9166.
47. M. Tenaillon, M. Sawkins, L. Anderson, J. Doebley, S. Stack, B. Gaut. (2002). Patterns of diversity and recombination along chromosome 1 of maize (*Zea mays* ssp. *mays* L.). *Genetics, 162,* 1401–1413.
48. M. Tenaillon, J. U'Ren, O. Tenaillon, B. Gaut. (2004). Selection versus demography: A multilocus investigation of the domestication process in maize. *Molecular Biology and Evolution, 21*(7), 1214–1225.
49. S. Whitt, L. Wilson, M. Tenaillon, B. Gaut, E. Buckler. (2002). Genetic diversity and selection in the maize starch pathway. *Proceedings of the National Academy of the Sciences of the United States of America, 99,* 2959–2962.
50. H. Wang, T. Nussbaum-Wagler, B. Li, Q. Zhao, Y. Vigouroux, M. Faller, K. Bomblies, L. Lukens, J Doebley. (2005). The origin of the naked grains of maize. *Nature, 436,* 714–719.
51. R. Wang, A. Stec, J. Hey, L. Lukens, J. Doebley. (1999). The limits of selection during maize domestication. *Nature, 398,* 236–239.
52. S. White, J. Doebley. (1999). The molecular evolution of *terminal ear1,* a regulatory gene in the genus *Zea. Genetics, 153,* 1455–1462.
53. S. Wright, I. Vroh Bi, S. Schroeder, M.Yamasaki, J. Doebley, M. McMullen, B. Gaut. (2005). The effects of artificial selection on the maize genome. *Science, 308,* 1310–1314.
54. E. Steig, E (1999). Mid-Holocene climate change. *Science, 386,* 1485–1487.
55. D. Rue. (1987). Early agriculture and early Postclassic Maya occupation in western Honduras. *Nature, 326,* 285–286.

C H A P T E R

3

Ancient DNA and the Integration of Archaeological and Genetic Approaches to the Study of Maize Domestication

VIVIANE R. JAENICKE-DESPRÉS* AND BRUCE D. SMITH†

*Evolutionary Genetics, Max Planck Institute for Evolutionary Anthropology, Leipzig, Germany

†Archaeobiology Program, National Museum of Natural History, Smithsonian Institution, Washington, D.C.

Glossary

Adenine One of the four bases that comprises DNA.

Allele An allele is one of the several alternatives forms of a gene occupying a given locus on a chromosome.

Diploid A cell having two copies of each chromosome.

DNA Deoxyribonucleic acid built up by the four bases guanine, cytosine, thymine, and adenine.

Gene Segment of DNA often coding for a protein.

Guanine One of the four bases that comprises DNA.

Heterozygous An individual is heterozygous for a particular locus when he or she carries two different alleles at this locus.

Homozygous An individual is homozygous for a particular locus when he or she carries two identical alleles at this locus.

Polymorphic A gene is polymorphic in a population when several alleles of this gene exist simultaneously in the population.

Promoter A region of DNA usually at the beginning of a gene that is needed to initiate transcription.

Transcription One of the steps in "gene expression." During the transcription the DNA is transcribed into RNA. This RNA is the template for protein synthesis that occurs in a step called "translation."

Transcription factor A protein that is involved in the transcription process of DNA into RNA.

Archaeology and genetics have, to varying degrees, independently shed light on the process of maize domestication. Archaeologists are limited to interpreting the phenotypic characteristics of archaeological cobs, whereas geneticists interpret the present-day genetic consequences of human and natural selection by analyzing modern maize and its wild ancestor teosinte. Innovative approaches in the study of ancient plant genetics now allow for the possibility of a more comprehensive understanding of the domestication process. This chapter illustrates how analysis of ancient **DNA** can be used to integrate archaeological and genetic approaches and addresses several different sets of questions concerning the early history of maize. Three vital **genes** in the domestication of maize are studied whose effects are not directly observable in archaeological corn cobs, and that influence maize plant architecture, as well as biochemical properties of starch and protein. We found that selection for these crucial characteristics in maize had occurred by 4400 BP. One of the genes studied also provides further support for the evolutionary or phylogenetic linkage between Southwestern landraces of maize and the Northern Flint landraces of the northeastern United States.

INTRODUCTION

Maize *(Zea mays* ssp. *mays),* one of the world's most important food crops, has undergone significant genetic change and diversification since it was first domesticated from a wild grass—teosinte *(Zea mays* ssp. *parviglumis)*—in southern Mexico more than 6300, and perhaps as long as 10,000 years ago [2, 43, 49]. As is the case with any domesticated plant, however, the transformation from wild progenitor to world crop did not occur during a single brief episode of change in the far-distant past. Domestication is, in many respects, much better viewed as a constantly unfolding and ongoing process. The long and complex history of genetic change that has occurred in maize over this span of more than six millennia encompasses a range of different traits or characters that reflect both the result of "unconscious selection"—adaptive responses by the crop to the new selective pressures associated with human planting, harvesting, and storage, as well as deliberate "methodical" human selection for desirable attributes [25]. Some of the more important changes that have occurred in maize since it was initially brought under domestication include modifications in vegetative structures (e.g., reduction in branching), seed head morphology (e.g., the number, placement, size, shape, and row number of cobs), seed characteristics (e.g., kernel shape, hardness, and color), and various starch and protein properties. When considering the domestication of any crop plant, including maize, it is of central importance to be able to establish both the temporal depth and the sequence of change in such characteristics that together define and document the long history of "improvement" of food crops over time as they become more reliable, higher in yield and nutritional value, and lower in cost in terms of harvest and processing time.

MORPHOLOGICAL AND MOLECULAR APPROACHES TO DOCUMENTING THE EARLY HISTORY OF MAIZE

Working with overlapping data sets, biologists and archaeologists have approached a better understanding of the early chapters of maize domestication from a number of different perspectives over the years. Much of the research on the early history of maize that is focused on the plant itself can be sorted into a simple four-cell set of what has today become interrelated, interdependent, and reciprocally illuminating approaches, with the four cells consisting of genetic research and morphological research on modern and on archaeological maize (i.e., morphology–modern; morphology–ancient; genetics–modern; and genetics–ancient).

For many years, efforts to gain a better understanding of the early evolution and diversification of maize were limited to general comparative morphological studies; first focused on modern maize, and then increasingly extended to include ancient maize recovered from archaeological contexts. The initial description and labeling of hundreds of different present-day landraces of maize throughout the Americas provided a foundation for broad-scale comparative consideration of their geographical and environmental settings, as well as documentation of variation in a wide range of cob and kernel attributes. These studies, in turn, allowed for the projection of spatial patterns of morphological variation in modern maize back into the past, and the formulation of proposed histories of maize evolution and diversification up through time as the geographical range of the crop plant expanded to the limits of cultivation in the Americas.

The general morphological characterizations or descriptive profiles of different modern varieties or landraces of maize have also been used as templates for the "identification" of ancient types of corn, based on similarities in cob and kernel morphology of archaeological specimens to the modern races (e.g., Chapalote, Nal-Tel, and Maiz de Ocho [41, 66]). Once identified on the basis of overall morphological similarity to modern template varieties, these ancient maize types could then be used, along with what was known regarding modern landraces, in more informed efforts to reconstruct the early history of this major crop plant by charting its geographical expansion and diversification up through time in the archaeological record. Morphological studies and general characterizations of present-day and historical landraces of maize thus formed the basis for the identification and profile characterization of maize varieties in the archaeological past, with both modern and ancient varieties then used to frame discussions of the early history of the crop plant.

Research on the role of maize in the economies of past societies has not, of course, been limited to gross morphological characterizations of ancient landraces. Recovered maize remains from archaeological contexts often consist of small cob fragments lacking kernels. This, in turn, limits assemblage analysis and larger regional and temporal comparative studies to a consideration of a small number of specific morphological characteristics of maize cobs (e.g., estimated cob row number). Valuable comparative morphological studies of maize remains in archaeological contexts can also be carried out on much smaller scales of analysis (e.g., consideration of the size and morphology of maize starch granules, phytoliths, and pollen preserved intact on processing tools and in other cultural contexts [47, 48, 51]).

Archaeologists also consider the role of maize in past economies and its relative importance in human diets within the larger context of reconstructing the overall way of life of the societies being studied, through analysis of material culture assemblages, settlements, storage facilities, the size and location of fields, and so on. In many regions, including the southwestern United States, for example, the relative abundance and ubiquity of cob and kernel remains in archaeological sites have been used as approximate measurements of the increasing dietary importance of maize up through time [69]. Similarly, changes in the size and abundance of maize processing tools such as manos and metates [22], and the appearance of specialized rooms used for grinding maize [8] are used as general indicators of the increasing economic importance of maize. In addition, in regions such as eastern North America, where C-4 carbon pathway plants such as maize are not a major indigenous food source, stable carbon isotope analyses of human burial populations can provide a much more direct measure of the changing dietary importance of maize by tracing temporal shifts in $\delta^{13}C$–values in human bone [1, 58]. Taken together, all of these different mutually reinforcing, morphology based approaches and associated archaeological context considerations have provided substantial information and insights regarding the early history of maize.

When combined with these existing approaches, molecular studies can expand access to a critically important, and up until now, unreachable general category of evidence of maize development and diversification: the timing and sequence of selection for characteristics that left no observable phenotypic signature in archaeological remains.

Genetic research on modern maize and teosinte, for example, has opened another significant potential window on the early history of this major crop. A number of different domestication genes that played an important role in the development of modern maize have been identified [63, 64, 67]. What has been lacking up until now, however, has been a clear understanding of the timing and sequence of selection of these genes.

Although analysis of archaeological maize has documented to some extent when some different crucial attributes of maize first appeared, these studies have been limited to a consideration of those few traits that can be easily observed on cob fragments (e.g., cob size and row number, glume, and cupule characteristics) [2, 3]. Left undocumented, however, are a number of central characteristics of domesticated maize that, until recently, have remained invisible and unrecognized in the archaeological record (e.g., traits affecting plant architecture, starch properties, kernel color, and growing habits). Given the continuing documentation of the modern maize genome and the associated recognition of additional genes that distinguish maize from its wild ancestor, the application of molecular approaches to archaeological maize, holds the promise of expanding and enriching our overall understanding of the long and complex history of maize.

Just as analysis of morphological variation in modern maize has set the stage for similar studies of archaeological specimens, so, too, does molecular analysis of archaeological maize depend on documentation of the modern maize genome. Once a specific gene, which has been the subject of selection during the long process of domestication, has been identified in modern maize, it can, in turn, be studied in archaeological maize, even if it left no recognizable morphological signature. Information regarding plant architecture, for example, can be obtained from analysis of DNA existing in cob fragments, even though the plant itself is not preserved [33]. Thus, genetic analysis of ancient maize, in theory, holds the promise of providing information on the timing and sequence of change in all of the genes that were subject to selection over the past 6000 years. In reality, however, analysis of selection of genes in archaeological maize depends not only on the successful identification of the genes in modern studies, but also on the availability of archaeological samples that contain DNA that can be extracted and amplified.

MOLECULAR LEVEL ANALYSIS OF ARCHAEOLOGICAL MAIZE: A CASE STUDY

This chapter offers a case study example of how analysis of selected genes in archaeological maize can significantly expand and improve our understanding of the overall timing and sequence of selection for different vital attributes in this domesticated crop plant over the past 6000 years. The three genes that are targeted are all of central importance in the unfolding transformation of the wild grass teosinte into maize: 1) *Teosinte branched-1 (tb1)* affects general plant architecture; 2) the *prolamin box-binding factor (pbf)* regulates the expression of storage proteins in the maize kernel; and 3) *sugary-1 (su1)* is involved in the starch biosynthesis pathway. Information gained during the present study regarding the timing and order in which these genes were selected is then combined with what has already been documented regarding selection for those morphological traits that can be directly observed in archaeological maize assemblages.

Monitoring for Selection of Preferred Attributes in Ancient Maize

Usually a single nuclear gene or part of a gene (sequence fragment) is represented in each **diploid** genome by two copies **(alleles),** one on each chromosome. When both alleles are identical and carry the exact same base–pair sequence, the individual is **homozygous** for this gene or sequence fragment. If the two alleles differ in their base–pair sequence the individual is **heterozygous** for the gene or sequence fragment. For any specific gene, there will be a number of different alleles present in the allele pool of a population. The greater the number of different alleles that are present for any particular gene, the greater the diversity of the gene or sequence fragment in the population. Alleles often not only differ in their sequence but can also vary slightly in their function. As a result, in domesticated crop plants, one allele can prove less advantageous than another from year to year, depending on annual changes in the surrounding natural and cultural environment.

Each year, early farmers were repeatedly selecting, either inadvertently or deliberately, for those alleles that were linked to certain preferred characteristics in maize. During this ongoing annual cycle of selection, the relative frequency of an advantageous allele increases, and simultaneously, the frequency of all other alleles for this gene or sequence fragment will decrease. Some alleles may even disappear entirely from the population. Thus, along with human cultivation and increased selection for certain attributes in domesticated crop plants comes a significant reduction in allele frequencies for particular genes when compared with wild teosinte populations. Such a reduction in diversity in the allele pool for certain genes as a result of domestication and human selection does not necessarily affect entire genes but, rather, can be restricted to a particular region of a gene (e.g., the regulatory **(promoter)** region such as in *tb1*).

As shown in Figure 3-1A, only the promoter region is reduced in diversity in *tb1*. The reduction of diversity does not extend to either the 3′ region [64] or to the 5′ region [6], indicating that no other neighboring genes are reduced in their diversity in association with reduced diversity in the promoter region.

As shown in Figure 3-1B, teosinte population allele pools include many different alleles within the promoter region of the *tb1* gene, whereas maize includes few. The full genetic diversity represented in wild populations of teosinte, however, was in all likelihood not present in those limited stands first targeted by humans for life cycle intervention and active management. Allelic diversity in *tb1* would have

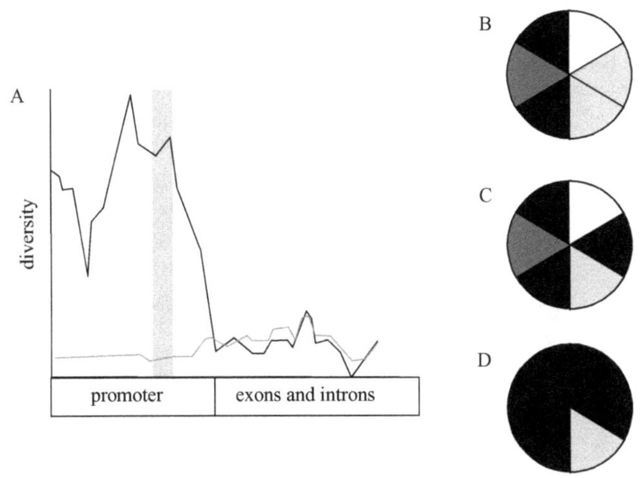

FIGURE 3-1 Reduction of diversity in selected genes. In **A,** the reduction of diversity is illustrated across the whole gene *tb1*. On the x-axis is the gene sequence with its regulatory promoter region, and its transcribed regions (exons and introns). On the y-axis the diversity is given (modified from [64]). Teosinte is colored black, whereas maize is colored gray. The gray bar indicates the 56-bp region in *tb1* analyzed in the ancient maize. Within this region the diversity (number of alleles) in teosinte is high and in maize low. Before an allele is selected; In **B,** several alleles exists in the population. With ongoing selection on one allele (black area); In **C,** the frequency of the selected allele increases, whereas other alleles get lost or decrease in frequency. After selection, in D, the selected allele is present in high frequency or fixed as the only allele left in the population.

been further reduced once new selection pressures associated with human planting and harvesting began. The frequency of the allele with selective advantages would have increased, whereas the frequency of other alleles would have decreased, in some cases to zero. After the new set of selection pressures associated with human management have been in effect for a period of time, the selected advantageous allele would be established at a high frequency or even fixed (Figures 3-1C and 3-1D). In the current study, we were interested in establishing the selection status of the three genes identified previously *(tb1, pbf, su1)* in archaeological maize assemblages (i.e., before, during, and after selection).

The three genes selected for this study are among the few that have so far been described as to function. *Tb1* is a **transcription factor** [10]. In maize, it carries an allele that represses the growth in axillary meristems and results in the unbranched plant architecture typical for maize. *Tb1* also contributes to the relocation of female ears and male tassels in maize plants in a pattern different from that seen in teosinte [13–15]. As part of a quantitative trait locus (QTL), *tb1* is not the only gene responsible for this set of changes but the one that is mainly responsible [15].

The *pbf* is involved in the control of storage protein expression in maize kernels [61, 62, 65]. Cereal storage proteins all have a conserved seven base pair (bp)-long sequence (P-box) which lies roughly 330-bp upstream of the **transcription** start. *Pbf* binds to this motif and acts as a transcription activator [61].

Su1 is involved in starch biosynthesis. Starch consists of the linear sugar amylose and the branched sugar amylopectin. *Su1* encodes a starch debranching enzyme expressed in kernels [52]. Debranching enzymes cleave the (1-6)-α-bonds in amylopectin [5] into short sugar molecules which can be turned into energy. Together with branching enzymes, *su1* determines the branching structure of amylopectin [34, 52]. The chain length of amylopectin and the ratio of amylose to amylopectin in starch are important for the gelatinization properties of starch [38, 67].

When an organism dies, its DNA begins to degrade. DNA recovered from archaeological contexts is often broken up into short fragments of only about 100-bp in length [30, 45]. As a result, when we studied the *tb1, pbf,* and *su1* genes in archaeological maize, we targeted sequence fragments that were still informative but short. For *tb1* we chose a 56-bp fragment in the promoter region of the gene. For *pbf* we amplified either a 94-bp fragment if the associated deletion was not present or, if the deletion was present, a 54-bp fragment that contained the **polymorphic** site [Jaenicke-Després, unpublished data]. For *su1* we analyzed a 60-bp fragment [33]. The amount of DNA present also decreases once an organism dies, and the extent to which surviving DNA is damaged increases [26, 46]. To ensure that the sequences recorded from archaeological maize were accurately determined, we amplified, cloned, and sequenced every fragment at least twice [26, 33]. As the amount of DNA in ancient samples decreases, and the number of starting molecules in an amplification process is reduced to low levels, allelic drop-out can occur [44]. In extracts consisting of small amounts of DNA, the two alleles for each sequence fragment will no longer be present in equal frequencies. As a result, amplification may start just from one of the two alleles, and the other allele will not be amplified at all. To address this potential problem, in those cases where we repeatedly observed two segregating alleles, we counted them both and assumed heterozygosity for the sequence fragment in question. At the same time, in those cases where just a single sequence was detected, the possibility of an allelic drop-out was recognized, and only one allele rather than two was counted.

The Archaeological Maize

A total of six maize cobs from Romero's and Valenzuela's Caves near Ocampo in Tamaulipas in northeastern Mexico [40, 57], and 11 maize cobs from Tularosa Cave in the Mogollon Highlands of New Mexico [42, 68] were initially selected for inclusion in the present study. Only 5 of the 6 cobs from the Ocampo Caves and 6 of the 11 Tularosa cobs, however, yielded DNA that could be successfully amplified. All of the cobs selected for analysis were uncarbonized, and with exception of a single specimen from Tularosa Cave, all lacked kernels. When the temporal context of each of these 11 cobs was established through direct, small-sample accelerator mass spectrometry (AMS) radiocarbon age determinations, they extend back almost 4000 years (Table 3-1) [33].

Tb1: Maize Plant Architecture 4400 Years Ago

Teosinte, the wild ancestor of maize, has a bush-like architecture with abundant long side branches, each of which is tipped by male tassel, and carries numerous small female grain spikes with 10 to 12 kernels each. Maize plants, in contrast, have a large dominant main stalk or axis, which carries a single terminal male tassel. Usually, two or three large cobs, each with up to 500 kernels are located on the end of short primary branches, close to the main axis. This striking difference between teosinte and maize in terms of branching structure, tassel, and cob placement is primarily a result of selection on the *tb1* gene. Acting in conjunction with other genes, *tb1* is responsible for about half of the

TABLE 3-1 Age Determinations, Archaeological Provenience, and Museum Source Information for 11 Analyzed Maize Cobs

DNA study sample inventory number	Radiocarbon lab sample number (β Analytic)	Dendrocalibrated age in calendar time scale years before present (BP)		^{14}C age in radiocarbon years before present (BP)	Archaeological site and horizontal/ vertical location	Museum collection and catalog number
		Intercept	1 sigma age range			
5	β 162533	660	710–560	710 ± 50	Tularosa Cave Square 2R2/Level 14	Field Museum 314984
6	β 162534	730	900–670	830 ± 50	Tularosa Cave Square 2R2/Level 14	Field Museum 314984
9	β 166754	750	900–680	850 ± 40	Tularosa Cave Square 2R1/Level 13	Field Museum 315040
3	β 160185	910	950–750	950 ± 50	Tularosa Cave no provenience	NMNH 246279
7	β 166753	1860	1920–1730	1900 ± 40	Tularosa Cave Square 2R2/Level 14	Field Museum 314984
14	β 166755	1870	1940–1800	1920 ± 40	Tularosa Cave Square 3R2/Level 11	Field Museum 315267
29	β 166756	2350	2370–2320	2340 ± 40	Romero's Cave Square S20/E15	Harvard Botanical Museum
28	β 85432	2735	2770–2455	2560 ± 60	Romero's Cave Square S25/W5/Level 6	Harvard Botanical Museum
51	β 166757	2750	2770–2720	2600 ± 40	Romero's Cave Square S25/W5/Level 3	Harvard Botanical Museum
31	β 85433	4300	4440–4125	3890 ± 60	Valenzuela's Cave Square E20/Level 2b	Harvard Botanical Museum
27	β 85431	4405	4510–4230	3930 ± 50	Romero's Cave Square S20/E10/Level 5	Harvard Botanical Museum

phenotypic plant architecture change that distinguishes maize from teosinte [14, 64].

In all 11 of the ancient cobs analyzed in the study, only a single allele, Tb1-M1, is recorded [33]. Although Tb1-M1 is also the only allele found in modern maize landraces, it had a much lower frequency of occurrence (36%) in teosinte [64]. It is important to keep in mind that our sample of archaeological maize dating to 4400 BP numbered only two cobs (Table 3-1), and because Tb1-M1 has a frequency of about 36% in modern teosinte, we cannot state with certainty that it was the only allele present. We believe, however, that there is a strong likelihood that Tb1-M1 was present in high frequency, and that modern maize plant architecture was established by 4400 BP, at least in the Ocampo Cave maize [33].

It is not surprising that the plant architecture characteristic of modern maize was already established by 4400 BP, given that maize had by then been a domesticated crop plant for 2000 years or more. Such a change in branching architecture, relocation, and reduction in the number of cobs and tassels reflects, in several respects, the automatic response by seed plants in general to the new and powerful set of selective pressures associated with deliberate human harvesting and planting of seed stock. As a result, Tb1-M1 would be expected to be one of the early attributes of *the adaptive syndrome of domestication* to be selected for under human management [23, 25, 59].

In the wild, the bush-like plant architecture of teosinte and the associated abundance of side branches, with their branch-end tassels and numerous small grain-spikes, has clear advantages in terms of a plant's likelihood of surviving to maturity and contributing seed to the next generation. The bush-like architecture of teosinte and the abundant side branch small grain-spike "packaging," along with associated staggered sequential tassel and individual seed maturation, reflects a strong selective advantage for broad dispersal of risk of plant predation and seed loss because of natural causes like predators or climate. Such diversification of risk was no longer necessary or advantageous, however, when humans began to intervene in the life cycle of some populations of teosinte. Once humans began to harvest and store seed stock for planting, selection during harvesting would have strongly favored Tb1-M1. Human harvests would have selected for larger, more visible, and more easily harvested seed packages containing uniformly mature kernels (i.e., corn cobs closely attached to the main axis) [23, 25, 59]. In addition, the tall, narrow plant architecture of maize would allow greater crop density and yields.

Pbf and *Su1*: The Development of Starch and Protein Properties

It has been suggested that maize may have been initially brought under cultivation not as food source but for its sweet juice [56]. Today, however, maize is primarily grown as a source of starch and protein. Because maize starch and protein are stored in kernels, which are only occasionally preserved in archaeological contexts, it has been difficult until now to study the early history of development of maize starch and protein structure. In the current study, we were able to directly access genetic information regarding *pbf* and *su1*, two genes that influence protein and starch quality and quantity. In contrast to *tb1*, which was represented by a single allele, several alleles were identified for both *pbf* and *su1* in the 11 archaeological maize cobs we analyzed.

The *pbf* allele Pbf-M1 (with the base **adenine** at the polymorphic site), the major allele in modern maize, was present in all 11 archaeological cobs [33]. Pbf-M3 (with the base **guanine** in the polymorphic site and additionally the 40-bp deletion), also existing in modern maize, was present in 3 of the 5 cobs from Tularosa Cave, and in 1 cob from the Ocampo Caves. These 4 cobs were heterozygous for Pbf-M1-/-Pbf-M3. Interestingly, the presence and relative frequency of these two alleles for *pbf* in the cobs analyzed from the Ocampo Caves is comparable with that which occurs in modern maize (Table 3-2). This observed heterozygosity in the ancient samples, together with the observation that no damaged base–pairs could be observed in the ancient sequences, suggests that in general DNA was preserved in the ancient samples well.

The gene *su1* was similarly represented by two alleles (Su1-M1 and Su1-M2) in the archaeological cobs analyzed. Both of these alleles often appear in modern maize but are rare in teosinte [33]. Although the samples from the Ocampo Caves exclusively carried Su1-M2, ancient maize from Tularosa Cave was more variable. Although all five of the Tularosa Cave maize cobs carried the allele Su1-M1 in at least one chromosome, the two cobs from the early occupation of the Cave, dating to ca. 1900–1800 BP (cobs 7 and 14) were both heterozygous. Although one of the early component cobs from Tularosa Cave was heterozygous for Su1-M1-/-Su1-M2, the other carried an allele (Su1-T1), which occurs today in teosinte populations but not in maize. Alleles such as Su1-T1, which are present in modern teosinte and ancient maize but not in modern maize, likely represent alleles selected against during the last 2000 years of maize development. Thus, the presence of Su1-T1 in one of the Tularosa cobs might indicate that the selection process in this ancient population was not yet completed. A much larger sample of maize from the early occupational component of Tularosa Cave will have to be analyzed, however, before the frequency of occurrence of Su1-T1 can be determined with greater precision. Because modern maize still carries some alleles in low frequency, however, it appears more likely, based on the limited data currently available, that selection for both *pbf* and *su1* was complete by 4400 years ago.

Although the functions of both the *pbf* and *su1* are only partly understood, it is possible to offer some insights

TABLE 3-2 Allele Frequencies in Modern and Ancient Maize and Teosinte. For modern maize, the allele frequencies are given for the all modern maize together and for each region (Southwest, Middle and South America) separately. Allele frequencies are given for the major alleles in *pbf* and *su1*. For all modern maize and teosinte we assumed that each plant contains two alleles per individual. For the ancient samples we assumed that allelic drop-out occurs, thus the number of alleles used is given in brackets after the frequencies and differs from the theoretical number of alleles.

	Modern Maize				Wild Ancestor	Archaeological Maize	
	Total modern	Southwest U.S.	Mexico	South America	Teosinte parviglumis	Ocampo	Tularosa
ind.	72	16	27	29	27	5	6
Pbf-M1	0.77	0.78	0.89	0.66	0.13	0.83 (6)	0.67 (9)
Pbf-M2	0.01	0.06	0	0	0.52	0	0
Pbf-M3	0.19	0.16	0.07	0.34	0.07	0.17 (6)	0.33 (9)
Pbf-rest	0.03	0	0.05	0	0.28	0	0
Su1-M1	0.30	0.72	0.22	0.14	0.07	0 (5)	0.75 (8)
Su1-M2	0.61	0.19	0.59	0.86	0.09	1.0 (5)	0.13 (8)
Su1-T1	0	0	0	0	0.04	0	0.12 (8)
Su1-rest	0.09	0.09	0.19	0	0.83	0	0

regarding their selection relatively early in the development of maize. As a transcription factor of genes expressing storage proteins, *pbf* is thought to have an effect on both the quantity and quality of proteins present in maize kernels [33]. Although *pbf* is a gene that leaves no as yet known observable physical or phenotypic trace on maize cobs, the present study indicates that farmers had probably already actively selected the Pbf-M1 allele by 4400 BP.

Su1 is a debranching enzyme and participates in the biosynthesis of starch. *Su1* mutants are reduced in their total starch concentration and also have altered chain length distribution in amylopectin [11, 34]. The influence of *su1* on the pasting properties of starch has not yet been specified in association studies [70], but the ratio of amylose to amylopectin and the branching pattern of amylopectin are known to be influenced by *su1* [38, 67] and, thus, affect the textural properties of tortillas [35, 36]. It is interesting that *su1* may influence the pasting properties of maize starch and that this property, which is of importance for the suitability of the corn for making tortillas, may have been selected shortly after the initial domestication of maize. Like *pbf*, *su1* is not expressed phenotypically, and the evidence for selection having been completed for both of these genes by 4400 BP clearly indicates that early farmers in the Americas were deliberate and sophisticated in their selection for a range of different desirable traits, some more easily monitored than others.

Population Substructure in the *Sugary-1* Gene

In addition to carrying out a comparative analysis of allele frequencies for *tb1*, *pbf*, and *su1* in modern and ancient maize, we were also interested in determining whether any patterns of population substructure could be observed for any of the three analyzed genes. For *tb1*, there was an obvious absence of any such substructure for the analyzed sequence fragment because all modern, as well as ancient, maize exclusively carried the allele Tb1-M1 [33]. In *pbf*, all modern maize had a high frequency for one allele (Pbf-M1) and low frequencies for other alleles [33, Table 2 this chapter]. No substructure was detected for *pbf* in modern maize. Nor could we determine a substructure in ancient maize for the distribution of *pbf* alleles; although, one needs to keep in mind that our sample number was too low to determine accurate allele frequencies.

The situation was different, however, for *su1*. The two major alleles present in modern landraces of maize exhibited a distinct geographical pattern of separation. The allele Su1-M1 occurs at a frequency of 72% in southwestern United States landraces of maize but only has a frequency of 22% and 14% in Mexican and South America landraces, respectively. In contrast, the Su1-M2 allele is present in low frequency in the Southwest (19%), and in much higher frequencies, relative to Su1-M1, in Mexican (59%) and South American (86%) landraces [33]. Interestingly, inbred maize lines derived from the landrace Northern Flint are similar to the southwestern landraces in that they also carry Su1-M1 in high frequency [67]. With a regional continuity dating back at least 800 years, Northern Flint was an important maize landrace that was primarily grown in northeastern North America [4, 12].

A similar pattern of geographic separation in *su1* to that observed in modern landraces of maize was also evident in the archaeological cobs studied. The Su1-M2 allele alone was recognized in the maize from the Ocampo Caves in

Tamaulipas, Mexico. In contrast, Su1-M1 had a frequency of 75% in the Tularosa Cave maize from New Mexico, whereas Su1-M2 had a frequency of only 13% (Table 6-2). Thus, the genetic substructure for the *su1* gene observed in modern maize appears to already have been established by 2000 BP, based on the ancient DNA data from Tularosa Cave, and perhaps as early as 4400 years ago, given the exclusive occurrence of Su1-M2 in the Ocampo Cave maize cobs. Again, however, these interesting initial results need to be confirmed through analysis of much larger assemblages of archaeological maize. The striking similarity in allele frequencies for *su1* between modern and archaeological maize from the Southwest and Northern Flint landraces in the northeastern United States, however, provides further evidence for a strong developmental link between the maize landraces of the two regions.

The distinctive isozyme composition of Northern Flint, which, when compared with other landraces, exhibits the greatest similarity to southwestern maize, indicated that Northern Flint was quite probably derived from southwestern maize [12]. A distinctive type of maize labeled Eastern Eight Row, which morphologically is similar to Northern Flint, first appears in archaeological sites from the Ohio Valley north at ca. 1000 BP [9]. The present study indicates a high frequency of the Su1-M1 allele in both modern maize landraces from the Southwest and modern inbred maize lines derived from Northern Flint, as well as in 2000-year-old archaeological maize from New Mexico, providing results consistent with the interpretation that Northern Flint is derived from Southwestern maize.

Three logical alternative explanations for the emergence of this geographically patterned genetic substructure in *su1* can be proposed. It could reflect a founder effect, indicating that even though both the Su1-M1 and Su1-M2 alleles were present in maize landraces of Mexico before 4400 years ago, the small specific landrace population of maize, which was the founder population for southwestern maize, carried Su1-M1 in high frequency. Alternatively, rather than reflecting a founder effect, if Su1-M1 and Su1-M2 had about the same relative fitness, the high frequency of Su1-M1 in southwestern and Northern Flint maize might be the result of random genetic drift that occurred over thousands of years subsequent to maize arriving in the Southwest and becoming semi-isolated from Mexican landraces. Finally, it is possible that rather than genetic drift producing a higher frequency for Su1-M1 in southwestern landraces of maize, the Su1-M1 allele may have held some selective advantage in terms of the cultural or natural environment of the region, resulting in a postarrival increased frequency for the allele. It is not yet clear which of these alternative possibilities played roles in the separation of Mexican and southwestern landraces of maize, but perhaps the answers will come with future analyses of maize assemblages from Mexico and the Southwest that adequately span the period from 5000 to 3000 BP when maize initially diffused into northern Mexico and the southwestern United States.

THE TIMING AND SEQUENCE OF SELECTION FOR KEY ATTRIBUTES IN MAIZE: COMBINING MORPHOLOGICAL AND MOLECULAR EVIDENCE

This study reports on the molecular analysis of three genes that were important in the early history of maize, but which left no morphological signature on cobs or other maize materials preserved in the archaeological record. When added to what is already known regarding other crucial attributes that can be directly observed in archaeological maize, the results of this analysis offer an initial indication of the great potential that exists for substantially expanding our understanding of the timing and sequence of early selection for different vital characters in maize by combining molecular and morphological data sets.

It is important to keep in mind, however, that although early maize continues to be recovered from sites being excavated in the southwestern United States and northern Mexico, there are, with few exceptions, not many recent or ongoing efforts to increase the number of maize assemblages available for analysis that are close to the southern Mexico location and the likely 10,000–6500 BP time frame of initial maize cultivation [50]. At the same time, little early maize has been recovered through excavation from archaeological sites in Mexico in more than 40 years. The early maize assemblages that were excavated in the 1950s and 1960s are not at present easily accessible for molecular analysis. The oldest maize samples yet described and directly dated were recovered from Guilá Naquitz, a small cave in the mountainous eastern valley of Oaxaca in southern Mexico. The two Guilá Naquitz cobs date to 6300 BP and are about two centimeters long [2, 17, 49]. About 200 kilometers north of Oaxaca, two caves in the Tehuacán Valley (Coxcatlan Cave and San Marcos Cave) have long occupational sequences and have yielded substantial assemblages of early maize, with some cobs dating to 5500 BP [39]. Seven hundred kilometers north of Tehuacán, in northeastern Mexico, two caves near Ocampo, Tamaulipas (Romero's Cave and Valenzuela's Cave) have also yielded early maize assemblages from their long occupational sequences [40, 57]. Five cobs from the Ocampo Caves were analyzed in the present study. Morphological analyses of the Guilá Naquitz and Tehuacán Cave maize assemblages provide evidence of strong selection for three distinctive characteristics early in the history of maize, all of which are classed under the heading of the *adaptive syndrome of domestication* and associated with automatic and unintended adaptive responses by domesticated plants to new human harvest and seed-bed selective pressures: loss of natural seed dispersal mechanisms, fewer but larger seed clusters, and loss of germination dormancy [24, 25, 59].

Loss of Natural Seed Dispersal Mechanisms

One of the central differences between maize and its wild ancestor, teosinte, and indeed a primary criterion in distinguishing domesticated seed plants in general from their wild relatives, is the loss of natural seed dispersal mechanisms. In teosinte, each of the ca. 10 to 12 individual kernels and encasing fruitcases that comprise a seed spike are separated by an abscission layer, which as they sequentially ripen, allows individual fruitcases to be dispersed onto the ground. In maize cobs, in contrast, kernels are tightly attached, with a resultant complete loss of any mechanism of natural seed dispersal. Such a dramatic shift from sequential automatic seed dispersal in teosinte to the total retention of all kernels on a maize cob reflects the drastic shift in selective pressures associated with human intervention in the life cycle of the cultivated crop plant. Loss of seed dispersal mechanisms is included as one of the central morphological indicators of the adaptive syndrome of domestication and is identified as an early and automatic response by domesticated plants to human harvest selective pressures. Total retention of seeds on the cob, although substantially increasing the probability that the kernels will be harvested by humans and subsequently planted the next spring, also reflects the complete reliance of the crop plant on humans as a replacement seed dispersal mechanism. Without humans to harvest, store, and plant maize kernels, the cultivated plant rapidly ceases to exist.

This loss of seed dispersal, which often is considered as a crucial morphological criteria for identifying a domesticated seed plant in the archaeological record, has clearly, and not surprisingly, been observed in the earliest maize cobs recovered from archaeological contexts in Oaxaca (6300 BP) and Tehuacán (5500 BP) [2, 3].

Fewer Larger Seed "Packages"

Human harvesting of seed plants also strongly favors those seed clusters that are easily seen and represent the greatest yield. Larger seed clusters situated at the ends of branches have a strong selective advantage in terms of a higher probability of being collected and contributing to the next season's seed stock. In maize, the selection for fewer larger seed clusters (cobs) involved both a doubling of spikelets per fruitcase, and the doubling of ranks of fruitcases. A teosinte "cob" is composed of a linear arrangement of triangular units called cupulate fruitcases. Each fruitcase has a single spikelet that develops into a single kernel. In contrast, the cupulate fruitcase in maize has two spikelets that develop two kernels. Although a doubling of spikelets was present in one, but not the other two of the Guilá Naquitz cobs, paired spikelet morphology was present in all of the Tehuacán maize, indicating that the doubling of spikelets had been selected by 5500 BP.

Although teosinte cobs carry kernels only on two sides of the axis, maize in general has at least 4 rows of kernels. The maize cobs from Guilá Naquitz exhibit 2 or 4 rows of kernels, and the early cobs from the Tehuacán Caves varied in the number of rows they carried from 4 to 10 [3]. By the time maize first appears in the archaeological sequence of the Ocampo Caves, however, all of the cobs carried at least 8 rows [41]. The doubling of rows appears to have undergone a slower process of selection than doubling of spikelets, suggesting separate developmental pathways.

Although the early selection for paired spikelets and doubling of rows changed cob morphology and increased cob size relative to teosinte, cob size did not exhibit a consistent trend of overall size increase following these initial changes. The Tularosa Cave maize assemblage exhibits considerable variability in cob size [42, Jaenicke-Després, unpublished data]. Cob size remains one of the most variable characters today in different landraces.

Loss of Germination Dormancy

Although the loss of natural seed dispersal mechanisms and seed compaction are the result of human harvest selective pressures, a third major trait that appears to have been selected for early in the history of maize—the softening and reduction of a hard protective structure in maize kernels could well reflect automatic adaptive response to seed-bed selective pressures. There are a number of different physical and chemical factors involved in ensuring that once dispersed onto the ground, seeds from wild plants remain dormant until the onset of the next growing season. In teosinte, as is the case with many other wild grasses, individual seeds are enclosed by an outer glume that becomes extremely hard at maturity. Birds and other potential seed predators that consume mature teosinte fruitcases can facilitate their dispersal, traveling some distance before depositing the often still-intact seeds. The hard glume, which encases teosinte grains and provides protection from predation, may also be involved in germination dormancy and would be strongly selected for in wild populations.

Once humans began to store and protect harvested seed from predation and sow them at the beginning of the next growing season the selective advantage of a hard glume or seed coat, in terms of both germination dormancy and protection from predation, disappeared. At the same time, a new and powerful set of selective pressures associated with seed-bed competition between newly sprouted seedlings was established once sustained annual planting of stored seed stock began. Those seeds with reduced germination dormancy and reduced seed coats germinate first and have both a considerable advantage in being able to outgrow the other seeds in a planting bed and a higher probability of contributing to the harvest. Although reduced germination dormancy and reduced seed coats or glumes is an automatic response by plants to seed-bed selective pressures and part of the adaptive syndrome of domestication, reduced glumes is also, of course, a desirable attribute for humans because it substantially reduces processing time and effort freeing the valuable starch and could have been the target of deliberate human selection early on in the history of maize.

In maize, the glume is softened and reduced in size, no longer enclosing the entire kernel but only the base. A single gene—*teosinte glume architecture-1*—is responsible for the softening and reduction of the glume in maize [16].

Although in the 6300-year-old maize cobs from Guilá Naquitz the glume was only partially reduced, all of the cobs in the Tehuacán Cave maize assemblages exhibited reduced and softened glume architecture, indicating that selection for glume reduction, similar to the selection for spikelet doubling, was still underway 6300 years ago and was complete by 5500 years ago [2, 3].

Terminal Seed Clusters and Uniform Ripening

Along with the three characters discussed earlier, which all reflect early and automatic adaptive responses by domesticated maize to human harvesting and planting of stored seed stock, one of the genes that was the focus of the current molecular study can also be identified as having been influenced by unconscious human selection and involved in the *adaptive syndrome of domestication*. The *tb1* gene, which is involved in the suppression of side branches and cob and tassel relocation, would have been subject to human harvest selection pressures, because it results in a reduction in cob number and a relocation of cobs to the end of branches, which facilitates easier harvesting. In addition, the change from multiple tassels in teosinte to a single tassel at the top of the main stem of the maize plant reflects a shortening and synchronization of pollenization and a resultant uniform ripening of seed, another attribute strongly selected for by human harvesting. Because all of the cobs from the Ocampo Caves included in the present study carried only the *tb1* allele known from modern maize, it is reasonable to conclude that modern maize plant architecture had already been selected by 4400 BP. Establishing when selection for *tb1* was ongoing requires molecular analysis of earlier maize assemblages, as they become available for study.

Improved Starch and Protein Quality

Interestingly, neither of the two other genes considered in our molecular analysis of the Ocampo and Tularosa Caves maize assemblages would have been subject to automatic adaptive response selection pressures. Both the *pbf* and *su1*, which influence protein and starch quality and quantity, would have been deliberately and consciously targeted for selection by human farmers. Because the Ocampo cobs included in the study all had only those genes that occur in modern maize, it is likely that both had also already undergone selection before 4400 BP.

Unfortunately, the number of cobs analyzed in the present study was too small to accurately establish the allele frequencies for either *pbf* or *su1* in the 4400-year-old Ocampo maize assemblage. As larger and older archaeological maize samples are analyzed for the *su1* gene, it will be interesting to see if both Su1-M1 and Su1-M2 were already present before 4400 years ago in Mexico, or if one of the alleles was selected at a different time in maize history.

Future Directions in Ancient DNA Analysis of Crop Plants

Genetic analysis of ancient specimens, including materials recovered from archaeological contexts, has dramatically improved in the past two decades [46]. The development of the polymerase chain reaction represented a major breakthrough in that it made it possible to amplify DNA from limited beginning sample amounts [54], opening up the analysis of DNA from extinct plants and animals and to address a wide range of deep time phylogenetic questions [7, 18, 31, 60].

Additionally, the analysis of multiple modern and ancient individuals of a single species helps to illuminate phylogeography and to reconstruct migration patterns [19] and speciation processes [28]. The analysis of ancient coprolites even allows the reconstruction to some extent of the diet and the environmental and climatic settings of past animal populations [27].

All of these areas of inquiry can be approached through analysis of either mitochondrial or chloroplast DNA, which are represented in each cell in several thousand copies, in stark contrast to the much more limited quantity of nuclear DNA available for study. This study, however, demonstrates that even though nuclear DNA is present in small amounts, the effort necessary to successfully extract and amplify it is worthwhile because nuclear genes provide information on a range of topics, including the timing and sequence of selection for vital characteristics in the domestication of plant and animal domesticates.

There are a number of challenges and limitations, however, that are inherent in the analysis of ancient DNA. Ancient DNA is often damaged and broken into small fragments [29, 45]. Damaged DNA either cannot be amplified at all or when amplified is subject to sequence changes that can only be detected by performing several independent amplifications for one sequence fragment [26]. The often small fragment size of ancient DNA also limits the information that can be gained from the genes. The study of longer, potentially more informative sequence fragments makes it necessary to amplify and sequence several overlapping sequences, which is costly, time-consuming, and extract consuming. Unsuccessful amplifications can be caused by an absence of sufficient amounts of DNA, damaged DNA that cannot be amplified, and by the presence of factors that inhibit the amplification reaction [21, 32, 37]. Large amounts of such inhibitors are present in many soils and are difficult to remove.

Finally, the possibility of sample contamination from modern DNA is a constant serious danger. When accidentally present in an ancient extract, modern DNA is much more easily amplified [20, 29, 55]. A number of basic precautions must be taken to minimize contamination risk and several controls must be performed throughout the entire ancient DNA process of analysis to monitor the possibility of contamination [46].

It is also important to keep in mind that extraction of ancient DNA is a destructive sampling technique. Only rarely can DNA be retrieved without sample destruction [53]. Often only small amounts of sample material are available for extraction. Several seeds, for example, may have to be pooled and extracted simultaneously to retrieve sufficient DNA for amplification, thus preventing analysis on an individual level and increasing the likelihood of contamination. Despite these difficulties, however, there are a wide range of significant questions that can only be addressed through ancient DNA analysis.

When combined with comparative morphological analysis of modern and archaeological maize, with archaeological studies providing broader natural and cultural context, and with comparative molecular analysis of modern maize and teosinte, molecular studies of ancient maize assemblages can play an important role in helping to illuminate the long and complex developmental history of maize. In contrast to these other approaches, ancient DNA analysis offers the opportunity to target specific genes that have been selected during the domestication process but that have left no easily observable signature on the archaeological remains and to establish the timing and sequence in which they were selected. As new genes are discovered and their function understood, future studies of ancient DNA in maize and a range of other domesticated plants and animals should substantially improve our understanding of their long and complex developmental history.

Acknowledgments

We thank Steve Nash, Jonathan Haas, D. Martin-Ross, and Gary Feinman, Field Museum; E.W. Wood, D. Pfister, and Wilma Wetterstrom, Harvard University Herbaria, for access to and permission to sample their archaeological maize collections, Svante Pääbo, John Doebley, and Ed Buckler for helpful discussion and the Bundesministerium für Bildung und Forschung and the Max Planck Gesellschaft for financial support.

References Cited

1. S. H. Ambrose. (1987). Chemical and isotopic techniques of diet reconstruction in Eastern North America. In: W. Keegan (Ed.), *Emergent horticulture economies of the Eastern Woodlands*. Southern Illinois University–Carbondale: Center of Archaeological Investigations. pp. 87–107.
2. B. F. Benz. (2001). Archaeological evidence of teosinte domestication from Guila Naquitz, Oaxaca. *Proceedings of the National Academy of Sciences of the United States of America, 98,* 2104–2106.
3. B. F. Benz, H. I. Iltis. (1990). Studies in archaeological maize I: The "wild" maize from San Marcos cave reexamined. *American Antiquity, 55,* 500–511.
4. W. L. Brown, E. Anderson. (1947). The northern flint corns. *Annals of the Missouri Botanical Garden, 34,* 1–22.
5. B. B. Buchanan, W. Gruissem, R. L. Jones. (2000). Biochemistry and molecular biology of plants. Rockville, MD: American Society of Plant Physiologists.
6. R. M. Clark, E. Linton, J. Messing, J. F. Doebley. (2004). Pattern of diversity in the genomic region near the maize domestication gene *tb1*. *Proceedings of the National Academy of Sciences of the United States of America, 101,* 700–707.
7. A. C. Cooper, Mourer-Chauvire, G. K. Chamber, A. von Haeseler, A. C. Wilson, S. Pääbo. (1992). Independent origins of New Zealand moas and kiwis. *Proceedings of the National Academy of Sciences of the United States of America, 89,* 8741–8744.
8. L. S. Cordell. (1984). *Prehistory of the Southwest.* Orlando, FL: School of American Research.
9. G. Crawford, D. G. Smith. (2003). Paleoethnobotany in the Northeast. In: P. Minnis, (Ed.), *People and plants in ancient Eastern North America.* Washington, D.C.: Smithsonian Institution Press. pp. 172–257.
10. P. Cubas, N. Lauter, J. Doebley, E. Coen. (1999). The TCP domain: A motif found in proteins regulating plant growth and development. *Plant Journal: For Cell and Molecular Biology, 18,* 215–222.
11. J. R. Dinges, C. Colleoni, A. M. Myers, M. G. James. (2001). Molecular structure of three mutations at the maize sugary1 locus and their allele-specific phenotypic effects. *Plant Physiology, 125,* 1406–1418.
12. J. Doebley, M. M. Goodman, C. W. Stuber. (1986). Exceptional genetic diversity of Northern Flint corn. *American Journal of Botany, 73,* 64–69.
13. J. Doebley, A. Stec. (1991). Genetic analysis of the morphological differences between maize and teosinte. *Genetics, 129,* 285–295.
14. J. Doebley, A. Stec, C. Gustus. (1995). Teosinte branched 1 and the origin of maize: Evidence for epistasis and the evolution of dominance. *Genetics, 141,* 333–346.
15. J. Doebley, A. Stec, L. Hubbard. (1997). The evolution of apical dominance in maize. *Nature, 386,* 485–488.
16. J. Dorweiler, A. Stec, J. Kermicle, J. Doebley. (1993). Teosinte glume architecture 1: A genetic locus controlling a key step in maize evolution. *Science, 262,* 233–235.
17. K. V. Flannery. (1986). *Guilá Naquitz.* New York: Academic Press.
18. A. D. Greenwood, J. Castresana, G. Feldmaier-Fuchs, S. Pääbo. (2001). A molecular phylogeny of two extinct sloths. *Molecular Phylogenetics and Evolution, 18,* 94–103.
19. E. A. Hadly, M. H. Kohn, J. A. Leonard, R. K. Wayne. (1998). A genetic record of population isolation in pocket gophers during Holocene climatic change. *Proceedings of the National Academy of Sciences of the United States of America, 95,* 6893–6896.
20. O. Handt, M. Krings, R. H. Ward, S. Pääbo, (1996). The retrieval of ancient human DNA sequences. *American Journal of Human Genetics, 59,* 368–376.
21. C. Hänni, T. Brousseau, V. Laudet, D. Stehelin. (1995). Isopropanol precipitation removes PCR inhibitors from ancient bone extracts. *Nucleic Acids Research, 23,* 881–882.

22. R. J. Hard, R. Mauldin, P. Raymond, G. R. Raymond. (1996). Mano size, stable carbon isotope ratios, and macrobotanical remains as multiple lines of evidence of maize dependence in the American Southwest. *Journal of Archaeological Method and Theory, 3,* 253–318.
23. J. R. Harlan, J. deWet, E. Price. (1973). Comparative evolution of cereals. *Evolution, 27,* 311–325.
24. J. R. Harlan. (1975). *Crops and man.* Madison, WI: American Society of Agronomy.
25. C. Heiser. (1988). Aspects of unconscious selection and the evolution of domesticated plants. *Euphytica, 37,* 77–85.
26. M. Hofreiter, V. Jaenicke, D. Serre, A. von Haeseler, S. Pääbo. (2001). DNA sequences from multiple amplifications reveal artifacts induced by cytosine deamination in ancient DNA. *Nucleic Acids Research, 29,* 4793–4799.
27. M. Hofreiter, H. N. Poinar, W. G. Spaulding, K. Bauer, P. S. Martin, G. Possnert, S. Pääbo. (2000). A molecular analysis of ground sloth diet through the last glaciation. *Molecular Ecology, 9,* 1975–1984.
28. M. Hofreiter, G. Rabeder, V. Jaenicke-Després, G. Withalm, D. Nagel, M. Paunovic, S. Pääbo. (2004). Evidence for reproductive isolation between cave bear populations. *Current Biology, 14,* 40–43.
29. M. Hofreiter, D. Serre, H. N. Poinar, M. Kuch, S. Pääbo. (2001). Ancient DNA. *Nature Reviews: Genetics, 2,* 353–359.
30. M. Höss, P. Jaruga, T. H. Zastawny, M. Dizdaroglu, S. Pääbo. (1996). DNA damage and DNA sequence retrieval from ancient tissues. *Nucleic Acids Research, 24,* 1304–1307.
31. M. Höss, A. Dilling, A. Currant, S. Pääbo. (1996). Molecular phylogeny of the extinct ground sloth Mylodon darwinii. *Proceedings of the National Academy of Sciences of the United States of America, 93,* 181–185.
32. M. Höss, S. Pääbo. (1993). DNA extraction from pleistocene bones by a silica-based purification method. *Nucleic Acids Research 21,* 3913–3914.
33. V. Jaenicke-Després, E. S. Buckler, B. D. Smith, T. M. Gilbert, A. Cooper, J. Doebley, S. Pääbo. (2003). Early allelic selection in maize as revealed by ancient DNA. *Science, 302,* 1206–1208.
34. M. G. James, D. S. Robertson, A. M. Myers. (1995). Characterization of the maize gene sugary 1, a determinant of starch composition in kernels. *The Plant Cell, 7,* 417–429.
35. J. Jane, J. Chen. (1992). Effect of amylose molecular size and amylopectin branch chain length on paste properties of starch. *Cereal Chemistry, 69,* 60–65.
36. J. Jane, J. Chen, L. F. Lee, A. E. McPherson, K. S. Wong, M. Radosavljevic, T. Kasemsuwan. (1999). Effects of amylopectin branch chain length and amylose content on the gelatinization and pasting properties of starch. *Cereal Chemistry, 76,* 629–637.
37. T. Kalmar, C. Z. Bachrati, A. Marcsik, I. Rasko. (2000). A simple and efficient method for PCR amplifiable DNA extraction from ancient bones. *Nucleic Acids Research, 28,* e67.
38. J. D. Klucincec, D. B. Thompson. (2002). Amylopectin nature and amylose-to-amylopectin ratio as influences on the behavior of gels of dispersed starch. *Cereal Chemistry, 79,* 24–35.
39. A. Long, B. F. Benz, D. J. Donahue, A. J. T. Jull, L. J. Toolin. (1989). First direct AMS dates on early maize from Tehuacán, Mexico. *Radiocarbon, 31,* 1030–1035.
40. R. S. MacNeish. (1958). *Preliminary archaeological investigations in the Sierra de Tamaulipas, Mexico.* Philadelphia: American Philosophical Society.
41. P. C. Mangelsdorf, R. S. MacNeish, W. C. Galinat. (1967). Prehistoric wild and cultivated maize. In: D. S. Byers, (Ed.), *Environment and subsistence. The prehistory of the Tehuacan Valley.* Austin: University of Texas Press.
42. P. S. Martin, J. B. Rinaldo, E. Bluhm, H. C. Cutler, R. Grange, Jr. (1952). *Mogollon cultural continuity and change. The stratigraphic analysis of Tularosa Cave and Cordova Cave.* Chicago: Chicago Natural History Museum.
43. Y. Matsuoka, Y. Vigouroux, M. M. Goodman, J. Sanchez, E. Buckler, J. Doebley. (2002). A single domestication for maize shown by multilocus microsatellite genotyping. *Proceedings of the National Academy of Sciences of the United States of America, 99,* 6080–6084.
44. P. A. Morin, K. E. Chambers, C. Boesch, L. (2001). Vigilant, quantitative polymerase chain reaction analysis of DNA from noninvasive samples for accurate microsatellite genotyping of wild chimpanzees *(Pan troglodytes verus). Molecular Ecology, 10,* 1835–1844.
45. S. Pääbo. (1989). Ancient DNA: Extraction, characterization, molecular cloning, and enzymatic amplification. *Proceedings of the National Academy of Sciences of the United States of America, 86,* 1939–1943.
46. S. Pääbo, H. N. Poinar, D. Serre, V. Jaenicke-Després, J. Hebler, N. Rohland, M. Kuch, J. Krause, L. Vigilant, M. Hofreiter. (2004). Genetic analyses from ancient DNA. *Annual Review of Genetics, 38,* 645–679.
47. L. Perry. (2002). Starch granule size and the domestication of manioc and sweet potato. *Economic Botany, 56,* 335–349.
48. D. R. Piperno. (1988). *Phytoliths analysis: An archaeological and geological perspective.* San Diego, CA: Academic Press.
49. D. R. Piperno, K. V. Flannery. (2001). The earliest archaeological maize *(Zea mays L.)* from highland Mexico: New accelerator mass spectrometry dates and their implications. *Proceedings of the National Academy of Sciences of the United States of America, 98,* 2101–2103.
50. D. R. Piperno, A. Ranere, E. Moreno, M. Lachniet, J. Iriate, M. E. Pohl, R. Dickau, I. Holst. (2004). *Preliminary results of investigations into maize history in the central balsas watershed, Mexico.* 69th Annual Meeting of the Society for American Archaeology, Montreal, Canada.
51. K. O. Pope, M. E. Pohl, J. G. Jones, D. L. Lentz, C. von Nagy, F. J. Vega, I. Quitmyer. (2001). Origin and environmental setting of ancient agriculture in the lowlands of Mesoamerica. *Science, 292,* 1370–1373.
52. A. Rahman, K. Wong, J. Jane, A. M. Myers, M. G. James. (1998). Characterization of *SU1* isoamylase, a determinant of storage starch structure in maize. *Plant Physiology, 117,* 425–435.
53. N. Rohland, H. Siedel, M. Hofreiter. (2004). Nondestructive DNA extraction method for mitochondrial analysis of mueseum specimen. *Biotechniques, 36,* 814–816.
54. R. K. Saiki, D. H. Gelfand, S. Stoffel, S. J. Scharf, R. Higuchi, G. T. Horn, K. B. Mullis, H. A. Erlich. (1988). Primer-directed enzymatic amplification of DNA with a thermostable DNA polymerase. *Science, 239,* 487–491.
55. D. Serre, A. Langaney, M. Chech, M. Teschler-Nicola, M. Paunovic, P. Mennecier, M. Hofreiter, G. Possnert, S. Pääbo. (2004). No evidence of neanderthal mtDNA contribution to early modern humans. *Public Library of Science Biology, 2,* 313–317.
56. J. Smalley, M. Blake. (2003). Archaeological evidence of maize stalk use in Mesoamerica and South America. *Current Anthropology, 44,* 682–703.
57. B. D. Smith. (1997). Reconsidering the Ocampo Caves and the era of incipient cultivation in Mesoamerica. *Latin American Antiquity, 8,* 342–383.
58. B. D. Smith. (1989). Origins of agriculture in Eastern North America. *Science, 246,* 1566–1571.
59. B. D. Smith. (2006). Documenting domesticated plants in the archaeological record. In: M. Zeder, B. Eve, B. D. Smith, (Eds.), *Documenting domestication: New archaeological and genetic paradigms.* University of California Press. pp. 15–24.
60. R. H. Thomas, W. Schaffner, A. C. Wilson, S. Pääbo. (1989). DNA phylogeny of the extinct marsupial wolf. *Nature, 340,* 465–467.
61. T. Ueda, Z. Wang, N. P. Pham, J. Messing. (1994). Identification of a transcriptional activator-binding element in the 27-kilodalton zein promoter, the -300 element. *Molecular and Cellular Biology, 14,* 4350–4359.
62. J. Vicente-Carbajosa, S. P. Moose, R. L. Parsons, R. J. Schmidt. (1997). A maize zinc-finger protein binds the prolamin box in zein gene promoters and interacts with the basic leucine zipper transcriptional

activator Opaque2. *Proceedings of the National Academy of Sciences of the United States of America, 94,* 7685–7690.
63. Y. Vigouroux, M. McMullen, C. T. Hittinger, K. Houchins, L. Schulz, S. Kresovich, Y. Matsuoka, J. Doebley. (2002). Identifying genes of agronomic importance in maize by screening microsatellites for evidence of selection during domestication. *Proceedings of the National Academy of Sciences of the United States of America, 99,* 9650–9655.
64. R. L. Wang, A. Stec, J. Hey, L. Lukens, J. Doebley. (1999). The limits of selection during maize domestication. *Nature, 398,* 236–239.
65. Z. Wang, T. Ueda, J. Messing. (1998). Characterization of the maize prolamin box-binding factor-1 (*PBF*-1) and its role in the developmental regulation of the zein multigene family. *Gene, 223,* 321–332.
66. E. J. Wellhausen, L. M. Roberts, E. Hernandez X. (1952). *Races of maize in Mexico.* Cambridge, MA: Bussey Institution, Harvard University.
67. S. R. Whitt, L. M. Wilson, M. I. Tenaillon, B. S. Gaut, E. S. Buckler. (2002). Genetic diversity and selection in the maize starch pathway. *Proceedings of the National Academy of Sciences of the United States of America, 99,* 12959–12962.
68. W. H. Wills. (1988). *Early prehistoric agriculture in the American Southwest.* Santa Fe, NM: School of American Research Press.
69. W. H. Wills. (1995). Archaic foraging and the beginning of food production in the American Southwest. In: T. Price, S. Gebauer, (Eds.), *Last hunters–first farmers.* Santa Fe, NM: School of American Research Press.
70. L. M. Wilson, S. R. Whitt, A. M. Ibanez, T. R. Rocheford, M. M. Goodman, E. S. Buckler. (2004). Dissection of maize kernel composition and starch production by candidate gene association. *The Plant Cell, 16,* 2719–2733.

C H A P T E R

4

Dating the Initial Spread of *Zea mays*

MICHAEL BLAKE

Department of Anthropology and Sociology, University of British Columbia, Vancouver, British Columbia, Canada

Glossary

AMS dating A radiocarbon dating technique using accelerator mass spectrometry that requires only minute samples of ancient material to directly count the amount of ^{14}C remaining.

Bioturbation Disturbance to the layers of archaeological deposits, and the materials contained within them, caused by biological organisms such as earthworms, beetles, and rodents. Bioturbation can result in older materials being found in younger layers and vice versa.

Coprolites Preserved feces, often found in dry cave sites, that contain food residues and other materials, including pathogens, and can help in identifying ancient subsistence patterns and health.

Isoclines A set of lines drawn on maps to indicate regions of similar value; for example, elevations (contours), precipitation, temperature, age, and population density.

Kriging A method used for interpolating or predicting values, such as elevations and so forth, for a spatial data set when only a limited number of known measurements is available. The kriging formula uses a variogram model that minimizes the statistical error of estimation by using a set of known data points near the point to be estimated.

New studies of *Zea mays* genetics suggest that all modern maize evolved from teosinte *(Zea mays* ssp. *parviglumis),* originating in the Río Balsas drainage of western Mexico. The processes whereby early occupants of the region interacted with teosinte; harvested its seeds, leaves, and stalks; and eventually transported or traded the plant and its descendants far beyond its natural range are not yet clearly understood. This chapter traces the spatial and chronological radiation of maize and all current archaeological and paleoethnobotanical evidence of early *Zea mays* from Mesoamerica, Central America, and South America, demonstrating that social uses of the plants' many products, including sugar, could have been as important in its early spread as were its nutritional uses.

INTRODUCTION

Two important discoveries have helped to reframe our understanding of *Zea*'s domestication and early spread. The first is the accumulating genetic evidence that maize *(Zea mays* ssp. *mays)* arose from an annual teosinte *(Zea mays* ssp. *parviglumis),* whose present-day range is centered in the Río Balsas region of western Mexico [50, 101], but extends west to Jalisco and southeast to Oaxaca [31]. The second is the direct accelerated mass spectrometry (AMS) radiocarbon dating of exceedingly small fragments of maize, establishing a reliable, absolute chronology for its initial appearance and eventual dispersal. Bruce Smith [87] has recently observed, these two lines of biological and archaeological evidence are providing scholars with exciting new approaches to their research on the questions of when, why, and how maize spread out of its homeland to new regions far beyond its zone of initial domestication.

Other important questions include what were the ways in which the first maize cultivators used the plant and how did these initial uses lead to its distribution throughout the Americas? These are key questions because they highlight the selective forces that cumulatively shaped the genetic transformation of teosinte from a localized western Mexican wild grass to the world's number one food crop. As with most domesticates, these forces were primarily social ones. They resulted from the active decisions of countless people who used the plant for many different purposes, ranging from food and drink to building materials, for thousands of years. Many of maize's uses may have been more important in the past than they are today, and many of today's uses are much more varied and elaborate than they were in the past. These long-term shifts in the characteristics and potential uses of maize suggest that its social significance must also have changed, both within its original homeland and in the new regions where it was eventually adopted.

Definitive answers to these questions are not yet possible—mainly because we know so little about early maize distribution. Besides the well-known examples from several dry caves in three separate regions (Tamaulipas, Tehuacán, and Oaxaca), as pointed out by Smith [87, p. 1325], few macrobotanical samples of maize have been recovered from Pre-Ceramic sites in Mexico. The only directly dated maize remains (using AMS radiocarbon dating) from Pre-Ceramic contexts are restricted to 12 samples from the Tehuacán caves [48], 6 from the Ocampo Caves in Tamaulipas, [44, 85, pp. 373–374] and 2 from Guilá Naquitz in Oaxaca [66].

Despite this small number of directly dated specimens, it has been possible for researchers to detect trajectories of morphological and genetic changes in assemblages of maize cobs from these sites and conclude that Pre-Ceramic peoples were intentionally selecting for several characteristics. Benz and Long [6, p. 463], in their study of Tehuacán maize, suggest that before 2500 BC (ca. 4450 BP) humans were initially interested in selecting maize ears with more kernels; and only later did they select ears with larger kernel sizes. After 2500 BC the rate of change in ear morphology slowed, and people were possibly more concerned with increasing the number of ears per plant. Genetic analysis of maize from the Ocampo caves indicates people were selecting for increased protein and starch quality, and that some specimens were similar to modern maize by 4450 BP (ca. 2500 BC) [44, p. 1207]. Jaenicke-Després and her colleagues [44] also discovered, however, that as recently as 2000 BP maize from New Mexico still had an allele of the gene *sugary-1 (su1)* in common with teosinte; and that probably prevented it from producing the high-quality starch found in modern maize—a trait that, for example, gives maize starch the sticky consistency necessary in making tortillas.

One factor that has made it difficult to answer questions about the spread of maize is that the direct **AMS dating** of Pre-Ceramic maize in the United States, and to a lesser extent in Mexico, is not matched by similar developments to the south. Instead, almost all of the indications of Pre-Ceramic maize south and east of Oaxaca and into Central and South America come from two types of microbotanical remains: pollen and phytoliths. These types of remains have generally been recovered from cores in lakes and swamps, and sometimes cave deposits, and have been dated indirectly by association with charcoal or other organic materials also recovered from the cores or caves. Researchers have relied on these assemblages of microscopic phytoliths and pollen to determine the presence or absence of maize in this vast region, because Pre-Ceramic caves and open-air sites are both rare and have yielded few dateable maize macroremains.

Here, I summarize the current studies of directly dated maize macrobotanical remains, map their distribution in the Americas, and compare these data with studies of indirectly dated maize microbotanical remains. The purpose of this comparison is to determine the extent to which these different views of maize tell us similar or differing stories about maize's initial spread. During the past few years it has become clear that directly dated macrobotanical remains have yielded significantly younger dates than indirectly dated microbotanical remains. What are the implications of these discrepancies and how might they influence our models of the origins of maize agriculture? To make an initial attempt at answering these questions, I first map maize's spread using the earliest directly dated maize macrobotanical remains in each region where samples exist and then map maize microbotanical remains by age, to show how these data provide a different picture of the spread of maize. Finally, I map the age distribution of stable carbon isotope values of human bone samples indicating a significant consumption of maize in the diet. These maps are meant to summarize our current knowledge of maize's distribution and fill in the details that are suggested by earlier maps, such as Peter Bellwood's [3, p. 147] excellent recent summary.

TEMPORAL FRAMEWORKS FOR *ZEA MAYS*' EARLY DISPERSAL

To understand the spread of maize it is necessary to accurately map its first occurrence in every region of the Americas. The two main methods for dating maize macrobotanical remains have been: (1) indirect dating, that is, by association with organic remains such as wood charcoal in archaeological deposits, and (2) the direct dating of maize macroremains using AMS radiocarbon dating (and occasionally conventional dating). Here, I concentrate on direct AMS dates using maize macroremains because indirect dates have often proven to be unreliable. Although individual indirect dates may be correct in some cases (that is, truly associated with a given maize sample, as Smith [85] has shown for some of the Ocampo cobs), in many other cases they are not. As noted by Long and colleagues [48] for the Tehuacán maize, it is common for more recent macrobotanical remains to work their way downward, as a result of both natural and cultural disturbances, into earlier deposits and thereby be incorrectly associated with older organic materials (in Chapter 29, Rivera describes this problem for several Chilean sites, and Smith [88] documents the vertical movement of AMS dated cucurbit remains from deposits at Coxcatlán Cave).

DIRECT DATING OF MAIZE

Table 4-1 presents the directly dated samples of maize used in this analysis. This is not a complete list of all directly dated maize in the Americas. Instead, I have included only the earliest sample from each site (or cluster of sites in a region), omitting later examples that will not shed additional light on the question of the initial spread of maize. So, for example, at Romero's Cave in Tamaulipas, Mexico, I include the earlier of two dates on maize cobs from that cave. Unless otherwise noted, all dates are presented in uncalibrated radiocarbon (^{14}C) years BP. Table 4-1 includes the radiocarbon sample's laboratory identification number where available, as well as the published source of the data.

The dates were then plotted on a map of the Americas using *Surfer 8.0,* a commercially available and widely used program. *Surfer* allows the dates to be plotted as **isoclines,** representing interpolated age ranges, and calculated using the program's various grid interpolation algorithms. For the maps produced here, I have gridded the data using *Surfer*'s **kriging** algorithm.

Figure 4-1 shows the distribution of the 30 direct dates (predominantly AMS) recorded for the earliest occurrences of macrobotanical maize remains. The **isoclines,** or "age contours," are set at 500-year intervals and show the broad trends of dispersal based solely on known, dated specimens. As Bruce Smith [85] pointed out, there are huge gaps in our regional coverage, and a great deal more needs to be done to both date existing collections and to recover more samples from known contexts.

The distribution of sites in Mesoamerica and North America, in Figure 4-1, shows a clear "hot spot" resulting from the earliest AMS dates from Guilá Naquitz Cave in Oaxaca (5420 ± 60 BP [Beta-132511], [66]) and San Marcos Cave in the Tehuacán Valley (4700 ± 110 BP [AA-3311], [48]). Radiating out from this is a series of successively younger maize dates with only a few from southern Mesoamerica and the bulk from northern Mexico and the American Southwest. This pattern suggests that maize spread rather slowly northward to Tamaulipas (Romero's Cave [3930 ± 50 BP, Beta-85431], [85]) and later to Chihuahua (Cerro Juanaqueña [2980 ± 50 BP, INS-3983], [35]). From northern Mexico, maize was brought into the American Southwest by about 3000 radiocarbon years ago (e.g., Fresnel Shelter, New Mexico [2945 ± 55 BP, AA-6402], [93]). By 1500 to 2000 BP, maize spread out of the American Southwest and northeast into the major tributaries of the Mississippi River (e.g., the Holding Site, near Cahokia, Illinois [2077 ± 70 BP, AA-8717], [72]). The genetic characteristics of maize in the American Northeast are most similar to southwestern maize, suggesting a direct link between the two regions [50].

In Mesoamerica, the next earliest maize, after the Tamaulipas samples, comes from Chiapas, Mexico. Seven AMS dates have been made on maize from a cluster of sites in the Mazatán region along the Pacific Coast [8, 20]. The youngest of these comes from Aquiles Serdán (3000 ± 65 BP [Beta-62920], [20]) and the earliest is from the site of San Carlos (3365 ± 55 BP [Beta-62911], [20]). The only other directly dated maize samples in Mesoamerica are even later than the ones from the American Southwest. One cob fragment comes from San Andrés in Tabasco (2565 ± 45 BP [AA-33923], [70]) and another from El Gigante rock shelter in Honduras (2280 ± 40 BP [Beta-159055], [78]).

It is surprising that no other early context (either Pre-Ceramic or Early Formative) maize samples have yet been directly dated in Mesoamerica or Central America. One site in the Arenal Reservoir region of Costa Rica is reported to have a maize kernel in association with wood charcoal that has been conventionally dated to 4450 ± 70 [10, 80]. However, we must be cautious in accepting this date as an indication of early maize cultivation until the kernel itself can be directly dated using AMS.

In South America (Figure 4-1) there have been few directly dated maize macrobotanical remains. Unfortunately, as with most of Central America, there are not yet any AMS-dated samples of early maize from Colombia or Peru. There are, however, seven directly dated samples from Ecuador, Chile, and Argentina. The two earliest samples come from the Ramaditas site in Chile (2210 ± 55 [GX-21725], Chapter 29), and Gruta del Indio in Argentina (2065 ± 40 [GrN-5396] a conventional date [32]). Gil [32] also reports four more recent AMS dates on maize from several sites in the southern Mendoza region, not far from Gruta del Indio (these range from 740 BP ± 40 BP to 1045 ± 45 BP). Also included in Figure 4-1 is the AMS-dated sample from the Loma Alta site in coastal Ecuador [59, p. 223]. The actual date has not yet been published, but the maize sample is thought to be associated with deposits dated to 3500 BP (or later).

This pattern for South America is surprising given the microbotanical evidence (discussed later) for the presence of maize in Central and South America at early dates. Early maize macroremains in the zone from 12° north latitude to

TABLE 4-1 Direct Dates on Maize Macrobotanical Samples

Country/Region	Site name	Dated material	^{14}C Method	Radiocarbon years BP	Sample ID number	Reference
United States						
New Mexico	Tornillo Shelter	maize cobs (8 pooled)	Conventional	3175 ± 240	GX-12720	[97, pp. 412–414; 35, p. 1664]
New Mexico	Bat Cave	maize	AMS	3010 ± 150	n/a	[102; 35, p. 1664]
New Mexico	Fresnal Shelter	maize	AMS	2945 ± 55	AA-6402	[93, p. 317; 35, p. 1664]
Arizona	Milagro	maize	AMS	2930 ± 45		[42; 35, p. 1664]
Arizona	Three Fir Shelter	maize	AMS	2880 ± 140		[83; 35, p. 1664]
Arizona	Fairbank	maize	AMS	2815 ± 80		[41; 35, p. 1664]
Arizona	Cortaro Fan	maize	AMS	2790 ± 60		[74; 35, p. 1664]
Arizona	West End	maize	AMS	2735 ± 75		[41; 35, p. 1664]
New Mexico	LA18091	maize	AMS	2720 ± 265	UGa-4179	[81; 35, p. 1664]
New Mexico	Jemez	maize	AMS	2410 ± 360		[1, p. 279]
New Mexico	Sheep Camp Shelter	maize	AMS	2290 ± 210	A-3396	[81, p. 79]
Illinois	Holding (Cahokia)	maize cob	AMS	2077 ± 70	AA-8717	[72, pp. 493–494]
New Mexico	Tularosa Cave	maize cob	AMS	1920 ± 40	Beta-166755	[44, p. 1207]
Tennessee	Icehouse Bottom	maize	AMS	1775 ± 100	Beta-16576	[19, p. 353]
Ohio	Harness Mound	maize	AMS	1730 ± 85	n/a	[72, p. 495; 84]
Mexico						
Oaxaca	Guilá Naquitz	maize cob	AMS	5420 ± 60	Beta-132511	[66, p. 2102]
Puebla	San Marcos Cave	maize cob	AMS	4700 ± 110	AA-3311	[48, p. 1037]
Tamaulipas	Romero's Cave	maize cob	AMS	3930 ± 50	Beta-85431	[85, p. 373]
Tamaulipas	Valenzuela's Cave	maize cob	AMS	3890 ± 60	Beta-85433	[85, p. 374]
Chiapas	San Carlos	kernel	AMS	3365 ± 55	Beta-62911	[20; 8, p.164]
Sonora	La Playa	maize	AMS	3000 ± ?		[49, p. 345]
Chihuahua	Cerro Juanaqueña	maize cob	AMS	2980 ± 50	INS-3983	[35, p. 1664]
Tabasco	San Andres	maize cob	AMS	2565 ± 45	AA-33923	[70, p. 1372]
Honduras						
Inland	El Gigante	maize cob	AMS	2280 ± 40	Beta-159055	[78]
Ecuador						
Coastal	Loma Alta	kernels	AMS	<3500	n/a	[59, p. 223]
Peru						
Coast	El Caral	cobs	AMS	unreported	n/a	Chapter 28
Argentina						
Mendoza	Gruta del Indio	maize	Conventional	2065 ± 40	GrN-5396	[32, p. 297]
Chile						
North coast	Tiliviche 1-b	maize cob	AMS	920 ± 32	AA-56416	Chapter 29 (Table 29-3)
	Guatacondo	maize cob	AMS	1865	UCLA-1698c	Chapter 29 (Table 29-4)
	Ramaditas	maize cob	AMS	2210 ± 55	GX-21725	Chapter 29 (Table 29-4)
	Rixhasca	kernel	AMS	1025	GX-21748	Chapter 29

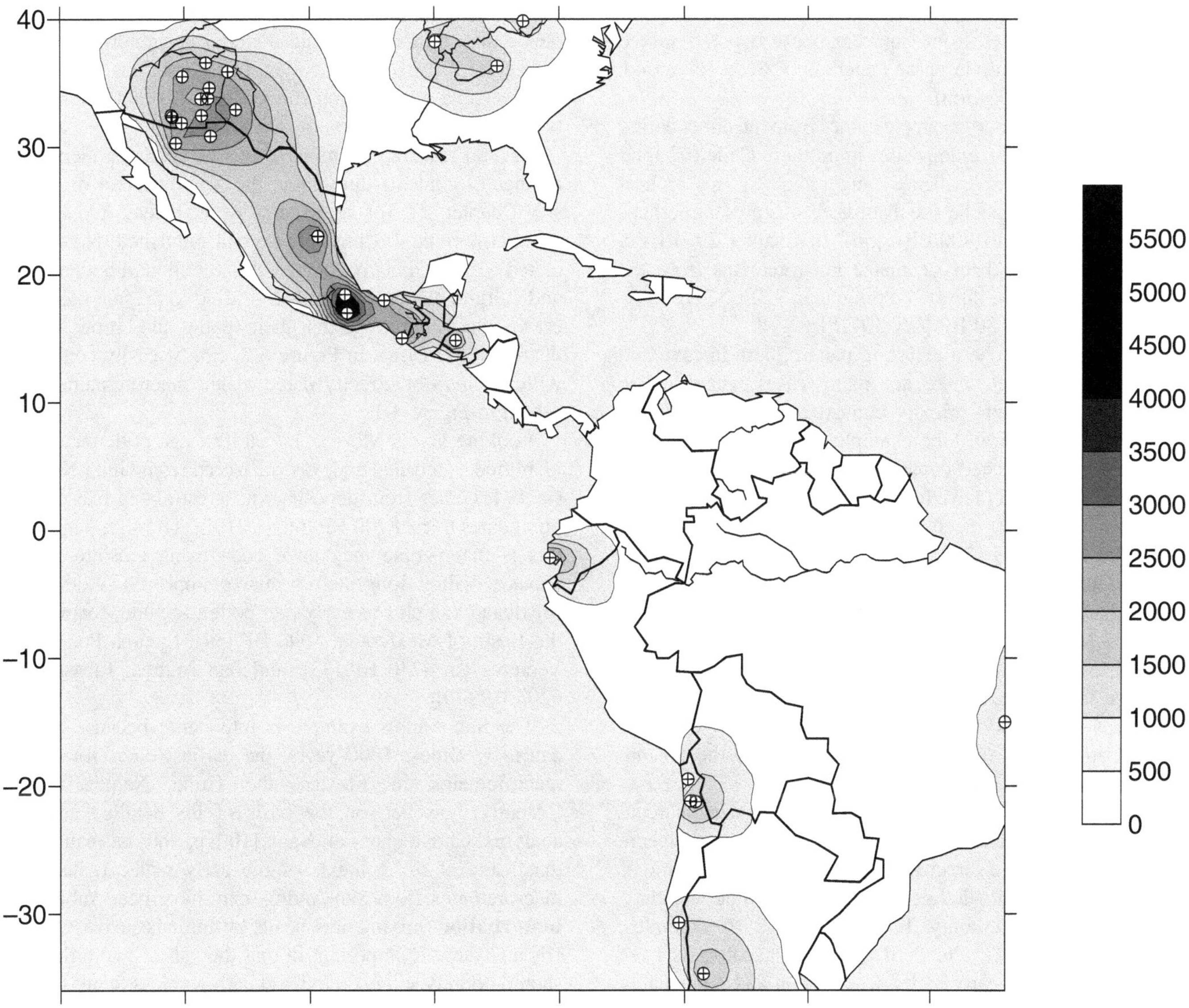

FIGURE 4-1 Direct dates on maize macrobotanical samples (using data from Table 4-1). Radiocarbon years BP.

approximately 20° south latitude do exist in archaeological collections, but few have been directly dated. The recovery and direct dating of maize macroremains from this zone should be a high priority for archaeologists and paleoethnobotanists because even one or two early specimens in the 3000 to 4000 year range would help to confirm the age estimates of maize based on indirect dating.

ON THE INDIRECT DATING OF MAIZE

Before AMS dating methods became widely available, most archaeologists relied on conventional dates of associated charcoal or other bulk carbon samples to date deposits. Even today, much of our knowledge of the antiquity of maize comes from such an approach. However, this is becoming a significant problem because the associations between the materials used for dating, such as wood charcoal, and the maize macrobotanical remains are not always secure. Long and colleagues [48] were the first to demonstrate this for the earliest maize from the Tehuacán Valley cave sites, where 4700-year-old maize made its way into deposits that, dated by association, were thought to be almost 2000 years older.

There are other good examples of early deposits containing younger maize. Deborah Pearsall [59, p. 223] notes that maize kernels from six different Valdivia 1 and 2 phase contexts at the Loma Alta site in coastal Ecuador (ca. 5000 BP) turned out to be much younger in age than were originally thought. One kernel was AMS dated to post 3500 BP and possibly comes from Chorrera phase deposits mixed with the Early Valdivia phase materials (J. Scott Raymond, personal communication).

Another cautionary example comes from the direct dating of maize from Pre-Ceramic sites in northern Chile (Chapter 29). Stratigraphic evidence suggested to researchers working at the Tiliviche site that maize was present perhaps as early as 7000 to 8000 BP [56]. In Chapter 29, Rivera shows that AMS dates on maize macroremains from the early deposits at the site are no more than 920 ± 32 BP [AA-56416] and 850 ± 30 BP [GX-30718].

This appears to be a recurring pattern, both in cave and open-air sites; yet, there are many cases where maize remains are in their primary context and can be reliably dated by association. One example where the association dates have been tested comes from coastal Chiapas, where my colleagues and I have found dozens of carbonized maize kernels and cob fragments in Early Formative period household trash deposits [7, 8, 20, 26]. In these deposits, which were dated by conventional radiocarbon methods on associated charcoal and by ceramic association, we were also able to use AMS dating of individual maize kernels and cob fragments. Six of seven AMS dates on maize macroremains (drawn from five Early Formative phases—Barra through Cuadros—ca. 3700 to 2800 BP) fell within their expected phase time ranges (at the one sigma level) and the seventh did so at the two sigma level.

A survey of radiocarbon dates on nonmaize charcoal from contexts containing early maize in the Americas, where we do not yet have direct AMS dates, helps fill in some of the gaps in our knowledge, but at the same time, creates a good deal of uncertainty. Table 4-2 shows 16 examples, ranging from Mexico to Peru, where archaeologists have hypothesized that early maize was present and presumably of some economic importance. Although some of these samples may turn out to be as old as suggested by indirect dating, I believe we should hold off on incorporating them into our distributional models until their ages can be confirmed with direct dating.

DATING THE EARLY DISTRIBUTION OF *ZEA* POLLEN

Pollen samples taken from sediments in lakes, swamps, and archaeological deposits provide an independent view of the presence or absence of *Zea* (both maize and teosinte) in the Americas. The dating of maize pollen grains recovered from core samples (long cylinders that penetrate deeply into the sediment layers) has usually been done indirectly (by association) with dateable charcoal or other organic remains found within the core. The same is true for dating pollen recovered from archaeological strata in both caves and open-air sites. In both cases, for the pollen analyst to have confidence that the dated organic material accurately estimates the age of the associated pollen, two conditions must hold: (1) the strata must be undisturbed; and (2) the pollen must be in its primary location of deposition.

Several research teams have recovered *Zea* pollen from a range of contexts throughout the Americas (for example, see Chapter 27 for a discussion of recent *Zea* pollen research). Table 4-3 presents several published dates associated with the first appearance of *Zea* pollen in both natural and cultural contexts. Here I have tried to present the earliest known examples so that their spatial and chronological distributions, shown in Figure 4-2, can be easily compared with the map of directly dated maize macroremains, presented in Figure 4-1.

Looking first at Mexico, the earliest *Zea* pollen has been identified as teosinte and was recovered from Guilá Naquitz Cave. It comes from deposits whose estimated phase midpoint dates to ca. 8240 BP [66, p. 2102]. The significance of this is that people may have been using teosinte in the Oaxaca Valley long before maize appeared. Additional Mexican examples of early *Zea* pollen include Zoapilco in the Basin of Mexico ca. 5090 BP [54]; Laguna Pompal in Veracruz ca. 4250 BP [33]; and San Andrés, Tabasco ca. 6208 BP [70].

The San Andrés example is interesting because it predates, by almost 1000 years, the earliest dates for maize macroremains in Mexico—the Guilá Naquitz cobs. Recently, however, on the basis of his detailed ceramic analyses, Christopher von Nagy [100, p. 960] has cautioned that some of the contexts of the early maize pollen and macroremains from San Andrés may have been subject to **bioturbation** (mixing of deposits by animals such as crabs). His analyses are important in that they show again that we should not rely solely on indirect dating unless there is some way of ensuring that the remains we are trying to date are in their primary context and are unequivocally linked to the material used to date them.

Table 4-3 summarizes *Zea* pollen dates from nine locations in Central America, including Guatemala, Belize, Honduras, Costa Rica, and Panama. These cases range in age from ca. 6860 BP at Cueva de los Ladrones in Panama [65] to ca. 2940 BP at Laguna Zoncho in Costa Rica [21]. One striking pattern in these data is that the most common dates for lake or swamp core sample locations is in the 4200 to 4700 BP range: Sipacate, Guatemala [52]; Cob Swamp, Belize [69]; Lake Yojoa, Honduras [75]; Laguna Martínez, Costa Rica [2]; and La Yeguada, Panama [64]. This pattern suggests that there was an expansion of maize planting during this period, correlating nicely with Benz and Long's [6, pp. 463–464] observation that maize cob size may have reached a size plateau around that 4450 BP.

Early examples of *Zea* pollen in South America come from four main locations, two locations in both Colombia

TABLE 4-2 Indirect Dates for *Zea* Macrobotanical Samples

Country/Region	Site name	Dated material	^{14}C Method	Associated *Zea*	Radiocarbon years BP	Sample ID number	Reference	Comments
Mexico								
Mexico	Zoapilco	n/a	Conventional	teosinte seeds	5090 ± 115	I-4405	[54, p.134]	
Tabasco	San Andres	wood charcoal	Conventional	charred fragments	>3340 ± 60	Beta-18198	[77, p. 103; 76, p. 187]	
Guatemala								
	Salinas la Blanca	assoc. charcoal	Conventional	cob impressions	2928 ± 105	Y-1150	[22, pp. 68–72]	50 cob impressions, Cuadros Phase, 1000–900 BC
Costa Rica								
	Arenal Reservoir	assoc. charcoal	Conventional	Kernel	4450 ± 70	n/a	[10, 80]	
Venezuela								
Amazonia	Parmana	ceramic seriation		Kernels	ca. 2550	phase	[73, p. 235; 60, p. 334]	phase midpoint 800–400 BC
Brazil								
Minas Gerais	Peruaçu Valley	palm nuts	Conventional cobs		990 ± 60	n/a	[30]	
Ecuador								
Coastal	La Ponga	assoc. charcoal	n/a	cob fragments and kernels	ca. 3150	n/a	[47, p. 118]	
Highland	La Chimba	assoc. charcoal	n/a		>2640	n/a	[59, pp. 230–231]	
Highland	Cotocollao	assoc. charcoal	n/a	cob fragments and kernels	ca. 3500	n/a	[59, p. 232]	
Highland	Nueva Era	assoc. charcoal	n/a	cob fragments and kernels	ca. 3000	n/a	[59, p. 232]	
Peru								
Inland	Chavin de Huantar	association	Conventional	two kernels (earliest one is from unit B2)	2190 ± 210	ISGS-510	[14, p. 254]	one Urabarriu (B2-b)
Inland	La Galgada	association	n/a	single cob at site	3130 ± 80	TX-4446	[34, p. 69; 89, p. 126]	phase endpoint
Casma	Las Haldas	assoc. charcoal	Conventional	maize	2990 ± 75	UGA-4526	[71, pp. 10, 30]	Early Horizon
	Casma PV32-1	unknown	Conventional	unidentified	6070 ± 70	GIF-6772	[9, p. 839]	
Coastal	Cardal	assoc. charcoal	n/a	unidentified	2925	27 14C dates	[15, p. 277]	phase midpoint 1150–800 BC
Supe	Caral	assoc. organics	Conventional	organics	ca. 3820	several dates	Chapter 28	Late Middle Period
	Los Gavilanes	unknown	Conventional	all plant parts	4140 ± 160	GX-5076	[9, p. 838]	

TABLE 4-3 Dates on Materials Associated with *Zea* Pollen Samples

Country/Region	Site name	Dated material	^{14}C Method	Associated *Zea*	Radiocarbon years BP	Sample ID number	Reference
Mexico							
Oaxaca	Guilá Naquitz	organic materials	Conventional	pollen (teosinte)	8240	estimated midpoint between ca. 9500–6980 ^{14}C years BP.	[79, p. 229; 66, p. 2102]
Tabasco	San Andrés	wood	AMS	pollen	6208 ± 47	AA38771	[70, p. 1372]
Mexico	Zoapilco	n/a	Conventional	pollen	5090 ± 115	I-4405	[54, pp. 132–137]
Veracruz	Laguna Pompal	pollen residue	AMS	pollen	4250 ± 70	CAMS-1770	[33, pp. 84–86]
Guatemala							
Pacific coast	Sipacate	wood	AMS	pollen	4600	not given	[52, 53]
Belize							
Caribbean coast	Cob Swamp	wood	Conventional	pollen	4610 ± 60	Beta-56775	[69, pp. 360–361]
Honduras							
	Lake Yojoa	wood	Conventional	pollen	<4770 ± 385	UGa-5380	[75, p. 178]
Costa Rica							
	Laguna Martínez	assoc. charcoal	AMS	pollen	4760 ± 40	not given	[2, p. 112]
	Lago Cote	assoc. charcoal	AMS	pollen	3630 ± 70	not given	[2, p. 112]
	Laguna Zoncho	wood	Conventional	pollen	2940 ± 50	Beta-115186	[21, p. 422]
Panama							
	Cueva de los Ladrones	n/a	n/a	pollen	6860 ± 90	n/a	[65, p. 873] (given as 4919 ± 90 BC)
	La Yeguada	n/a	n/a	pollen	4200	n/a	[64]
	Gatun Lake	charcoal?	n/a	pollen	ca. 4000	in core above UCLA-11354	[61, p. 17]
Colombia							
	Hacienda El Dorado	n/a	n/a	pollen	6680	n/a	[12]
	Hacienda Lusitania	n/a	n/a	pollen	5150 ± 180	n/a	[51]
Ecuador							
Amazonia	Lake Ayauchi	assoc. charcoal	AMS	pollen	4570 ± 70	Beta-20956	[17, p. 304]
	Lake San Pablo	n/a	n/a	pollen	4000	phase	[57, p. 421]

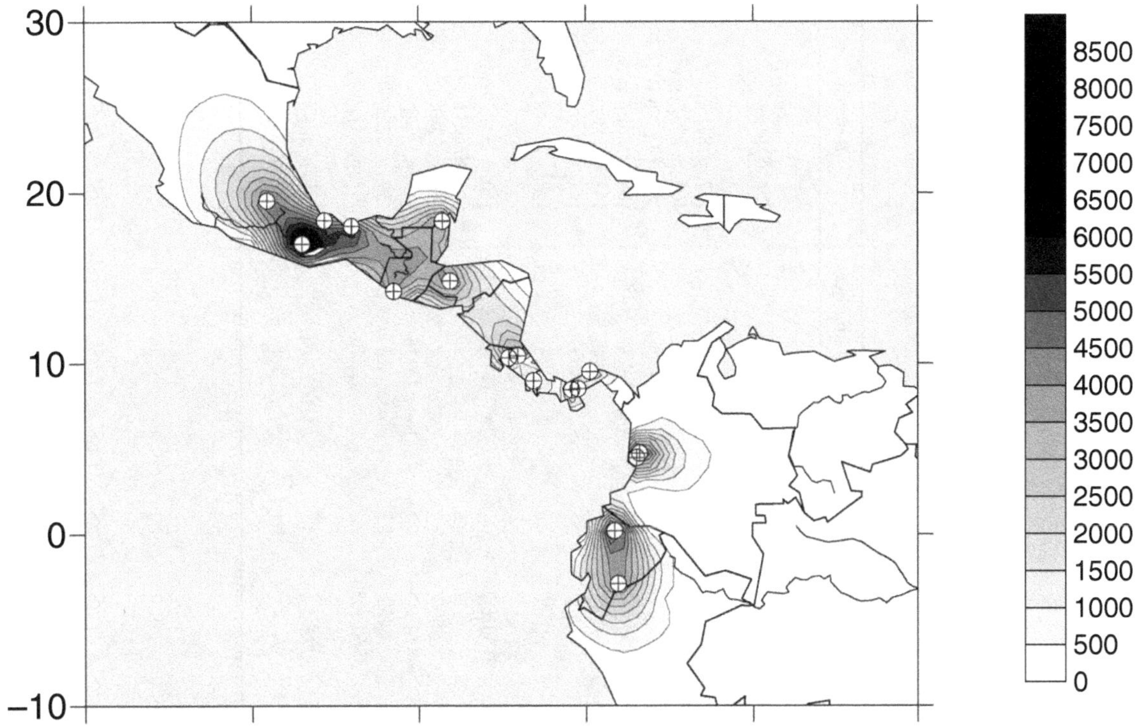

FIGURE 4-2 Indirect dates for *Zea* pollen samples (using data from Table 4-3). Radiocarbon years BP.

and Ecuador. The earliest Colombian sample is from Hacienda El Dorado and dates to ca. 6680 BP [12]. At Hacienda Lusitania, Monsalve [51] reported a date of ca. 5150 BP. Both of these are similar to the earliest dates for *Zea* pollen from Central America. In Ecuador, the pollen from lake cores is 1000 to 2000 years more recent: Lake Ayauchi, ca. 4570 [17] and Lake San Pablo ca. 4000 [57].

DATING THE EARLY DISTRIBUTION OF MAIZE PHYTOLITHS

Maize phytoliths have been recovered from many sites in the Americas, but here I present information on the earliest reported examples from Central and South America. To date, researchers rarely seek out maize phytoliths in sites in Mexico; the phytoliths have not been routinely recovered. In the few instances where researchers did look for them in early deposits, as in zones C through B1 at Guilá Naquitz Cave "no phytoliths characteristic of either teosinte fruitcases or maize cobs were present" [66, p. 2102].

Thanks to the work of Deborah Pearsall and Dolores Piperno and their colleagues, phytolith identification and analysis has become an important tool in the search for evidence of the early spread of maize. It has proven particularly useful in the neotropical regions of the Americas where maize macrobotanical remains are poorly preserved—sometimes providing the only evidence of the presence of maize in early sites. Where phytoliths have been found in association with other artifacts and macrobotanical remains in undisturbed deposits, they have helped to provide independent lines of evidence for the early use of maize. At the same time, where microbotanical remains such as phytoliths and pollen provide the only evidence of maize use, it has been difficult to resolve debates about the timing of maize's introduction into a region [90, 91].

This problem has arisen partly because maize phytoliths from archaeological soils cannot be directly dated; therefore, they must be dated indirectly by association with dateable materials. Phytoliths recovered from soil matrix samples are subject to mixing and movement, perhaps even more so than macrobotanical remains. This certainly poses the risk that microscopic phytoliths from younger contexts may have migrated into earlier deposits, erroneously giving the impression of maize cultivation long before it took place. New techniques of recovering phytoliths trapped in residues on cooking pots and in the dentition of ancient skeletons have had a twofold increase on the contextual reliability of phytoliths in our interpretations of early maize use [91; Chapter 30]. First, we can rule out postdepositional transport of such phytoliths and second, they can be reliably indirectly dated from the residues, ceramic styles, or human remains in which or on which they are embedded.

Table 4-4 presents the indirect dates of phytoliths associated with 14 early contexts in Central and South America, and Figure 4-3 shows their mapped distributions. In

TABLE 4-4 Dates on Materials Associated with *Zea* Phytolith Samples

Country/Region	Site name	Dated material	^{14}C Method	Associated *Zea*	Radiocarbon years BP	Sample ID number	Reference
Guatemala							
Pacific coast	Sipacate	wood	AMS	phytoliths	4600	n/a	[52, 53]
Panama							
	Monte Oscuro	bulk sediment	AMS	phytoliths	7500 ± 70	Beta-74292	[67, p. 81]
	Cueva de los Ladrones	unknown		phytoliths	6860 ± 90	n/a	[65, p. 873] (given as 4919 ± 90 BC)
	Gatun Lake	charcoal?		phytoliths,	4750 ± 100	UCLA-1354	[61, p. 15]
	Aguadulce	unknown		phytoliths	4500	n/a	[61, 62]
	La Yequada	unknown		phytoliths	4200	n/a	[64]
	Lake Wodehouse	unknown		phytoliths	3900	n/a	[63]
Ecuador							
Coast	Vegas site	assoc. shell	Conventional	phytoliths	7150 ± 70	Tx-3314	[60, p. 332; 92, pp. 618, 621]
Coast	Loma Alta	unknown		phytoliths	5000	n/a	[59, p. 224]
Amazonia	Lake Ayauchi	assoc. charcoal	AMS	phytoliths	4570 ± 70	Beta-20956	[17, p. 304]
Coast	Real Alto	unknown		phytoliths	4450	phase: Valdivia 2	[59, p. 225; 60, p. 330]
Coast	La Emerenciana	food residue on sherd	AMS	phytoliths in residue	3860 ± 50	Beta-125107	[91, p. 44]
Coast	La Emerenciana	charcoal	Conventional	phytoliths	3775 ± 165	SMU-2563	[91, p. 46]
Highland	Cotocollao	unknown		phytoliths	ca. 3500	n/a	[59, p. 232]

Guatemala, the earliest example, ca. 4600 BP, comes from the Sipacate region of the Pacific Coast [52]. Six examples come from both core samples and from archaeological sites in Panama. Four of these are in the 3900 to 4750 BP range and two are much older, in the 6860 to 7500 BP range [61, 62, 63, 67].

Seven samples are presented for six sites in Ecuador. These are Lake Ayauchi at ca. 4570 BP [17], the Las Vegas site ca. 7150 BP [60, 92], Real Alto ca. 4450 BP [60], Loma Alta ca. 5000 BP [59], La Emerenciana ca. 3775 BP [91], and Cotocollao ca. 3500 BP [59]. In addition to these samples, Staller and Thompson [91] report an AMS date on residues from a cooking vessel that contained maize phytoliths embedded in the residues: 3860 ± 50 BP [Beta-125107]. This latter date is particularly important because, as pointed out earlier, there is almost as much certainty that the AMS date is actually measuring the age of the maize phytoliths as there would be if they were able to directly date maize macroremains from the same context (although none have been recovered).

In comparing the early pollen distribution map (see Figure 4-2) with the early phytolith distribution map (see Figure 4-3) the pattern is similar in many ways. Both pollen and phytoliths occur at early dates—in fact, several samples from Mexico, Panama, Colombia, and Ecuador predate (indirectly) the earliest known maize macroremains from Guilá Naquitz by as much as 2100 years. This is hard to accept because the Guilá Naquitz maize is primitive—any more primitive and it would still have been close in morphology to teosinte [5, 91]. Both maps also show that there are discontinuities in the distribution of the earliest examples. If maize spread south at the time that the pollen and phytoliths suggest it did, then the people using it must have avoided or maneuvered around certain regions on their way south. Many centuries later, people must have spread it back to the north, filling in areas that had been bypassed in earlier times. Another possible interpretation is that maize did spread evenly or continuously to the south throughout the Americas, but we simply have not yet found the evidence. With regard to this last point, we are reminded once again of the need for directly dated remains so that remixing, secondary deposition, and bioturbation can be ruled out.

DATING THE EARLY DISTRIBUTION OF MODERATE-TO-HIGH STABLE CARBON ISOTOPE RATIOS

Smalley and Blake [82] summarized some of the published stable carbon isotope data for more than 600 individual human remains recovered from dozens of well-dated sites from Mesoamerica to South America. The pattern they observed was one of gradually increasing stable carbon ratios spanning the period from the first appearance of maize in Central Mexico to the time of the Spanish Conquest. In most regions, with three important exceptions, the shift to

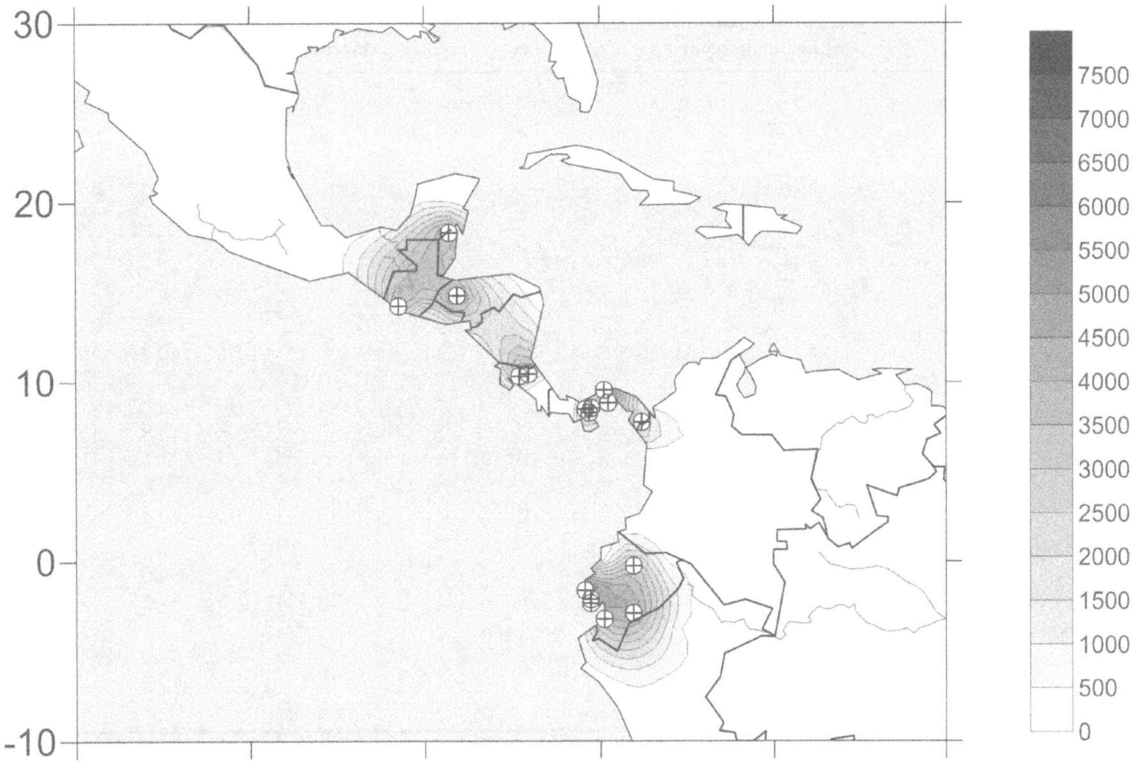

FIGURE 4-3 Indirect dates for *Zea* phytolith samples (using data from Table 4-4). Radiocarbon years BP.

higher stable carbon ratios, reflecting significant maize consumption in the diet, did not occur until after about 3000 years ago. That is, even where maize had been present in macrobotanical and microbotanical remains for as long as 5000 years, it was not generally a dietary staple until after 3000 BP, and in many regions, such as the Andean region, not until much later than that [94].

Table 4-5 presents the earliest occurrences of moderate-to-high stable carbon ratios (that is, greater than −15.0‰) in human remains [82, pp. 685–686]. People who eat little or no maize (or other C4 plants) have stable carbon isotope ratios in the −20‰ to −15‰ range, whereas moderate-to-high maize (or other C4 plant) consumption may be reflected in stable carbon ratios between −15.0‰ and −6‰. These values depend on a complex mix of foods in the diet and cannot simply be reduced to maize eating or not. However, most researchers do agree that the greater the reliance on maize in the diet, the higher the stable carbon ratio. Unfortunately, however, other plants and animals besides maize also produce high stable carbon isotope ratios. So, although we can say that individuals with low stable carbon isotope ratios are unlikely to have consumed much maize, we cannot say that all those with high stable carbon ratios did consume maize. Plants such as cacti and agaves, known as crassulacean acid metabolism (CAM) plants, also produce high stable carbon ratios. Furthermore, many other C4 plants, besides maize, were consumed by peoples in the New World, and these can equally lead to elevated stable carbon ratios. *Setaria,* or foxtail millet, is one example of an economically important C4 grass seed that was consumed during the Archaic period, whereas CAM plants such as agaves and cacti were even more abundant in the archaeological deposits at Guilá Naquitz and Coxcatlán. Several species of marine fish and crustaceans constitute another food type that can lead to high stable carbon isotope ratios. I consider these in the following section.

Figure 4-4 maps the dates of the moderate-to-high stable carbon values presented in Table 4-5. It shows several clear hot spots that may indicate high consumption of C4 or CAM plants or marine foods. The earliest occurrence of a "high" stable carbon ratio (−13.3‰) comes from the Tehuacán Valley's El Riego Phase (midpoint ca. 7900 BP, [25]). This individual, and one from the succeeding Coxcatlán Phase (midpoint ca. 6250 BP, [25]) with an even higher stable carbon ratio of −6.1‰, are unlikely to have been eating much maize, since the earliest maize at Tehuacán dates to ca. 4700 BP. A more plausible explanation for the high stable carbon ratios for these two individuals is that they were consuming other C4 plants such as *Setaria* and CAM plants such as agave. As Farnsworth and colleagues [25, p. 112] and Callen [18] point out, *Setaria* seeds were found in large quantities in more than 70% of the **coprolites** recovered

TABLE 4-5 Earliest Dates for Regions Where Human Stable Carbon Isotope Samples (del^{13}C) Are Greater Than −15‰

Country/Region or Site	Radiocarbon years BP for phase midpoint	N	Average del^{13}C ratio ‰	Reference[a]
Mexico				
Chiapas, Mazatán region	2800	4	−14	[7]
Chiapas, Tlacuachero	4400	2	−9.8	[7]
Puebla, Tehuacán Valley	7900	1	−13.3	[25]
Guatemala				
Peten	2100	16	−10.2	[99]
Belize				
Lowlands	2650	28	−12.9	[95]
Honduras				
Copan	1450	87	−9.7	[99]
Panama				
La Mula	2750	6	−11.7	[55]
Cerro Mangote	5750	16	−13.7	[55]
Ecuador				
Coast, Valdivia	3000	10	−12.3	[98]
Highlands, La Florida	1725	32	−11.3	[96]
Peru				
Highlands, Mantaro Valley	454	29	−15	[36, 37]
Coast, Puerto Moorin (V-66)	2000	not given	−11	[24]
Venezuela				
Parmana	1600	3	−10.3	[73]

[a]See J. Smalley, M. Blake (2003). Sweet beginnings: Stalk sugar and the domestication of maize. *Current Anthropology, 44,* 82; see p. 685 for full details of sample calculation.

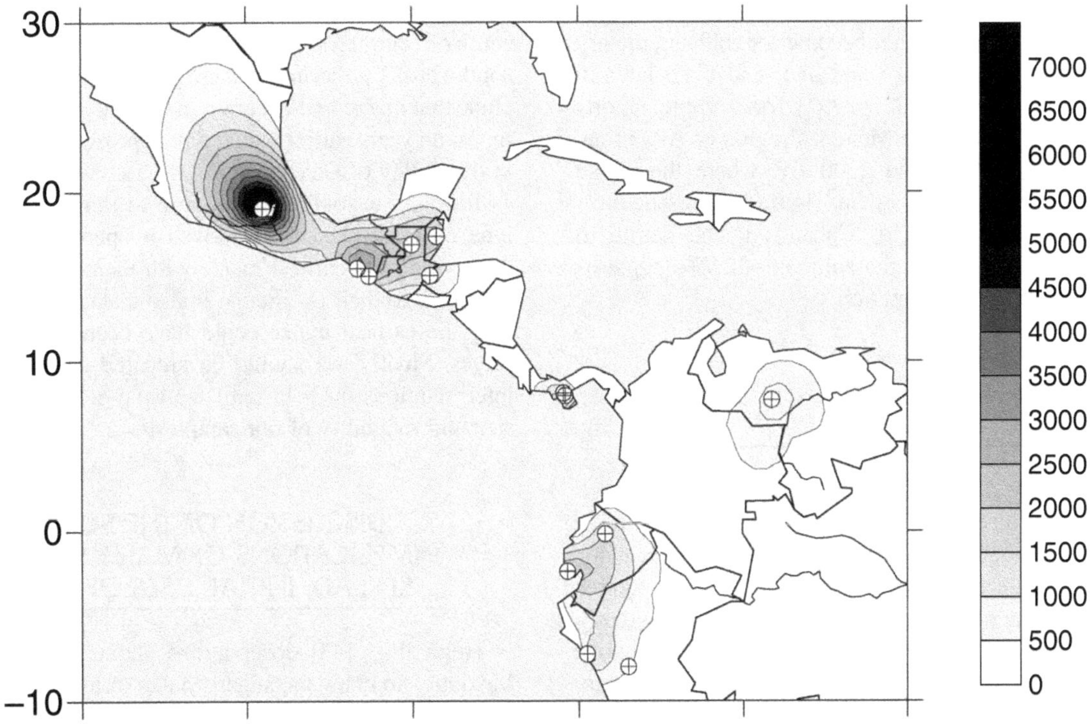

FIGURE 4-4 Dates for moderate-to-high stable carbon isotope values (del^{13}C) greater than −15‰ (using data from Table 4-5). Radiocarbon years BP.

from El Riego Phase and Coxcatlán Phase deposits. Macrobotanical remains from the cave deposits also show that CAM plants—especially agave and prickly pear cactus—were also frequently consumed. Therefore, the high stable carbon isotope ratio at such an early date in central Mexico (Figure 4-4) is unlikely to have resulted from maize consumption.

Another early case of high stable carbon isotope ratios is recorded for two Archaic period (Chantuto B Phase, midpoint ca. 4400 BP) individuals from the Soconusco region of Chiapas [7]. The mean stable carbon ratio for these two individuals is −9.8‰, but as Chisholm and Blake (Chapter 12) discuss, this could have been a result of a high marine diet, not necessarily C-4 plants such as maize. One possible explanation is that these people were specializing in shrimp and other marine resource harvesting, and that their stable carbon isotope levels reflect that specialization, not maize farming. The stable nitrogen isotope values (del^{15}N) ratios for these two individuals are much lower than would be expected for a marine diet and more in line with a C4 plant diet. It is not possible, with current evidence, to distinguish between these two possible interpretations.

A similar pattern holds for the Cerro Mangote site in Panama. Norr [55] reported on 16 individuals with a mean stable carbon ratio of −13.7‰. They date to Pre-Ceramic period (5000–2500 BC—phase midpoint estimated at ca. 5750 BP) and show stable carbon ratios that are intermediate between a C3 and C4 diet, but are also in the range of a marine diet [55, pp. 218–219]. The stable nitrogen isotope values (del^{15}N) are higher than would be expected for marine diet, however, and this leads Norr to suggest "a diet comprised of predominantly terrestrial fauna and maize."

All of the other examples (from Mexico to Peru) of moderate-to-high stable carbon isotope ratios occur after ca. 3000 BP. This pattern suggests that with the possible exception of Pacific coastal Panama, few people were significantly dependent on maize as their primary staple until about 3000 years ago. Even after 3000 BP there are examples of individuals who did not appear to rely heavily on maize. For example, Brady and colleagues [11] report two individuals recovered from the Cueva del Río Talgua in northeastern Honduras; one of the individuals dated to the Classic Period (1385 ± 75 BP, WG-286), yet had a stable carbon isotope ratio of only −20.02‰ (J. Brady, personal communication). This is slightly lower than an Early Formative period individual from the same cave (3110 ± 85 BP, WG-285), whose stable carbon isotope ratio was −18.62‰ (J. Brady, personal communication). Neither of the individuals could have eaten much maize, even though maize consumption was generally high throughout the Maya region by the Classic Period [95].

In South America, with the exception of people living in Coastal Ecuador, maize did not become a significant dietary component until after 2000 years ago, and even later in many places [e.g., 16]. Gil [32, p. 297], for example, reports several individuals from the Mendoza region of Argentina, dating between 200 BP and 2300 BP, where the stable carbon isotope ratios are less than −15.0‰. Only one individual, from the Agua de los Caballos I site, dating to between 250 and 350 BP, had a value of −12.7‰, suggesting significant maize consumption.

COMPARING THE DIFFERENT LINES OF EVIDENCE

Admittedly the data for any single category of evidence as presented earlier are still sparse. Furthermore, the individual lines of evidence are often neither in geographical nor chronological agreement. In all cases, they need further refinement and the sample sizes need to be increased greatly to improve their reliability. However, it is worth summarizing and comparing the patterns among the different data sets.

First, the direct AMS dates on maize macroremains provide a benchmark against which the other categories of data may be compared. This has been especially useful in examining the indirectly dated maize macroremains and questioning their reliability. Because there is, in most cases, no reason not to directly date existing maize macroremains using AMS dating, it makes little sense to continue using indirect dates. One or two dates from well provenanced samples will quickly resolve any outstanding chronological issues. This, of course, is not possible where maize macroremains have not been recovered, and as Deborah Pearsall [58] has pointed out, charcoal preservation in some tropical environments may continue to make it unlikely that maize macrofossils will soon be recovered.

Second, when comparing maize microfossil and macrofossil evidence, it must be noted that both are subject to post-depositional mixing and disturbance. Kent Flannery [27, pp. 271–272; 29, p. 290] has pointed this out several times—and it is worth restating his caution—that we must be careful in making interpretations about the presence of early agriculture, let alone its importance, based solely on small samples of botanical remains, be they macro, micro, or molecular specimens.

Figures 4-1, 4-2, and 4-3 clearly show how small samples can influence and distort our perspective. The distribution of early direct dates shows the gradual progression of the spread of maize in Mesoamerica and the American Southwest. But moving southward, into Central and South America, there is a huge gap with no directly dated macroremains. Conversely, the pollen and phytolith data, if taken at face value, would lead us to think that maize domestication occurred first in the tropical lowlands far to the east and south of central Mexico and then later spread northward and southward. Furthermore, these data might lead us to conclude that maize first occurred in Central America one to two thousand years earlier than it first appeared in primitive form in the Valley of Oaxaca.

It makes a good deal of sense to think of exactly what type of plant was being moved by people as they either transported the earliest maize with them, or as they traded or gave it to their neighbors. It also makes sense to consider how the earliest maize could have been used in its initial stages. Finally, we should be prepared to put most of our interpretations on hold until we have greatly increased the size and reliability of our samples.

DISCUSSION OF THE SOCIAL IMPLICATIONS OF MAIZE'S EARLY SPREAD: INITIAL USES OF MAIZE

Hugh Iltis' [43] observations about people in western Mexico who chew the sugary stalks of teosinte—much like sugarcane—prompted John Smalley and me to consider the following possibility: that one of the initial primary uses of teosinte in the wild may have been to extract the sugary juice in its stalk [82]. We initially became interested in this process of shifting uses of maize after reading ethnographic accounts of maize beer production using the sugary juice of the maize stalk among peoples such as the Tarahumara in northern Mexico [e.g., 4, 46]. We also considered the possibility that teosinte was planted by people in zones outside its natural range to increase the availability of a source of sugar, either for brewing beverages containing alcohol or for casual eating.

Teosinte's initial domestication may have come about as Flannery [28] described: People practiced agriculture to reduce the effect of annual variations in natural factors such as rainfall, which could otherwise affect the productivity of desirable plants. People's attempts to intervene in the natural life cycle of teosinte by planting it in more favorable locations; tending it by watering and weeding; and selecting larger, healthier seeds for replanting would have eventually led to its domestication. The persistence of this intervention over several generations must have required a great deal of interest in the plant—at least to the point where cobs began to appear and people could start selecting for attributes that produced larger kernels and cobs [5, 66]. People's earliest interest in maize may have had to do with its nutritious greens and sugary stalk, and people tended it and moved it to new locations for those reasons [43, 68]. Later, when small cobs developed, people may still have been interested in the greens and stalk, but it also became easier to collect the seeds in cob form. Later still, the nutritional value of the dry kernels became accessible once the hard casing of teosinte seeds began to disappear from domesticated populations.

If early Mesoamericans grew teosinte to make alcohol from its stalk or to provide fresh sweet greens, or both, then they were doing so in the social context of food consumption that goes well beyond simple nutritional value of the grain. They were using the teosinte plant for its potential in facilitating social interaction. In fact, as Henry Bruman [13] pointed out 60 years ago in his doctoral dissertation on traditional alcohol production in Mexico, the maize stalk (and by extension the teosinte stalk) was just one among many sources of sugar and carbohydrates that could be turned into alcohol and consumed in the ritual cycles that mark the Mesoamerican calendar. Alcohol production, as an essential ingredient of social interactions, would have been an instrumental part of the broader process of feasting behavior that has become of great interest in recent years [23, 45]. Social gatherings, such as feasts, may have encouraged the production and consumption of alcohol, and provided a stimulus to both the initial domestication and the rapid expansion of maize beyond its natural habitat. Archaic period Mesoamericans, involved in the cultivation and harvesting of many different plants, could have easily and quickly incorporated new genera such as maize into their embryonic systems of agriculture.

This hypothesis—that the initial spread and use of *Zea* (either teosinte or early maize) was prompted by the value of the stalks in producing large quantities of fermentable juice—may help to make sense of the several conflicting lines of evidence and observations that have arisen in the past few years, as well as those presented earlier in the chapter. On the one hand, these discrepancies may be explained simply as gaps in our data or the result of differential preservation of macroremains and microremains of maize. On the other hand, it is possible that maize was used differently in earlier times than it was in later times, after it had undergone a series of genetic transformations leading to the plant that we know today. The maize stalk sugar hypothesis helps to explain some of the discrepancies observed in the archaeological and paleoethnobotanical records, and it gives new ways to incorporate social processes into our explanations. For example, the widespread early occurrence of *Zea mays* pollen and phytoliths in the studies presented in Tables 4-3 and 4-4 (and in studies in this volume) could have resulted from the rapid and early spread of *Zea* (either teosinte or early maize) into the tropical lowlands of southern Mesoamerica, Central America, and perhaps even Colombia and Ecuador. Its lack of a well-defined cob in these early stages, and perhaps its use for sugar and sweet greens rather than as a storable grain, could help explain why so few macroremains have been found from these early contexts [58].

The widespread occurrence of maize in ceremonial contexts in South America (Chapter 32), and the apparent use of the grain in *chicha* production, shows that the growing, harvesting, and consumption of maize was not homogeneous throughout the Americas. Its uses varied at different times and in diverse places. Numerous peoples transformed maize to suite their needs, just as maize eventually transformed the many peoples who used it [38, 39].

One promising new avenue of research lies in the linguistic evidence for the spread of maize throughout the Americas. Jane Hill's [40; Chapter 46] research on the correspondence of terms for maize in the broad Uto-Aztecan language group suggests that Proto–Uto-Aztecan speaking peoples brought maize with them as they spread from Mexico northward into the American Southwest. The pan-American glottochronological study by Cecil Brown (Chapter 47) indicates, however, that the terms for maize differ among Uto-Aztecan speakers possibly because they may have been using maize long before it became an economically significant food source. However, Brown shows that for other groups, such as the Mayan speakers, the terms for maize are so consistently similar that its spread must have coincided with its rapid adoption as an economically significant food staple—perhaps at the same time that Mayan speakers colonized the regions that today form their homelands. Studies such as these will help us link our knowledge of the domestication and spread of maize with much broader questions of agricultural and linguistic dispersals on a world-wide scope, as outlined by Bellwood [3].

When viewed on such a broad scale, the evidence for the dating of the spread of maize indicates that we need to consider a range of social and economic processes, such as a growing demand for alcohol that was possibly used in feasts and rituals or an increased need for reliable staple food crops, in our explanations for its initial spread throughout the Americas. Testing such interpretations in reliable and convincing ways will only proceed when we increase the sample of accurately dated remains.

Acknowledgments

This chapter is a revised version of a paper presented at the 69th Annual Meeting of the Society for American Archaeology, March 31 to April 4, 2004. I thank John Smalley who was coauthor of the S.A.A. version of the paper. My thanks to John E. Staller and R. G. Matson for their many comments and suggestions about this topic over the past two years. Bruce Benz, Bob Hard, and Marcus Winter kindly offered many helpful suggestions to improve an earlier draft of the paper. I also thank J. Scott Raymond and James Brady for making some of their unpublished data available.

References Cited

1. K. R. Adams. (1994). A regional synthesis of *Zea mays* in the prehistoric American Southwest. In: S. Johannssen, C. A. Hastorf, (Eds.), *Corn and culture in the prehistoric New World*. Boulder, CO: Westview Press. pp. 273–302.

2. M. R. Arford, S. P. Horn. (2004). Pollen evidence of the earliest maize agriculture in Costa Rica. *Journal of Latin American Geography, 3,* 108–115.

3. P. Bellwood. (2005). *First farmers: The origins of agricultural societies.* London: Blackwell.

4. W. C. Bennett, R. M. Zingg. (1976). *The Tarahumara: An Indian tribe of northern Mexico* (Reprinted). Glorieta, NM: Rio Grande Press.

5. B. F. Benz. (2001). Archaeological evidence of teosinte domestication from Guilá Naquitz, Oaxaca. *Proceedings of the National Academy of Sciences of the United States of America, 98,* 2104–2106.

6. B. F. Benz, A. Long. (2000). Early evolution of maize in the Tehuacán Valley, Mexico. *Current Anthropology, 41,* 459–465.

7. M. Blake, B. S. Chisholm, J. E. Clark, B. Voorhies, M. W. Love. (1992). Prehistoric subsistence in the Soconusco region. *Current Anthropology, 33,* 83–94.

8. M. Blake, J. E. Clark, B. Voorhies, G. Michaels, M. W. Love, M. E. Pye, A. A. Demarest, B. Arroyo. (1995). Radiocarbon chronology for the Late Archaic and Formative periods on the Pacific Coast of Southeastern Mesoamerica. *Ancient Mesoamerica, 6,* 161–183.

9. D. Bonavia, A. Grobman. (1989). Preceramic maize in the Central Andes: A necessary clarification. *American Antiquity, 54,* 836–840.

10. J. E. Bradley, T. Vieja. (1994). An Archaic and Early Formative site in the Arenal region. In: P. D. Sheets, B. R. McKee, (Eds.), *Archaeology, volcanism, and remote sensing in the Arenal region, Costa Rica.* Austin: University of Texas Press. pp. 73–86.

11. J. E. Brady, C. Begley, J. Fogarty, D. J. Stierman, B. Luke, A. Scott. (Talgua Archaeological project: A technical assessment. Accessed on September 7, 2005, at: http://www.calstatela.edu/academic/anthro/Talgua_pre.htm.

12. W. Bray, L. Herrera, M. C. Schrimpff, P. Botero, J. G. Monsalve. (1987). The ancient agricultural landscape of Calima, Colombia. In: W. Denevan, K. Mathewson, G. Knap, (Eds.), *Pre-hispanic agricultural fields in the Andean region.* Oxford, UK: British Archaeological Reports International Series. pp. 443–481.

13. H. J. Bruman. (2000). *Alcohol in ancient Mexico.* Salt Lake City: University of Utah Press.

14. R. L. Burger. (1984). *The prehistoric cccupation of Chavín de Huántar, Peru. University of California Publications in Anthropology 14.* Berkeley: University of California Press.

15. R. L. Burger, L. Salazar-Burger. (1991). The second season of investigations at the initial period center of Cardal, Peru. *Journal of Field Archaeology, 18,* 275–296.

16. R. L. Burger, N. J. van der Merwe. (1990). Maize and the origin of Highland Chavín civilization: An isotopic perspective. *American Anthropologist, 92,* 84–95.

17. M. B. Bush, D. R. Piperno, P. A. Colinvaux. (1989). A 6,000 year history of Amazonian maize cultivation. *Nature, 340,* 303–305.

18. E. O. Callen. (1967). Analysis of Tehuacan Coprolites. In: D. S. Byers, (Ed.), *The prehistory of the Tehuacan Valley: Volume I. Environment and subsistence.* Austin: University of Texas Press. pp. 261–289.

19. J. Chapman, G. D. Crites. (1987). Evidence for early maize (*Zea mays*) from the Icehouse Bottom Site, Tennessee. *American Antiquity, 52,* 352–354.

20. J. E. Clark. (1994). *The development of Early Formative rank societies in the Soconusco, Chiapas, Mexico.* Ph.D. dissertation, University of Michigan–Ann Arbor.

21. R. M. Clement, S. P. Horn. (2001). Pre-Columbian land-use history in Costa Rica: A 3000-year record of forest clearance, agriculture and fires from Laguna Zoncho. *The Holocene, 11,* 419–426.

22. M. D. Coe, K. V. Flannery. (1967). Early cultures and human ecology in South Coastal Guatemala. In: *Smithsonian contributions to anthropology,* Vol. 3. Washington, D.C.: Smithsonian Institute.

23. M. Dietler. (1990). Driven by drink: the role of drinking in the political economy and the case of early Iron Age France. *Journal of Anthropological Archaeology, 9,* 352–407.

24. J. E. Ericson, M. West, C. H. Sullivan, H. W. Krueger. (1989). The development of maize agriculture in the Viru Valley, Peru. In: T. D. Price, (Ed.), *The chemistry of prehistoric human bone.* Cambridge, UK: Cambridge University Press. pp. 68–104.

25. P. Farnsworth, J. E. Brady, M. J. DeNiro, R. S. MacNeish. (1985). A reevaluation of the isotopic and archaeological reconstruction of diet in the Tehuacán Valley. *American Antiquity, 50,* 102–116.

26. V. L. Feddema. (1993). *Early Formative subsistence and agriculture in Southeastern Mesoamerica.* Master's Thesis, Department of Anthropology and Sociology, Vancouver: University of British Columbia.

27. K. V. Flannery. (1973). The origins of agriculture. *Annual Review of Anthropology, 2,* 271–310.

28. K. V. Flannery. (1986). The research problem. In: K. V. Flannery, (Ed.), *Guilá Naquitz: Archaic foraging and early agriculture in Oaxaca, Mexico.* Orlando, FL: Academic Press. pp. 3–18.

29. K. V. Flannery. (2002). Editorial Comment. *Antiquity, 76,* 289–291.

30. F. de O. Freitas. (2001). Estudo genético-evolutivo de amostras modernas e arqueológicas de milho *(Zea mays mays, L.)* e feijão *(Phaseolus vulgaris, L.).* Ph.D. Dissertation, Escola Superior de Agricultura Luiz de Queiroz. University of São Paulo, Brazil.

31. K. Fukunaga, J. Hill, Y. Vigouroux, Y. Matsuoka, J. Sanchez G., K. Liu, E. S. Buckler, J. Doebley. (2005). Genetic diversity and population structure of teosinte. *Genetics, 169,* 2241–2254.

32. A. F. Gil. (2003). *Zea mays* on the South American periphery: Chronology and dietary importance. *Current Anthropology, 44,* 295–300.

33. M. Goman, R. Byrne. (1998). A 5000-year record of agriculture and tropical forest clearance in the Tuxtlas, Veracruz, Mexico. *The Holocene, 8,* 83–89.

34. T. Grieder. (1988). Radiocarbon measurements. In: T. Grieder, A. B. Mendoza, C. E. Smith, Jr., R. M. Malina, (Eds.), *La Galgada, Peru: A preceramic culture in transition.* Austin: University of Texas Press, pp. 68–72.

35. R. J. Hard, J. R. Roney. (1998). A massive terraced village complex in Chihuahua, Mexico, 3000 years before present. *Science, 279,* 1661–1664.

36. C. A. Hastorf. (1998). The use of paleoethnobotanical data in prehistoric studies of crop production, processing, and consumption. In: C. Hastorf, V. Popper, (Eds.), *Current paleoethnobotany: Analytical methods and cultural interpretations of archaeological plant remains.* Chicago: University of Chicago Press. pp. 119–144.

37. C. A. Hastorf. (1991). Gender, space and food in prehistory. In: J. Gero, M. Conkey, (Eds.), *Engendering archaeology: Women and prehistory.* Oxford, UK: Blackwell Press. pp. 132–159.

38. C. A. Hastorf. (1994). Cultural meanings: Introduction to part four. In: S. Johannssen, C. A. Hastorf, *Corn and culture in the prehistoric New World.* Boulder, CO: Westview Press. pp. 395–398.

39. C. A. Hastorf. (1999). Cultural implications of crop introductions in Andean prehistory. In: C. Gosden, J Hather, (Eds.), *The prehistory of food: Appetites for change.* London: Routledge. pp. 35–58.

40. J. H. Hill. (2001). Proto-Uto-Aztecan: A community of cultivators in central Mexico? *American Anthropologist, 1003,* 913–934.

41. B. B. Huckell. (1990). *Late preceramic farmer-foragers in Southeastern Arizona: A cultural and ecological consideration of the spread of agriculture into the arid Southwestern United States.* Ph.D. Dissertation, Department of Arid Lands Resource Sciences, University of Arizona, Tucson.

42. B. B. Huckell, L. Huckell, S. K. Fish. (1995). *Investigations at Milagro, a late preceramic site in the Eastern Tucson Basin.* Technical Report 94-5. Tucson, AZ: Center for Desert Archaeology.

43. H. H. Iltis. (2000). Homeotic sexual translocations and the origin of maize (*Zea mays, Poaceae*): A new look at an old problem. *Economic Botany, 54,* 7–42.
44. V. Jaenicke-Després, E. S. Buckler, B. D. Smith, M. Thomas, P. Gilbert, A. Cooper, J. Doebley, S. Pääblo. (2003). Early allelic selection in maize as revealed by ancient DNA. *Science, 302,* 1206–1208.
45. J. Jennings, K. L. Antrobus, S. J. Atencio, E. Glavich, R. Johnson, G. Loffler, C. Luu. (2005). Drinking beer in a blissful mood: Alcohol production, operational chains, and feasting in the ancient world. *Current Anthropology, 46,* 275–303.
46. J. G. Kennedy. (1978). *Tarahumara of the Sierra Madre: Beer, ecology and social organization.* Arlington Heights, IL: AHM Publishing.
47. R. D. Lippi, R. McK. Bird, D. M. Stemper. (1984). Maize recovered at La Ponga, an early Ecuadorian site. *American Antiquity, 49,* 118–124.
48. A. Long, B. F. Benz, D. J. Donahue, A. J. T. Jull, L. J. Toolin. (1989). First direct AMS dates on early maize from Tehuacán, Mexico. *Radiocarbon, 31,* 1035–1040.
49. R. G. Matson. (2003). The spread of maize agriculture into the U.S. Southwest. In: P. Bellwood, C. Renfrew, (Eds.), *Examining the farming/language dispersal hypothesis.* Cambridge, UK: McDonald Institute for Archaeological Research, University of Cambridge. pp. 341–356.
50. Y. Matsuoka, Y. Vigouroux, M. M. Goodman, J. Sanchez G., E. Buckler, J. Doebley. (2002). A single domestication for maize shown by multilocus microsatellite Genotyping. *Proceedings of the National Academy of Sciences of the United States of America, 99,* 6080–6084.
51. J. Monsalve. (1985). A Pollen Core from the Hacienda Lusitania. In: W. Bray, (Ed.), *Pro Calima: Archälogische-ethnologisches Projekt im Westlichen Columbien/Sudamerika No. 5.* Vereingung Pro Calima, Bern. pp. 40–44.
52. H. Neff, B. Arroyo, J. G. Jones, D. M. Pearsall, D. E. Freidel. (2002). Nueva evidencia pertinente a la ocupación temprana del sur de Mesoamérica. Paper Presented at the XII Encuentro Internacional: Los Investigadores de la Cultura Maya, Campeche, November 10–14, 2002. University of Campeche.
53. H. Neff, D. M. Pearsall, J. G. Jones, B. Arroyo de Pieters, D. E. Freidel. (In press). Climate change and population history in the Pacific Lowlands of Southern Mesoamerica. *Quaternary Research.*
54. C. Niederberger. (1979). Early sedentary economy in the Basin of Mexico. *Science, 203,* 131–142.
55. L. Norr. (1995). Interpreting dietary maize from bone stable isotopes in the American tropics: The state of the art. In: P. W. Stahl, (Ed.), *Archaeology in the lowland American tropics: Current analytical methods and recent applications.* Cambridge, UK: Cambridge University Press. pp. 198–223.
56. L. Nuñez. (1986). Evidencias arcaicas de maíces y cuyes en Tiliviche: Hacia el semisedentarismo en el litoral fértil y quebradas del norte de Chile. *Chungara, 16/17,* 25–48.
57. D. M. Pearsall. (1999). The impact of maize on subsistence systems in South America: An example from the Jama River Valley, Coastal Ecuador. In: C. Gosden, J. Hather, (Eds.), *The prehistory of food: Appetites for change.* London: Routledge. pp. 419–437.
58. D. M. Pearsall. (2000). *Paleoethnobotany: A handbook of procedures* (2nd ed.). San Diego, CA: Academic Press.
59. D. M. Pearsall. (2003). Plant food resources of the Ecuadorian Formative: An overview and comparison to the Central Andes. In: J. S. Raymond, R. L. Burger, (Eds.), *Archaeology of Formative Ecuador: A symposium at Dumbarton Oaks, 7 and 8 October 1995.* Washington, D.C.: Dumbarton Oaks Research Library and Collection. pp. 213–257.
60. D. M. Pearsall, D. R. Piperno. (1990). Antiquity of maize cultivation in Ecuador: Summary and reevaluation of the evidence. *American Antiquity, 55,* 324–337.
61. D. R. Piperno. (1985). Phytolithic analysis of geological sediments from Panama. *Antiquity, 59,* 13–19.
62. D. R. Piperno. (1988). *Phytolith analysis: An archaeological and geological perspective.* San Diego, CA: Academic Press.
63. D. R. Piperno. (1994). Phytolith and charcoal evidence for prehistoric slash-and-burn agriculture in the Darien rainforest of Panama. *The Holocene, 4,* 321–325.
64. D. R. Piperno, M. B. Bush, P. A. Colinvaux. (1990). Paleoenvironments and human occupation in Late-Glacial Panama. *Quaternary Research, 33,* 108–116.
65. D. R. Piperno, K. H. Clary, R. G. Cooke, A. J. Ranere, D. Weiland. (1985). Preceramic maize in Central Panama: Phytolith and pollen evidence. *American Anthropologist, 87,* 871–878.
66. D. R. Piperno, K. V. Flannery. (2001). The earliest archaeological maize (*Zea mays* L.) from Highland Mexico: New accelerator mass spectrometry dates and their implications. *Proceedings of the National Academy of Sciences of the United States of America, 98,* 2101–2103.
67. D. R. Piperno, J. G. Jones. (2003). Paleoecological and archaeological implications of a Late Pleistocene/Early Holocene record of vegetation and climate from the Pacific coastal plain of Panama. *Quaternary Research, 59,* 79–87.
68. D. R. Piperno, D. M. Pearsall. (1998). *Origins of agriculture in the Lowland Neotropics.* San Diego, CA: Academic Press.
69. M. D. Pohl, K. O. Pope, J. G. Jones, J. S. Jacob, D. R. Piperno, S. D. De France, D. L. Lentz, J. A. Gifford, M. E. Danforth, J. K. Josserand. (1996). Early agriculture in the Maya Lowlands. *Latin American Antiquity, 7,* 355–372.
70. K. O. Pope, M. D. Pohl, J. G. Jones, D. L. Lentz, C. von Nagy, F. J. Vega, I. R. Quitmeyer. (2001). Origin and environmental setting of ancient agriculture in the lowlands of Mesoamerica. *Science, 292,* 1370–1373.
71. S. Pozorski, T. Pozorski. (1987). *Early settlement and subsistence in the Casma Valley, Peru.* Des Moines: University of Iowa Press.
72. T. J. Riley, G. R. Walz, C. J. Bareis, A. C. Fortier, K. E. Parker. (1994). Accelerator mass spectrometry (AMS) dates confirm early *Zea mays* in the Mississippi River Valley. *American Antiquity, 59,* 490–498.
73. A. C. Roosevelt. (1980). *Parmana: Prehistoric maize and manioc subsistence along the Amazon and Orinoco.* New York: Academic Press.
74. B. J. Roth. (1989). *Late Archaic settlement and subsistence in the Tucson Basin.* Ph.D. Dissertation, Department of Anthropology, University of Arizona, Tucson.
75. D. J. Rue. (1989). Archaic Middle American agriculture and settlement: Recent pollen data from Honduras. *Journal of Field Archaeology, 16,* 177–184.
76. W. Rust, B. W. Leyden. (1994). Evidence of maize use at Early and Middle Preclassic La Venta Olmec Sites. In: S. Johannssen, C. A. Hastorf, (Eds.), *Corn and culture in the prehistoric New World.* Boulder, CO: Westview Press. pp. 181–202.
77. W. Rust, R. Sharer. (1988). Olmec settlement data from La Venta, Tabasco, Mexico. *Science, 242,* 102–104.
78. T. E. Scheffler. (El Gigante rock shelter: Archaic Mesoamerica and transitions to settled life. Reports submitted to FAMSI. Accessed on September 7, 2005 at: www.famsi.org/reports/00071/index.html.
79. J. Schoenwetter, L. D. Smith. (1986). Pollen analysis of the Oaxaca archaic. In: K. V. Flannery, (Ed.), *Guilá Naquitz: Archaic foraging and early agriculture in Oaxaca, Mexico.* Orlando, FL: Academic Press. pp. 179–218.
80. P. D. Sheets. (1994). The proyecto prehistórico Arenal: An introduction. In: P. D. Sheets, B. R. McKee (Eds.), *Archaeology, volcanism, and remote sensing in the Arenal region, Costa Rica.* Austin: University of Texas Press. pp. 1–23.

81. A. H. Simmons. (1986). New evidence for the early use of cultigens in the American Southwest. *American Antiquity, 51,* 73–89.
82. J. Smalley, M. Blake. (2003). Sweet beginnings: Stalk sugar and the domestication of maize. *Current Anthropology, 44,* 675–703.
83. F. E. Smiley. (1994). The agricultural transition in the northern Southwest: Patterns in the current chronometric data. *Kiva, 60,* 165–189.
84. B. D. Smith. (1992). *Rivers of change.* Washington, D.C.: Smithsonian Institution Press.
85. B. D. Smith. (1997). Reconsidering the Ocampo caves and the era of incipient cultivation in Mesoamerica. *Latin American Antiquity, 8,* 342–383.
86. B. D. Smith. (1998). Research: Origins of agriculture—between foraging and farming. *Science, 279,* 1651–1652.
87. B. D. Smith. (2001). Documenting plant domestication: The consilience of biological and archaeological approaches. *Proceedings of the National Academy of Sciences of the United States of America, 98,* 1324–1326.
88. B. D. Smith. (2005). Reassessing Coxcatlan cave and the early history of domesticated plants in Mesoamerica. *Proceedings of the National Academy of Sciences of the United States of America, 102,* 9438–9445.
89. C. E. Smith. (1988). Floral remains. In: T. Grieder, A. B. Mendoza, C. E. Smith, Jr., R. M. Malina, (Eds.), *La Galgada, Peru: A preceramic culture in transition.* Austin: University of Texas Press. pp. 125–151.
90. J. E. Staller. (2003). An examination of the palaeobotanical and chronological evidence for an early introduction of maize *(Zea mays L.)* into South America: A Response to Pearsall. *Journal of Archaeological Science, 30,* 373–380.
91. J. E. Staller, R. G. Thompson. (2002). A multidisciplinary approach to understanding the initial introduction of maize into coastal Ecuador. *Journal of Archaeological Science, 29,* 33–50.
92. K. E. Stothert. (1985). The preceramic Las Vegas culture of coastal Ecuador. *American Antiquity, 50,* 613–637.
93. M. D. Tagg. (1996). Early cultigens from Fresnal shelter, Southeastern New Mexico. *American Antiquity, 61,* 311–324.
94. R. H. Tykot, J. E. Staller. (2002). The importance of early maize agriculture in coastal Ecuador: New data from La Emerenciana. *Current Anthropology, 43,* 666–677.
95. R. H. Tykot, N. J. van der Merwe, N. Hammond. (1996). Stable isotope analysis of bone collagen and apatite in the reconstruction of human diet: A case study from Cuello, Belize. In: M. V. Orna, (Ed.), *Archaeological chemistry: Organic, inorganic and biochemical analysis.* ACS Symposium Series 625. Washington, D.C.: American Chemical Society. pp. 355–365.
96. D. H. Ubelaker, M. A. Katzenberg, L. G. Doyon. (1995). Status and diet in precontact Highland Ecuador. *American Journal of Physical Anthropology, 97,* 403–411.
97. S. Upham, R. S. MacNeish, W. C. Galinat, C. M. Stevenson. (1987). Evidence concerning the origin of maiz de ocho. *American Anthropologist, 89,* 410–419.
98. N. J. van der Merwe, J. A. Lee-Thorp, J. S. Raymond. (1993). Light, stable isotopes and the subsistence base of formative cultures at Valdivia, Ecuador. In: J. B. Lambert, G. Grupe, (Eds.), *Prehistoric human bone: Archaeology at the molecular level.* Berlin: Springer-Verlag. pp. 63–97.
99. N. J. van der Merwe, R. H. Tykot, N. Hammond, K. Oakberg. (2000). Diet and animal husbandry of the Preclassic Maya at Cuello, Belize: Isotopic and zooarchaeological evidence. In: S. Ambrose, A. Katzenberg, (Eds.), *Biogeochemical approaches to paleodietary analysis.* New York: Kluwer Academic/Plenum Publishers. pp. 23–38.
100. C. von Nagy. (2003). *Of meandering rivers and shifting towns: Landscape evolution an community within the Grijalva delta.* Ph.D. Dissertation, Department of Anthropology, Tulane University, New Orleans.
101. S. R. Whitt, L. M. Wilson, M. I. Tenaillon, B. S. Gaut, E. S. Buckler, IV. (2002). Genetic diversity and selection in the maize starch pathway. *Proceedings of the National Academy of Sciences of the United States of America, 99,* 12959–12962.
102. W. H. Wills. (1988). *Early prehistoric agriculture in the American Southwest.* Santa Fe, NM: School of American Research.

CHAPTER

5

Toward a Biologically Based Method of Phytolith Classification

GREG LADEN

Department of Anthropology, University of Minnesota, Minneapolis, Minnesota

Glossary

Allelic variation Variation in morphological features of shape or size that is caused by developmental or physical process directly controlled by genes.

Phytolith A silica formation or body formed among plant cells as part of a poorly understood physiological process. Phytoliths can preserve morphological information that can, in the absence of the plant tissues themselves, be used to identify plant taxa, sometimes to the level of species or even variety, in archaeological materials.

Rondel phytoliths A form of phytoliths especially common in grasses.

Reaction norm As used here, variation in morphological features of shape or size that is *not* caused by genetic variation but, rather, by variations in the environmental context of development of an organism.

Variable/variate As used here, a variable is a type of measurement, such as the shape of part of a rondel phytoliths, whereas variate is the specific observation subservient to the variable (e.g., round, square, triangular).

A previously developed and applied method of classifying maize **rondel phytoliths** into ethnographically documented distinct varieties has been shown to be effective in a wide range of contexts. However, the underlying factors that result in successful clustering have not been examined in detail. This analysis is a first attempt to demonstrate whether variation in the data that are collected represent primarily genetic factors, as opposed to reaction norms or observer error, and to assess whether some **variables** (or **variates** among those variables) contribute to a greater or lesser degree to the clustering process. This analysis shows that the different variables used in the method have roughly similar behavior of variability and may represent primarily **allelic variation.** However, there appear to be underlying multivariate relationships in the data that could not be discovered in this preliminary study, which suggests the need for a more extensive analysis using much larger samples (primarily consisting of multiple samples within each of several maize varieties).

INTRODUCTION

Previous work by Robert Thompson and colleagues [3, 5] (as reported and exemplified in various publications and in Chapters 30 and 37 of this volume) has revealed a relationship between patterns of morphological features of maize rondel phytoliths and ethnographically distinct maize varieties. The genetics that distinguish the "domesticated" form of maize (*Zea mays* L.) from its wild progenitor teosinte *(Zea mays* ssp. *parviglumis)* are now believed to involve relatively few genes [1, 2], with one of those genes being teosinte glume architecture (tga-1), which, in turn, is thought to affect silica deposition. It is, therefore, reasonable to hypothesize that the classification of maize varieties by **phytolith** characteristics is based on and reflects true taxonomic (phylogenetic) patterning.

Cluster analysis of morphological trait presence or trait absence data of populations of 100 phytoliths is successful in differentiating among known varieties and has been useful in classifying archaeological phytolith samples into known extant (ethnographically distinct) varieties. However, cluster analysis is best viewed as an exploratory technique that involves few or no assumptions about the underlying causes of variation, clustering, or difference among assemblages, other than that there may be statistically valid clusters and that an archaeological sample may fit to one of the extant samples. The present analysis is an attempt to examine the underlying patterns of variation that result in what appears to be a successful clustering methodology, as a first step in ultimately discovering which morphological traits are the best indicators of genetic variation in maize rondel phytoliths.

Any method that uses these morphological characteristics as a total set incorporates nonbiological causes of variation with equal weight as biological causes. This, then, is a first step in developing clustering, classification or tree-building schema that are biologically sensible rather than merely exploratory.

THE RAW DATA AND ITS PRESUMED MEANING

The raw data collected on several samples by Thompson, Lusteck, and colleagues consist, in part, of the presence or absence of shape-related (nonsize dependent) morphological features on rondel phytoliths. By using assemblages of 100, it is believed that sample sizes are large enough to average out nonallelic variation, and subsequent manipulation of the data require less arithmetic transformation. A premise of this method is the presumption that some of the variation in these traits is caused by underlying genetic variation, but it is also reasonable to suggest that developmental **reaction norms,** and observer variation will contribute additional variation. In fact, it can be unequivocally stated that an exhaustive list of sources of variation will consist of:

1. Allelic variation affecting rondel phytolith growth;
2. Norm of reaction (developmental) caused variation; and
3. All other variation including, and especially, error introduced because of the application of the technique (vagaries of microscopy, observer error, inconsistency in classification, etc.)

The data collected by Thompson and Lusteck in a typical analysis can be described as follows:

1. *Face morphology.* This is a letter code (A, B, C, or D) expressing basic morphology of the face of a rondel phytolith.
2. *Face size relationship,* which is the relationship between the size of the first (larger) and second (smaller) face.
3. *Complexity* of the edge of each face, with "1" referring to the simplest morphology, and "7" referring to a convoluted and decorated morphology; and
4. *Geometric shape* of each face (i.e., oval, square, triangular, etc.)

Thompson's [4] method combines these variables in a single code, so that, for example, a rondel that is entire (in its edge), with similar-sized faces of rectangular shape and no indentations or decorations would be called "A1A 1 1." By combining the different variables into a single code, an exhaustive set of unique identifiers is generated, and counts of phytoliths per code represent a morphological profile of an assemblage. Of the many logically possible variants of traits, a large number are rarely or never observed, such that fewer than half of the potential shape-related features are ever observed at any frequency. Of these, some traits are rare (seen only once or twice in an assemblage of 100 phytoliths), whereas others are more common.

Cluster analysis of these data readily group different samples from the same known variety and distinguish among varieties remarkably well, and, thus, presumably this technique of data recording and cluster analysis should work well for classifying archaeological samples. However, despite the fact that this method works well, it is still true that variation in the natural system that is being studied can caused by multiple sources, only one of which is genetic (allelic), whereas others are not. For this reason, the present analysis seeks to partition at least some of the variance into categories of greater or lesser relevance to genetic variation in an effort to (ultimately, not necessarily in this study) increase both the certainty of specific classifications (of samples) and to perhaps amend the methodology. The former is important because, as stated earlier, cluster analysis is mainly an exploratory technique and should ultimately lead to more systematic understanding of the variables and their real meaning. The latter is important because, although the technique works well at this point, as the number of ethnographic and archaeological samples goes up, it will actually become more difficult rather than easier to match archaeological samples to known varieties because a denser sampling of the actual range of ethnographic sampling will be differentiated by increasingly subtle or overlapping characteristics.

EXPLORING GENETIC VERSUS NONGENETIC VARIATION

In the absence of a detailed model for genetic control over development, it is possible to estimate the relationship between allelic variation and reaction norm (environmental variation) by comparing variation in data within ethnographic variety samples and among them. This is based on two assumptions. First, there is a closer genetic relationship among samples from one variety than among varieties (by definition) so that pure genetic variation should be small within varieties and larger among them. Second, although reaction norms may be larger among samples because the samples could come from different environments, the differential variation within and among samples should be much less than for allelic variation. Furthermore, if the ethnographic varieties are all sampled from the same general region, with similar ecological conditions, it is likely that reaction norms would vary randomly across the entire set of samples. In the present study, all of the ethnographic samples are from highland Bolivia and, therefore, are assumed to vary randomly among samples rather than being clustered by ethnographic variety. (This assumption would be further supported were it known that growing practices were similar across sampling areas. Beyond this, experimentally growing the maize varieties in controlled conditions would be the ideal strategy, perhaps to be pursued in a future study.)

For the present study, the within-variety versus among-variety variation for each of the components that make up these morphological profiles is assessed separately. This requires first generating "parsed variables" from the original morphological profile data set to examine them separately. In this way, the four variables listed earlier (face morphology, face shape relationship, etc.) are examined separately. The average values for each parsed variable within and among varieties provides a set of absolute residuals for each individual sample. The distribution of these residuals should be smaller in magnitude (lower average residual value) within varieties than among varieties because of sample size effects. However, the relative difference in these measures across variables should indicate the relative degree to which a particular measure indicates variation that is allelic versus other (reaction norm and error).

This is shown in Figure 5-1. Variation within samples is typically about half of that among samples, and all but one of the different variables have about the same relationship of within-variety versus among-variety variation. The exception is geometric shape, which may, therefore, be thought of as a more discriminatory variable than the others. Most importantly, however, the similarity of patterns of variation among the parsed variables indicates that it is unlikely that any one type of observation is contributing in any usual way to the cluster analysis.

There seems to be no exceptional degree of covariance among the parsed variables when analyzed at the finest level possible. This was examined with the use of several tech-

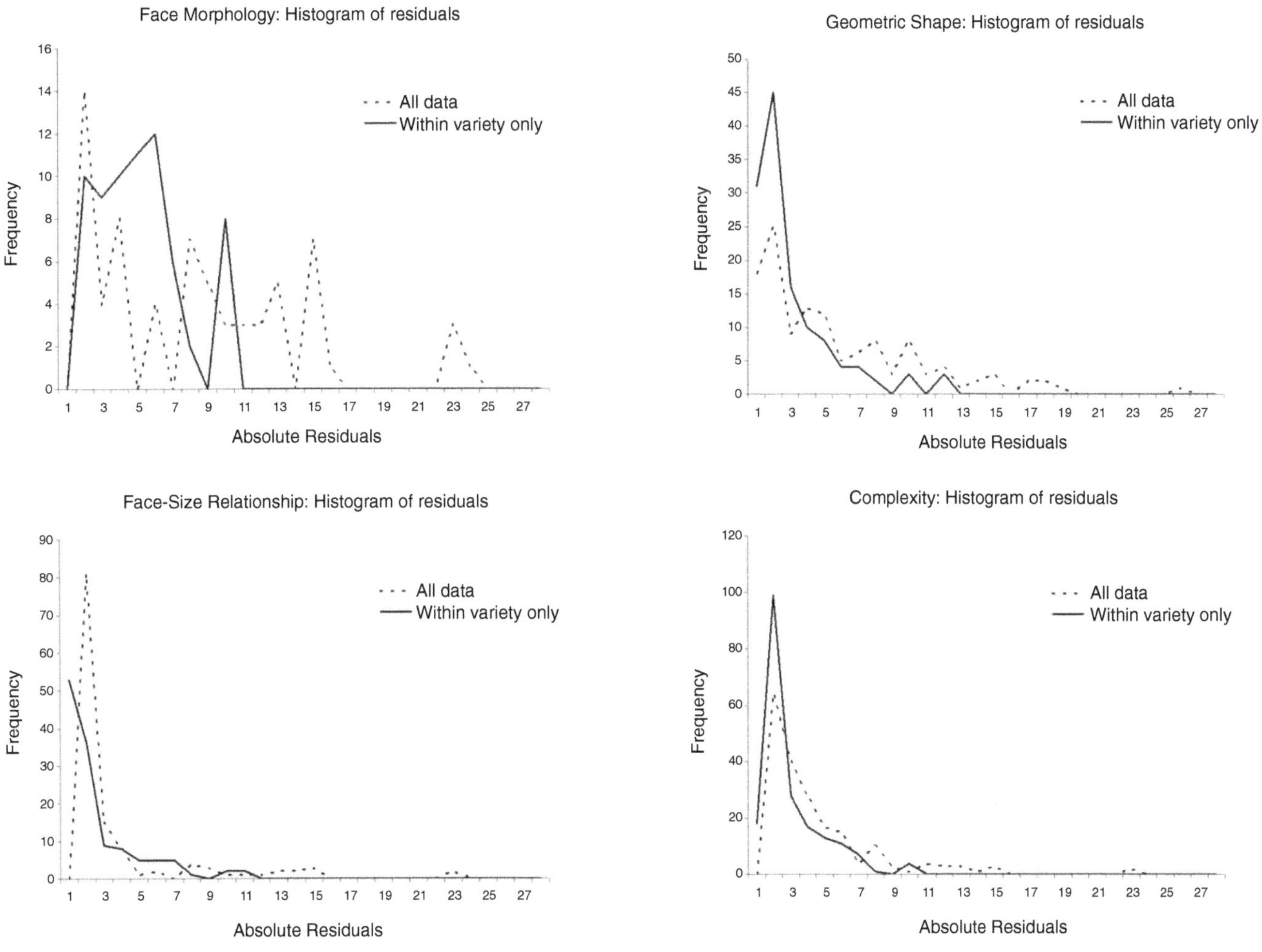

FIGURE 5-1 Comparison of observations within eight Bolivian ethnographically collected varieties of maize versus among varieties. These are histograms of residuals calculated by subtracting the observed value (number of variates per sample of 100 phytoliths) from the average across either one variety or across all varieties.

niques; however, any results must be considered preliminary because at this time the number of cases (maize varieties, $n = 8$, or samples of maize varieties $n = 18$) is far smaller than the number of different variables or unique variates of those variates of those variables, and each of the varieties contain only two or three samples, all of which limits confidence in the analysis. Briefly, the distribution of Pearson's r-values as a scaled indicator of covariance appears bimodal and may be composed of a set of random bivariate factors with a mean of about 0 and a set of negatively correlated factors with a mean of about –0.8. Factor analysis; other forms of covariate analysis; and, in particular, principle component analysis have been tried and have not shown that any one morphological characteristic can be eliminated or combined with another, although individual variates do correlate. (Notably, the pattern of covariance does seem to be related to which varieties are examined, but there is not enough data at this time to explore this further. If this turns out to be a general pattern, there may be different sets of maize varieties that would be best differentiated using different classification or clustering algorithms.)

A possible concern of the total morphological method used by Thompson [4] and Lusteck is that of the many variables that are used, there is some conflation of less-well-behaved and better-behaved variables into the total morphological sample. Although this is contraindicated by the pattern of residuals summarized in Figure 9-1, it is possible that among the parsed-out variables, one or more would do a better job at classifying samples, whereas others would do a poor job. For example, the comparison of residuals shown in Figure 5-1 suggests that among the parsed variables, geometric shape would do the best job at discriminating among varieties.

To explore this, K-means cluster analysis was applied to each set of data that were used to generate Figure 5-1. These are shown in Figure 5-2. Each parsed variable is used to

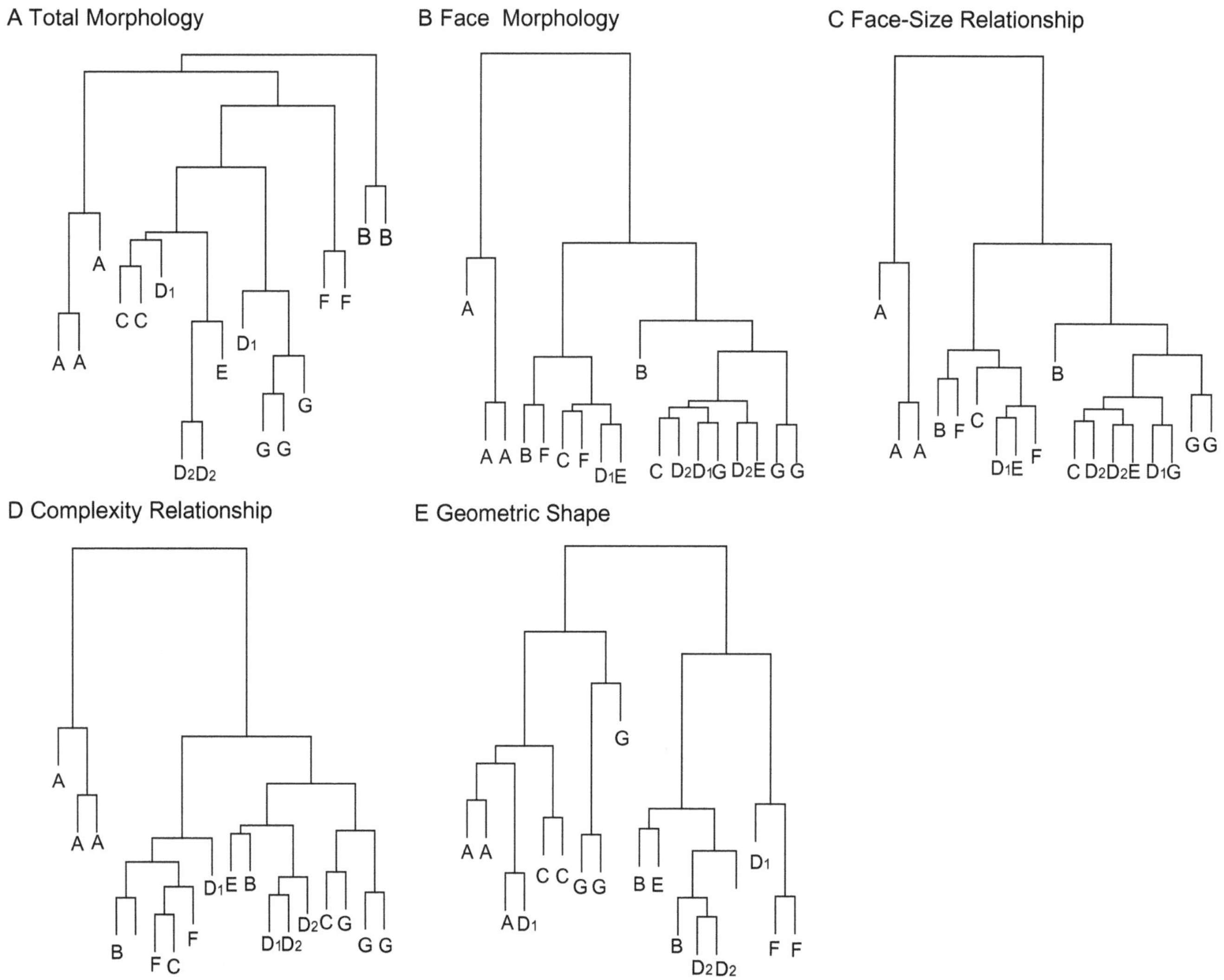

FIGURE 5-2 A series of K-means cluster analysis of the Bolivian ethnographically collected maize samples using the entire morphological pattern.

TABLE 5-1 Residuals and Ranked Differences of Bolivian Maize Varieties (a, b, c, d1, d2, e, f, g)

			Absolute residual (difference from average for each sample)								**Ranked difference from average (1 = most different). First, second, and third ranks are shaded.**							
	Variate	**Average value**	**a**	**b**	**c**	**d1**	**d2**	**e**	**f**	**g**	**a**	**b**	**c**	**d1**	**d2**	**e**	**f**	**g**
Face morphology	a	68	34	2.7	5.7	3.7	4.3	5.7	18	3.4	1	8	3	6	5	3	2	7
	b	66	32	7.3	5.8	2.8	4.7	5.3	16	0.5	1	3	4	7	6	5	2	8
	c	32	35	2.6	5.6	4.6	2.4	6.1	16	2.9	1	7	4	5	8	3	2	6
	d	33	33	6.8	5.3	1.3	3.2	5.3	18	0.7	1	3	4	7	6	4	2	8
Face size relationship	AA	0.3	0.3	0.3	0.3	0.3	0.3	0.3	1.8	0.3	2	2	2	2	2	2	1	2
	AB	65	34	4.9	3.9	3.9	3.1	6.9	17	0.1	1	4	5	5	7	3	2	8
	AD	3.1	0.2	1.6	2.4	0.4	0.6	0.6	3.1	3.2	8	4	3	7	5	5	2	1
	CB	1.6	1.4	2.4	1.9	1.1	1.6	1.6	1.6	0.1	6	1	2	7	3	3	3	8
	CC	0.1	0.6	0.1	0.1	0.1	0.1	0.1	0.1	0.1	1	2	2	2	2	2	2	2
	CD	30	32	4.9	7.4	3.4	4.1	4.4	14	2.2	1	4	3	7	6	5	2	8
	DD	0.1	0.1	0.1	0.1	0.9	0.1	0.1	0.1	0.1	2	2	2	1	2	2	2	2
Complexity	s1	15	3.2	3.2	13	1.2	5.2	7.7	13	5.2	6	6	1	8	4	3	1	4
	s2	12	1.5	8.4	5.6	2.4	4.9	7.4	12	4.1	8	2	4	7	5	3	1	6
	s3	103	17	18	20	2.9	14	14	20	32	5	4	3	8	6	7	2	1
	s4	36	2.4	12	1.5	3.5	17	2	12	36	6	4	8	5	2	7	3	1
	s5	33	18	5.9	0.9	2.6	9.9	3.4	32	3.3	2	4	8	7	3	5	1	6
	s6	0.5	0.1	0.5	0.5	0.5	0.5	0.5	0.5	0.9	8	2	4	2	4	4	4	1
	s7	0	0	0	0	0	0	0	0	0	1	1	1	1	1	1	1	1
Complexity	c1	68	34	2.7	5.7	3.7	4.3	5.7	18	3.3	1	8	3	6	5	3	2	7
	c2	65	33	4.6	4.6	3.6	3.4	6.6	17	0.2	1	4	4	6	7	3	2	8
	c3	1.2	0.5	2.8	1.3	0.7	1.2	1.2	1.2	0.2	7	1	2	6	3	3	3	8
	c4	13	5.3	1.8	5.3	4.2	1.7	4.8	3.8	4.7	1	7	1	5	8	3	6	4
	c5	3.2	3.8	0.8	0.7	2.7	0.8	0.3	2.7	0.2	1	4	6	2	4	7	2	8
	c6	5.8	14	1.3	2.8	1.3	0.8	5.3	2.3	0.5	1	5	3	5	7	2	4	8
	c7	9.2	12	0.2	3.3	4.7	0.8	3.8	7.2	8.2	1	8	6	4	7	5	3	2
	c8	14	1.1	5.4	3.9	4.1	3.1	3.4	4.4	8.8	8	2	5	4	7	6	3	1
	c9	3.2	3.8	0.8	1.2	2.7	0.8	0.8	2.7	0.2	1	5	4	2	5	5	2	8
	c10	7	15	2	3.5	2	2	6.5	3.5	0	1	6	3	5	6	2	3	8

generate a different cluster tree, and these are compared with the cluster tree generated from the total morphological sample. Interestingly, although the total morphological sample does a good job at differentiating among known varieties, none of the individual parsed variables does as well. Certain varieties are classified together and distinct from other varieties more consistently than others, but all of the trees based on parsed variables are erroneous in one way or another.

Breaking down the "total morphological pattern" by the individual components (parsed variables) that are used to generate the morphological codes into constituent parts appears to *lose* critical information. Multivariate analysis should restore this information by allowing through PCA, or other methods, the reformation of the morphological categories; however, attempts to do so have failed to yield satisfactory results so far. There seems to be an enigma in that parsing out the total morphological variables into their known constituent parts, and even recombining them, using standard statistical techniques produces results that are less accurate than leaving the data in their original form.

To explore this more closely, a more detailed, brute force analysis is represented in Table 5-1. For each variate of each variable, the average value for each maize variety is compared with the average for all varieties combined expressed as absolute differences. These values are then ranked with respect to relative magnitude of difference across varieties. This is meant to be a rough but potentially useful indicator of the possible contribution of a particular variable to differentiating among varieties, as well as the degree to which a particular variety may stand separate from others in a cluster analysis. If one variety was the most different from the average (as variety A seems to be in many respects) then one would expect the first rank differences to be clustered there (as they indeed are). However, if one considers all first, second, and third ranked relationships across the entire table, it becomes clear that there are some variables that should differentiate each of the eight maize varieties in the collection.

CONCLUSIONS

1. There is less variability within samples of a given variety of maize than there is among varieties, so the assertion that it is possible to tell varieties apart in both ethnographic and archaeological samples is supported.

2. The topological relationships of a cluster tree of varieties *based on the total morphological data* as generated by Thompson and Lusteck in various studies using any of a number of different clustering techniques, "make sense" in that varieties are grouped on branches of the tree without overlap or ambiguity.

3. Parsing out the elements of the morphological data, such as face shapes, complexity, and so on, shows that some varieties are distinguished by only one or two components of the total morphological data. However, cluster analysis using these parsed-out data produces trees in which the varieties are grouped somewhat like the tree on the basis of the original assemblage data but with frequent "errors" in classification. This suggests that there are as yet undiscovered subtleties in the relationship among variates and variables, which is worthy of further investigation with larger samples.

4. To some degree, variation is concentrated among the samples so that some varieties (especially "A") are "more different" and more easily classified and classified on the basis of different criteria than are other varieties.

References Cited

1. J. Dorweiler, J. Doebley. (1997). Developmental analysis of teosinte glume architecture 1: A key locus in the evolution of maize *(Poaceae)*. *American Journal of Botany, 84,* 1313–1322.
2. V. Jaenicke-Després, E. S. Buckler, B. D. Smith, T. M. Gilbert, A. Cooper, J. Doebley, S. Pääbo. (2003). Early allelic selection in maize as revealed by ancient DNA. *Science, 302,* 1206–1208.
3. J. E. Staller, R. G. Thompson. (2002). A multidisciplinary approach to understanding the initial introduction of maize into coastal Ecuador. *Journal of Archaeological Science, 29,* 33–50.
4. R. G. Thompson. (2006). Tracing the diffusion of maize into South America through phytolith analysis of food residues. In: M. A. Zeder, E. Emshwiller, D. Bradley, B. D. Smith (Eds.), *Documenting domestication: New Genetic and archaeological paradigms*. Berkeley: University of California Press. pp. 82–96.
5. R. G. Thompson, S. Mulholland. (1994). The identification of corn in food residues on utilized ceramics at the Shea Site (32CS101). *Phytolitharian Newsletter, 8,* 7–11.

PART II

Histories of Maize:

Domestication and Production

CHAPTER

6

Early Agriculture in Chihuahua, Mexico

ROBERT J. HARD*, A. C. MACWILLIAMS†, JOHN R. RONEY‡, KAREN R. ADAMS§, WILLIAM L. MERRILL¶

*Department of Anthropology, University of Texas at San Antonio, San Antonio, Texas

†Department of Archaeology, University of Calgary, Calgary, Alberta, Canada

‡Albuquerque Field Office, Bureau of Land Management, Albuquerque, New Mexico

§Crow Canyon Archaeological Center, Cortez, Colorado

¶Department of Anthropology, Smithsonian Institution, Washington, D.C.

Glossary

Altithermal Proposed middle Holocene hot and dry period for the western United States.

Cerros de trincheras sites Hilltop sites with terraces, walls, and other stone features.

Fluvial terraces Landforms resulting from riverine deposition and erosion.

Proto-Uto-Aztecan language Hypothetical language ancestral to modern Uto-Aztecan languages.

Tarahumara Indigenous people of southern Chihuahua, Mexico.

Northwestern Mexico is critical for understanding the spread and adoption of farming as it lies between Mesoamerica, where maize farming began, and the American Southwest where maize had a tremendous impact on local societies. Recent research in the state of Chihuahua reveals new information regarding the ecological and cultural contexts of the spread and adoption of maize farming. Substantial levels of farming were underway in northwestern Chihuahua by ca. 1300 BC, and maize was present by 200 BC in the Sierra Madre Occidental of southwestern Chihuahua, and tentatively by 1700 BC in south-central Chihuahua. The pace at which maize was integrated into local economies across the American Southwest and northwestern Mexico in the millennia following its initial introduction around 2000 BC is generally gradual but was more rapid in some locales including northwestern Chihuahua and southern Arizona.

INTRODUCTION

Understanding the spread of maize agriculture from Mesoamerica to northern Mexico and the American Southwest is fundamental to addressing a series of theoretical issues related to the impetus and consequences of the adoption of farming in this region and around the world. Here, we review recent research on the earliest maize in the northern Mexican state of Chihuahua and assess its relevance to current issues including the timing, spread, adoption, and consequences of agriculture.

EARLY AGRICULTURE

Chihuahua sits between Mesoamerica where maize agriculture originated and the American Southwest where agriculture had a profound impact on most local societies. Compared with Chihuahua, both of these areas are well known archaeologically and provide frameworks for considering the processes involved in the arrival and adoption of maize agriculture in northern Mexico.

Despite the lively debate about the details of the timing, geography, and biology of early agriculture in Mesoamerica, researchers agree that the major cultigens—maize, beans, and squash—originated in tropical Mesoamerica and then spread to other parts of the New World [10, 33, 68, 69, 70, 90, 96]. The Guilá Naquitz site in Oaxaca has yielded maize dating to 4300 CAL BC, the earliest directly dated maize encountered thus far, but molecular evidence suggests that maize was first domesticated somewhat earlier along the Río Balsas of Michoacán and Guerrero [112, p. 2102]. An earlier initial domestication of maize in Mesoamerica also is suggested by rapidly accumulating, yet still controversial, pollen and phytolith evidence, which raises the possibility that maize arrived in Central and South America between 5500 and 3000 BC [111, 116, p. 1371; 134]. By about 3500 CAL BC, maize had spread northeastward to Puebla's Tehuacán Valley, and it was present in the northern Mexican state of Tamaulipas by 2400 CAL BC [80, p. 1039; 84, 130, 131, p. 1326]. Between ca. 1500 and 1000 BC maize farming populations occupied permanent villages across much of southern tropical Mexico and temperate western Mexico and complex societies were present in some locales by ca. 1000 BC [16, 17, 37, 42, 55, 91, 104, 113, 114, 132].

In the American Southwest, the Early Agricultural Period is defined as the interval before the widespread use of ceramic technology during which domesticated plants were introduced and became widely accepted, although its span flexes as new data accumulate (Table 6-1) [64, 65]. The Early Agricultural Period cross-cuts the traditional Middle and Late Archaic Periods, which are used in this chapter as purely temporal units without regard to the presence or absence of agriculture. The date for the initial arrival of maize on the Colorado Plateau is now firmly placed at about 2100 CAL BC with a cluster 10 accelerated mass spectrometry (AMS) maize dates [62]. These results confirm previous reports that maize is in multiple locations by 2000 BC [43, p. 118; 44, pp. 253; 66, 147]. By 1600 CAL BC, maize is widespread in the region, and by AD 100 it is nearly ubiquitous [49, pp. 1661; 83, 147]. The ending date of the Early Agricultural Period should perhaps be when ceramic vessel technology is well integrated into the economy rather than simply "pre-ceramic" as some ceramic objects are now known to be present during the Early Agricultural Period [81, p. 786]. A number of ceramic objects including figurines, beads, and sherds of small fired ceramic vessels have been found in a several Tucson Basin sites dating as early as 2100 BC.

Two fundamental processes characterize the Early Agricultural Period: 1) the introduction of agriculture and 2) the formation of an agricultural economy. The introduction of agriculture by any means defines the time when people first have access to domesticated grains, such as maize, and some information regarding how to produce a harvest. In an extensive review of published schemes for partitioning these processes, Bruce D. Smith [132, Figure 7] delineates three adaptive patterns: 1) low level food production without domesticates, 2) low level food production with domesticates, and 3) agriculture. The agriculture pattern includes both primary dependence on domesticates as well as subsistence strategies where domesticated plants comprise 30–50% of the diet.

Here we employ Smith's [132] scheme, using the term "agricultural economy" as a synonym for his term "agriculture" to designate the state in which hunter–gatherer adaptations have shifted in settlement, seasonal movement, social organization, and material culture to accommodate farming. Over much, but not all, of the American Southwest, the Early Agricultural Period appears to have involved low level food production with domesticates, as there is little evidence of major, agriculture-driven changes in settlement or population. Farming seems to have contributed a relatively minor portion of the diet, but may have been important in reducing risks or enhancing access to nondomesticated resources [31, 56, p. 118; 102, 148, 150]. The availability of maize did not necessarily mean, however, that all groups accepted it at the same pace or to the same extent [36, 47, 101]. For decades the standard assumption for the American Southwest was that low level food production with domesticates continued for several thousand years before the formation of agricultural settlements, and furthermore that such settlements did not appear until the middle of the first millennium AD [30, 56, 129]. A series of remarkable discoveries in southern Arizona, northwestern Chihuahua, and northwestern Sonora now challenge this traditional model.

TABLE 6-1 Time Periods

Temporal unit	Age	Notes	Refs
Early Agricultural Period	2100 BC–AD 200	Earliest evidence of maize to widespread use of ceramics; cross-cuts Middle and Late Archaic Periods	62, 64, p. 16
Middle Archaic Period	3500 BC–1500 BC	1200 or 1500 BC is frequently used	65
Late Archaic Period	1500 BC–AD 200		64, 65
San Pedro Phase	1500 BC–800 BC	1200 or 1500 BC	64

Doubts about this model were first raised by Bruce B. Huckell's [63, 65] work in southeastern Arizona, and then by other excavations of large, Late Archaic Period (Table 33-1) sites the Tucson Basin consisting of numerous pit-houses and storage features that were associated with the emergence of elaborate material culture, irrigation, and limited use of ceramics [43, 45, 81–83]. These settlements are thought to be reoccupied settlements with occupations consisting of no more than perhaps about 30 persons with increasing sedentism later in time, although debate around these issues is ongoing [45, 81, 127]. Irrigation agriculture may have also have been underway by 1000 BC in the Zuni area of western New Mexico, although the cultural context of this adaptation is unknown [28].

At the multicomponent site of La Playa in northwestern Sonora maize is common along with wild plants in Early Agricultural Period roasting pits and 160 excavated human remains are associated with that period, representing 83% of the total excavated burial population [24, pp. 15–16, 103]. Preliminary analyses indicate that dental caries occurred at rate similar to that found among the Early Agricultural Period Tucson populations. La Playa occupants were also involved in a significant levels of farming and as well as foraging economy similar to those in southern Arizona [24, pp. 14–17, 103]. Some Basketmaker II occupations on the Colorado Plateau were substantially dependent on maize by ca. 100 BC [47, 93, 95]. In northwestern Chihuahua we have investigated Cerro Juanaqueña and a series of related **cerros de trincheras sites** (hills with constructed terraces and walls) whose occupants relied on farming for a significant part of their economy from 1350–1300 CAL BC until about 1100 CAL BC [1, 50, 51, 120]. We will discuss our findings in more detail in this chapter.

In sum, it now appears that in various locales approximately 1000 years of low level food production with domesticates, from ca. 2000 BC to ca. 1300–1000 BC, precedes a level of farming that can be considered an agricultural economy in a few locations. However, over most of the American Southwest, low level food production with domesticates persisted for 2000–2500 years. In these areas the standard model still appears to be applicable with settled, farming communities arising only after about AD 100–500 [47, 126, 149]. In some regions, such as the southern Jornada Mogollon near El Paso, Texas, this transition does not occur until after AD 900, reflecting more than 3000 years of low level food production with maize present in that region beginning in the Late Archaic Period [47, 145].

THE INTRODUCTION OF MAIZE

Since the 1940s, most scholars have believed that it took a period of several thousand years for successive groups of hunter–gatherers to spread maize northward from central Mexico to the American Southwest. Several routes of diffusion have been proposed: the uplands of the Sierra Madre Occidental, the north–south trending river valleys of Chihuahua, Arizona, and New Mexico, and the Sonoran coastal plain [6, p. 249, 56, p. 113, 65]. However, the early and widespread occurrence of maize in the American Southwest, northwestern Mexico, and northeastern Mexico suggests that maize may have diffused along one or more broad fronts [9].

Migrations northward by groups of farmers is usually cast as an alternative explanation to group-to-group diffusion of maize. Most migration theorists suggest that local people quickly adopted the newcomers' maize. Claudia F. Berry and Michael S. Berry [11] initiated recent debates by proposing that the San Pedro Phase in the southern Southwest began around 1200 BC with the arrival of farmers from Mexico [64]. This phase spans about 1200–800 BC and is recognized on both sides of the international border [135]. Matson [93, 94] also suggests the San Pedro Phase and other, related cultures represent migrants from Mexico, whose descendants included Western Basketmaker II occupants on the Colorado Plateau.

Jane H. Hill [58–61] attributes both the original domestication of maize and its spread from central Mexico into the northern American Southwest to growing populations of **Proto-Uto-Aztecan** farmers who expanded northward, out competing local hunters and gatherers. In large part she bases her view on reconstructed agricultural terms that she proposes formed part of Proto-Uto-Aztecan and similar language–agricultural dispersal models from Europe and other parts of the world [7, 119]. Basketry, projectile point, and architectural styles, as well as a dental study have all been marshaled to support a northward migration of farmers [94]. Mitochondrial DNA and blood serum evidence do not readily support Hill's original arguments, but a scenario in which only Proto-Uto-Aztecan males migrated northward, bringing language and maize, and intermarrying with local females has been proposed as a way of accommodating this evidence within a migrationist framework [89, 133]. Nonetheless, Hill's reconstructions of Proto-Uto-Aztecan agricultural terms and migrations have been questioned [22], and alternative models that argue that Uto-Aztecan populations spread in the opposite direction, from northern California [105] or the American Southwest and northern Mexico into central Mexico [39, 100] remain viable.

Regardless if maize agriculture was introduced to northern Mexico and the American Southwest by diffusion or migration, ecological factors must be taken into consideration if the spread and role of farming among societies in this region is to be understood. Several relevant ecological models have been proposed. Resource imbalance models suggest that local hunters and gatherers are "pushed" to adopt agriculture when environmental changes or increasing populations create resource shortages [12, 14, 46, 52, 67, 71,

72, 102]. In contrast, "pull" models propose that floodplains and other settings provide attractive niches for farming for both migrant farmers and local hunter–gatherers [58, 65, 93, 102].

Combinations of factors have also been proposed to explain the northward diffusion of maize from Mesoamerica, and more recently John P. Carpenter and colleagues [23] suggest that a reduction in wild resources brought on by **Altithermal** aridity (ca. 5500–2500 BC) pressured Proto-Uto-Aztecan populations to split into northern and southern branches and to abandon the lower elevations in the southern American Southwest and northwestern Mexico [73]. As climatic conditions ameliorated these populations reoccupied the desert regions of the American Southwest, Chihuahua, and Sonora, and expanded northward into areas attractive for agriculture. Other models based on optimal foraging theory are neither push nor pull but emphasize the trade-offs between domesticated and nondomesticated resources [5, 31, 52]. Of course, as Matson [93] notes, the pace for the northward expansion of maize agriculture was at least partially dependent on genetic adaptability and changes within the plant itself.

EARLY AGRICULTURE IN CHIHUAHUA

The northern Mexican state of Chihuahua includes segments of two of the routes that have been proposed for the transmission for maize agriculture between Mesoamerica and the American Southwest: the uplands of the Sierra Madre Occidental and the north–south trending river valleys east of this mountain range. Both areas would also likely have been involved if maize farming spread northward along broad fronts rather than more delimited routes and thus appear crucial for understanding early maize agriculture. Nonetheless, little is known about the chronology and processes associated with the adoption and expansion of farming in these and surrounding areas within Chihuahua because only limited research has been completed there.

To begin eliminating this gap in our knowledge, we have been investigating early agriculture in three areas of Chihuahua that include two distinct ecological zones. The zones are the Basin and Range province and, to the west, the Sierra Madre Occidental (Figure 6-1 and Table 6-2). Basin elevations increase toward the west, and mean annual precipitation increases moving south and west in Chihuahua. The first investigated area is in the high basin and range country of northwestern Chihuahua where our work focused on the Río Casas Grandes Valley with its wide floodplains and excellent arable land. Here rain-fed farming (more commonly known as dry farming) is marginal and modern irrigation is widespread (see Figure 6-1). The second area is in south-central Chihuahua, also in the high basin and range country where we have worked primarily along the perennial Río San Pedro and Río Satevo, both tributaries of the Río Conchos, in the vicinity of the town of Santa María de Cuevas. Also in south-central Chihuahua, but about 100 kilometers further south, we have examined sites in the vicinity of Parral where the perennial drainages are the Río Primero, Río Santa Bárbara, and Río Sapién, all tributaries of the Río Florida. Most of these midsize drainages and their tributaries contain wide floodplains and well-defined **fluvial terraces.** Rain-fed farming is more widespread in alluvial soils, and modern irrigation farming is common. The third area is the rugged Sierra Tarahumara of southwestern Chihuahua within the Sierra Madre Occidental ecological zone (see Figure 6-1 and Table 6-2). The Sierra Tarahumara is now inhabited by Mestizo and **Tarahumara** (Rarámuri), both of whom routinely use dry farming [97].

PALEOENVIRONMENT

Research addressing paleoenvironment in Chihuahua is relatively recent, but in broad terms the results of this research correspond to findings from southern Arizona and New Mexico [51, 52]. Warmer temperatures in the early Holocene (9000–7000 BC) brought about major environmental changes. By the middle Holocene (7000–2000 BC), vegetation shifts occurred in response to combinations of increased summer temperatures and increased aridity. The late Holocene (2000 BC–present) brings the establishment of modern vegetation but with fluctuating moisture levels.

General warm dry middle Holocene conditions increases in moisture are reflected against uncorrelated episodes of lake filling and marsh development in Chihuahua and surrounding areas [25, 38, 99, 108, 141, 142]. By 3000–1600 BC

TABLE 6-2 Ecological Zones

Zone	Elevation (m)	Precipitation (mm/yr)	Biotic community	Vegetation	References
High basin and range	150–1800	325–450	Plains and semidesert grassland	Shrublands, grasslands, and woodlands	21, 122, 144
Sierra Madre Occidental	1800–2800	500–850	Madrean evergreen woodland and Petran Montane conifer forest	Conifer and oak forests	21, 32

a regime was underway consisting of high magnitude floods, moist conditions, dramatic lake high-stands, and cold lake water temperatures [34, 57, 99, p. 169; 110, p. 202; 140, 143, 146]. Southwestern United States high magnitude flooding decreased markedly between 1600–200 BC, and overall moisture regimes remained high at ca. 1000 BC [106, 109, p. 1177].

The following millennium witnessed a slow drying trend. In the Babícora Basin of central Chihuahua, modern, semi-arid conditions were established [99, 108]. These drying conditions were not necessarily widely synchronous, as moist conditions continued to ca. AD 1 in some locations [115, 140, 141]. The absence of high magnitude floods facilitated floodplain building along the Río Casas Grandes until ca. 200 BC, but they resumed after AD 1 [34, 106]. However, mesic (damp) conditions did not commence again until ca. AD 500 [34, 106, 139, Figure 7.9].

The general picture is one of a warm and dry middle Holocene followed by widespread moisture increases late in the middle Holocene to the beginning of the late Holocene, a return to somewhat arid conditions following 1000 BC, then mesic conditions returning at ca. AD 500. These overall trends, however, are sometimes contradicted by studies of specific areas within the region. Further paleoenvironmental work in Chihuahua is required if the complex relationships between climatic shifts and the spread of farming are to be understood.

PREVIOUS RESEARCH IN CHIHUAHUA

In the past, research on early agriculture in Chihuahua has been directed in large part toward enhancing an understanding of the spread of maize agriculture to the American Southwest. Working in Cave A near Norogachi, located in the eastern Sierra Tarahumara (Figure 6-1).), Robert M. Zingg [151, p. 212; 152, pp. 1–43] found lithics and preserved organic material, including corn, but no ceramics. He concluded that the local sequence began with the Río Fuerte Basketmakers who farmed maize but lacked beans and pottery. Significantly, in the context of current migration debates, Zingg [152, pp. 18–31] shows a number of similarities between the Rio Fuerte Basketmaker material and that of the Colorado Plateau Western Basketmaker, particularly in weaving technology [cf. 94].

Two decades later, Robert H. Lister [79] found a few lithics and three maize cobs in the lower strata of Swallow Cave in northwestern Chihuahua, below the ceramic levels. Although lacking temporal controls, Zingg's [152] and Lister's projects provided qualified evidence that maize agriculture extended into the Sierra Madre Occidental quite early. They also fostered the idea that the Sierra Madre Occidental was a corridor by which agriculture spread north, a conclusion reinforced by the discovery of early agricultural sites in the highlands of New Mexico, such as Bat Cave and Tularosa Cave [31, 56].

Sporadic, largely unpublished, reconnaissance surveys beginning in the 1930s gradually demonstrated that there were scattered groups of hunter–gatherers and small-scale farmers in the basin and range country of southern and central Chihuahua [20, 73, 92, 107, 123]. With a few exceptions [73, 74, 85], however, work in southern Chihuahua has focused on the Sierra Tarahumara [4, 27, 28, 78, 138, 151, 152]. Collectively this work indicates that the Sierra Tarahumara has been occupied at least since the Middle to Late Archaic Period (see Table 33-1) but does not provide additional insights into early agriculture. In fact, little work regarding early agriculture was conducted in Chihuahua until we began a 4-year excavation program (1997–2001) at several sites in northwestern Chihuahua [49, 51, 120, 121] followed by two seasons of reconnaissance and testing (2002–2003) in south-central and southwestern Chihuahua [86–88].

NORTHWESTERN CHIHUAHUA

Our excavations focused on Cerro Juanaqueña, which lies at the confluence of the Río Casas Grandes and Río San Pedro about 50 kilometers east of the Sierra Madre Occidental. We also tested three other intermediate-sized cerros de trincheras sites—Cerro Torres, Cerro Vidal, and Cerro el Canelo—and conducted limited surface investigations at another nine small hilltop sites in the area [121]. All 13 of the sites are located on isolated volcanic hills distributed among a series of wide flood basins aligned along a 70-kilometer stretch of the Río Casas Grandes (see Figure 33-1). Two of the small sites are found along the Río Santa Maria, the next river valley to the east.

Cerro Juanaqueña is by far the largest of the 13 sites as it contains 100 rock rings and 50 terraces that include 8 kilometers of constructed walls. (Rock rings are alignments of rocks in circular to oval patterns.) The terraces are distributed in two groups on the 140 meters high basalt hill. More than 300 of the terraces are within a 6 hectare area on the top and upper slopes of the hill. Almost every bit of this area has been modified through clearing stones and constructing terraces, and artifact density is typically high. Most of the remaining terraces occur lower down on the lower west flank of the hill about 20 to 40 meters above the floodplain.

Dating of the four excavated sites is based on a series of 22 AMS dates on maize cupules, kernels, and cobs plus another five dates on short-lived species including three on ocotillo *(Fouquieria)* stems, one on a wild gourd *(Cucurbita digitata)* seed and one on a saltbush (*Atriplex* sp.) stem [49–51, 120]. (AMS radiocarbon dates use the accelerator mass spectrometry method that allows small samples to be

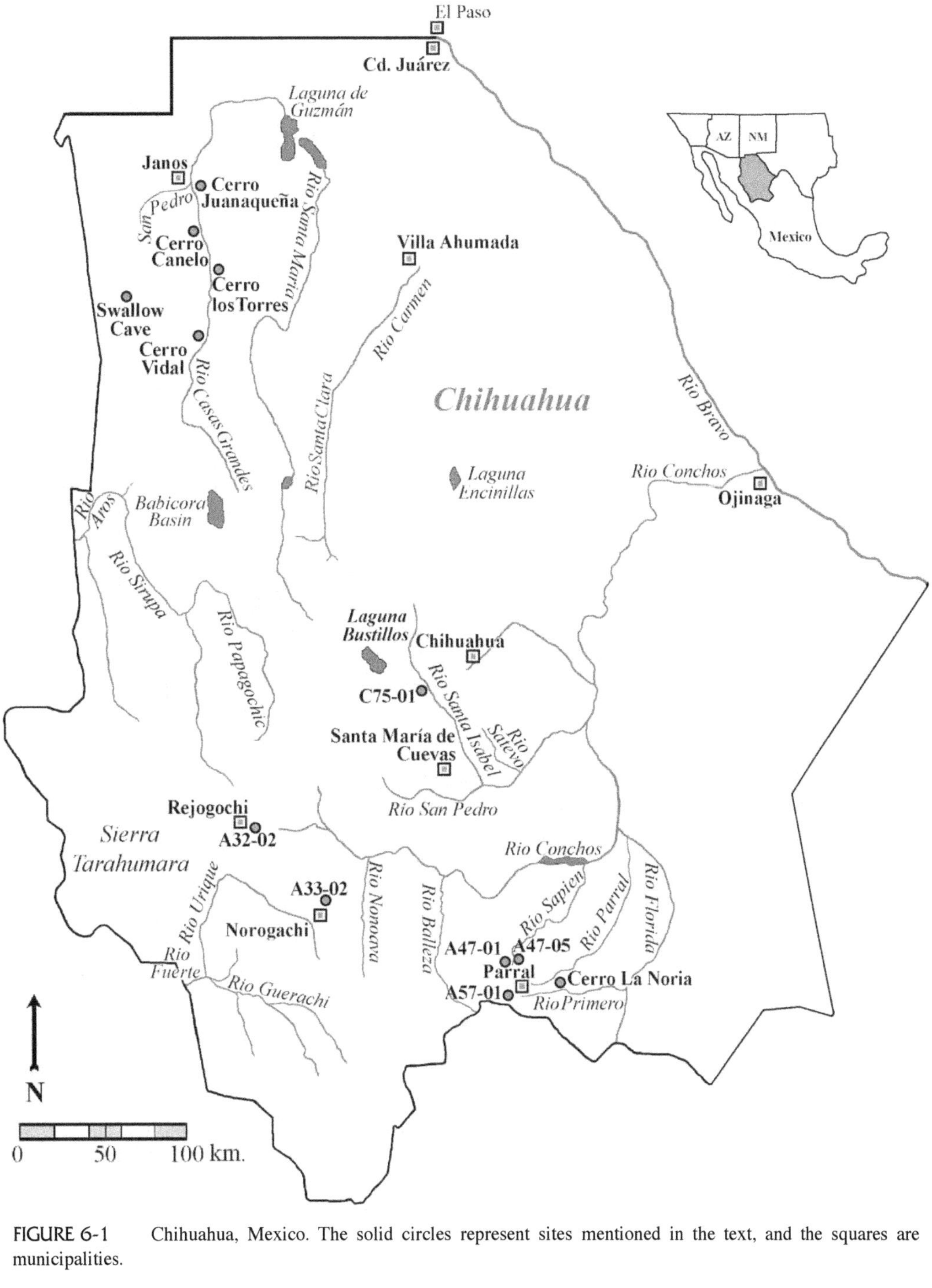

FIGURE 6-1 Chihuahua, Mexico. The solid circles represent sites mentioned in the text, and the squares are municipalities.

analyzed.) These dates suggest contemporaneous occupations at around 1300 CAL BC for Cerro Juanaqueña, Cerro los Torres, and Cerro el Canelo. Based on OxCal probability calculations, Cerro Juanaqueña itself was only occupied for about 200 years, and occupation at the other two sites may have been equally brief [18, 50, 53, 120]. None of the upper terraces at Cerro Juanaqueña show evidence of residential use following this occupation, but ^{14}C dates from two of the lower terraces at Cerro Juanaqueña do document a limited, later reoccupation between 400 and 100 CAL BC. Cerro Vidal also revealed an occupation contemporaneous with that of the second occupation of some of the Cerro Juanaqueña lower terraces.

Excavations demonstrate that the terraces were constructed by first piling the larger basalt cobbles to form an arc-shaped berm. The term "berm" indicates the terraces were constructed by loosely mounding rocks rather than stacking them as vertical walls were generally not constructed. The pocket between the berm and the hill slope was then filled with smaller cobbles, and fine sediment was used to cover the surface. The resulting terraces at Cerro Juanaqueña average about 17 meters in length and have an average surface area of 51.5 square meters. Altogether, a calculated 31,000 cubic meters of stone and sediment were incorporated into these constructions at Cerro Juanaqueña alone, and we estimate that they represent about 30 person-years of labor [54].

The terraces appear not to have been built for agricultural purposes. If the 3.6 hectares of total surface area behind all 550 terraces were planted in maize, the harvest would feed only six people for a year, assuming rainfall typical of higher elevations in the Sierra Madre [54]. In terms of labor cost per hectare of agricultural land, they would have been as costly to construct as Maya raised fields [54]. Further, they do not contain attributes typical of agricultural terraces, such as water control devices or a forward slope.

Our excavations revealed that some terraces contained midden deposits with ample burned and unburned animal bone, charcoal, and lithic debris [49–51, 120]. The 689 basin metates and 666 manos, most of which are on or near terrace surfaces, indicate that substantial domestic effort was expended to process seeds and grains. The basin metates particularly exhibited heavy wear with many being exhausted. In addition, 99 circular rock rings or alignments, with a mean diameter of 2.5 meters, were constructed on or near the terrace surfaces, some of which contained compacted living surfaces. Metates were encountered on the floors of two of these circular features, one of which also contained an informal hearth. This evidence leads us to conclude that most of the terraces were constructed as house and activity area platforms and that the rock rings often served as houses [51, 120].

At Cerro Juanaqueña and many other northwestern Chihuahua cerro de trincheras settlements, perimeter berms encircle large portions of the sites [54, 120]. The berm at Cerro Juanaqueña lies along its northern, eastern, and southern perimeters and consists of 22 linked, individual terraces forming a 400 meters long alignment. In contrast to typical platform terraces that run perpendicular to the slope, the perimeter berm on the northern and southern sides of the site runs down or parallel to the downward slope. Cerro los Torres is surrounded by an impressive perimeter berm; a similar feature defines much of the exterior boundary of Cerro el Canelo, and Cerro Vidal is enclosed by a perimeter berm on three sides while the fourth side consists of a near vertical slope. At least five of the other, smaller cerros de trincheras sites also have perimeter berms.

These perimeter berms appear to have been planned to accommodate definite spatial needs, with each constructed in a single or a few coordinated episodes. As such they suggest a level of population aggregation, organized decision making, and possibly leadership that has not otherwise been noted for this period [54, 120]. We estimate that the average population at Cerro Juanaqueña ranged between 100 and 300 people, over a span of about two centuries. We base this estimate on a consideration of several different lines of evidence, including the extent of occupied areas at specific points in time and levels of food processing as revealed through the wear documented on the ground stone basin metates [51, 120].

The archaeobotanical material recovered from these trincheras sites indicates that maize agriculture was an important economic pursuit. Fifty-one percent of 157 flotation samples were examined, and 15 of 17 terraces at Cerro Juanaqueña contained fairly well-preserved charred maize [1]. Sixteen round cobs had between 8 and 12 rows of kernels, with a mean of 10.6 rows. Kernels were of flint and flint–flour types. These traits of the Cerro Juanaqueña maize are generally consistent with a landrace of popcorn, having small, rounded, kernels, and flinty endosperm, with some lesser portion of floury endosperm. They do not overlap with those of Chapalote, an early and long persistent maize in the American Southwest, which is described as a 12–14 rowed flint or popcorn maize with isodiametric (globular) kernels marked with husk striations [2].

Amaranth (*Amaranthus* spp.) seeds were also present, and based on the presence of thin seed coats, may have been domesticated [40, 41]. In addition, a total of 16 wild plants apparently figured in the subsistence economy. These included plants of disturbed habitats, such as goosefoot (*Chenopodium* spp.), horse purslane *(Trianthema cf. portulacastrum),* chia (*Salvia* spp.), globemallow (*Sphaeralcea* spp.), lovegrass (*Eragrostis intermedia),* and some unidentified grasses (*Gramineae* spp.). Other plants used in the past include those that prefer mesic locations, such as bulrush (*Scirpus* spp.), mesquite (*Prosopis* spp.), and walnut (*Juglans* spp.), and those that can be found on drier landscapes, such as barrel cactus *(Ferocactus),* lemonade berry *(Rhus aromatica),* a wild curcurbit *(Cucurbita digitata),* and others [1]. Amaranth and wild *Chenopodium* seeds were the most commonly recovered resources after maize, and the other wild plants preserved at lower ubiquity levels, together revealing that both domesticated and wild plant foods were important in subsistence.

The most likely location for successful farming at Cerro Juanaqueña would have been the floodplain at the base of the site. A downstream channel constriction creates a flood basin forming a large expanse of alluvial sediment that is periodically susceptible to summer monsoonal flooding. This basin would have provided fairly good farm land, although drought and declining nitrogen levels may have been obstacles to farming [1]. Despite extensive trenching in the Río Casas Grandes floodplain we did not identify any evidence of irrigation canals. Of course, floodplains sometimes receive more water than plants need or can sustain, but the risk of occasional crop loss could have been offset by frequent successes.

The faunal assemblage consisted of 2550 individual specimens (NISP), and 75% of these were rabbits and hares, with jackrabbits being by far the most numerous. About 5% were pronghorn and deer, with the balance including rodents and reptiles, many of which were probably intrusive. Birds and fish also were present, and small fish appear to have been an important resource [124].

The available evidence strongly suggests that agriculture, rather than low level food production with domesticates, was practiced in northwestern Chihuahua by 1300 CAL BC. We also have concluded that Cerro Juanaqueña and the other Late Archaic Period cerros de trincheras sites probably were constructed for defensive purposes [51, 53, 76]. The isolated, steep, high hilltop locations are easily defended, the sites are compact, and they include possibly defensive perimeter berms. Moreover, terrace distributions suggest that all directions could be monitored from them, and many of the sites are visible from one to another, a characteristic of defensive locations [76]. Through a cross-cultural examination of 40 ethnographic groups that live on elevated landforms, we have shown that it is rare for nonindustrial populations to live on hilltops without compelling reasons to do so [53]. Sixty-five percent (26) of cases include defense as an explanation for hilltop living and another 20% of cases (8) can be accounted for by an absence of alternative, more convenient residential areas. The remaining cases can be attributed to environmental or farming conditions related to the avoidance of flooding or enhanced access to food resources and fields.

SOUTH-CENTRAL CHIHUAHUA

We have also conducted reconnaissance and tested sites in the basin and range province of south-central Chihuahua for the purpose of identifying early agricultural sites. It is apparent that the entire area was inhabited during the mid-Holocene interval when maize agriculture spread north. Middle Archaic and Late Archaic Period sites are relatively common, with evidence of occupation found on fluvial terraces, alluvial fans, cerros de trincheras, and in rockshelters. To date our work in this area has focused on two distinct types of sites: rockshelters with "D"-shaped terraces built in front of them and cerros de trincheras.

D-Shaped Terrace Sites

Rockshelters in the basin and range country with built-up terraces behind D-shaped walls are a phenomenon first noted by Richard H. Brooks [19]. The four that we have recorded are habitation sites are within 1.5 kilometers of fluvial terraces suitable for agriculture and have substantial assemblages suggestive of heavy use, evidenced by cobble manos, ample flaked stone debris, and faunal material. The constructed terraces follow cliff bases for 20 to 25 meters, defining the straight side of an uppercase D, and arc out about 10 to 15 meters from cliffs, suggesting a notable labor investment. We tested three of these rockshelters during 2003 [87, 88]. Work on the first was quickly abandoned due to pervasive rodent disturbance, but the other two proved more informative.

At Cueva de los Indios (C75-01), located 40 kilometers west of Ciudad Chihuahua, terrace deposits extended to 2.3 meters below surface (see Figure 33-1). The chipped stone assemblage includes at least 25 late Middle Archaic and Late Archaic Period projectiles points and abundant bifacial reduction flakes. The faunal assemblage is remarkable both for its quantity of more than 300 NISP and its composition: 80% of the identified elements are tortoise and turtle [125]. Ceramic and Recent Period components primarily rest above the Archaic materials, and a Plainview point base reveals that some Paleoindian materials are deep in the terrace fill.

Sitio Pienso (A47-05) is another D-shaped terrace rockshelter site located northwest of Parral (see Figure 33-1). More than one dozen projectile points from this site derive from the late Middle Archaic to Late Archaic Periods. A 1 by 2 meter excavation produced more than 4000 pieces of flaked stone, 25% of it from biface reduction, and not a single ceramic sherd. A diverse faunal assemblage includes turtle, tortoise, jackrabbit, cottontail, and mule deer [125]. Cobble manos, a small grinding slab, charcoal, and fire-cracked rock were also present. A recent radiocarbon date of 170 ± 40 BP (Beta 185632) on charred tree bark of an unknown species seems obviously intrusive, while an older date of 2770 ± 40 BP (Beta 182382) on a charred walnut shell is in line with the material culture assemblage. Maize was not recovered in our limited excavations or in flotation samples. (Here, we report all radiocarbon dates in uncalibrated years before present, with "present" being 1950 by international convention.)

At this point, it is impossible to determine if initial occupation of these sites predates the arrival of agriculture in the area, but if so they connote predisposition to accepting agriculture. Wirt H. Wills [147, p. 41] suggests that predisposition involves repetitive seasonal movement, limited geographical range, and resource storage capacity. Similarly, Fish and Fish [36, p. 86] suggest that experience manipulating wild plants may be an important preagricultural adaptation. The investment in terrace construction and dense assemblages at these sites are congruent with logistically organized settlement, as the rockshelters are far from being typical forager camps [13]. Such a strategy may imply wild plant intensification and decreased residential mobility [15]. Locations close to fluvial terraces may be related to the use of wild and domestic resources or both. The generalization that hunter–gatherers will move to the least portable resource and reach the others with task groups [13, 147, p. 43] may be quite applicable in this instance.

The projectile point assemblages from these sites are based largely on Middle and Late Archaic Texas types found in the Lower Pecos and Río Grande areas [136, 137]. They suggest that the principal initial Archaic Period occupations at these Chihuahua sites were no earlier than 2500–2000 BC, with continued use through the Late Archaic Period, with minimal evidence of Late Paleoindian presence.

Cerros de Trincheras

A total of nine cerros de trincheras sites are now known from the central basin and range province in south-central Chihuahua, and some of these hills have evidence of Early Agricultural Period occupation. These cerros de trincheras are consistently between 50 and 70 meters high with good views of valleys and are close to arable floodplains. They have from 5 to 40 terrace walls built with stacked rocks and circular rock outlines of structures. Nusbaum [107] first recorded several of these sites and Gerry R. Raymond [117, 118] tested three of them in 2000. One of Raymond's samples from Cerro Prieto de Santa Bárbara (site A57-01) produced a single maize radiocarbon date of about 1700 CAL BC (3410 ± 40 BP, Beta 174996) and another one from Cerro La Noria produced a maize date of about AD 100 CAL (1950 ± 50, Beta 174994) [118].

We have since re-examined one of these sites, Cerro Prieto de Santa Bárbara, mapped another site, and both mapped and tested three others. On four of these sites, we found at least one dart point that is indistinguishable from well-defined Archaic dart points in Arizona, New Mexico, and Texas, along with more recent arrow points and ceramics. Points from Cerro Prieto include two late Middle Archaic points, one is a Langtry or a variant, and the other is similar to Almagre or Augustin forms. Late Archaic Period points from this site include an untyped one and a Shumla point. Extensive bifacial reduction is evident on those hills with sizable samples of lithics, a pattern consistent with Archaic rather than Ceramic Period assemblages. This is particularly true of Cerro las Flojeras (A47-01) located within direct sight of three of the D-shaped terrace sites. These observations provide limited corroboration for the 1700 BC radiocarbon date at Cerro Prieto, although we have yet to obtain a second date in this early time range.

Cerros de trincheras of southern Chihuahua are important for several reasons. As discussed previously, in northwestern Chihuahua Cerro Juanaqueña and a series of these sites date to the Late Archaic Period and have abundant maize. The south-central Chihuahua cerros de trincheras are not as extensively developed as Cerro Juanaqueña, but they are similar to some of the other smaller, Late Archaic Period hilltop sites along the Río Casas Grandes [121]. This pattern of Early Agricultural Period cerros de trincheras extends as far north as the multicomponent Tumamoc Hill site in the Tucson Basin, and continues south into Mexico for an unknown distance [35]. Unlike cerros de trincheras in northwestern Chihuahua, those in the south-central region have substantial Ceramic Period occupations that date to ca. AD 500–1000, making their interpretation more complicated.

In summary, sites or components that likely date to the window when agriculture first arrived are widespread in the higher basin and range country of Chihuahua. These sites occur in diverse settings and in most instances imply some stabilization in settlement, which is potentially tied to the arrival of maize agriculture. Only archaeological excavation and radiocarbon dating of appropriate material will firmly establish the timing and location of early maize in this region and the degree to which local economies were dependent on agriculture.

THE SIERRA TARAHUMARA

R. G. Matson [93, p. 209] suggests that Tarahumara maize farming in the deep canyons *(barrancas)* within the Sierra Madre Occidental was the basic, original pattern for maize cultivation in the area, and that other planting patterns must have emerged from this foundation. Colonial Tarahumara agricultural strategies included floodwater and dry farming using hillsides, mesas, alluvial terraces, and floodplains and irrigation in some mission communities [29, p. 80; 75, 98, 128]. Although the broken terrain limits agricultural land, today the Tarahumara typically farm on medium to small upland valley bottoms as well as practicing limited slash-and-burn bean farming. Narrow river floodplains and small seep fields are farmed occasionally as well [3, 8, 48].

Five sites including two rockshelters, date to the Middle to Late Archaic Period (see Table 6-1, Figure 6-1). The first of these sites is a rockshelter known as Sowérare (A32-02), which is perched on a ledge below the base of one cliff and above a second cliff, in the Río Urique watershed near the community of Rejogochi. Part of the site has been thoroughly plundered, but deep midden deposits remain intact. We have excavated a single 1 m unit in the midden to a depth of more than 120 centimeters below surface. Under spoil from the looted areas, the actual midden deposits contained dense lithics, rare ground stone, and charcoal. No maize was found, but two samples of wood charcoal yielded radiocarbon dates of 800 CAL BC (2660 ± 40 BP, Beta 185630) and 1700 CAL BC (3500 ± 40 BP, Beta 182381) that are in stratigraphic order and are chronologically consistent with the midden assemblage and dart points from the site.

We have also tested a sizable cave locally known as Ganóchi (A33-02) near Norogachi, that was probably originally recorded by Zingg [151, 152]. A single 1 m unit replicated the pattern of lithics and maize below ceramics in deep cave fill as reported by Robert H. Lister [79] at Swallow Cave. The upper 80 centimeters contained abundant artifacts, including pottery and uncharred bits of cord and basketry, with remnants of a mud and grass container. The few lithics included a high proportion of biface reduction flakes. A charred maize sample from 180 centimeters, just above bedrock, dated to 400–200 CAL BC (2280 ± 40 BP, Beta 185631).

At this point it is reasonable to argue that the Sierra Tarahumara was occupied no later than the late Middle Archaic Period and potentially much earlier. Sites are widely distributed and are in locations presently well suited to small-scale subsistence agriculture. Overall, they are small, even by the modest standards of south-central Chihuahua, and our limited work has produced no indication of montane cerros de trincheras or the D-shaped terrace sites.

DISCUSSION

Maize is now documented at 2100 BC in the American Southwest, 1300 BC in the basin and range country of northwestern Chihuahua, tentatively at 1700 BC in the basin and range region of south-central Chihuahua, and at no later than 200 BC in the Sierra Tarahumara of southwestern Chihuahua. The most likely interpretation is that maize spread quickly along a broad front, from central Mexico into northwestern Mexico and then into the American Southwest, rather than along single route [9]. The spread of maize into the American Southwest is now typically argued to be either the result of migration or group-to-group diffusion. However, we suspect the processes involved are not as straightforward as this dichotomy suggests. For example, people who adopt agriculture may partially shift territorial range on a seasonal basis to incorporate arable land or occupy previously unoccupied niches, obscuring the migration–diffusion dichotomy [15].

Floodplains are potentially productive locales for farming that could attract migrants to these valleys [93]. Identifying factors that might push migrant farmers northward from further south in Mexico is more difficult. However, because the widespread formation of settled farming communities in Mesoamerica did not occur until after 1500 BC, Mexican populations may not have been at a level that would have encouraged a migration that could account for maize in the American Southwest by 2000 BC. On the other hand, it appears that the population of northwestern Chihuahua increased or became more concentrated around 1300 BC, based on the appearance of cerros de trincheras settlements in the Río Casas Grandes Valley and the evidence for intergroup conflict [51, 53, 77, p. 361, 94]. Local populations were also using less built cerros de trincheras sites in south-central Chihuahua during the Early Agricultural Period, but the nature and extent of this use and the interrelationships among the sites are currently unknown. By this time it appears that populations were rising in Mesoamerica and complex societies were coalescing coincident with improved environmental conditions for farming in northern Mexico and the American Southwest. Both the Sierra Tarahumara and south-central Chihuahua were widely inhabited before this, during the later mid-Holocene (ca. 2500–2000 BC); however, we lack the relevant data to know if and how these various circumstances were related.

Maize was a prominent component of the local economy at Cerro Juanaqueña and related sites in northwestern Chihuahua maize, qualifying it as agricultural following Bruce D. Smith's [132] scheme. We know that farming was important in southern Arizona and apparently in northwestern Sonora as well. The data currently available for south-central and southwestern Chihuahua are insufficient to allow a more complete reconstruction of subsistence other than to say maize was present in the Early Agricultural Period.

In south-central Chihuahua there are indications of increased settlement stability on a modest scale from the late Middle Archaic Period while in the Sierra Tarahumara sites from the late Middle Archaic Period onward are present but relatively small. The deep deposits at both Sierra Tarahumara tested cave sites probably reflect repeated, but not populous, occupation, and evidence for settlement stability, such as cerros de trincheras sites or D-shaped terrace rockshelter sites, is also currently absent. In contrast, the Río Casas Grandes Valley of northwestern Chihuahua seems to have been populated to a level such that warfare and aggregation associated with farming were ongoing by 1300 BC, but the nature and extent of occupations before this time are unknown. Our future research will continue to address the variability in the ecological context and spread of farming in Chihuahua and its implications for the timing, spread, adoption, and consequences of agriculture.

References Cited

1. K. A. Adams, K. J. Hanselka. (2006). Plant use in the Late Archaic Period. In: R. J. Hard, J. R. Roney, (Eds.), *Early farming and warfare in Northwest Mexico.* Manuscript in preparation, to be submitted to University of Utah Press.
2. K. R. Adams. (1994). A regional synthesis of *Zea mays* in the prehistoric American Southwest. In: S. Johannessen, C. A. Hastorf, (Eds.), *Corn and culture in the prehistoric New World.* Boulder, CO: Westview Press. pp. 273–302.
3. K. R. Adams, W. L. Merrill. (2004). Effects of Colonialism on Rarámuri agriculture and subsistence: Searching for archaeological analogs for early maize farming strategies. Paper presented at the 33rd Gran Quivira Conference, October 8–10 2004, Las Cruces, New Mexico.
4. R. Ascher, F. J. Clune Jr. (1960). Waterfall Cave, Southern Chihuahua, Mexico. *American Antiquity, 26,* 270–274.

5. K. R. Barlow. (2002). Predicting maize agriculture among the Fremont: An economic comparison of farming and foraging in the American Southwest. *American Antiquity, 67,* 65–88.
6. R. L. Beals. (1943). Relations between Meso America and the Southwest. In: R. G. Granados, (Ed.), El norte de Mexico y el sur de Estados Unidos. Tercera reunión de mesa redonda sobre problemas antropológicos de México y Centro America. México, D.F.: Castillo de Chapultepec. pp. 245–251.
7. P. Bellwood. (2002). Farmers, foragers, languages, genes: The genesis of agricultural societies. In: P. Bellwood, C. Renfrew, (Eds.), *Examining the farming/language dispersal hypothesis.* Cambridge, MA: McDonald Institute for Archaeological Research, University of Cambridge. pp. 17–28.
8. W. C. Bennett, R. M. Zingg. (1935). The Tarahumara: An Indian tribe of Northern Mexico. Chicago: The University of Chicago Press.
9. B. F. Benz. (1999). On the origin, evolution, and dispersal of maize. In: M. Blake, (Ed.), *Pacific Latin America in prehistory: The evolution of Archaic and Formative cultures.* Pullman, WA: Washington State University Press. pp. 25–38.
10. B. F. Benz. (2000). A long, prehistoric maize evolution in the Tehuacan Valley. *Current Anthropology, 41,* 459–465.
11. C. F. Berry, M. S. Berry. (1986). Chronological and conceptual models of the Southwest Archaic. In: C. J. Condie, D. D. Fowler, (Eds.), *Archaeology of the American Southwest.* University of Utah Anthropological Papers, No. 110. Salt Lake City: University of Utah. pp. 253–327.
12. L. R. Binford. (1968). Post-Pleistocene adaptations. In: S. R. Binford, L. R. Binford, (Eds.), *New perspectives in archeology.* Chicago: Aldine. pp. 313–341.
13. L. R. Binford. (1980). Willow smoke and dogs' tails: Hunter-gatherer settlement systems and archaeological site formation. *American Antiquity, 45,* 4–20.
14. L. R. Binford. (2001). *Constructing frames of reference: An analytical method for archaeological theory building using ethnographic and environmental data sets.* Berkeley: University of California Press.
15. M. Blake. (1999). Introduction to the archaeology of Pacific Latin America. In: M. Blake, (Ed.), *Pacific Latin America in prehistory: The evolution of Archaic and Formative cultures.* Pullman, WA: Washington State University Press. pp. 3–10.
16. M. Blake, J. E. Clark. (1999). The emergence of hereditary inequality: The case of Pacific Coastal Chiapas, Mexico. In: M. Blake, (Ed.), *Pacific Latin America in prehistory: The evolution of Archaic and Formative cultures.* Pullman, WA: Washington State University Press. pp. 55–74.
17. M. Blake, B. S. Chisholm, J. E. Clark, B. Voorhies, M. W. Love. (1992). Prehistoric subsistence in the Soconusco Region. *Current Anthropology, 33,* 83–94.
18. R. C. Bronk. (1999). OxCal Program v3.3, Radiocarbon Accelerator Unit. University of Oxford. http://www.rlaha.ox.ac.uk/oxcal/oxcal.htm. Accessed April 23, 2003.
19. R. H. Brooks. (1971). Lithic traditions in Northwestern Mexico, Paleo-Indian to Chalchihuites. Ph. D. Dissertation. Boulder: Department of Anthropology, University of Colorado.
20. D. E. Brown, C. H. Lowe. (1980). Biotic communities of the Southwest. USDA Forest Service General Technical Report—RM-78. Washington, D.C.: U.S. Government Printing Office.
21. B. J. Buck, H. C. Monger. (1999). Stable isotopes and soil geomorphology as indicators of Holocene climate change, Northern Chihuahuan Desert. *Journal of Arid Environments, 43,* 357–373.
22. L. Campbell. (2002). What drives linguistic diversification and language spread? In: In: P. Bellwood, C. Renfrew, (Eds.), *Examining the farming/language dispersal hypothesis.* Cambridge, MA: McDonald Institute for Archaeological Research, University of Cambridge. pp. 467–475.
23. J. P. Carpenter, G. Sanchez, M. E. Villalpando. (2002). Of maize and migration: Mode and tempo in the diffusion of *Zea mays* in Northwest Mexico and the American Southwest. In: S. H. Schlanger, (Ed.), *Traditions, transitions, and technologies: Themes in Southwestern Archaeology.* Proceedings of the 2000 Southwest Symposium. Boulder: University Press of Colorado. pp. 245–256.
24. J. P. Carpenter, G. Sanchez, M. E. Villalpando. (2003). Sonora precerámica: del Arcaico y del surgimiento de aldeas agrícolas. *Arqueología, 29,* 5–29.
25. P. J. Castiglia. (2002). Late Quaternary climate history of the pluvial Lake Palomas system, northern Chihuahua, Mexico. Master's Thesis Thesis. Albuquerque: University of New Mexico.
26. E. Chacón Soria. (2001). Proyecto Minero El Sauzal, Informe al Consejo de Arqueología. México, D.F.: Insituto Nacional de Antropología e Historia.
27. H. Cutler. (1960). Cultivated plant remains from Waterfall Cave, Chihuahua. *American Antiquity, 26,* 277–279.
28. J. E. Damp, S. A. Hall, S. J. Smith. (2002). Early irrigation on the Colorado Plateau near Zuni Pueblo, New Mexico. *American Antiquity, 67,* 665–676.
29. S. M. Deeds. (2003). Defiance and deference in Mexico's colonial north: Indians under Spanish colonial rule in Nueva Vizcaya. Austin: University of Texas Press.
30. H. W. Dick. (1965). *Bat cave.* Monograph 27. Santa Fe, NM: School of American Research.
31. M. W. Diehl. (1997). Rational behavior, the adoption of agriculture, and the organization of subsistence during the Late Archaic Period in the Greater Tucson Basin. In: C. M. Barton, G. A. Clark, (Eds.), *Rediscovering Darwin: Evolutionary theory and archeological explanation.* Archeological Papers of the American Anthropological Association No. 7. Arlington, VA: American Anthropological Association. pp. 251–265.
32. Direccion General de Geografia del Territorio Nacional. (1981). Carta de Precipitation Total Anual, 1 : 100,000, Chihuahua. México, D.F.: Secretaria de Programacion y Presupuesto.
33. J. F. Doebley, A. Stec, L. Hubbard. (1997). The evolution of apical dominance in maize. *Nature, 386,* 485–488.
34. L. L. Ely. (1997). Response of extreme floods in the Southwestern United States to climatic variations in the Late Holocene. *Geomorphology, 19,* 175–201.
35. P. R. Fish, S. K. Fish, A. Long, C. Miksicek. (1986). Early corn remains from Tumamoc Hill, southern Arizona. *American Antiquity, 51,* 563–572.
36. S. K. Fish, P. R. Fish. (1994). Prehistoric desert farmers of the Southwest. *Annual Review of Anthropology, 23,* 83–108.
37. K. V. Flannery. (1976). Interregional religious networks. In: K. V. Flannery, (Ed.), *The early Mesoamerican village.* New York: Academic Press. pp. 329–333.
38. H. L. Fleischhauer, Jr., W. J. Stone. (1982). *Quaternary geology of Lake Animas, Hidalgo County, New Mexico.* Circular 174. Socorro, NM: New Mexico Bureau of Mines and Mineral Resources.
39. C. S. Fowler. (1983). Some lexical clues to Uto-Aztecan Prehistory. *International Journal of American Linguistics, 49,* 224–257.
40. G. Fritz. (2006). Amaranth and chenopod seeds from Early Agricultural Cerros de Trincheras Sites. In: R. J. Hard, J. R. Roney. *Early farming and warfare in Northwest Mexico.* Manuscript in preparation, to be submitted to University of Utah Press.
41. G. Fritz, K. Adams, R. Hard, J. R. Roney. (1999). Evidence for cultivation of Amaranthus sp. (Amaranthaceae) 3,000 years ago at Cerro Juanaqueña, Chihuahua. Paper presented at the 22nd Annual Conference of the Society of Ethnobiology, March 11, 1999. Oaxaca, Mexico.
42. S. Gorenstein, M. S. Foster. (2000). West and northwest Mexico: The ins and outs of Mesoamerica. In: M. S. Foster, S. Gorenstein, (Eds.), *Greater Mesoamerica: The archaeology of west and north-*

west Mexico. Salt Lake City: The University of Utah Press. pp. 3–20.

43. D. A. Gregory. (1999). Data integration and synthesis. In: D. A. Gregory, (Ed.), *Excavations in the Santa Cruz River floodplain: The Middle Archaic component at Los Pozos*. Anthropological Papers No. 20. Tucson, AZ: Center for Desert Archaeology. pp. 85–123.
44. D. A. Gregory. (2001). An evaluation of Early Agricultural Period chronology in the Tucson Basin. In: D. A. Gregory, (Ed.), *Excavations in the Santa Cruz River floodplain: The Early Agricultural Period component at Los Pozos*. Anthropological Papers No. 21. Tucson, AZ: Center for Desert Archaeology. pp. 237–254.
45. D. A. Gregory, M. W. Diehl. (2002). Duration, continuity, and intensity of occupation at a Late Cienega Phase settlement in the Santa Cruz River floodplain. In: S. H. Schlanger, (Ed.), *Traditions, transitions, and technologies: Themes in southwestern archaeology*. Proceedings of the 2000 Southwest Symposium. Boulder: University Press of Colorado. pp. 200–223.
46. R. J. Hard. (1986). Ecological relationships affecting the rise of farming economies: A test from the American Southwest. Ph.D. Dissertation. Albuquerque: Department of Anthropology, University of New Mexico.
47. R. J. Hard, R. P. Mauldin, G. R. Raymond. (1996). Mano size, stable carbon isotope ratios, and macrobotanical remains as multiple lines of evidence of maize dependence in the American Southwest. *Journal Archaeological Method and Theory, 3,* 253–318.
48. R. J. Hard, W. L. Merrill. (1992). Mobile agriculturalists and the emergence of sedentism: Perspectives from Northern Mexico. *American Anthropologist, 94,* 601–620.
49. R. J. Hard, J. R. Roney. (1998). A massive terraced village complex in Chihuahua, Mexico, 3000 years before present. *Science, 279,* 1661–1664.
50. R. J. Hard, J. R. Roney. (2004). Late Archaic Period hilltop settlements in Northwestern Chihuahua, Mexico. In: B. Mills, (Ed.), *Society and politics in the Greater Southwest*. Boulder: University of Colorado Press. pp. 276–294.
51. R. J. Hard, J. R. Roney. (2006). Early farming and warfare in Northwest Mexico, with contributions by K. R. Adams, G. J. Fritz, K. Hanselka, L. Nordt, K. M. Schmidt, B. J. Vierra. Manuscript in preparation, to be submitted to University of Utah Press.
52. R. J. Hard, J. R. Roney. (2005). The transition to farming on the Rio Casas Grandes and in the Southern Jornada Mogollon region in the North American Southwest. In: B. J. Vierra, (Ed.), *Current research on the Late Archaic across the Borderlands*. Austin: University of Texas Press. pp. 141–186.
53. R. J. Hard, J. R. Roney. (In press). Cerros de trincheras in Northwestern Chihuahua: The arguments for defense. In: S. Fish, P. Fish, E. Villalpando, (Eds.), *Enduring Borderlands traditions: Trincheras sites in time, space and society*. Tucson: University of Arizona Press.
54. R. J. Hard, J. E. Zapata, B. K. Moses, J. R. Roney. (1999). Terrace construction in Northern Chihuahua, Mexico: 1150 B. C. and modern experiments. *Journal of Field Archaeology, 26,* 129–146.
55. C. A. Hastorf, S. Johannessen. (1994). Becoming corn-eaters in prehistoric America. In: S. Johannessen, C. A. Hastorf, (Eds.), *Corn and culture in the prehistoric New World*. Boulder, CO: Westview Press. pp. 427–444.
56. E. W. Haury. (1962). The greater American Southwest. In: R. J. Braidwood, G. R. Willey, (Eds.), Courses toward urban life. Viking Fund Publications in Anthropology, Number 32. New York: Wenner-Gren Foundation for Anthropological Research, Inc. pp. 106–131.
57. C. V. Haynes. (1981). Geochronology and paleoenvironments of the Murray Springs Clovis Site, Arizona. *National Geographic Society Research Reports, 13,* 243–251.
58. J. H. Hill. (2001a). Proto-Uto-Aztecan: A community of cultivators in Central Mexico? *American Anthropologist, 103,* 913–934.
59. J. H. Hill. (2001b). Dating the break-up of Southern Uto-Aztecan. In: J. L. Moctezuma Zamarrón, J. H. Hill, (Eds.), *Avances y balances de lenguas yutoaztecas: Homenaje a Wick R. Miller*. Mexico City: Instituto Nacional de Antropología e Historia. pp. 345–357.
60. J. H. Hill. (2002a). Toward a linguistic prehistory of the Southwest: "Azteco-Tanoan" and the arrival of maize cultivation. *Journal of Anthropological Research, 58,* 457–475.
61. J. H. Hill. (2002b). Proto-Uto-Aztecan cultivation and the northern devolution. In: P. Bellwood, C. Renfrew, (Eds.), *Examining the farming/language dispersal hypothesis*. Cambridge, MA: McDonald Institute for Archaeological Research, University of Cambridge. pp. 331–340.
62. E. K. Huber. (2005). Early maize at the Old Corn Site (LA 137258). In: E. K. Huber, C. R. Van West, (Eds.), *Archaeological data recovery in the New Mexico Transportation Corridor and First Five-Year Permit Area. Fence Lake Coal Mine Project, Catron County, New Mexico*. Synthetic Studies and Summary, Draft final report. Technical Series 84, Vol. 4. Tucson, AZ: Statistical Research. pp. 36.1–36.31.
63. B. B. Huckell. (1988). Late Archaic archaeology of the Tucson Basin: A status report. In: W. H. Doelle, P. R. Fish, (Eds.), *Recent research on Tucson Basin prehistory: Proceedings of the second Tucson Basin Conference*. Anthropological Papers No. 10. Tucson, AZ: Institute for American Research. pp. 57–80.
64. B. B. Huckell. (1995) *Of Marshes and Maize. Preceramic Agricultural Settlements in the Cienega Valley, Southeastern Arizona*. Anthropological Papers 59. Tucson, AZ: University of Arizona Press.
65. B. B. Huckell. (1996). The Archaic prehistory of the North American Southwest. *Journal of World Prehistory, 10,* 305–373.
66. B. B. Huckell, L. W. Huckell, M. S. Shackley. (1999). McEuen Cave. *Archaeology Southwest, 13,* 12.
67. R. L. Hunter-Anderson. (1986). *Prehistoric adaptation in the American Southwest*. Cambridge, MA: Cambridge University Press.
68. H. H. Iltis. (1987). Maize evolution and agricultural origins. In: T. Soderstrom, K. Hilu, C. Campbell, M. Barkworth, (Eds.), *Grass systematics and evolution*. Washington, D.C.: Smithsonian Institution Press. pp. 195–213.
69. V. Jaenicke-Després, E. S. Buckler, B. D. Smith, M. Thomas, P. Gilbert, A. Cooper, J. Doebley, S. Pääbo. (2003). Early allelic selection in maize as revealed by ancient DNA. *Science, 302,* 1206–1208.
70. S. Johannessen, C. A. Hastorf, (Eds.). (1994). *Corn and culture in the prehistoric New World*. University of Minnesota Publications in Anthropology No. 5. Boulder, CO: Westview Press.
71. A. L. Johnson. (1997). Explaining variability in the pace and pattern of cultural evolution in the North American Southwest: An exercise in theory building. Ph. D. Dissertation. Dallas, TX: Southern Methodist University.
72. A. L. Johnson. (2004). A method for anticipating patterns in archaeological sequences: Projecting duration for the transition to agriculture in Mexico. In: L. Webster, M. McBrinn, (Eds.), *Archaeology without borders: Contact, commerce and change in the U.S. Southwest and Northwestern Mexico*. Proceedings of the 2004 Southwest Symposium. Boulder: under review University Press of Colorado.
73. J. C. Kelley. (1956). Settlement patterns in North-Central Mexico. In: G. R. Willey, (Ed.), *Prehistoric settlement patterns in the New World*. Viking Fund Publications in Anthropology 23. New York: Wenner Gren Fund for Anthropological Research. pp. 128–139.
74. J. H. Kelley, J. D. Stewart, A. C. MacWilliams, L. C. Neff. (1999). A west-central Chihuahuan perspective on Chihuahua Culture. In: C. F. Schaafsma, C. L. Riley, (Eds.), *The Casas Grandes world*. Salt Lake City: University of Utah Press. pp. 63–77.
75. J. G. Kennedy. (1978). *Tarahumara of the Sierra Madre*. Arlington Heights, IL: AHM Publishing Corporation.

76. S. A. Leblanc. (1999). *Prehistoric warfare in the American Southwest.* Salt Lake City: University of Utah Press.
77. S. A. Leblanc. (2002). Conflict and language dispersal: Issues and a New World example. In: P. Bellwood, C. Renfrew, (Eds.), *Examining the farming/language dispersal hypothesis.* Cambridge, MA: McDonald Institute for Archaeological Research, University of Cambridge. pp. 357–365.
78. S. Lewenstein. (1993). Proyecto arqueológico-ethnoarqueológico Sierra Tarahumara, 1991: Los Sitios, Su Entorno, (Herencia Mogollon). In: M. T. Cabrero G., (Ed.), *II Coloquio Pedro Bosch-Gimpera.* Universidad Nacional Autónoma de México, México, D.F.: Instituto de Investigaciones Anthropológicas. pp. 496–510.
79. R. H. Lister. (1958). Archaeological excavations in the Northern Sierra Madre Occidental, Chihuahua and Sonora, Mexico. In: R. H. Lister, (Ed.), *Archaeological excavations in the northern Sierra Madre Occidental, Chihuahua and Sonora, Mexico.* University of Colorado Studies, Series in Anthropology No. 7. Boulder: University of Colorado Press. pp. 1–95.
80. A. Long, B. F. Benz, D. J. Donahue, A. J. T. Jull, L. J. Toolin. (1989). First direct AMS dates on early maize from Tehuacán, Mexico. *Radiocarbon, 31,* 1035–1040.
81. J. B. Mabry. (1998). *Conclusions, archaeological investigations of early village sites in the Middle Santa Cruz Valley. Part II: Analysis and Synthesis.* Anthropological Papers 19. Tucson, AZ: Center for Desert Archaeology. pp. 757–791.
82. J. B. Mabry. (2002). The role of irrigation in the transition to agriculture and sedentism in the Southwest. In: S. H. Schlanger, (Ed.), *Traditions, transitions, and technologies: Themes in Southwestern archaeology in the year 2000.* Boulder: University of Colorado Press.
83. J. B. Mabry, D. L. Swartz, H. Wöcherl, J. J. Clark, G. H. Archer, M. W. Lindeman. (1997). *Archaeological investigations of early village sites in the Middle Santa Cruz Valley: Description of the Santa Cruz Bend, square hearth, stone pipe, and canal sites.* Anthropological Papers No. 18. Tucson, AZ: Center for Desert Archaeology.
84. R. S. MacNeish, M. W. Eubanks. (2000). Comparative analysis of the Rio Balsas and Tehuacan models for the origin of maize. *Latin American Antiquity, 11,* 3–20.
85. A. C. MacWilliams. (2001). Archaeology in Laguna Bustillos Basin, Chihuahua, Mexico. Ph. D. Dissertation. Tucson: Department of Anthropology, University of Arizona.
86. A. C. MacWilliams, R. J. Hard, J. R. Roney, K. R. Adams, W. L. Merrill. (2004). Observations on the ceramic period of Southern Chihuahua. Paper presented at the 69th Annual Meeting of the Society for American Archaeology, March 31–April 4, 2004, Montreal, Canada.
87. A. C. MacWilliams, R. J. Hard, J. R. Roney, K. R. Adams, W. L. Merrill. (2004). The search for early farming sites in Northwestern Mexico. Report to the National Geographic Society committee for Research and Exploration, Grant Number 7399-03.
88. A. C. MacWilliams, R. J. Hard, J. R. Roney, K. R. Adams, W. L. Merrill. (2004). The setting of early agriculture in southern Chihuahua, Mexico. In: L. Webster, L. McBrinn, (Eds.), *Archaeology without borders: Contact, commerce and change in the U.S. Southwest and Northwestern Mexico.* Proceedings of the 2004 Southwest Symposium. Boulder: Under review, University Press of Colorado.
89. R. S. Malhi, H. M. Mortensen, J. A. Eshleman, B. M. Kemp, J. G. Lorenz, F. A. Kaestle, J. R. Johnson, C. Gorodezky, D. G. Smith. (2003). Native American mtDNA prehistory in the American Southwest. *American Journal of Physical Anthropology, 120,* 108–124.
90. P. C. Mangelsdorf. (1986). The origin of corn. *Scientific American, 255,* 80–86.
91. J. Marcus, K. V. Flannery. (2004). The coevolution of ritual and society: ^{14}C dates from Ancient Mexico. *Proceedings of the National Academy of Sciences of the United States of America, 101,* 18257–18261.
92. G. J. Marrs. (1949). Problems arising from the surface occurrence of archaeological material in Southeastern Chihuahua, Mexico. Master's Thesis. Albuquerque: Department of Anthropology, University of New Mexico.
93. R. G. Matson. (1991). *The origins of southwestern agriculture.* Tucson: University of Arizona Press.
94. R. G. Matson. (2002). The spread of maize agriculture into the U.S. Southwest. In: P. Bellwood, C. Renfrew, (Eds.), *Examining the farming/language dispersal hypothesis.* Cambridge, MA: McDonald Institute for Archaeological Research, University of Cambridge. pp. 341–356.
95. R. G. Matson, B. Chisholm. (1991). Basketmaker II subsistence: Carbon isotopes and other dietary indication from Cedar Mesa, Utah. *American Antiquity, 56,* 444–459.
96. Y. Matsuoka, Y. Vigouroux, M. M. Goodman, J. Sanchez G., E. Buckler, J. Doebley. (2002). A single domestication for maize shown by multilocus microsatellite genotyping. *Proceedings of the National Academy of Sciences of the United States of America, 99,* 6080–6084.
97. W. L. Merrill. (1988). *Rarámuri souls: Knowledge and social process in Northern Mexico.* Washington, D. C.: Smithsonian Institution Press.
98. W. L. Merrill. (1993). Conversion and colonialism in Northern Mexico: The Tarahumara response to the Jesuit mission program, 1601–1767. In: R. W. Hefner, (Ed.), *Conversion to Christianity: Historical and anthropological perspectives on a Great Transformation.* Berkeley: University of California Press. pp. 129–163.
99. S. E. Metcalfe, A. Bimpson, A. J. Courtice, S. L. O'Hara, D. M. Taylor. (1997). Climate change at the monsoon/westerly boundary in Northern Mexico. *Journal of Paleoliminology, 17,* 155–171.
100. W. R. Miller. (1983). Uto-Aztecan languages. In: A. Ortiz, (Ed.), *Southwest handbook of North American Indians.* Vol. 10. W. C. Sturtevant, (Gen. Ed.). Washington, D.C.: Smithsonian Institution. pp. 113–124.
101. P. E. Minnis. (1985). Domesticating people and plants in the Greater Southwest. In: R. I. Ford, (Ed.), *Prehistoric food production in North America.* University of Michigan Anthropological Papers 75. Ann Arbor, MI: University of Michigan. pp. 309–340.
102. P. E. Minnis. (1992). Earliest plant cultivation in the Desert Borderlands of North America. In: C. W. Cowan, P. J. Watson, (Eds.), *The origins of agriculture: An international perspective.* Washington, D.C.: Smithsonian Institution Press. pp. 121–141.
103. C. Montero, J. Carpenter, E. Barnes, A. Rohn, J. Watson. (2004). Early Agricultural Period burials at La Playa (SON F. 10:03), Sonora, Mexico: Further preliminary results. Paper prepared for the 69th Annual Meeting of the Society for American Archaeology. March 31–April 4, 2004, Montreal, Canada.
104. J. B. Mountjoy. (2000). Prehispanic cultural development along the Southern Coast of West Mexico. In: M. S. Foster, S. Gorenstein, (Eds.), *Greater Mesoamerica: The archaeology of west and northwest Mexico.* Salt Lake City: The University of Utah Press. pp. 81–106.
105. M. J. P. Nichols. (1983–1984). Old California Uto-Aztecan. *Journal of the Steward Anthropological Society, 15,* 23–46.
106. L. Nordt. (2003). Late Quaternary fluvial landscape evolution in Desert Grasslands of Northern Chihuahua, Mexico. *Geological Society of America Bulletin, 115,* 596–606.
107. D. Nusbaum. (1940). Gila Pueblo Detail Sheet for GP U: 13:2 (CHIH). Document on File. Tucson: Arizona State Museum Archives.
108. J. R. Ortega-Ramirez. (1995). Los paleoambientes holocenicos de la laguna de babicora, Chihuahua, Mexico. *Geofisica Internacional, 34,* 107–116.

109. J. R. Ortega-Ramirez, A. Valiente-Banuet, J. Urrutia-Fucugauchi, C. A. Mortera-Gutierrez, G. Alvarado-Valdéz. (1998). Paleoclimatic changes during the Late Pleistocene-Holocene in Laguna Babícora, near the Chihuahuan Desert, Mexico. *Canadian Journal of Earth Science, 35,* 1168–1179.
110. M. R. Palacios-Fest, A. L. Carreño, J. Ortega-Ramirez, G. Alvarado-Valdéz. (2002). A Paleoenvironmental reconstruction of Laguna Babícora, Chihuahua, Mexico, based on ostracode paleoecology and trace element shell chemistry. *Journal of Paleolimnology, 27,* 185–206.
111. D. M. Pearsall, K. Chandler-Ezell, A. Chandler-Ezell. (2003). Identifying maize in neotropical sediments and soils using cob phytoliths. *Journal of Archaeological Science, 30,* 611–627.
112. D. R. Piperno, K. V. Flannery. (2001). The earliest archaeological maize *(Zea mays L.)* from highland Mexico: New accelerator mass spectrometry dates and their implications. *Proceedings of the National Academy of Sciences of the United States of America, 98,* 2101–2103.
113. M. D. Pohl, K. O. Pope, J. G. Jones, J. S. Jacob, D. R. Piperno, S. D. de France, D. L. Lentz, J. A. Gifford, M. E. Danforth, J. K. Josserand. (1996). Early agriculture in the Maya Lowlands. *Latin American Antiquity, 7,* 355–372.
114. H. P. Pollard. (2000). Tarascan external relationships. In: M. S. Foster, S. Gorenstein, (Eds.), *Greater Mesoamerica: The archaeology of west and northwest Mexico.* Salt Lake City: the University of Utah Press. pp. 71–80.
115. V. J. Polyak, Y. Asmerom. (2001). Late Holocene climate and cultural changes in the Southwestern United States. *Science, 294,* 148–151.
116. K. O. Pope, M. E. D. Pohl, J. G. Jones, D. L. Lentz, C. von Nagy, F. J. Vega, I. R. Quitmyer. (2001). Origin and environmental setting of ancient agriculture in the Lowlands of Mesoamerica. *Science, 292,* 1370–1373.
117. G. R. Raymond. (2001). La investigación de algunos sitios al Sur de Chihuahua. In: R. J. Hard, J. E. Zapata, J. R. Roney, (Eds.), *Una investigación arqueológica de los sitios cerros con trincheras del arcaico tardío en Chihuahua, México. Informe al Consejo de Arqueología. Instituto Nacional de Antropología e Historia.* Center for Archaeological Research, UTSA, Special Report, No. 27-S. San Antonio: University of Texas at San Antonio. pp. 37–44.
118. G. R. Raymond, J. R. Roney, A. C. MacWilliams, R. J. Hard, K. R. Adams, W. L. Merrill. (2003). Looking for early maize between the Southwest and Mesoamerica: A field report from Southwest Chihuahua, Mexico. Paper presented at the Annual Meeting of the Society for American Archaeology, April 9–13, 2003. Milwaukee, Wisconsin.
119. C. Renfrew. (2002). "The Emerging Synthesis": The Archaeogenetics of farming/language dispersals and other spread zones. In: P. Bellwood, C. Renfrew, (Eds.), *Examining the farming/language dispersal hypothesis.* Cambridge, MA: McDonald Institute for Archaeological Research. University of Cambridge, pp. 3–16.
120. J. R. Roney, R. J. Hard. (2002). Early agriculture in Northwestern Chihuahua. In: S. Schlanger, (Ed.), *Traditions, transitions, and technologies: Themes in Southwestern Archaeology.* Boulder: University of Colorado Press. pp. 163–180.
121. J. R. Roney, R. J. Hard. (2004). A review of cerros de trincheras in Northwestern Chihuahua. In: G. E. Newell, E. Gallaga, (Eds.), *Surveying the archaeology of Northwest Mexico.* Salt Lake City: University of Utah Press. pp. 127–148.
122. J. Rzedowski. (1983). *Vegetación de México.* México: Editorial Limusa.
123. E. B. Sayles. (1936). *An archaeological survey of Chihuahua, Mexico.* Medallion Papers 22. Globe, AZ: Gila Pueblo Foundation.
124. K. M. Schmidt. (2006). Faunal remains. In: R. J. Hard, J. R. Roney. *Early farming and warfare in Northwest Mexico.* Manuscript in preparation, to be submitted to University of Utah Press.
125. K. M. Schmidt. (2004). Faunal remains from test excavation at Early Agricultural Sites in Southern Chihuahua, Mexico. Manuscript on file. San Antonio: Center for Archaeological Research, University of Texas, San Antonio.
126. K. G. Schollmeyer, C. G. Turner II. (2004). Dental caries, prehistoric diet, and the Pithouse to Pueblo Transition in Southwestern Colorado. *American Antiquity, 69,* 569–582.
127. M. R. Schurr, D. A. Gregory. (2002). Fluoride dating of faunal materials by ion-selective electrode: High resolution relative dating at an Early Agricultural Period site in the Tucson Basin. *American Antiquity, 67,* 281–299.
128. T. E. Sheridan, T. H. Naylor. (1979). *Rarámuri: A Tarahumara colonial chronicle.* Flagstaff, AZ: Northland Press.
129. A. H. Simmons. (1986). New evidence for the early use of cultigens in the American Southwest. *American Antiquity, 51,* 73–89.
130. B. D. Smith. (1997). Reconsidering the Ocampo Caves and the era of incipient cultivation in Mesoamerica. *Latin American Antiquity, 8,* 342–383.
131. B. D. Smith. (2001). Documenting plant domestication: The consilience of biological and archaeological approaches. *Proceedings of the National Academy of Sciences of the United States of America, 98,* 1324–1326.
132. B. D. Smith. (2001). Low-level food production. *Journal of Archaeological Research 9,* 1–43.
133. D. G. Smith, J. Lorenz, B. K. Rolfs, R. L. Bettinger, B. Green, J. Eshleman, B. Schultz, R. Malhi. (2000). Implications of the distribution of Albumin Maskapi and Albumin Mexico for New World prehistory. *American Journal of Physical Anthropology, 111,* 557–572.
134. J. E. Staller. (2003). An examination of the palaeobotanical and chronological evidence for an early introduction of maize *(Zea mays L.)* into South America: a response to Pearsall. *Journal of Archaeological Science, 30,* 373–380.
135. M. N. Stevens, R. J. Sliva. (2002). Empire points: An addition to the San Pedro phase lithic assemblage. *Kiva, 67,* 297–326.
136. E. S. Turner, T. R. Hester. (1993). *A field guide to stone artifacts of Texas Indians.* Houston, TX: Gulf Publishing.
137. S. A. Turpin. (1991). Time out of mind: The radiocarbon chronology of the Lower Pecos River Region. In: S. A. Turpin, (Ed.), *Papers on Lower Pecos prehistory.* Studies in Archeology No. 8. Austin, TX: Texas Archeological Research Laboratory, University of Texas. pp. 1–49.
138. R. A. Tyson. (1985). The Chihuahua mummy research project at the San Diego Museum of Man. In: R. A. Tyson, D. V. Elerick, (Eds.), *Two mummies from Chihuahua, Mexico: A multidisciplinary study.* San Diego Museum Papers, No. 19. San Diego, CA: San Diego Museum of Man. pp. 1–8.
139. T. R. Van Devender. (1990). Late Quaternary vegetation and climate of the Sonoran Desert, United States and Mexico. In: J. L. Betancourt, T. R. Van Devender, P. S. Martin, (Eds.), *Packrat middens: The last 40,000 years of biotic change.* Tucson: University of Arizona Press. pp. 104–133.
140. T. R. Van Devender. (1995). Desert grassland history: changing climates, evolution, biogeography, and community dynamics. In: M. P. McClaran, T. R. Van Devender, (Eds.), *The desert grassland.* Tucson: The University of Arizona Press. pp. 68–99.
141. T. R. Van Devender, R. D. Worthington. (1977). The herpetofauna of Howell's Ridge Cave and the paleoecology of the Northwestern Chihuahuan Desert. In: R. H. Wauer, D. H. Riskind, (Eds.), *Transactions of the symposium on the biological resources of the Chihuahuan Desert Region: United States and Mexico.* No. 3, National Park Service Transactions and Proceedings Series. Washington, D.C.: U.S. Department of the Interior. pp. 85–106.
142. M. R. Waters. (1989). Late Quaternary lacustrine history and paleoclimatic significance of pluvial Lake Cochise, Southeastern Arizona. *Quaternary Research, 32,* 1–11.

143. M. R. Waters, C. V. Haynes. (2001). American Southwest. *Geology, 29,* 399–402.
144. R. C. West. (1964). The natural regions of Middle America. In: R. C. West, (Ed.), *Natural environment and early cultures: Handbook of middle American Indians.* Vol. I. Austin: The University of Texas Press. pp. 363–383.
145. M. E. Whalen. (1994). Moving out of the Archaic on the edge of the Southwest. *American Antiquity, 59,* 622–638.
146. D. E. Wilkins, D. R. Currey. (1999). Radiocarbon chronology and $\delta^{13}C$ analysis of Mid- To Late-Holocene aeolian environments, Guadalupe Mountains National Park, Texas, USA. *The Holocene, 9,* 363–371.
147. W. H. Wills. (1988). *Early prehistoric agriculture in the American Southwest.* Santa Fe, NM: School of American Research Press.
148. W. H. Wills. (1995). Archaic foraging and the beginning of food production in the American Southwest. In: T. D. Price, A. B. Gebauer, (Eds.), *Last hunters, first farmers.* Santa Fe, NM: School of American Research. pp. 215–242.
149. W. H. Wills. (2001). Pithouse architecture and the economics of household formation in the prehistoric American Southwest. *Human Ecology, 29,* 477–500.
150. W. H. Wills, B. B. Huckell. (1994). Economic implications of changing land-use patterns in the Late Archaic. In: G. J. Gumerman, (Ed.), *Themes in Southwest prehistory.* Santa Fe, NM: School of American Research. pp. 33–52.
151. R. M. Zingg. (1939). *A reconstruction of Uto-Aztekan History, Contributions to Ethnography: II University of Denver.* New York: G. E. Stechert and Company.
152. R. M. Zingg. (1940). *Report on archaeology of Southern Chihuahua, Contributions of the University of Denver.* Denver, CO: Center of Latin American Studies.

CHAPTER

7

El Riego and Early Maize Agricultural Evolution

BRUCE F. BENZ*, LI CHENG†, STEVEN W. LEAVITT‡, AND CHRIS EASTOE§

*Biology Department, Texas Wesleyan University, Ft. Worth, Texas

†Laboratory of Tree-Ring Research, University of Arizona, Tucson, Arizona

‡Laboratory of Tree-Ring Research, University of Arizona, Tucson, Arizona

§Department of Geosciences, University of Arizona, Tucson, Arizona

Glossary

Darwin A metric developed by J. Haldane and embellished by G. G. Simpson in the 1940s for comparing the amount of morphological change in paleontological assemblages through time.

Evolution Genetic changes in a population through time or across space. Human mediated evolution is documented by morphological change of a cultigen in the absence of simultaneous environmental change.

Scaled evolutionary rates Adjusted rates in Darwins employed to facilitate comparisons of rates derived from divergent long and short time spans.

Stable isotopes Isotopes found in organic molecules whose abundance in organic material is used as a proxy measure for documenting atmospheric and environmental conditions.

Morphological analyses of maize from the East Niche of the El Riego Oasis in the Tehuacan Valley document temporal variation in human selection for maize ear traits. Changing intensities of human selection are inferred from the absence of apparent environmental changes over the past 4000 years.

INTRODUCTION

The search for the origins of agriculture and the beginnings of complex society in the New World led R. S. MacNeish to excavate numerous rock shelters, overhangs, and open sites in the Tehuacan Valley during the 1960s. Two small inter-connected rock shelters, located a short distance from the community of El Riego, were excavated by MacNeish and A. Garcia Cook. The two shelters within the larger cave showed different amounts of accumulated human refuse. The East Niche faces southwest. It is the larger of the two rock shelters, measuring 15 meters wide, that is a maximum of 7 meters deep under the overhang. The accumulation of human refuse under the overhang measured nearly 2 meters thick. The West Niche faces southeast and is considerably smaller and had less accumulated refuse. The human refuse in both rock shelters showed sharp stratigraphic discontinuities in the field. These discontinuities allowed MacNeish and Garcia Cook to reconstruct the human activities represented in each stratigraphic layer or zone based on the distribution of artifacts, features, and ecofacts that each contained. An abundance of maize cobs were recovered from excavation though no absolute dates—radiocarbon dates—were obtained from either rock shelter [14]. The maize remains were not subject to the same detailed analysis by Mangelsdorf and colleagues as were those recovered from San Marco or Coxcatlan [13]. We have taken advantage of the abundance of maize recovered from the East Niche and the apparent stratigraphic integrity of this rock shelter's deposits to document the rate of morphological change in maize represented throughout these archaeological deposits.

Maize is a model organism for describing the domestication process. Remains of its ears, stems, and roots are common in the dry rockshelter deposits of the Tehuacan Valley and other regions of the New World. Together this archaeological evidence of maize's local evolutionary history documents human behavior because the ear's morphological change over time is a mirror of human effort to create and perpeutate a reliable food source. Human selection focused on relatively few traits that are archaeologically visible over the five and half thousand years maize has been subject to domesticating selection. Grain size increased over time, as did the ear and the ancillary structures associated with it. These trends have been well-known since MacNeish and colleagues completed excavation of rockshelters in Tamaulipas and Tehuacan [1, 13, 14]. The rate at which these morphological changes took place was hardly mentioned by those who had the opportunity to examine the archaeological maize from Tehuacan. Interested archaeologists apparently thought the trends in ear size were germane to understanding of cultural developments in central Mexico, though none consulted the collections themselves. Following the development of accelerator mass spectrometry (AMS) dating, the Tehuacan maize assemblages availed themselves to a novel approach to study maize **evolution.** This approach [3, 5] sought to document the amount of morphological change in maize over time, adopting a time tested paleontological method [6–8] to ascertain the relative rate of change in the major lineages of fossil forms during the past 500 million years. This method shows great promise, although it is not without assumptions that some are unwilling to accept.

Our first application of Haldane's evolutionary rate calculation focused on the early maize from the Tehuacan Valley [5]. Morphological comparison of some of the earliest cobs from San Marcos and Coxcatlan showed subtle morphological differences; they were dated by AMS, and their ages were calibrated and tested for contemporaneity. A total of 20 AMS dated specimens were used to describe the morphological change in 26 cobs covering a time span of almost 3700 years. Skeptics argue that mixing of archaeological strata or the movement of cobs through bioturbation severely limits comparison and makes inferences about the tempo of maize's morphological change meaningless. This criticism is shortsighted, if not mistaken, even if well-intentioned. All specimens used in that study from Coxcatlan had associated AMS dates—six in all—and whose contemporaneous ages were averaged. And since those findings were published, one additional contemporaneous date has been obtained that corroborates the earliest of Coxcatlan maize dates. The initial findings about the Tehuacan maize can be corroborated and should be considered as "open to alternative and more focused analyses," [17, p. 9443] but they are hardly meaningless.

The intent of Haldane's development of the paleontological metric, the **Darwin,** was to reconcile the paleontological record with the powerful predictive ability of population genetics and systematics. Morphological variation through time suggests that the population will wander through an adaptive landscape unless genetic drift, nonrandom mating, mutation, migration, or natural selection effects directional change. The variable nature of the earliest maize in Tehuacan suggested at first glance that it was a heterogeneous population that humans were propagating but perhaps not intensively manipulating. Dating individual specimens, each of which could be precisely described in morphological terms, provided an opportunity to test the heterogeneous population hypothesis using the Darwin as a proxy for the magnitude of change when subject to selection for a specific length of time (1000 Darwins refers to an 0.1% change per generation, which, over 1000 generations, will result in significant changes). The magnitude of change documented interesting behaviors but by itself was not the only estimate of population manipulation derived [3, 5].

Perhaps more important than magnitude of change was determination of whether human selection intensity could be referenced in terms of effective population size. Domesticating selection can be understood as the degree to which a desirable allele becomes more abundant than the less desirable allele during successive generations. The more intense the domesticating selection, the more common the desirable allele becomes in relation to the less desirable allele. This type of selection can be effective if the population that remains for planting from one generation to the next is small and overrepresents the desirable allele. Estimates of the size of the domesticating bottleneck were derived for estimating whether genetic drift or natural selection was the decisive factor. These estimates were modeled by Lande [12]. We used them to evaluate whether genetic drift was an important factor in domesticating maize.

DOMESTICATION AND AGRICULTURE

It is important to make the distinction between domestication and agriculture when describing the evolutionary changes in *Zea mays* over the past 6000 years. Domestication refers to the genetic modification of teosinte, which rendered it dependent on humans for dispersal. Cereals all tend to follow a similar pattern of domestication in that in most or all species, the dispersal mechanism was the first trait modified by humans [9]. Continued human selection followed by focusing on growth habit, harvestability, competitive ability, production or yield, and edibility. The nonshattering rachis of all three cobs from Guila Naquitz indicates that *Zea* was already dependent on humans for dispersal; hence, it can appropriately be referred to as domesticated [2, 4]. The single genetic change in *Tga1* that made the grain easier to harvest is another such domestication gene, although when this mutation appeared has not yet been determined from archaeological research. Its phenotype confronts significantly greater challenges dispersing its seed than that of its wild ancestor because the grain is unprotected [18]. Other morphological changes in the ear and cob, such as an increase in grain size and larger ears, followed after dependency on humans was well established. Establishment of these morphological modifications is associated with an adaptation to seed production or competition associated with seedling competition [9]. Morphological variation of grain, cob, and ear are expected to manifest itself once agriculture is clearly established, because early cultivators were involved in mass selection of open-pollinated populations when saving seed from harvest to planting. Directional selection would be expected to arise among agricultural populations that are separated by reproductive barriers when cultivators isolate populations and impose limits on the range of morphological variation they are willing to save and perpetuate from year to year.

Maize shows considerable morphological variation between populations once agriculture selection was firmly established as well. Such variation in cob and rachis morphology is widely apparent in the Americas and has been widely used to describe distinct extant landraces in the Americas [15]. Human selection of cultigens, domesticated species, or races, focused on variable traits that are subject to human preference and environmental factors that impose constraints on plant development. Differences in ear or grain size can, thus, be expected to result from human desire and environmental conditions. One trend apparent to Mangelsdorf, MacNeish, and Galinat in their study of the Tehuacan maize collections is that ear and grain size increased from the time "wild" maize appeared to the time when racial variation was evident in uppermost stratigraphic layers in the assemblages [14, pp. 147–164].

We examine the maize assemblage from the East Niche of the El Riego locality to ascertain whether human selection or environmental variation is responsible for morphological changes over time. If morphological, specifically climate, variation is caused by human selection, evolutionary change should not vary with proxy measures for environmental, specifically climate, variation. Tests of this hypothesis are made using AMS dating, **stable isotopes,** and corn cob morphology. Dating individual cobs provides a temporal framework for describing changes in cob morphology and variation $^{13}C/^{12}C$, D/H, and $^{18}O/^{16}O$ istope ratios as possibly related to environmental changes involving soil moisture, relative humidity, and temperature (see [20]). Morphological variation left unexplained by variation in these stable isotopes related to climate can be attributed to human selection.

METHODS

Study of the El Riego collections was initiated because previous analyses [13] indicated that the site offered a long, relatively uninterrupted stratigraphic sequence of human occupations, many of which were laid down before maize was domesticated. More importantly, upper strata of the East Niche contained an abundance of relatively recent postcontact evidence of human occupation that included abundant maize cobs. The fact that no radiocarbon determinations were previously obtained by MacNeish [13] for the El Riego locality suggested that development of an absolute chronology would be a worthy undertaking. It also important to realize that MacNeish and Garcia Cook's [13] analysis of the El Riego niches focused on the stratigraphic zones (Tc35w–West Niche) or layers (Tc35e–East Niche) as distinct archaeological components; each of which was taken to represent occupation by a single cultural group for varying spans of time. These discrete cultural units promised great utility for describing how maize evolved under agricultural selection.

Because maize remains were so abundant in the rockshelters, we specifically sought excavation units with long, successive stratigraphic sequences for study. Excavation descriptions indicate distinct strata were recognized in the field and excavation proceeded in an orderly fashion, which ensured re-collection of materials from each square in a top to bottom fashion that focused on well-defined stratigraphic deposits or layers [13, pp. 20–21]. The excavators' field catalog indicates excavation proceeded in arbitrary levels, whereas published accounts indicate natural stratigraphy was followed during excavation. The number of excavated levels listed in the field catalog and the published stratigraphic profiles [13, p. 18] suggest that arbitrary levels of 20 cm were used where natural stratum thickness exceeded 20 cm. There appears to be close correspondence between the number of excavation levels and stratigraphic zones or layers in the East Niche. Only 25% of excavation units had more than seven excavation levels. As a result we selected maize for study only from excavation units with nine excavation levels that also contained the greatest and most uniform number of corn cobs per level throughout the stratigraphic profile.

Analytical priority was given to two excavation units. S2E3 (catalog no. 10) and N2E1 (catalog no. 32) provided maize remains for morphological study and radiocarbon and stable isotope analyses because both units satisfied the aforementioned selection criteria. Specimens from all layers within an excavation unit were arranged on the tabletop before analysis and inspected to determine whether fragments might be refitted. All specimens in each layer were photographed and labeled before analysis. All specimens in each layer were measured using a vernier caliper and a

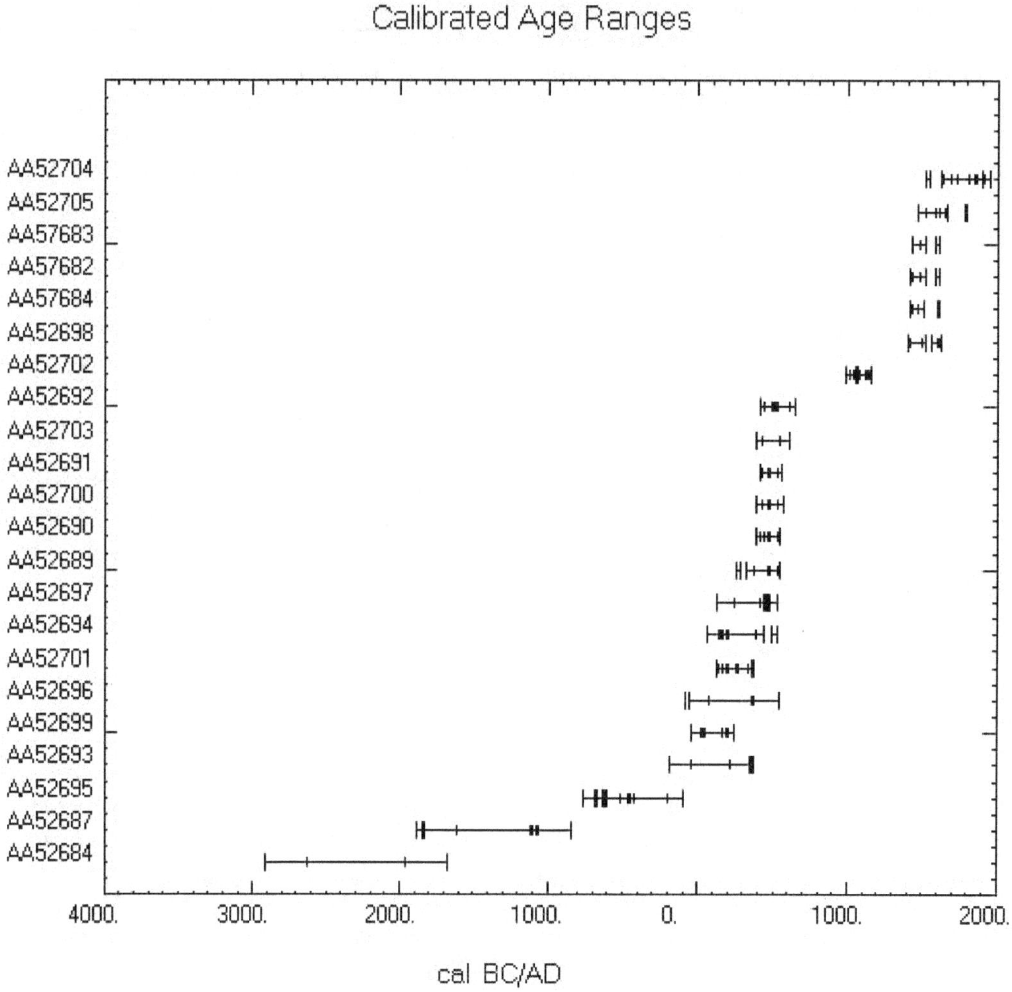

FIGURE 7-1 Calibrated radiocarbon dates from East Niche of the El Riego Oasis.

binocular dissecting microscope equipped with an ocular micrometer. Morphological measurements were obtained as described [4]. Measurements of rachis size attest to varying amounts of productive surface area of the cob. Kernel or grain size is inferred from rachis dimensions. Cupule width is a proxy for kernel width, whereas rachid length measures kernel thickness. Row number varies inversely with cupule width but accounts for variation in kernel width dimensions only when rachis diameter does not vary.

The cost of radiocarbon dating, stable isotope determination, and DNA analysis dictated that only three specimens per level or zone could be sampled. The small size of many cobs also dictated that no more than 20 milligrams could be taken from each cob. The three largest specimens in each level were selected for radiocarbon and stable isotope analysis. Stable isotopes composition was measured at the Laboratory of Isotope Geochemistry, University of Arizona, using established protocols.

$$\delta \text{ (in ‰ units)} = [(H/L)_{sample} / (H/L)_{standard}] - 1 \times 1000$$

where H/L refers to the ratio of heavy to light isotope (i.e., $^{13}C/^{12}C$ for $\delta^{13}C$, $^{18}O/^{16}O$ for $\delta^{18}O$, and D/H for δD). Holocellulose was extracted from samples using sodium chlorite for delignification in a modified Jayme-Wise process [after 19] and used directly for $\delta^{13}C$ and $\delta^{18}O$ analysis. For δD, the holocellulose was nitrated to remove exchangeable hydrogen atoms [22] before analysis. Radiocarbon dating was initiated by combusting samples to CO_2 and reducing them to graphite [21] for accelerator targets used in age determinations NSF-University of Arizona AMS facility.

Rates of morphological change were calculated from morphological measurements of individually dated specimens, and when possible, averages of contemporaneous corn populations. Rate determination and scaling were performed according to published methods [6–8, 10, 12].

Calibration of AMS dates and testing for contemporaneity were accomplished with CALIB 5.0 using the decadal northern atmospheric calibration dataset [16].

TABLE 7-1 El Riego AMS Dates on Maize Cobs Used to Establish Antiquity and to Test for Contemporaneity and Averaging. One and two sigma ranges were calibrated with CALIB 5.0 [16]. Contemporaneity is indicated by the letters in column at right. Averages calculated according to Reimer and Stuiver's [16] recommendation.

Laboratory number	Site/Catalog no.-level	^{14}C radiocarbon age	1 s range calibration BP	2 s range calibration BP	Average calibrated date group[a]
AA52704	Tc35e 10-2a	195 ± 61	128–233	54–317	i
AA52705	Tc35e 10-2b	294 ± 41	358–430	385–472	h
AA52706	Tc35e 10-2e	135 ± 41	68–118	169–282	i
AA52701	Tc35e 10-3a	1777 ± 45	1688–1740	1592–1820	d
AA52702	Tc35e 10-3b	986 ± 36	902–936	891–961	f
AA52703	Tc35e 10-3c	1562 ± 55	1404–1520	1339–1560	e
AA52698	Tc35e 10-4a	429 ± 48	457–526	426–538	g
AA52699	Tc35e 10-4b	1900 ± 62	1776–1900	1706–1988	d
AA52700	Tc35e 10-4e	1585 ± 45	1415–1473	1373–1563	e
AA52694	Tc35e 10-5a	1759 ± 89	1564–1741	1512–1884	d
AA52695	Tc35e 10-5b	2290 ± 100	2147–2367	2042–2545	e
AA52696	Tc35e 10-5c	1800 ± 130	1591–1869	1407–2002	d
AA52697	Tc35e 10-5e	1709 ± 72	1540–1703	1508–1816	d
AA52690	Tc35e 10-6a	1598 ± 35	1417–1468	1405–1556	e
AA52691	Tc35e 10-6e	1582 ± 35	1416–1469	1394–1541	e
AA52692	Tc35e 10-6f	1520 ± 62	1345–1421	1306–1529	e
AA52693	Tc35e 10-6g	1920 ± 110	1726–1989	1601–2134	d
AA52687	Tc35e 10-7b	3107 ± 220	3057–3560	2782–3781	b
AA52689	Tc35e 10-7c	1630 ± 49	1485–1568	1404–1627	e
AA52685	Tc35e 10-8f	3210 ± 490	2837–3994	2310–4729	b
AA52684	Tc35e 10-9b	3850 ± 240	3906–4569	3617–4864	a
AA62686	Tc35e 32-7a	1805 ± 39	1703–1813	1685–1826	d
AA62685	Tc35e 32-8a	1625 ± 39	2147–2367	2024–2545	e
AA62684	Tc35e 32-8c	1845 ± 39	1725–1823	1699–1876	d

[a]Averaged dates: a: 4240 BP, b: 3120 BP, c: 2290 BP, d: 1810 BP, e: 1590 BP, f: 920 BP, g: 480 BP, h: 380 BP, i: 150 BP.

RESULTS: CALIBRATING AND AVERAGING AMS DATES

Radiocarbon dates were obtained for 24 maize cobs from two excavation units in 9 separate stratigraphic layers. More than 30 samples were submitted for AMS dating, but samples from the 2 lowest layers yielded insufficient carbon to obtain a date. The cobs in these two lower strata—layers 8 and 9—were spongy and brittle. The cupules of these specimens were filled with a dry, white mycelium suggesting that they were water-logged while they laid buried in the deposits overlying the travertine floor [13, p. 17]. Dates range from 3850 ± 250 to 191 ± 61 14C years BP (all dates hereafter refer to calibrated dates before the present). Dates generally follow the expected stratigraphic sequence, with recent dates in the upper strata, and the oldest dates occurring at the bottom (Table 7-1). However, considerable stratigraphic mixing (movement of older specimens up and later specimens down) through the profile has occurred. Although there should be little doubt that the excavators accurately recognized and excavated distinct strata, one is left wondering whether such bidirectional migration of plant material throughout the profile indicates that such movement would have likewise displaced other cultural materials between stratigraphic zones. Although this indicates that the reconstruction of cultural activities and cultural historical reconstructions hypothesized by MacNeish and colleagues resides on relative dating methods, such as seriation and intersite comparisons of assemblages, the AMS dates provide us with an absolute chronological basis of morphological comparison largely independent of stratigraphic mixing.

Calculation of contemporaneity involves statistical testing to determine whether sample ages are estimating the same date [16]. Our approach assumes that corn cobs with contemporary sample ages belong to the same cultivated crop population. Such testing led to recognition of 9 sample ages for the 24 dates, 3 of which result from sample age averages (Table 7-1). The nine sample ages identify distinct depositional periods that roughly approximate the following archaeological phase assignments [11, p. 5] for the Tehuacan Valley: a) Purron; b) Ajalpan; c) and d) Sta. Maria; e) Palo Blanco; f), g), and h) Venta Salada; and i) an early nineteenth-century occupation. These averaged sample ages do not coincide exactly with the phase assignments made on the basis of artifact content [13, Table 1] nor do the non-averaged ages. Specifically, these AMS dates indicate MacNeish and colleagues failed or were unable to recognize the Middle and Late Formative occupations of the East Niche. Nevertheless, they did correctly identify two of three upper zones with phases representative of the Classic and Post-Classic periods in Tehuacan. Based on these dates, the East Niche was most intensively occupied during two or three separate occasions during the Sta. Maria phase (Classic) and again during the Venta Salada (Post-Classic) phase. Overall, the time span represented by the maize dates extends occupation of the East Niche of the El Riego Oasis more than 2000 years earlier than previous estimates.

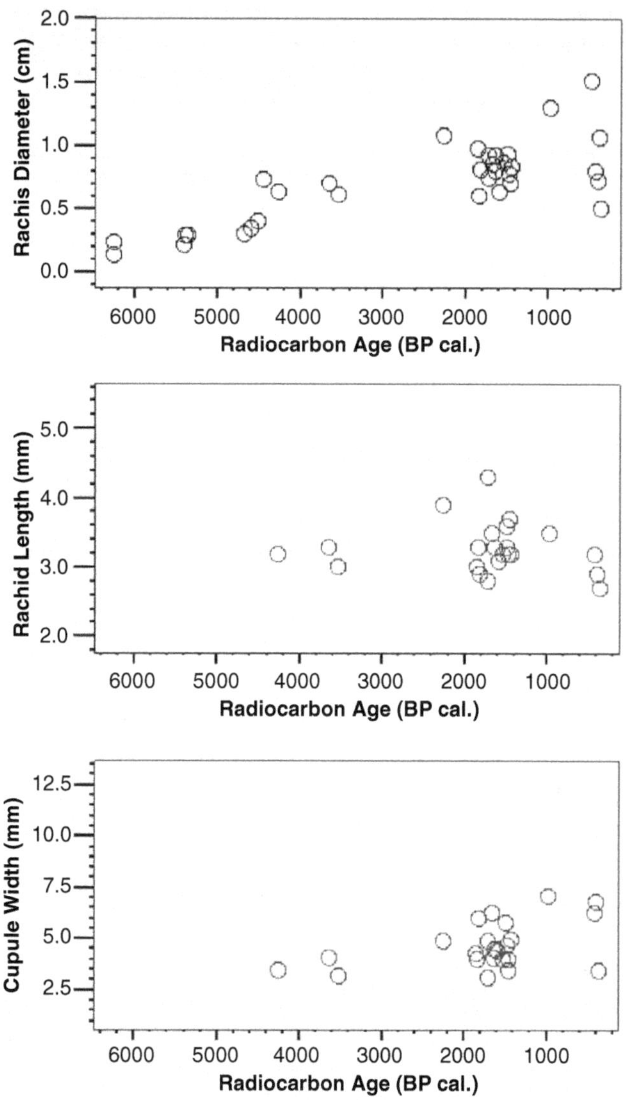

FIGURE 7-2 Morphological attributes of dated maize from El Riego East Niche (Tc 35e) in the Tehuacan Valley. Cupule width is correlated ($r^2 = 0.22$; $p < 0.06$) with radiocarbon age. Row Number (not shown) is inversely correlated ($r^2 = 0.41$; $p < 0.05$) with cupule width.

RESULTS: MORPHOLOGICAL TRENDS AND RATES

The corn cobs from Tc35e demonstrate increasing grain width through time. Cupule width increases only slightly during the first 1000 years of the chronological sequence represented in the East Niche. The subsequent Sta. Maria phase witnesses a dramatic increase in grain width (Figures 7-2 and 7-3). This rise is followed by a period of increasing row numbers (not shown) and concomitant decrease in kernel width until the middle Palo Blanco Phase (Classic Period). Grain size increases dramatically again during the Venta Salada phase (Early Post-Classic) except for a brief decrease in size around 480 years ago. The width of the cupule increases again at the end of the sequence.

Rachid length or grain thickness varies little over the 4000 years at El Riego (see Figures 7-2 and 7-3). Grain thickness remains static during the first 1000 years then increases dramatically during the period between 3120 and 1590 BP Rachid length varies considerably during the succeeding Classic, Post-Classic, and Colonial periods.

Like cupule width, rachis diameter changes little during the Middle to Late Formative transition. The diameter of the rachis increases steadily through the 3120 to 1590 BP period. There appears to have been a rapid increase in diameter until 980 BP after which the rachis narrows through the Post-Classic and Colonial periods.

Combining the three already described dimensions of the maize cob from the East Niche leads us to suggest that little change in ear productivity occurred during the 4200 to 3100 BP time period. Major changes in productivity are first evident around 2290 BP and stabilize or slightly decrease during the succeeding 700 years until about 1600 BP. The latter part of the sequence suggests selection took a variety of directions, some of which may have involved increasing grain number by reducing kernel thickness and increasing row number, that is, toward an ear with many rows of thin narrow kernels.

RESULTS: EVOLUTIONARY RATES

Rates of morphological change in cobs from the East Niche are greater overall than they were from earlier time periods (see Table 7-2). [5]. The highest rates appear late in

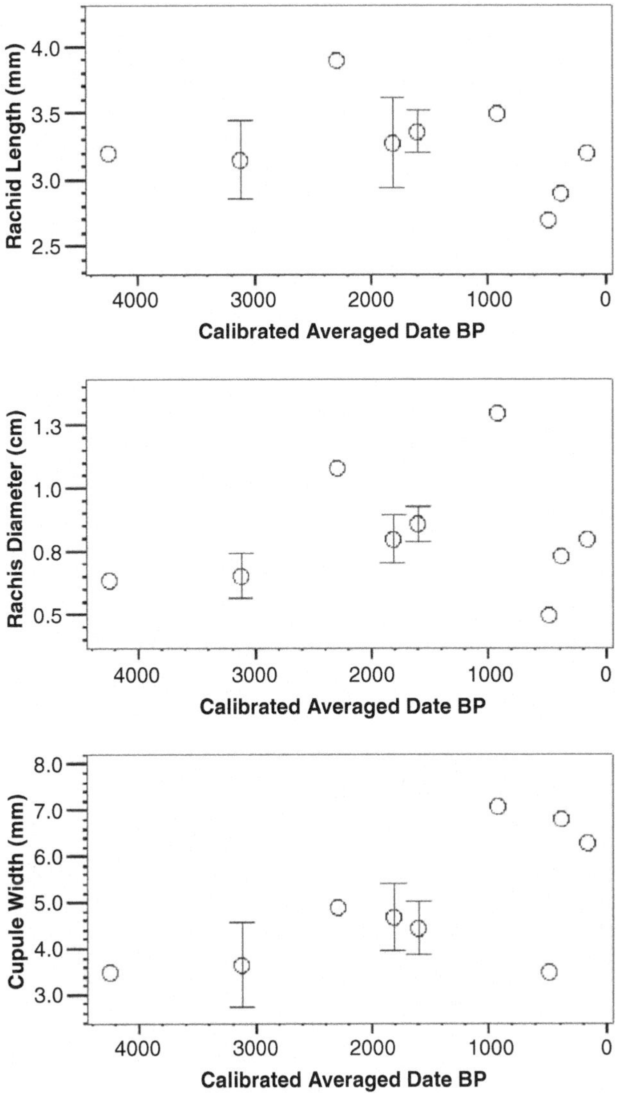

FIGURE 7-3 Three morphological attributes of dated maize from El Riego East Niche in the Tehuacan Valley. Cupule width and kernel row number are strongly correlated with calibrated averaged dates ($r^2 = 0.22$; $p < 0.05$ and $r^2 = 0.26$; $p < 0.05$ respectively). Ears apparently decreased in row number as kernel width increased over time. Error bars represent two standard errors.

the sequence, roughly around 920 BP. They remain high until 480 BP, when they fall to levels reminiscent of the time period between 1590 and 920 BP.

Scaled evolutionary rates also exceed most rates from earlier time period except for the first transitions described for the San Marcos maize assemblage (Figure 5-4; cf. Benz and Long [5], Figure 1). The changes attested to by these rates suggest morphological changes on the order of 0.08 to 0.3% per year or perhaps slightly greater. The latter such change would be perceptible by cultivators during their lifetimes. Overall individual rates of change when scaled according to the time interval are not vastly different from one another (Figure 7-5). Rates of change for three periods are remarkable. Morphological rates for the earliest transition (4240 to 3120 BP) appear to be less than those for the 920 to 480, the 48 to 320 BP transitions as well as those describing the amount of change required to create either of two extant races from the last archaeological populations existing 150 years ago. The apparent difference in evolutionary rates between the Formative and Post-Classic periods suggest that much greater interest and investment was made to modify maize's productive capacity during much later times than during the Middle to Late Formative at El Riego. Even though there is a little perceptible change in evolutionary rates from the Formative up to the Post-Classic, rates increase steadily during this time period. This steady rise suggests that gradual intensification was occurring around ear selection throughout the sequence at El Riego.

RESULTS: STABLE ISOTOPE DETERMINATIONS

$\delta^{13}C$ determinations were obtained for more samples than could be dated. Only dated determinations, 21 in all, are included here for consideration. Fewer $\delta^{18}O$ and δD determinations could be obtained from the maize samples, particularly the earliest samples. Although whole-tissue sample weights were low but well within the range commonly used, sample quality (e.g., insufficient holocellulose content) is probably the reason why no result was forthcoming for the earliest samples because other samples analyzed from the same strata did not yield AMS dates either.

One $\delta^{13}C$ observation, from a corn cob obtained from the basal-most stratum of the East Niche, yielded a significantly lower value than any other samples (Figure 5-6). This sample's content is believed to be aberrant because no other samples submitted from this stratum have yielded AMS dates or stable isotope results to date. Apart from this observation, remaining samples are virtually invariant over the subsequent 3000 year time frame.

$\delta^{18}O$ determinations from El Riego show considerable variation over time (see Figure 5-6). Values obtained from samples dating around 3120 BP are most varied and cannot be distinguished from values obtained from samples from later time periods. The lack of systematic temporal variation suggests that water availability for maize cultivation did not change over the period of time represented by these samples.

δD determinations were difficult to obtain for many of the samples. The single sample dating around 4300 BP appears to be different than samples dating around 1600 BP. Otherwise values show little systematic temporal variation (see Figure 7-6).

Variability of $\delta^{18}O$ and δD are surprising because we anticipated they would yield evidence of climatic variation. Such variation is not apparent in this set of samples. Addi-

TABLE 7-2 Rates of Change in Darwins for Four Morphological Traits of Corn Cobs from the East Niche of El Riego, Tehuacan, and Two Morphologically Distinctive Extant Races from Central and Western Mexico

Transitions (from Table 5-1)	Duration in calendar years	Rachis diameter	Row number	Cupule width	Rachid length
a–b	1120	80	119	141	27
b–c	830	522	142	215	201
c–d	480	625	888	91	361
d–e	220	328	508	249	110
e–f	670	617	135	701	61
f–g	440	2171	350	1607	589
g–h	100	3784	3365	6642	715
h–i	230	398	0	332	428
i-Arrocillo	150	559	1421	3197	386
i-Tabloncillo	150	2062	3530	3000	1520
a–i	4090	55	82	144	0

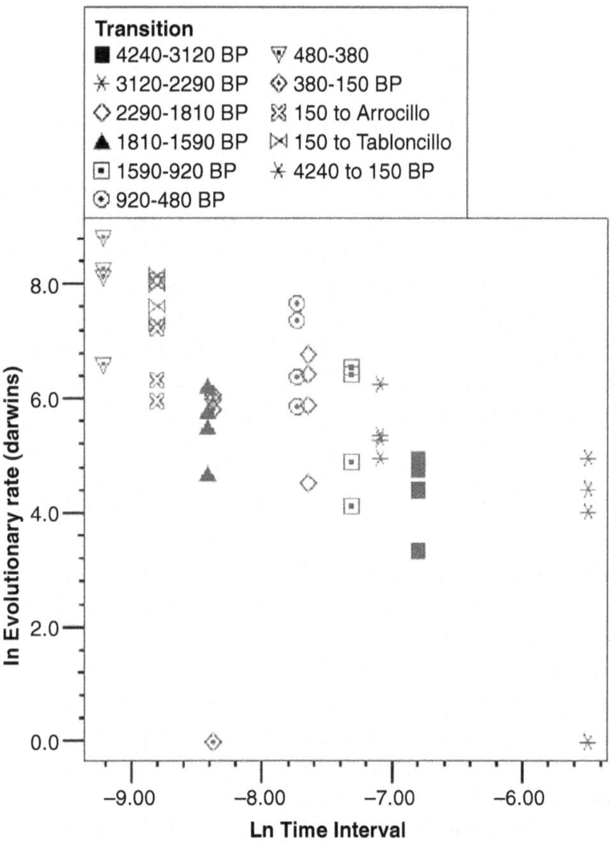

FIGURE 7-4 Scaled evolutionary rates for four morphological attributes of maize over 11 dated transitions from the East Niche from El Riego Oasis, Tehuacan. Time interval on x-axis shows transitions with shorter durations on the left while longer transitions are on the right. Ln Time Interval refers to time in millions of years (1000 years is 1×10^{-6} or –6.0, 100 years is 1×10^{-9}). Rates are scaled to correct for longer transition durations that appear to have lower rates and to facilitate comparison. Rates of six over a 1000-year transition can produce dramatic evolutionary change.

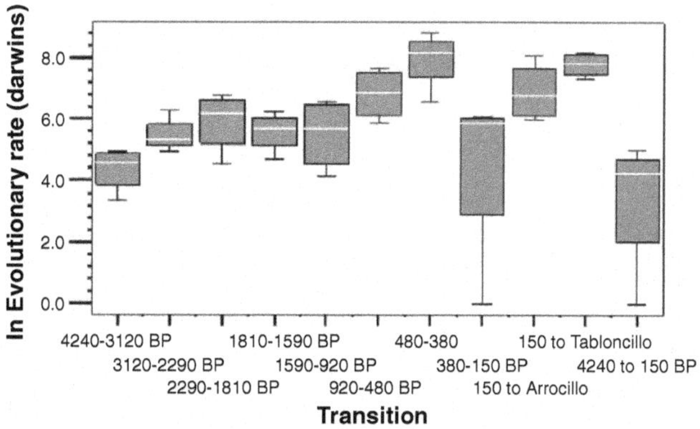

FIGURE 7-5 Box plot of evolutionary rates by transition. Each box represents the median rate (white horizontal line) in scaled Darwins and interquartile ranges. By the Middle Formative, evolutionary rate had increased to the point where morphological change averaged around 0.02% per year. This rate of change remained constant until the mid-Classic when it began to rise significantly achieving a 0.1% per year increase and then a 0.3% increase around the time of Spanish contact. The overall rate of change in corn cob size during the 4200 years represented by the El Riego sequence appears to be less than or equal to 0.02% per year. This of course averages all increase and decrease in morphological change over time. Rate of change increases significantly from 4240 through 150 BP ($r^2 = 0.47$; $F = 8.8$; $p < 0.05$, and $r^2 = 0.40$; $F = 6.7$; $p < 0.05$ when rates and ln rates were transformed to ranks).

tional samples are currently undergoing analysis that we hope adds to current evidence that permit us to address hypotheses of climate-induced environmental change in Tehuacan.

DISCUSSION AND SUMMARY

This study assumes that there is genetic continuity of maize populations through time. By this we mean that vari-

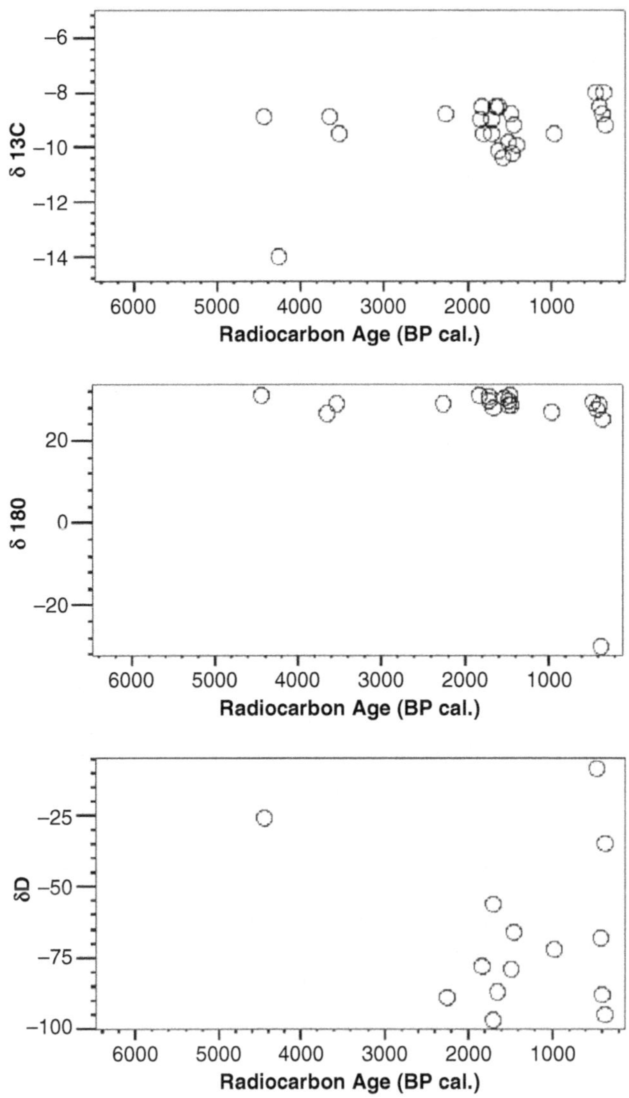

FIGURE 7-6 $\delta^{13}C$, $\delta^{18}O$, and δD results obtained from dated maize cobs excavated from the East Nishe of El Riego Oasis.

eties cultivated during the Palo Blanco Phase are assumed to be descendants of the populations cultivated during the Purron Phase. This assumption facilitates summary because it enables interpretation that the changes in maize morphology cultivated over the course of human occupation of the El Riego Oasis East Niche is because of either human selection or environmental variation over the past 4500 years. Although the inability to make this assumption does not necessarily negate interpretations that human selection alone or in part explains changing rates in maize ear evolution, it permits us to infer that changes in human population behaviors associated with varying intensities of human selection are the result of local constraints on maize production, not the result of introductions of foreign germplasm. Although this latter possible explanation is worth considering, data currently available do not permit us to test its worthiness, although appropriate data will be available in the future.

Another assumption this analysis relies on is that a small number of individual corn cobs accurately reflect the morphology and changes in shape and size of the populations they represent. Although it is always difficult to accept a single individual as representative, we take solace in knowing that all dated specimens represent the largest of all cobs found in each of the strata. Nevertheless, there is always chance for bias in relying on small samples, so, as always, our interpretations should be viewed as working hypotheses.

The presence of maize cultivators occupying the El Riego Oasis during the Purron and Ajalpan phases occupation in the Tehuacan Valley is not surprising, although it was previously undocumented. This is probably the result of stratigraphic mixing that introduced younger materials into older strata. Maize was under cultivation at least 1500 years before in other parts of the valley where soil moisture was less abundant. Preservation of material in the lower levels of the East Niche complicated radiometric and stable isotope determinations of material in the lowermost strata, although we suspect that future analyses will yield specimens of contemporaneous ages.

In addition to the maize found in Purron Phase deposits, it was also subsequently deposited in the East Niche on at least eight other occasions. Another Formative Period occupation (Ajalpan Phase) left a large number of corncobs whose morphology is similar to that left by previous occupants. The occupation following the Ajalpan, dating to the last centuries of the Late Formative, shows significant changes in ear morphology from the first two. Maize apparently changes character during the ensuing occupations, whose material date around 1800 and 1600 BP and produced ears with thin kernels and narrow rachises. These occupations, during the Palo Blanco phase, appear to have had the greatest archaeological effect on the site because corncobs of these ages are distributed more widely throughout the stratigraphic profile than any others. Four subsequent occupations of different ages are evident in the dated material from the East Niche. Each is denoted by a single dated corncob and show widely divergent morphologies. Trends are suggested but must be considered cautiously in light of limited data.

Rates of morphological change during the Formative, Classic, and Post-Classic phases recognized in Tehuacan are generally greater than or equal to similar rates calculated for Late Archaic through Early Formative maize from San Marcos and Coxcatlan. The rates of morphological change, if single ears are representative, steadily increase beginning with the Formative occupations. This steady increase documents an overall increase in rachis diameter (see Figure 7-3) through the beginning of the Post-Classic. Human selection intensities appear to have increased, with a focus on increasing the productive capacity of the ear. It is likely that cultivators noted a significant increase in ear size during their own lifetimes.

Evidence for environmental effects on the variation in maize ear size and shape and the increasing rates of morphological change through time is lacking. Systematic temporal variation in stable isotopes of carbon, oxygen, and hydrogen is lacking from the corn cobs in the East Niche of the El Riego Oasis.

Acknowledgments

This research was supported by NSF grant BSC-91490. We gratefully acknowledge the Consejo de Arqueología, of the Mexican Instituto Nacional de Antropología e Historia for permission to study the collections described herein and to A. Montufar and R. Zavala for making the collections available for study. D. Webster made useful comments on an earlier draft for which we are grateful.

References Cited

1. B. Benz. (1999). On the origin, evolution and dispersal of maize, In: M. Blake, (Ed.), *Development of agriculture and emergence of formative civilizations in Pacific Central and South America. The Prehistory of the Pacific Basin.* Seattle, WA: WSU Press. pp. 25–38.
2. B. Benz. (2001). Archaeological evidence of teosinte somestication from Guilá Naquitz, Oaxaca. *Proceedings of the National Academy of Science of the United States of America, 98,* 2104–2106.
3. B. Benz. (In press). Evidencia arqueológica sobre la evolución del maíz, In: A. Casas, B. Rendón, (Coords.), *Procesos de evolución de plantas bajo domesticación en Mesoamérica.*
4. B. Benz, H. H. Iltis. (1990). Studies in archaeological maize I: The "wild" maize from San Marcos cave reexamined. *American Antiquity, 55,* 500–511.
5. B. Benz, A. Long. (2000). Early evolution of maize in the Tehuacan Valley, Mexico. *Current Anthropology, 41,* 459–465.
6. P. Gingerich. (1983). Rates of evolution: Effects of time and temporal scaling. *Science, 222,* 159–161.
7. P. Gingerich. (1984). Reply. *Science, 226,* 995–996.
8. J. Haldane. (1949). Suggestions as to quantitative measurement of rates of evolution. *Evolution 3,* 51–56.
9. J. Harlan, J. DeWet, E. Price. (1973). Comparative evolution of cereals. *Evolution, 22,* 311–325.
10. A. Hendry, M. Kinnison. (1999). The pace of life: Measuring rates of contemporary microevolution. *Evolution, 53,* 1637–1653.
11. F. Johnson, R. MacNeish. (1972). Chronometric dating. In: F. Johnson, (Ed.), *The prehistory of the Tehuacan Valley,* Vol. 4. Austin: University of Texas Press. pp. 3–58.
12. R. Lande. (1976). Natural selection and random genetic drift in phenotypic evolution. *Evolution, 30,* 314–334.
13. R. MacNeish, A. Garcia Cook. (1972). Excavations in the locality of the El Riego Oasis. In: R. MacNeish, M. Fowler, A. Garcia Cook, F. Peterson, A. Nelken-Terner, J. Neely, (Eds.), *The prehistory of the Tehaucan Valley,* Vol. 5. Austin: University of Texas Press. pp. 14–65.
14. P. C. Mangelsdorf. (1974). *Corn: Its origin, evolution and improvement.* Cambridge, MA: Belknap Press of Harvard University Press.
15. H. Perales, B. Benz, S. Brush. (2005). Maize diversity and ethnolinguistic diversity in Chiapas, Mexico. *Proceedings of the National Academy of Sciences of the United States of America, 102,* 940–954.
16. M. Stuiver, P. Reimer. (1993). Extended 14C database and revised CALIB radiocarbon calibration program. *Radiocarbon, 35,* 215–230.
17. B. Smith. (2005). Reassessing Coxcatlan cave and the early history of domesticated plants in Mesoamerica. *Proceedings of the National Academy of Sciences of the United States of America, 102,* 9438–9944.
18. H. Wang, T. Nussbaum-Wagler, B. Li, Q. Zhao, Y. Vigouroux, M. Faller, K. Bomblies, L. Lukens, J. Doebley. (2005). The origin of the naked grains of maize. *Nature, 436,* 714–719.
19. S. W. Leavitt, S. R. Danzer. (1993). Method for batch processing small wood samples to holocellulose for stable-carbon isotope analysis. *Analytical Chemistry, 65,* 87–89.
20. D. McCarroll, N. J. Loader. (2004). Stable isotopes in tree rings. *Quaternary Science Reviews, 23,* 771–801.
21. P. J. Slota, Jr, A. J. T. Jull, T. W. Linick, L. J. Toolin. (1987). Preparation of small samples for ^{14}C accelerator targets by catalytic reduction of CO_2. *Radiocarbon, 29,* 303–306.
22. L. S. L. Sternberg. (1989). Oxygen and hydrogen isotope measurements in plant cellulose analysis. H. F. Linksens, J. F. Jackson, (Eds.), *Plant fibres: Modern methods of plant analysis V. 10.* New York: Springer-Verlag. pp. 88–99.

CHAPTER

8

The Maize Revolution

A View from El Salvador

ROBERT A. DULL

Department of Geography and the Environment, University of Texas at Austin, Austin, Texas

Glossary

Teosinte A tropical grass from the genus *Zea*, which is the presumed progenitor of domesticated maize *(Zea mays* ssp. *mays)*.

Phytoliths Microscopic siliceous bodies that form in cell walls of some terrestrial plants.

Plant macrofossils Plant fossils that are large enough to be visible without magnification.

Fossil pollen analysis The study of morphologically distinct microscopic pollen grains to characterize past vegetation assemblages.

Milpa A traditional Mesoamerican agricultural field, most typically planted with maize *(Zea mays* ssp. *mays)*.

CAL BP Calendar years before present. This is the standard age reporting format in the natural–physical sciences and for most cross-disciplinary research.

The Middle American isthmus has been occupied by people since the late Pleistocene. The earliest human inhabitants of this region were nomadic hunter–gatherers who traveled in small bands, seasonally exploiting the resources from a wide range of microenvironments. By approximately 5000 Calendar years BP (ca. 3000 BC), incipient agriculture allowed people throughout Middle America to begin relying less on wild foods. Sedentary living became commonplace for the first time as subsistence economies gradually shifted from those based on hunting and foraging to those centered on food production. Crucial to this shift was the domestication and spread of maize *(Zea mays* ssp. *mays)*. The modern-day country of El Salvador is located approximately 1200 kilometers southeast of the Balsas River basin, the region of Mexico where maize was most likely first domesticated. This chapter reviews the microfossil and macrofossil evidence for the arrival, spread, and cultivation of the genus *Zea* in El Salvador.

INTRODUCTION

The history and geography of maize cultivation in the Americas has benefited greatly from research developments achieved over the past several years. Although a lively debate continues regarding the chronology and geography of maize domestication [33, 36, 39, 40, 51, 52], some general conclusions can now be drawn from the data that are available. There is indisputable AMS-dated macrofossil evidence for maize in Mexico by ca. 6250 **CAL BP** [34]. A case has been made that maize was present as early as 7000 BP in the lowland tropics of Panama and Ecuador [35]; this thesis, however, is not currently widely accepted due to questions regarding maize phytolith taxonomy [13, 39, 40, 51], incongruous radiocarbon dating results [18, 48], and a complete lack of corroborating pollen or plant macrofossil evidence.

The most widely accepted paradigm for the geography of maize domestication derives from studies reconstructing racial phylogeny using plant morphological and genetic evidence [7, 12, 24, 25, 31]. These studies have demonstrated that the closest living ancestor to domesticated maize is the Balsas **Teosinte** *(Zea mays* ssp. *parviglumis)*, a *Zea mays* subspecies found in the Rio Balsas drainage basin in the southern Mexican state of Guerrero. Indeed this theory is most consistent with the *Zea* pollen and macrofossil evidence now available. Matsuoka and colleagues [31] used

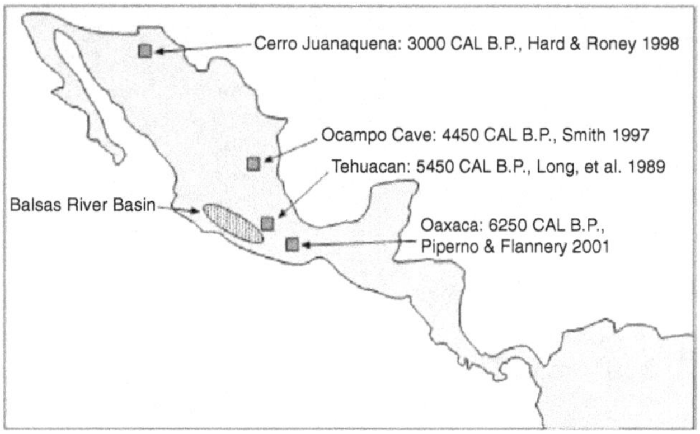

FIGURE 8-1 Map of maize *(Zea mays* ssp. *mays)* macrofossil sites from central and northern Mexico. The Rio Balsas basin is the presumed site of the maize-from-teosinte domestication event.

known mutation rates in *Zea* to calculate a maize from teosinte divergence date of ca. 9188 CAL BP (95% confidence limits: 5689–13,093 CAL BP).

Accepting that maize was most likely domesticated in the Balsas drainage basin sometime between ca. 9200 and 6250 CAL BP (ca. 7250–4300 BC), we can trace its spread north and south using data derived both from archaeological contexts and from lake sediment cores. The spread of maize within central Mexico and north toward the U.S. border is currently much better understood than the southern spread, because the aridity of northern Mexico and the southwestern United States has facilitated the excellent preservation of maize plant macrofossils in those regions (Figure 8-1). Maize macrofossils have been dated by accelerated mass spectrometry (AMS) in the central Mexico valleys of Oaxaca (ca. 6250 CAL BP) and Tehuacan (ca. 5450 CAL BP), in the Gulf Coast region of Tamaulipas (ca. 4450 CAL BP), and in the Sonoran desert borderlands of northern Mexico (ca. 3000 CAL BP) [23, 29, 34, 50]. Collectively these dates provide a working chronology of the spread of maize north from central Mexico that is broadly accepted by the scientific community.

The spread of maize south from Mexico to Central and South America is still not as well understood as its northward dispersal, and the subject is certainly more contentious. The current debate about the spread of maize into Central and South America hinges on methodological issues regarding maize microfossils—pollen and **phytoliths.** Because environments conducive to the preservation of maize macrofossils are rare in the humid tropics of Central America, and because the utility of phytoliths for identifying and dating maize remains controversial, much of our understanding of the dispersal of maize horticulture southeast into Central America comes from fossil pollen analysis-stratigraphic pollen records extracted from wetlands (lakes and marshes). Pollen preservation in tropical lake sediments is generally excellent [9], especially for anemophilous (wind pollinated) taxa such as maize.

Our understanding of the spread of maize horticulture southeastward from Mexico has improved considerably in recent years (Figure 8-2). Pollen studies record the arrival of *Zea* (and most likely the domesticate *Zea mays* L.) on the Gulf Coast of Veracruz and Tabasco, Mexico, between 5000 and 7000 CAL BP [21, 38, 49], in the Maya lowlands of Guatemala and Belize between 5500 and 4600 CAL BP [37, 54], and in north-central Honduras by ca. 4500 CAL BP [41]. These sites, however, lie well to the north of the assumed Pacific dispersal route [43]. The only well-dated early pollen records from the Pacific coast region of Central America have come from Costa Rica, where maize pollen first appears at ca. 5500 CAL BP [2].

El Salvador is situated astride the volcanic axis of Central America on the Pacific coast just southeast of Guatemala (see (Figure 8-2). Almost nothing is known about the early and middle Holocene archaeology of the country, but several large-scale archaeological studies have been carried out on late Holocene sites (Pre-Classic through Post-Classic) in central and western El Salvador [47]. It is from these sites and from lake sediment records in the same regions that we can cull the information that allows us to begin to reconstruct the history of maize cultivation in El Salvador.

MAKING SENSE OF FOSSIL *ZEA* POLLEN FROM EL SALVADOR AND BEYOND

Tripsacum, Zea mays (teosinte), and *Zea mays* ssp. *mays* (maize) are the three New World grasses that have the largest long axis pollen diameters. There is substantial overlap in the size ranges of *Tripsacum* and teosinte pollen and in those of teosinte and maize pollen (Table 8-1). Despite common practice by some researchers, discriminating maize pollen from teosinte in the fossil record has proved impossible based on morphological characters such as exine spinule density and the annulus diameter to long axis diameter ratio [22, 48, 53]. It has been found that the long axis measurement alone is the only way to unambiguously separate maize from teosinte, and this is true exclusively in the case of large (>92 μ) maize pollen specimens [5, 48, 55].

Primitive corn might have produced pollen grains in the size range of modern teosinte [5, 18, 30]; even smaller (<50 μ) Poaceae pollen grains from middle Holocene contexts have been identified as maize [38, 42], although no modern studies have clearly demonstrated this possibility. A positive correlation between cob size and pollen size in the archaeological record may lend some support these claims of small diameter maize [20, 30], even if identifications based on spinule density and major axis–pore diameter

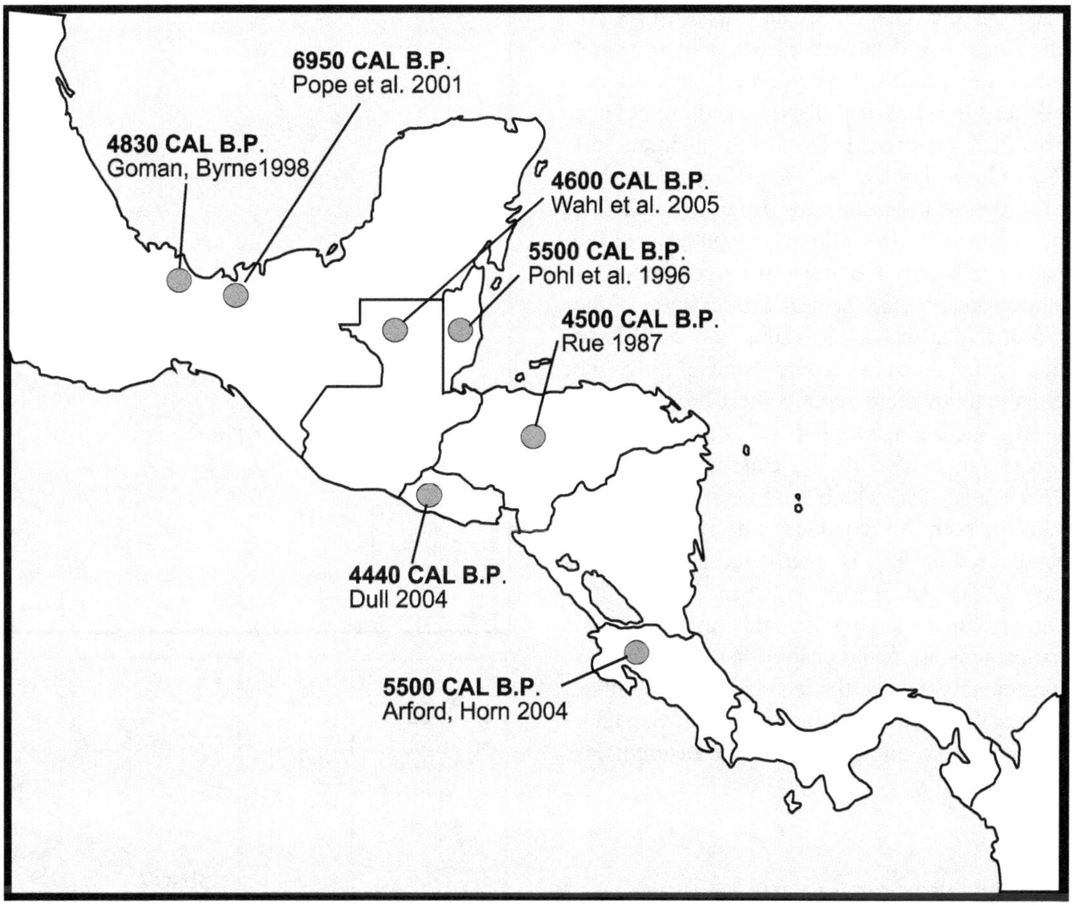

FIGURE 8-2 Earliest published claims for *Zea* pollen from sites in Mesoamerica. All are based on stratigraphic interpolation of calibrated ^{14}C dates on associated organic material. *Zea* pollen identification criteria vary among researchers, and therefore these data must be considered tentative until standardization and consensus is achieved.

TABLE 8-1 Pollen Long Axis Size Ranges for Three Neotropical Grasses

Study	Maize range (μ)	Teosinte range (μ)	*Tripsacum* range (μ)
Tsukada, 1964 [51]	85.0–135.0	60.0–92.0	40.0–70.0
Whitehead and Langham, 1965 [55]	58–98.6	46.4–87.0	33.5–56.55
Barghoorn and colleagues, 1954 [4]	[a]72–141.7 [b]64.1–130.8	[a]74.0–102.0 [b]65.9–90.8	[a]33.6–64.0 [b]29.9–57.0

[a]Uncorrected glycerin measurements.
[b]These numbers reflect a correction multiplication factor of .89 for samples mounted in glycerin as prescribed by [48].

ratios do not. However, if larger maize pollen grain diameters have evolved under the process of human selection during the middle to late Holocene, it appears that it would be impossible to distinguish maize from teosinte during earliest phase of the domestication process (e.g., before ca. 6000 BP). This does not mean that the application of stratigraphic palynology to the question of New World agricultural origins is a futile endeavor. Even if the earliest maize had pollen grain diameters in the 50–90 μ size range (that of modern teosinte), the initial occurrence of such grains together with other agricultural indicators (i.e., increases in charcoal and agricultural weeds) could signal the emergence of maize agriculture.

What is certain about the published body of literature on fossil *Zea* pollen is that there is no single standard for maize pollen identification, which is accepted and practiced by all pollen analysts. The result of these multiple *Zea* pollen identification practices is a confusing corpus of maize pollen claims, some substantiated and some not. Many researchers in the archaeological community have unwittingly embraced new claims of increased maize antiquity from pollen studies without realizing that these claims were based on questionable *Zea* pollen identification criteria.

It is my goal in this chapter to take a conservative approach to the *Zea* pollen data from El Salvador so as not to further confuse the published record. Teosinte does not occur in El Salvador today, or if it does, it has not yet been reported. However, *Z. luxurians* is found in Guatemala just west of the Salvadoran border, and *Z. nicaraguensis* has recently been discovered at coastal sites fringing the Gulf of Fonseca in Nicaragua, just a few kilometers southeast of the Salvadoran border [12, 26]. The modern biogeography of these two teosinte species—and the near extinct status of the latter—supports the idea that El Salvador may well have hosted teosinte populations at some time during the Holocene. Pollen identifications reported here are restricted solely to long axis measurements [14–16]. Those Poaceae grains in the size range 65–90 μ are classified in a size category for the genus *Zea,* which could represent either teosinte or maize or both. A second category of *Zea* grains with long axes exceeding 90 μ is interpreted as cultivated maize exclusively. Poaceae grains between 35–65 μ are placed in a category called "large grass"; this category could include *Tripsacum,* teosinte, and perhaps other tropical grass genera, but presumably not cultivated maize. Of course, some of the *Zea* pollen identified from sites in El Salvador was folded, and therefore no accurate long measurements can be provided.

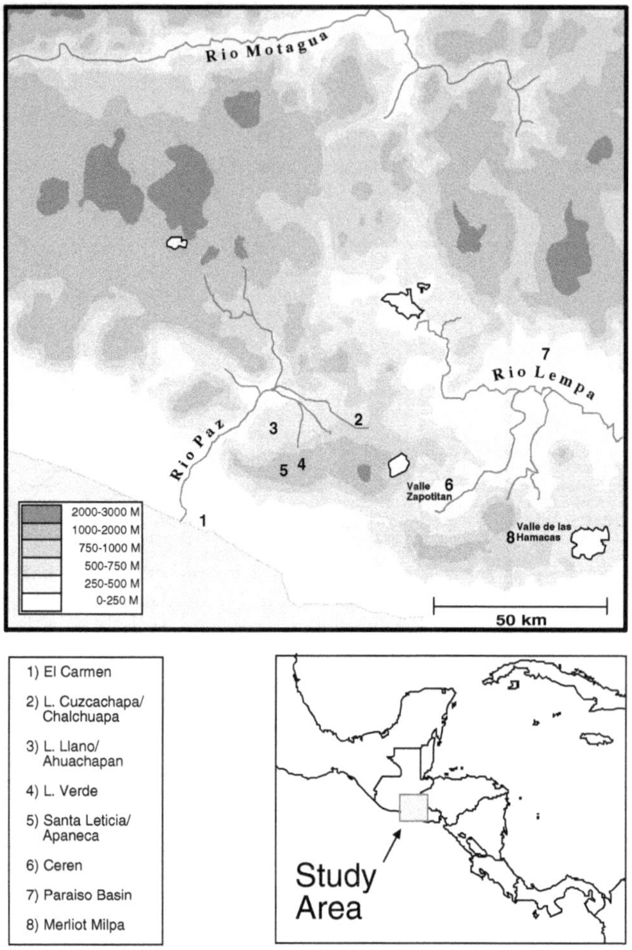

FIGURE 8-3 Salvadoran maize macrofossil and *Zea* pollen sites.

PREHISTORIC MAIZE FROM WESTERN EL SALVADOR

The earliest pollen and macrofossil evidence for *Zea* in El Salvador has been discovered in the westernmost regions of the country (Figure 8-3). For purposes of discussion the west is broken up here into three geographic zones: the Pacific plain, the Rio Paz basin, and the Sierra de Apaneca–Llamatepec highlands.

The Pacific Coastal Plain

The first evidence of pottery and sedentary living in El Salvador comes from the Pacific coast littoral site of El Carmen (see Figure 8-3). El Carmen is the easternmost locale in a string of sedentary habitation sites located along the Pacific coastal plain of southern Mexico (Chiapas), Guatemala, and El Salvador during the Early Formative Period [8]. The earliest occupation levels at El Carmen date to ca. 3660 CAL BP (3430 ± 90 ^{14}C yr. BP) [3, 8]. Corn cob fragments were found at El Carmen in the deepest deposits, representing the earliest macrofossil evidence for maize agriculture in El Salvador [3]. The Bostan ceramic phase at El Carmen predates Tok phase ceramics from Chalchuapa, and is apparently closely related to Locona-Ocós pottery of the Soconusco region [11]. The period of occupation at El Carmen apparently lasted just over 200 years [3, 11].

The Rio Paz Basin

The settlement data available indicate that the Rio Paz Valley in western El Salvador was one of the first inland valleys in El Salvador to be settled by sedentary agriculturalists [44, 45]. Situated on a natural migration route from the Pacific (see Figure 8-3), the Chalchuapa Archaeological Zone in the upper Rio Paz Basin was settled sometime before 3200 CAL BP. Chalchuapa is also the site of some of the earliest developments of social rank–complexity in the southern Maya realm. The evidence for pre-3200 CAL BP occupation at Chalchuapa consists of sparse obsidian blades and chipped stone debris, along with clay figurine and pottery fragments, which could be related to Ocos or Barra phases (ca. 3600–3200 CAL BP) on the Pacific coast [45]. Pollen data from Laguna Cuzcachapa support this specula-

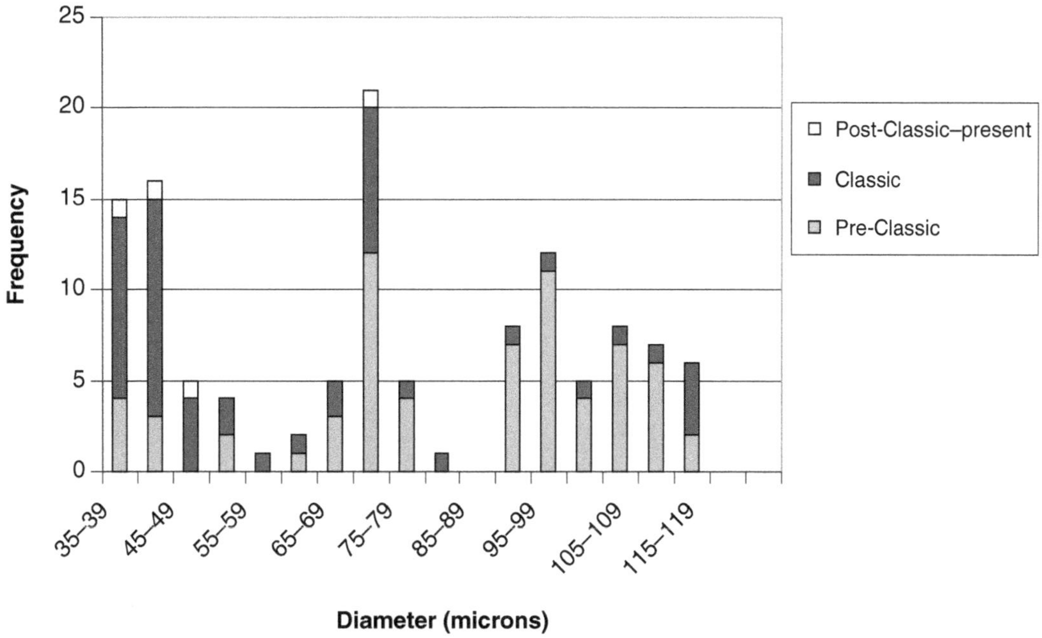

FIGURE 8-4 Poaceae pollen greater than 35 μ size frequency histogram from Laguna Cuzcachapa Chalchuapa. *Zea* pollen was identified in routine pollen counts of 300 nonaquatic pollen grains. Folded and torn *Zea* grains were excluded from the analysis.

tion, indicating the presence of maize agriculture by ca. 3710 CAL BP [17], suggesting that maize cultivation here was coeval with that at El Carmen. The maize pollen assemblage at Chalchuapa is quite remarkable both because of the high concentration levels of *Zea* pollen and because of the large size of the grains that occur over a period spanning nearly four millennia. Many of the grains exceed 100 μ in length, even in the early Pre-Classic strata (Figure 8-4). Pollen was present in the deepest (ca. 3710 CAL BP) stratum analyzed, but no further probing was possible with the equipment available. Given the dense concentration of maize pollen in the basal sediments, it is quite likely that maize cultivation began some centuries before ca. 3710 CAL BP.

Approximately 20 kilometers west of Chalchuapa (see (Figure 8-3). the site of Laguna Llano yielded *Zea* pollen in the size range exclusive to maize dating to ca. 2910 CAL BP [16]. Poor pollen preservation in the deeper sediments at this site prevented the documentation of the initial arrival of maize.

The Sierra de Apaneca–Llamatepec Highlands

Laguna Verde, a crater lake situated at 1600 masl, has yielded the earliest evidence of the genus *Zea* from anywhere in El Salvador (see Figure 8-3). The 8000-year continuous pollen record from Laguna Verde records the first occurrence of *Zea* pollen in the lake at ca. 4440 CAL BP [15]. The date is based on a linear interpolation of calibrated AMS ^{14}C dates on **plant macrofossils** above (ca. 4030 CAL BP) and below (ca. 5830 CAL BP) the horizon where *Zea* pollen first appears. *Zea* pollen was found in several samples between ca. 4440 and 1520 CAL BP. The record of *Zea* pollen from this site is distinctly different from those found at Laguna Cuzcachapa and Laguna Llano 900 meters below in the Rio Paz basin. The *Zea* pollen at Laguna Verde is present in much lower concentrations than the Rio Paz basin sites, and the grains that are present are relatively small in size (diameter range = 65–82 μ, $n = 7$).

No archaeological sites have been reported within the Laguna Verde crater, however the Pre-Classic site of Santa Leticia is located just 4 kilometers southwest at 1400 masl (see (Figure 8-3). Human occupation at Santa Leticia has been dated as far back as ca. 2350 CAL BP [10]. A small collection of paleobotanical remains has been reported from Apaneca [10, 32]. The most common plant fragments identified in the Pre-Classic and Classic Period middens were maize *(Zea mays* ssp. *mays)* cupules. The maize has been identified as Dzit Bacal, an 8- to 10-rowed dent landrace associated with Classic Maya populations elsewhere in Mesoamerica both prehistorically and today [33].

PREHISTORIC MAIZE FIELDS FROM CENTRAL EL SALVADOR

Although the earliest evidence for maize in El Salvador comes from the western part of the country, the central region of the country has yielded an uncommonly well-preserved series of fossil maize fields buried under volcanic tephra. Linear fields, characterized by 10- to 15-centimeter high ridges spaced roughly one meter apart have been found throughout the San Salvador Valley [1, 6], the Zapotitán Valley [46, 57], and the Rio Lempa bottomlands/Paraiso basin [18] (see Figure 26-3). These ridge and furrow fields date to three separate episodes of volcanism: ca. 2780 CAL BP, ca. 1520 CAL BP, and ca. 1290 CAL BP. The first and the last of these are distinguished by the presence of maize plant impressions excavated from their surfaces. At least 18 fields have now been identified underlying widespread tephra deposited by the ca. 1520 CAL BP Tierra Blanca Joven eruption of the Ilopango caldera; none, however, contains well-preserved macrofossils [1, 17]. In total, more than two dozen pre-Hispanic ridge and furrow fields have now been identified in El Salvador. Although similar ridge and furrow fields have been reported elsewhere in Mesoamerica and throughout the Americas, these food productions systems have received little scholarly attention [26].

Valle de las Hamacas (San Salvador)

Four ridge and furrow fields dating to ca. 2780 CAL BP have been identified in the vicinity of Antiguo Cuzcatlan and Merliot, in the Valle de las Hamacas, San Salvador [1, 6]. Ridge spacing on these fields ranges from 56–84 centimeters, and the average furrow depth measured was 11 centimeters [1]. One of the three fields—the "Merliot Milpa" (see (Figure 8-3) —is remarkable in that it contains abundant plant macrofossil impressions, all of which have been identified as maize (Figures 26-5, 26-6).

FIGURE 8-5 Close-up photograph of ridge and furrow features from the ca. 2830 CAL BP Merliot Milpa. See color insert. (Courtesy of Robert A. Dull)

FIGURE 8-6 Maize leaf cast impression excavated from the Merliot Milpa. This is one of dozens that were excavated from furrows on the field surface. See color insert. (Courtesy of Robert A. Dull)

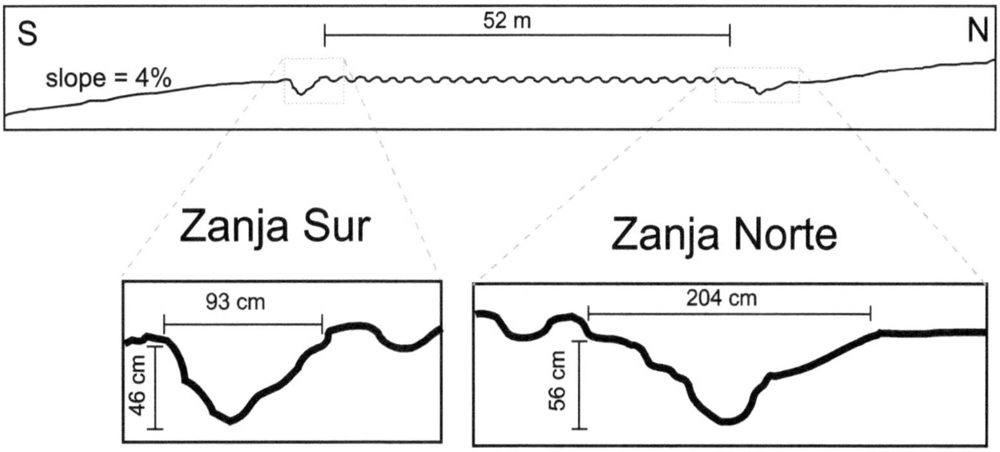

FIGURE 8-7 Diagram of Merliot Milpa exposure (not to scale). Fifty-two meter long ridge and furrow milpa surface is flanked by ditches ("zanjas") on both ends. The flat field surface marks a break in the hillslope surface gradient, which is about 4%.

This 52-meter long ridge and furrow field is also distinguished by the deep (about 50 cm) ditches that were excavated at both ends of the field surface (Figure 8-7). The Merliot **Milpa** represents the earliest corn field with in situ macrofossil impressions found anywhere in the Americas.

The Zapotitán Basin

The archaeological site of Cerén in the Zapotitán Valley of central El Salvador contains exquisitely well-preserved prehistoric maize milpas [28, 46, 56, 57]. Although roughly 1500 years younger in age than the Merliot Milpa [17], these ridge and furrow fields are nearly identical in morphology with ridge spacing ranging from 60–80 centimeters, and furrow depths ranging from 10–20 centimeters [57]. Plaster-poured casts of corn stalks reveal the ca. 1290-year-old plants exactly as they stood in the fields where they were planted (Figure 26-8). Maize cob measurements from Cerén give an average row number of 13.1, which is intermediate between Chapalote Nal-Tel and Tepecintle maize landraces, although kernel dimensions (width = 6.3 mm, thickness = 4.3 mm) and lack of denting indicate a close affinity with Chapalote Nal-Tel [28].

CONCLUSIONS

The history of maize cultivation in El Salvador is still only partially understood. The earliest evidence of *Zea* pollen in El Salvador comes from Laguna Verde in the Sierra de Apaneca. The 4440-year-old (ca. 2490 BC) *Zea* pollen grain from Laguna Verde is only 65 μ long, suggesting that it was produced by either teosinte or an early race of cultivated maize. By ca. 3700 CAL BP (1750 BC) there is unambiguous evidence for maize cultivation in western El Salvador—both along the southwestern Pacific coast at El Carmen and in the Rio Paz basin at Chalchuapa. The evidence collected to date both in El Salvador and in other Central America countries (Belize and Costa Rica = ca. 5500 BP) suggests that maize was probably present at least 1000 years earlier in El Salvador than the 4440 BP pollen from Laguna Verde would imply, but no physical evidence has yet been discovered in El Salvador to support this thesis.

Two generations of fossil maize fields have been reported from central El Salvador. The fields are located respectively in San Salvador (ca. 2780 CAL BP, or 830 BC) and the Zapotitán Valley (ca. 1290 CAL BP, or AD 660). These maize milpas represent a dryland farming strategy that was employed by the Pre-Hispanic residents of El Salvador for more than two millennia. It is remarkable that maize fields separated in time by 1500 years retained such similar morphological characteristics. In both cases maize plants were planted on linear ridge features interspersed with furrows. Both generations of maize fields contained deeper ditch features that were presumably used to facilitate drainage. It is hypothesized that milpa dry farming was probably a common food production strategy used throughout Mesoamerica, but that few relicts exist today because of erosion and subsequent land use.

FIGURE 8-8 Plaster cast of maize stalks (**A**) and excavated maize milpa surface (**B**) from Cerén dating to ca. 1290 CAL BP. See color insert. (Courtesy of Robert A. Dull)

References Cited

1. P. Amaroli, R. Dull. (1999). Milpas prehispánicas en El Salvador. *Proceedings of the XII Simposio de Investigaciones Arqueológicas en Guatemala*. pp. 639–650.
2. M. R. Arford, S. P. Horn. (2004). Pollen evidence of the earliest maize agriculture in Costa Rica. *Journal of Latin American Geography, 3,* 108–115.
3. B. Arroyo. (1995). Early ceramics from El Salvador: The El Carmen site. In: W. K. Barnett, J. W. Hoopes, (Eds.), *The emergence of pottery: Technology and innovation in Ancient Societies*. Washington, D.C.: Smithsonian Institute Press. pp. 199–208.
4. E. S. Barghoorn, M. K. Wolf, K. H. Clisby. (1954). Fossil maize from the Valley of Mexico. *Botanical Museum Leaflets, Harvard University, 16,* 229–240.
5. G. W. Beadle. (1981). Origin of corn: Pollen evidence. *Science, 213,* 890–892.
6. G. Bello Suazo. (1991). Rescate arqueológico en Antiguo Cuscatlán: Informe preliminar. *Mesoamerica, 12,* 115–121.
7. B. Benz. (1986). Racial systematics and the evolution of Mexican maize. In: L. Manzanilla, (Ed.), *Studies in the Neolithic and urban revolutions*. Mexico: B.A.R.
8. M. Blake, J. E. Clark, B. Voorhies, G. Michaels, M. W. Love, M. E. Pye, A. A. Demarest, B. Arroyo. (1995). Radiocarbon chronology for the Late Archaic and Formative periods on the Pacific coast of southeastern Mesoamerica. *Ancient Mesoamerica, 6,* 161–183.
9. M. B. Bush. (1995). Neotropical plant reproductive strategies and fossil pollen representation. *The American Naturalist, 145,* 594–609.
10. A. A. Demarest. (1986). *The archaeology of Santa Leticia and the rise of Maya civilization*. New Orleans: Tulane University Middle America Research Institute.
11. A. A. Demarest, B. Arroyo, P. Amaroli. (1989). *Proyecto El Carmen*. Nashville, TN: Vanderbilt University.
12. J. Doebley. (1990). Molecular systematics of *Zea* (Gramineae). *Maydica, 35,* 143–150.
13. W. E. Doolittle, C. D. Frederick. (1991). Phytoliths as indicators of prehistoric maize *(Zea mays* susp. *mays, Poaceae)* cultivation. *Plant Systematics and Evolution, 177,* 175–184.
14. R. A. Dull. (2001). *El bosque perdido: A cultural-ecological history of Holocene environmental change in western El Salvador geography*. Berkeley: University of California.
15. R. A. Dull. (2004). An 8000-year record of vegetation, climate, and human disturbance from the Sierra de Apaneca. El Salvador. *Quaternary Research, 61,* 159–167.
16. R. A. Dull. (2004). A Holocene record of Neotropical savanna dynamics from El Salvador. *Journal of Paleolimnology, 32,* 219–231.
17. R. A. Dull, J. R. Southon, P. Sheets. (2001). Volcanism, ecology and culture: a reassessment of the Volcán Ilopango TBJ eruption in the southern Maya realm. *Latin American Antiquity, 12,* 25–44.
18. W. R. Fowler, H. H. Earnest. (1985). Settlement patterns and prehistory of the Paraiso Basin of El Salvador. *Journal of Field Archaeology, 12,* 19–32.

19. G. J. Fritz. (1994). Are the first American farmers getting younger? *Current Anthropology, 35,* 305–309.
20. W. C. Galinat. (1961). Corn's evolution and its significance for breeding. *Economic Botany, 15,* 320–325.
21. M. Goman, R. Byrne. (1989). A 5000-year record of agriculture and tropical forest clearance in the Tuxtlas, Veracruz, Mexico. *The Holocene, 8,* 83–89.
22. C. A. Grant. (1972). A scanning electron microscopy survey of some Maydae pollen. *Grana, 12,* 77–184.
23. R. J. Hard, J. R. Roney. (1998). A massive terraced village complex in Chihuahua, Mexico, 3000 years before present. *Science, 279,* 1661–1664.
24. H. H. Iltis. (1983). From teosinte to maize: The catastrophic sexual transmutation. *Science, 222,* 886–894.
25. H. H. Iltis. (2000). Homeotic sexual translocations and the origin of maize *(Zea mays, Poaceae)*: A new look at an old problem. *Economic Botany, 54,* 7–42.
26. H. H. Iltis, B. F. Benz. (2000). *Zea nicaraguensis (Poaceae),* a new teosinte from Pacific coastal Nicaragua. *Novon, 10,* 382–390.
27. T. W. Killion. (1992). The archaeology of settlement agriculture. In: T. W. Killion, (Ed.), *Gardens of prehistory: The archaeology of settlement agriculture in greater Mesoamerica.* Tuscaloosa: University of Alabama Press. pp. 1–13.
28. D. L. Lentz, M. P. Beaudry-Corbett, M. L. R. D. Aguilar, L. Kaplan. (1996). Foodstuffs, forests, fields, and shelter: A paleoethnobotanical analysis of vessel contents from the Ceren site, El Salvador. *Latin American Antiquity, 7,* 247–262.
29. A. Long, B. Benz, J. Donohue, A. Jull, L. Toolin. (1989). First direct AMS dates on early maize from Tehuacán, Mexico. *Radiocarbon, 31,* 1035–1040.
30. P. C. Mangelsdorf, E. S. Barghoorn, U. C. Banerjee. (1978). Fossil pollen and the origin of corn. *Harvard University Botanical Museum Leaflets 26,* 238–255.
31. Y. Matsuoka, Y. Vigouroux, M. M. Goodman, J. Sanchez, G. E. Buckler, J. Doebley. (2002). A single domestication for maize shown by multilocus microsatellite genotyping. *Proceedings of the National Academy of the Sciences, 99,* 6080–6084.
32. C. H. Miksicek. (1986). Appendix 2. Paleobotanical identifications. In: A. Demarest, (Ed.), *The archaeology of Santa Leticia and the rise of the Maya civilization.* New Orleans: Middle American Research Institute. pp. 199–200.
33. D. M. Pearsall, K. Chandler-Ezell, A. Chandler-Ezell. (2004). Maize can still be identified using phytoliths: Response to Rovner. *Journal of Archaeological Science, 31,* 1029–1038.
34. D. R. Piperno, K. V. Flannery. (2001). The earliest archaeological maize (*Zea mays* L.) from highland Mexico: New accelerator mass spectrometry dates and their implications. *Proceedings of the National Academy of Sciences of the United States of America, 98,* 2101–2103.
35. D. R. Piperno, D. M. Pearsall. (1998). *The origins of agriculture in the Lowland Neotropics.* San Diego: Academic Press.
36. D. R. Piperno, D. M. Pearsall, R. A. Benfer, Jr., L. Kealhofer, Z. Zhao, Q. Jiang. (1999). Phytolith morphology. *Science, 283,* 1265–1267.
37. M. D. Pohl, K. O. Pope, J. G. Jones, J. S. Jacob, D. R. Piperno, S. D. deFrance, D. L. Lentz, J. A. Gifford, M. E. Danforth, J. K. Josserand. (1996). Early agriculture in the Maya lowlands. *Latin American Antiquity, 7,* 355–372.
38. K. O. Pope, M. E. D. Pohl, J. G. Jones, D. L. Lentz, C. von Nagy, F. J. Vega, I. R. Quitmyer. (2001). Origin and environmental setting of ancient agriculture in the lowlands of Mesoamerica. *Science, 292,* 1370–1373.
39. I. Rovner. (1999). Phytolith analysis. *Science, 283,* 487.
40. I. Rovner. (2004). On transparent blindfolds: comments on identifying maize in Neotropical sediments and soils using cob phytoliths. *Journal of Archaeological Science, 31,* 815–819.
41. D. J. Rue. (1987). Early agriculture and early Postclassic Maya occupation in western Honduras. *Nature, 326,* 285–286.
42. W. F. Rust, B. W. Leyden. (1994). Evidence of maize use at Early and Middle Preclassic La Venta Olmec sites. In: S. Johannessen, C. A. Hastorf, (Eds.), *Corn and culture in the prehistoric New World.* Boulder, CO: Westview Press. pp. 181–202.
43. C. O. Sauer. (1941). The personality of Mexico. *Geographical Review, 31,* 353–364.
44. R. J. Sharer. (1974). The prehistory of the southeastern Maya periphery. *Current Anthropology, 15,* 165–176.
45. R. J. Sharer. (1978). *The prehistory of Chachuapa, El Salvador.* Philadelphia: The University of Pennsylvania Press.
46. P. D. Sheets. (1982). Prehistoric agricultural systems in El Salvador. In: K. Flannery, (Ed.) *Maya subsistence.* New York: Academic Press. pp. 99–117.
47. P. D. Sheets. (1984). The prehistory of El Salvador: An interpretive summary. In: F. W. Lange, D. Z. Stone, (Eds.), *The archaeology of Lower Central America.* Albuquerque: New Mexico Press. pp. 85–112.
48. A. Sluyter. (1997). Analysis of maize *(Zea mays* subsp. *mays)* pollen: Normalizing the effects of microscope-slide mounting media on diameter determinations. *Palynology, 21,* 35–39.
49. B. D. Smith. (1995). *The emergence of agriculture.* New York: Scientific American Library.
50. B. D. Smith. (1997). Reconsidering the Ocampo caves and the era of incipient cultivation in Mesoamerica. *Latin American Antiquity, 8,* 342–383.
51. J. E. Staller. (2003). An examination of the paleobotanical and chronological evidence for an early introduction of maize (*Zea mays* L.) into South America: a response to Pearsall. *Journal of Archaeological Science, 30,* 373–380.
52. J. E. Staller, R. G. Thompson. (2002). A multidisciplinary approach to understanding the initial introduction of maize into coastal Ecuador. *Journal of Archaeological Science, 29,* 33–50.
53. M. Tsukada, J. R. Rowley. (1964). Identification of modern and fossil maize pollen. *Grana Palynologica, 5,* 406–412.
54. D. Wahl, R. Byrne, T. Schreiner, R. Hansen. (In press). Holocene vegetation change in the northern Peten and its implications for Maya prehistory. *Quaternary Research.*
55. D. R. Whitehead, E. J. Langham. (1998). Measurement as a means of identifying fossil maize pollen. *Bulletin of the Torrey Botanical Club, 92,* 7–20.
56. J. Wiseman. (1998). The art of gardening: Eating well at a Mesoamerican Pompeii. *Archaeology, January–February,* 12–16.
57. C. J. Zier. (1980). A Classic-period Maya agricultural field in western El Salvador. *Journal of Field Archaeology, 7,* 65–74.

CHAPTER

9

Pre-Columbian Maize Agriculture in Costa Rica

Pollen and Other Evidence from Lake and Swamp Sediments

SALLY P. HORN

Department of Geography, The University of Tennessee, Knoxville, Tennessee

Glossary

AMS radiocarbon date A date obtained using a method of radiocarbon dating in which the amount of ^{14}C in a sample is measured directly (rather than by waiting for decay events to occur, as in standard radiocarbon dating), using an accelerator mass spectrometer. The method greatly reduces the quantity of organic material required for radiocarbon dating, making it possible to date single small leaf fragments or pieces of wood or charcoal weighing as little as 5 milligrams.

Calibration of radiocarbon dates The adjustment of dates in conventional radiocarbon (^{14}C) years to calendar years, using calibration datasets constructed by measuring the radiocarbon age of tree rings of known age or other independently dated samples. Calibration is made necessary by fluctuations over time in the amount of radiocarbon present in the atmosphere.

Pollen grains Microscopic structures carrying the male gametes of seed plants, which are shed from pollen cones of gymnosperms or from anthers of flowering plants. Transfer of pollen grains to the female reproductive structures of plants by wind, insects, or other means leads to pollination and seed formation. Pollen grains are produced in great abundance by many plants and when deposited in lakes or wetlands may be preserved in sediments for thousands of years, or longer. They can be extracted from sediment cores obtained from such sites, and identified and counted to reveal changes in local or regional vegetation over time. Identification of pollen grains to family, genus, or in some cases species is made possible by their distinctive shapes, sizes, and surface sculpturing patterns.

Sediment core A cylindrical sample of mud from a lake, bog, or other wetland obtained using a hand-operated sediment corer or a machine-driven drilling rig. Most soft sediment cores obtained from hand-operated corers for paleoecological study are about 5 centimeters in diameter, although larger diameter cores are sometimes collected.

Stable carbon isotope values In addition to the radioactive isotope ^{14}C, carbon also has two naturally occurring stable isotopes: ^{12}C and ^{13}C. The proportions of these two isotopes in a sample is expressed in a per mil (‰) notation relative to a standard known as the Vienna-Pee Dee belemnite (VPDB) marine-carbonate, as a $\delta^{13}C$ value. The $\delta^{13}C$ value of a sample is calculated as: $1000 \cdot [(R_{sample}/R_{standard}) - 1]$, where $R = {}^{13}C/{}^{12}C$. Plants that use the C_4 photosynthetic pathway, such as maize, produce organic matter with more positive $\delta^{13}C$ values than do plants that use the more common C_3 photosynthetic pathway.

Pollen grains and other evidence of Pre-Columbian maize agriculture preserved in **sediment cores** from lakes and swamps in Costa Rica complement and extend the archaeological record. Most lakes and swamps below 1500 meters elevation that preserve Late Holocene pollen assemblages contain some maize pollen. The large size of maize pollen grains compared with most other pollen types facilitates their discovery using low power scans of microscope slides. The oldest maize pollen grains in Costa Rica were deposited about 3550 BC in Laguna Martínez in the northwest lowlands and appear to constitute the oldest botanical or archaeological evidence of maize cultivation in the country. In some younger lakes on the Caribbean slope and in southern Pacific Costa Rica, maize pollen is present in the oldest sediments recovered and extends through the European period with hardly a break.

INTRODUCTION

Botanical and genetic studies by John Doebley and collaborators [5, 14, 15, 41] have identified an annual species of teosinte *(Zea mays* subsp. *parviglumus)* native to mid-elevations of the Balsas River Valley of Mexico as the progenitor of domesticated maize *(Zea mays* subsp. *mays).* This aspect of the ancestry of maize presently enjoys wide acceptance [50, 58–60], despite the existence of a rival model [17, 40]. But other aspects of the nature, geography, and timing of maize domestication, and the subsequent spread of the crop throughout the Americas, continue to generate debate [50, 58, 66].

This chapter contributes to advancing the understanding of the southward spread of maize out of Mexico by summarizing pollen evidence for the presence of maize at different times and places in the area of southern Central America that is now the country of Costa Rica. The data come from analyses of pollen assemblages and other components of lake and swamp sediments that my students and collaborators and I have carried out to document long-term vegetation, climate, and human land-use histories at a range of sites in Costa Rica. I compare our maize pollen data and other sedimentary evidence of maize agriculture with the occurrence of maize macrofossils, pollen, and phytoliths from archaeological excavations in Costa Rica, and with prehistoric dietary information derived from stable isotope analyses of human skeletal tissues. Archaeological data other than the bone analyses are taken mainly from a review by Aida Blanco and Giselle Mora [7] of archaeobotanical evidence of wild and cultivated plant species at archaeological sites in Costa Rica.

MAIZE POLLEN IDENTIFICATION AND DISPERSAL AND ASSOCIATED PALEOECOLOGICAL EVIDENCE

Pollen analysts generally distinguish pollen of the genus *Zea* from that of other grass genera based on the length of the maximum diameter of the grain, which can be examined using light microscopy. The maximum diameter of the annulus or thickened ring of tissue surrounding the pore may also be instructive. Additional features may be observable using scanning electron microscopy [18], but this is generally not practical for routine identification and counting of pollen samples. Pollen grains of cultivated maize *(Zea mays* subsp. *mays)* overlap in size with those of the teosintes (*Zea mays* subsp. *parviglumis* H. H. Iltis & Doebley, *Zea perennis* (Hitchc.) Reeves & Mangelsd. and other *Zea* L. spp.; [57]), complicating the reconstruction of maize agriculture from sedimentary pollen at sites where pre-Columbian populations of teosinte are known or suspected. In the modern period, wild populations of teosinte have been discovered as far south as the Gulf of Fonseca region in the Pacific lowlands of northernmost Nicaragua [30]. If we assume that Pre-Columbian populations of teosinte reached no farther south than do modern populations, then the maize–teosinte identification issue disappears in southern Central America and South America. However, analysts must still take care to rule out other potential sources of large grass pollen grains, including more distant maize relatives in the tribe Maydeae, and certain aquatic grasses and bamboos. This requires reference collections of modern grasses from the area of interest, ideally produced from herbarium specimens that have been identified by experts [71]. As indicated later, the reference and fossil grains need to be prepared in similar fashion, and mounted on microscopes slides using the same mounting medium.

Work by Donald Whitehead and E. James Langham [73] in the 1960s established size ranges for maize, teosinte, and a third grass in the same tribe, gamma grass or *Tripsacum.* Based on modern specimens mounted in silicone oil, Whitehead and Langham reported diameters for cultivated maize (12 races) of 58.00–98.60 µm; for teosinte (10 races) of 46.40–87.00 µm, and for *Tripsacum dactyloides* (one diploid and one tetraploid race) of 33.35–56.55 µm. Beatriz Ludlow-Wiechers and collaborators [39] found certain Mexican races of maize with maximum diameters up to 120 µm. These and other papers [57, 74] on size characteristics of pollen grains of maize and its relatives reveal the strong effects that different mounting media (and to some extent, different processing methods) and slide thickness can have on grain size. Andrew Sluyter [57] conducted experiments to facilitate comparison of sizes of maize pollen grains mounted in silicone oil, glycerine jelly, and a new type of acrylic resin mounting medium. He suggested a conversion factor of 0.92 between glycerine and silicone mounts for modern maize samples measured within one day, and 0.89 for 30 days, indicating an expansion of 11–12%, but other studies have suggested that glycerine jelly can lead to swelling of grains by 20% [39] or even as much as 50% [43]. The reliability of pollen-based reconstructions of the history of maize in southern Central America and South America would be enhanced by controlled studies in which fossil pollen samples from sediment cores across the region are prepared and measured in the same lab using the same chemical procedures, mounting medium, microscope, and measuring system. I am in the process of seeking collaborators and funding for such a study, which could be designed to also shed light on possible changes in maize pollen diameter over time, as revealed by frequency distributions of the diameters of large grass pollen grains in samples [57; for examples see 10 and Chapter 26]. Such a study requires the isolation of large numbers of maize grains from sediment samples, but it may be possible to do this using sieving protocols developed to concentrate cultigen pollen in European sediment profiles [8].

Most of the maize pollen grains that we have identified in Costa Rican lake and swamp sediments range from 63–90 μm in diameter, based on specimens prepared using standard preparation techniques (KOH, HCl, HF, Acetolysis [6]) and mounted on microscope slides in silicone oil (2000 cs viscosity). However, we occasionally find much larger grains (up to 125 μm in diameter). These measurements are within the reported size ranges of maize pollen mounted in silicone oil [39, 57, 74]. Our maize grains do not overlap in size with the *Tripsacum* species studied by Whitehead and Langham [73], nor with any other genus of wild grasses (including aquatic grasses and bamboos) native to Costa Rica, based on our examination of reference pollen samples and of grain sizes reported in the literature. Our reference pollen sample of *Tripsacum intermedium* revealed a few grains up to 66 μm (most were <58 μm), but surface sculpturing on these grains under light microscopy is coarsely scabrate to verrucate, in contrast to the granular sculpturing on maize grains; moreover, whether this species actually occurs in Costa Rica is apparently in some doubt [21]. Our Costa Rican maize grains do overlap in size with various teosintes, with the sterile maize–teosinte hybrid *Tripsacum andersonii* and with the introduced Asian grass *Coix lacryma-jobi* (also in the grass tribe Maydeae), but there is no evidence that these grasses were present in Costa Rica during prehistoric times.

A caveat must be mentioned. Robert Dull (personal communication) has questioned whether we can be certain that some of our ancient maize pollen is not, in fact, teosinte pollen. He has suggested that it may be inappropriate to use the modern, known biogeography of the teosintes to infer their prehistoric distribution. Perhaps one or more of the teosintes once grew in Costa Rica, but populations were eliminated by post-European landscape alteration, including the introduction of cattle, or by other environmental changes resulting from human or natural factors. As we learn more about the modern distributions and history of maize and its relatives, we may need to reconsider our identifications of Costa Rican maize pollen. If modern, apparently native populations of teosinte are discovered in Costa Rica, or if macrofossil evidence from archaeological sites reveals the prehistoric presence of teosinte, then we would need to conclude that the pollen we have identified as cultivated maize might also include (or perhaps entirely comprise) teosinte pollen. But at present we have no such evidence, so we interpret large grass pollen grains that overlap in size with both maize and teosinte pollen as representing maize.

The large size of maize pollen grains compared with most other types is an advantage for documenting their presence in sediment samples, because it is possible to search for them on microscope slides at low power magnification. My students and I have found it convenient to count maize pollen along with our regular pollen counts (made along transects at 400X magnification, to a total of 250–300 pollen grains excluding spores and indeterminate grains), and then to completely scan multiple slides at 100× magnification to determine the presence or absence of maize. If control spores are added to preparations and their numbers on slides can be counted or reliably estimated, it may be possible to convert counts of maize grains made by scanning to percentage counts or concentrations. As discussed earlier, it may also be possible to prepare special samples in which maize grains are concentrated by sieving out smaller grains and other organic debris.

Although the relatively large size of maize pollen is an advantage when it comes to documenting the presence of the pollen type, this size difference affects dispersal of the grain to core sites and may lead to its underrepresentation in pollen records. Maize is wind pollinated, but its comparatively heavy pollen grains are unlikely to travel far under normal atmospheric conditions; Gilbert Raynor and collaborators [52] found that 95% of maize pollen grains fall within 60 meters of the parent plant. Gerald Islebe and co-authors [31] suggested that maize pollen may be common in lake sediments only when the crop is grown on the shore of the lake.

The interpretation of maize pollen is strengthened when core samples reveal other evidence of agricultural activity. Pollen assemblages containing maize often show high percentages of other grass pollen along with Asteraceae and Amaranthaceae pollen that may derive from agricultural weeds [55]. The maize-bearing samples might also show depressed tree pollen percentages in comparison with other samples, and it is common to find microscopic charcoal in the pollen assemblages and macroscopic charcoal in sieved samples from the same core intervals [1]. Phytoliths of *Zea* and agricultural weeds can provide additional microfossil evidence of maize cultivation [48, 49].

Chad Lane and Claudia Mora and I have discovered that **stable carbon isotope values** of lake and swamp sediments provide an additional proxy for maize agriculture [36]. Whereas tropical rainforest trees and understory herbs that grow beneath the rainforest canopy are predominantly C_3 plants, maize uses the less common C_4 photosynthetic pathway, as do many herbs and weeds of more open ground that might be expected in prehistoric corn fields. Plants using the C_3 photosynthetic pathway typically produce organic matter with $\delta^{13}C$ values between −35‰ and −20‰ (VPDB), whereas plants using the C_4 photosynthetic pathway typically produce organic matter with $\delta^{13}C$ values between −14‰ and −10‰ [4, 47]. The organic matter that accumulates beneath plants and that washes into lakes or swamps preserves the distinctive $\delta^{13}C$ values associated with these differing photosynthetic pathways, allowing paleoecologists to potentially infer the past abundance of C_3 versus C_4 plants in a watershed based on the $\delta^{13}C$ values in core samples. In cores from rainforest sites that are dominated by C_3 plants in undisturbed conditions, forest clearance for maize agriculture should result in more positive stable carbon isotope values in sedimented organic matter. This increase would result primarily from the increased relative abundance of C_4 plants in the watershed (e.g., maize, agricultural weeds, and other herbs of open ground). The presence of a more open canopy at the site could also influence stable carbon isotope values by reducing what is known as the canopy effect, which leads understory herbs to more strongly discriminate against $\delta^{13}C$ in photosynthesis under shaded conditions [72].

MAIZE POLLEN IN ARCHAEOLOGICAL REGIONS OF COSTA RICA

Three broad archaeological regions are recognized in Costa Rica (Figure 9-1) [62]. The Guanacaste–Nicoya archaeological region in northwestern Costa Rica is a part of the Greater Nicoya archaeological subarea, which also includes parts of western Nicaragua. The Diquís archaeological region along the southern Pacific coast is part of the larger subarea known as the Greater Chiriquí, which includes the Panamanian province of Chiriquí to the east. Separating these two areas is the Central Highlands–Atlantic Watershed archaeological region. Sediment cores from lakes and swamps document Pre-Columbian maize in all three archaeological regions. Table 9-1 summarizes data on the temporal distribution of maize pollen in lacustrine sediments from sites shown in Figure 9-1 For comparison, Table 9-1 also includes summary data on maize macrofossils,

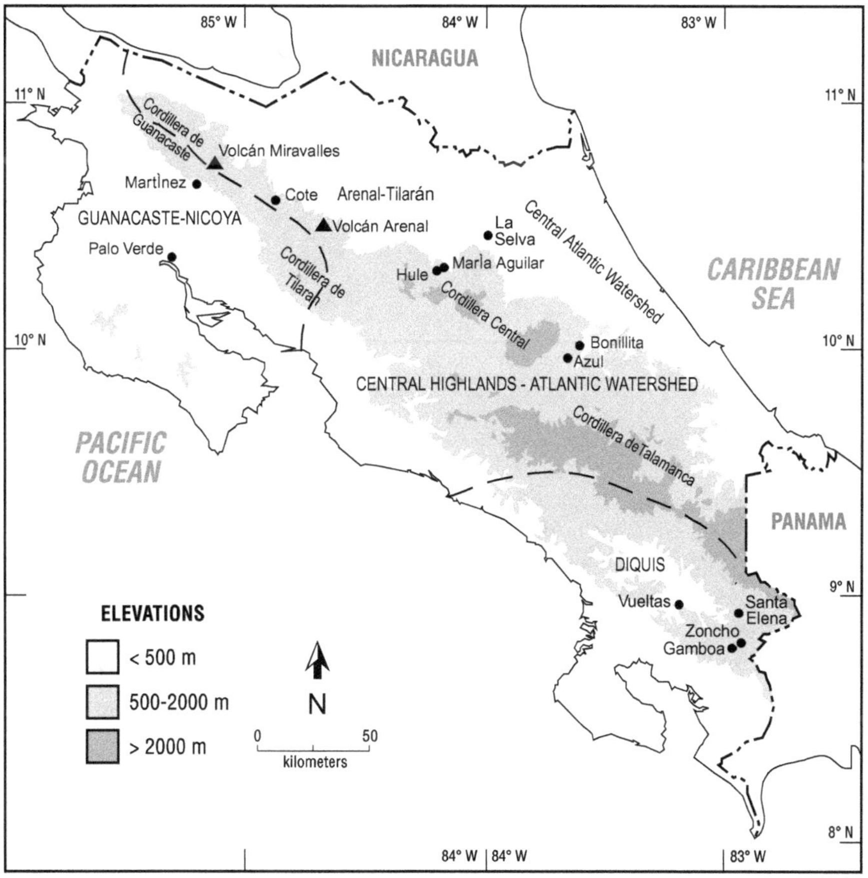

FIGURE 9-1 Map showing archaeological regions in Costa Rica and locations of sites mentioned in text. Archaeological region boundaries are based on maps in M. J. Snarskis. (1981). The archaeology of Costa Rica. In: E. P. Benson, (Ed.), *Between continents/between seas: Precolumbian art of Costa Rica.* New York: Harry N. Abrams Publishers. pp. 15–84.

TABLE 9-1 Summary of Sedimentary and Archaeological Evidence for the Presence of Maize in Selected Archaeological Regions and Subregions of Costa Rica. Regionalization and periodization follows Snarskis (1984) and Blanco and Mora (1994). X = evidence of maize from region or subregion. Classes of evidence: SP = maize pollen in sediment cores; M = maize macrofossils from archaeological excavation; AP = maize pollen in archaeological excavation; APh = maize phytoliths in archaeological excavation, B = stable isotope evidence from bone collagen suggesting maize consumption.

Archaeological region and core Site	Earliest maize (cal yr BC/AD) for region and site	Evidence of maize in sediment records and archaeological sites				References
		III: 4000 BC–1000 BC	IV: 1000 BC–AD 500	V: AD 500–AD 1000	VI: AD 1000–AD 1550	
Central Highlands–Atlantic Watershed	3140 BC	X	X	X	X	
Central Atlantic Watershed subregion						
Laguna Bonillita	670 BC		SP	SP	SP	46, 37
Laguna Bonilla	uncertain			?	SP	46
Laguna Azul	AD 1390				SP	this chapter
Cantarrana Swamp	AD 920			SP	SP	32, 33, 34
Machita Swamp	650 BC		SP, I	SP, I	SP, I	25, 36
Laguna María Aguilar	1030 BC		SP			26
Archaeological sites	AD 420		M	M	M	7, 61
Tilarán–Arenal subregion						
Lago Cote	2000 BC	SP				2
Archaeological sites	3140 BC	M	M, AP, APh	M, AP, APh	M, AP, APh	7, 56
Central Valley subregion						
Archaeological sites	300 BC		M, APh, B	M, APh, B	M, APh, B	7, 44
Diquís	1330 BC	X	X	X	X	
Laguna Zoncho	1330 BC	SP, I	SP, I	SP, I	SP, I	12, 36
Laguna Santa Elena	AD 140		SP	SP	SP	1
Laguna Vueltas	AD 850			SP, I	SP, I	28, 69
Laguna Gamboa	590 BC		SP			this chapter
Archaeological sites	500 BC			M	M	7
Guanacaste–Nicoya	3550 BC	X	X	X	X	
Laguna Martínez	3550 BC	SP	SP	SP	SP	3
Archaeological sites	AD 300		Ap, B	M, B	M, B	7, 44, 45

pollen, and phytoliths recovered from archaeological excavations in Costa Rica, and stable isotope evidence of maize consumption from studies of bone collagen [7, 44, 45, 56, 61]. Newer findings likely exist, and my hope is that this table, incomplete in terms of archaeological evidence, will spur new effort to discover, document, and synthesize maize evidence from archaeological sites.

In this chapter I present and discuss important aspects of the pollen records summarized in Table 9-1. I include two representative pollen diagrams that show the presence of maize pollen grains within the context of Late Holocene pollen and microscopic charcoal assemblages. Additional pollen diagrams have been or will be presented elsewhere. Pollen, charcoal, and sediment analyses reported here were carried out in the Paleoenvironmental Research Laboratory at the University of Tennessee, which I co-direct with Professor Ken Orvis. Isotope analyses were carried out in the Stable Isotope Laboratory directed by Professor Claudia Mora. Lake and swamp cores were collected primarily with a square-rod piston corer [75] and a Colinvaux-Vohnout locking piston corer [13], operated at our lake sites from an anchored floating platform constructed of plywood or planks lashed across two inflatable boats or four truck tire inner tubes.

The Central Highlands–Atlantic Watershed Archaeological Region

My students and I have recovered and analyzed sediment cores from six lakes and two swamps in the Central Highlands–Atlantic Watershed region that preserve Pre-Columbian pollen assemblages (see Table 9-1, Figure 9-1). The sites range in elevation from about 42 to 770 meters and are located mostly in the Central Atlantic Watershed subregion of this extensive archaeological region. Seven of them contain maize pollen.

Laguna Bonillita (9°59′36″ N, 83°36′47″ W, 450 m elevation) is a 6.0 ha lake located in the Río Reventazón Valley, on the lower eastern slope of Volcán Turrialba (see Figure 9-1). Sediment cores from both Laguna Bonillita and its larger (30.8 ha) sister, Laguna Bonilla, contain maize pollen [46], but owing to dating problems with the Bonilla sediments I focus here on the Bonillita record.

The central portion of Laguna Bonillita is 20 meters deep [24], but we anchored our coring platform at a water depth of 6.1 meters, in a gently sloping portion of the basin on the northeastern end of the lake. From this site about 50 meters from shore, we recovered a 9.5-meter-long sediment core. This is the longest core from a modern lake that we have recovered in Costa Rica, but it is not the oldest core, as the sediment in it accumulated fairly rapidly. Elliott Burden and coauthors [9] suggested that cores collected in nearshore areas of lakes within archaeological areas are more likely to yield evidence of human presence than are cores from sites further offshore. The abundance of maize pollen in the Bonillita core (up to 5% of the pollen sum; Figure 9-2), and the presence of large Cucurbitaceae pollen that may represent a cultivated species, may in part reflect our nearshore coring location. In future work at other lakes we plan paired investigations of nearshore and more centrally located cores to explore the effect of distance from shore on the preservation of large cultigen pollen and other signs of human influence.

The Bonillita pollen diagram reveals the clear importance of maize throughout the time period spanned by the core. Maize is present in the lowest sample analyzed (in sediments dated to 2560 ± 60 ^{14}C yr BP, or about 670 BC), and in nearly every sample above with the exception of the most recent three samples at the top of the core. (Note: **Calibration of radiocarbon dates** was performed using version 5.0 of the CALIB program [68], and the IntCal04 data set [53]. The single BC or AD calendar age provided for each radiocarbon date cited here represents the weighted mean of the calibration probability distribution [70], rounded to the nearest 10 years.) Stable carbon isotope values in bulk sediments from the Bonillita core also indicate maize agriculture [37]. These data provide new context for archaeological interpretations. William Kennedy [35] examined two archaeological sites near the lakes that include burial mounds, petroglyphs, and late Period IV (El Bosque phase) artefacts. Based on his reconnaissance (full excavations have yet to take place), Kennedy interpreted the sites as temporary habitation sites of people who relied for subsistence on hunting, fishing, and the cultivation of tubers. The pollen results from Laguna Bonillita argue instead that these or other sites in the area were permanently occupied during early Period IV (La Montaña phase) by people whose subsistence activities included maize agriculture. These results contrast greatly with the interpretations of Kennedy, who followed Doris Stone [67] in regarding maize as a postconquest introduction in the Central Atlantic Watershed. They instead support the alternative regional interpretation of Michael Snarskis [63], who suggested that maize cultivation was in full swing during the El Bosque period (and for which time period maize macrofossils also exist [7]). From the maize pollen in Laguna Bonillita, we know that maize farmers were present in the Central Atlantic Watershed subregion even earlier.

The Bonillita sediments also contain variable amounts of microscopic charcoal that may signal agricultural burning; and high percentages of Asteraceae and Amaranthaceae pollen that may derive from agricultural weeds, along with pollen of the Cucurbit *Sechium* (chayote), which may represented a cultivated species. The sharp decline in *Myrsine* pollen above about 5 meters depth in the core may be due to forest clearance by indigenous people and the preferential use of *Myrsine floridiana* for constructing houses [46]; fluctuations in percentages of the weedy secondary tree *Cecropia* may also relate to episodes of forest clearance.

Additional evidence for Pre-Columbian maize agriculture in the Río Reventazón Valley comes from sediments of *Laguna Azul* (9°57′22″ N, 83°39′03″ W, 630 m elevation), located some 6 km upslope from Laguna Bonillita, near the major, complex Period V–VI archaeological site of Guayabo de Turrialba [62]. The Laguna Azul story is a sad one for paleoecology and archaeology. When Kurt Haberyan and I first visited the site in 1991, it was an attractive small lake surrounded by agricultural fields [24]. However, when I returned to core it in 1992, the process had already begun of transforming this natural lake into a concrete-lined holding tank for a small hydroelectric project. Surrounding vegetation had been cut and thrown into the lake and was rotting, and the odor of gases bubbling up from the bottom of the lake was so nauseatingly strong that we had to cut short our field time. We were only able to recover a short core of near-surface sediments, and a single 1-meter core section beginning about 3 meters below the sediment–water interface at the time (which had been raised by the deposition of vegetation in the lake and inwashing of soil from newly unprotected slopes). An AMS radiocarbon date of 530 ± 50 ^{14}C yr BP (about AD 1390) on leaf fragments sieved from near the middle of the section shows that we did recover Pre-Columbian sediments, and unpublished pollen analysis by John Rodgers, Lisa Northrop, and myself has revealed the presence of La Cabaña phase (Period VI) maize cultivation at the site. However, the opportunity for more detailed reconstruction at the site was lost when the lake was destroyed before it could be properly sampled.

The La Selva Biological Station (see Figure 9-1) is a world famous research site located on the lower Caribbean slope and adjacent coastal plain of northern Costa Rica (10°26′ N, 83°59′ W, 30–150 m elevation). Owned and operated by the Organization for Tropical Studies, the station hosts thousands of scientists, students, and other visitors each year, who come to study and observe tropical

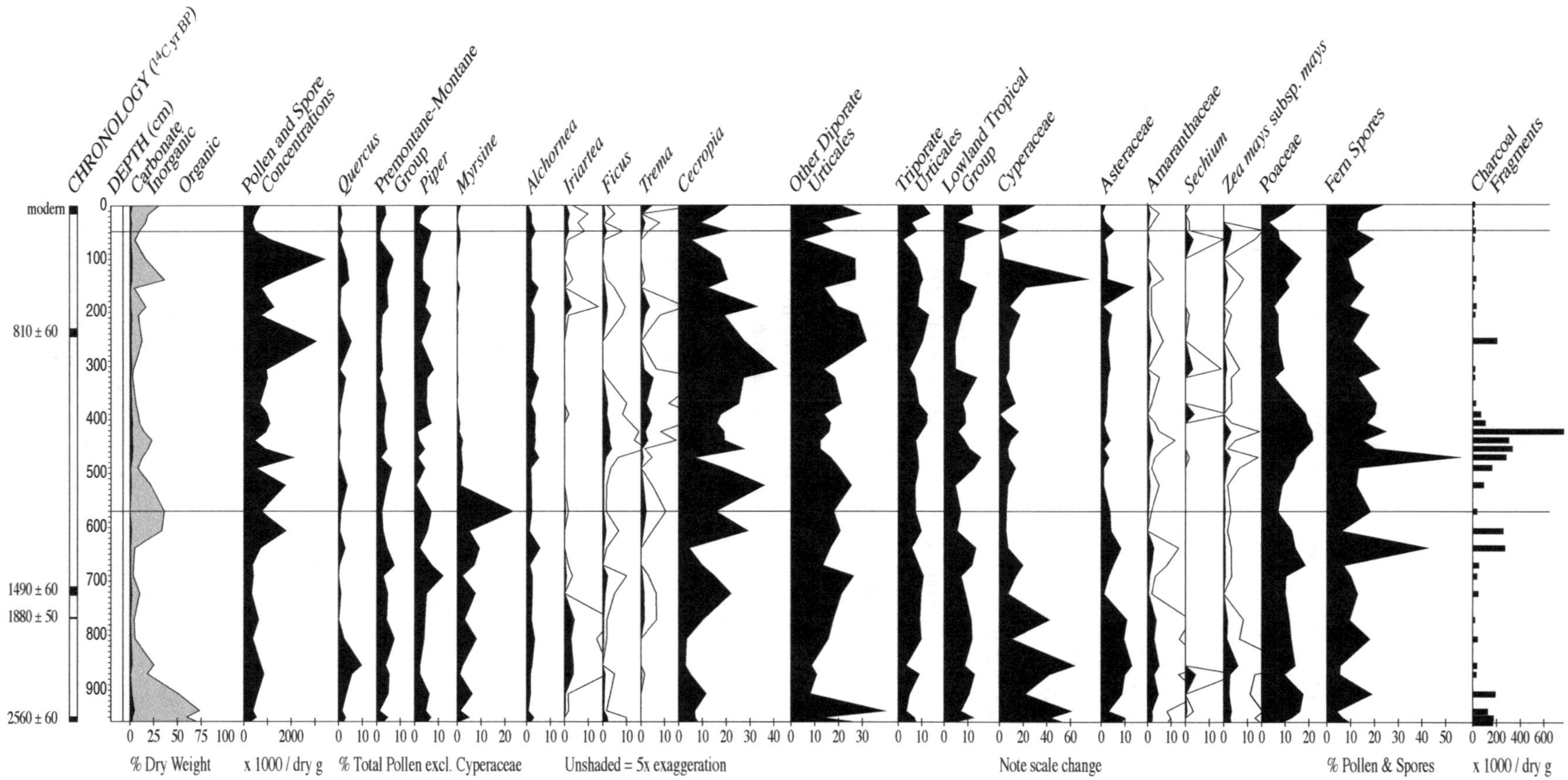

FIGURE 9-2 Pollen diagram from Laguna Bonillita, Costa Rica, showing sediment composition based on loss-on-ignition (carbonate, noncarbonate inorganic, and organic content), pollen and spore concentrations, percentage curves for major taxa and composite groups, charcoal profile, and pollen zones. The premontane–montane group includes *Alnus, Hedyosmum, Weinmannia, Alfaroa-Oreomunnia, Malpighia, Myrica, Podocarpus,* and *Ulmus*. The lowland tropical group includes *Anacardium, Apocynaceae, Ilex,* Bignoniaceae, Bombacaceae, *Cordia, Tournefortia, Bursera,* Ericaceae, *Acalypha, Croton, Hura, Sapium,* Flacourtiaceae, *Hippocratea,* Mimosoideae, other Fabaceae, *Bunchosia, Brysonima,* Malvaceae, Melastomataceae/Combretaceae, *Virola,* Myrtaceae, Arecaceae other than *Iriartea,* Rhamnaceae, Rosaceae, Rubiaceae, *Zanthoxylum,* Sapindaceae, Sapotaceae/Meliaceae, *Guazuma,* and *Mortoniodendron*. Cyperaceae pollen is excluded from the pollen sum used for calculating percentages for the composite groups and all individual taxa other than itself, but is included in the pollen concentration curve and in the pollen sum used to plot fern spore percentages. Charcoal concentrations are expressed on a dry weight basis and are for fragments greater than or equal to 25μm in length scanned during the pollen counts. Revised slightly from Figure 4 in L. A. Northrop, S. P. Horn. (1996). Pre-Columbian agriculture and forest disturbance in Costa Rica: Paleoecological evidence from two lowland rainforest lakes. *The Holocene, 6,* 289–299; copyright 1996 Edward Arnold (Publishers) Ltd., used with permission of the publisher.

rainforest ecology [42]. Mature forests of the station were once thought to be pristine or virgin, but local archaeological research, together with our sediment coring and paleoecological analyses, have revealed that the site has a long history of human occupation and land use that includes maize agriculture on fertile alluvial soils bordering the Puerto Viejo river [25, 33], and prehistoric fires of likely human origin on both terrace and residual soils [27 and unpublished data]. Two sediment cores from swamps on alluvial terraces contain prehistoric maize pollen. From the *Cantarrana Swamp* we recovered a 5.9-meter-long sediment core that spans the past ca. 3200 years, based on AMS and standard radiocarbon dates on wood, leaves, and charcoal in the core. Lisa Kennedy [32] discovered maize pollen at a depth of 216 centimeters in the core during standard pollen counts conducted at 400X magnification for her M.S. thesis research on paleoecological evidence of environmental history at La Selva. To check for additional maize grains and to determine the temporal distribution of maize pollen in the core, we subsequently conducted low power (100X magnification) scans of one to several microscope slides from over 60 additional levels in the core [33]. We initially reported that maize pollen was present between 104–224 centimeters; however, further research revealed maize to be present over the wider interval of 92–312 centimeters [33]. Most of the pollen samples with maize also contain microscopic charcoal, suggesting fires associated with agriculture. Extrapolating from calibrated radiocarbon dates on charcoal within and below this section suggests that the "maize interval" at the Cantarrana Swamp corresponds to the period from about AD 920 to AD 1670 [34]. This time period falls mainly within the subregional La Cabaña archaeological phase (Period VI in Table 27-1), which began about AD 1000 and extended to the conquest in AD 1550, but also overlaps with the latest part of the subregional La Selva phase from AD 500 to AD 1000 (Period V in Table 9-1).

The abundance of microscopic charcoal in lower samples from the Cantarrana core suggested that the absence of maize pollen in these samples might only indicate the local absence of maize cultivation. Was the charcoal found below the "Cantarrana maize interval" evidence of agricultural activity elsewhere at La Selva? To test this possibility, we cored the nearby "Research" and "Machita" swamps (unofficial names). Pollen preservation was poor in the two cores we obtained from the *Research Swamp,* possibly because the site dries out at intervals. However, the Research Swamp sediments do contain charcoal fragments that may indicate prehistoric agricultural activity; one charcoal sample from the core yielded an AMS radiocarbon date of 2710 ± 50 ^{14}C yr BP (about 870 BC).

We also recovered two cores from the *Machita Swamp,* located at the head of the Quebrada Leonel, about 1 km northwest of the Cantarrana swamp. Pollen is absent below 70 centimeters depth in the 3.0-meter-long core we recovered from near the edge of the swamp, probably (as at the Research Swamp) due to periodic drying of the deposits and consequent oxidation of pollen grains. But pollen is present throughout the 3.4-meter-long core we raised in 1992 from near the center of the swamp. Preservation is not as good in these samples as in our samples from the Cantarrana swamp, and we have not attempted complete pollen counts. We have instead focused on the presence or absence of maize as determined by scanning multiple slides from 31 levels in 1992 core and 6 in the 1997 core. In the 1997 core, we found maize pollen at 40 centimeters, and in the 1992 core, we found maize pollen between 40 and 248 centimeters (Figure 9-3). There were some gaps in maize occurrence from 40–201 centimeters, but between 201–248 centimeters nearly every sample we examined contained maize pollen [25]. Our sediment cores from the Machita Swamp also preserve stable carbon isotopic ($\delta^{13}C$) signatures that we interpret as additional evidence of past maize cultivation in the watershed [36]. An AMS radiocarbon date on charcoal fragments sieved from the same interval of the core that contains the lowest maize pollen yielded a date of 2540 ± 60 ^{14}C yr BP (about 650 BC). We have not obtained a radiocarbon date on the uppermost maize grains in the Machita sediments, but extrapolating from the date presented earlier would suggest that the most recent maize grains might have been deposited about AD 1560. This date is consistent with the distribution of maize pollen in the Cantarrana sediments, again pointing to maize agriculture during the La Cabaña archaeological phase (Period VI). The deepest maize grains in the core would correspond to the early part of Period IV (subregional La Montaña phase, 1000–300 BC), and the presence of maize in intervening samples would suggest maize cultivation also took place at the site during the later part of Period IV (subregional El Bosque Phase, 300 BC–AD 500), and during Period V (subregional La Selva phase, AD 500–1000). These data extend the known time frame of human occupancy of La Selva forests. Almost all artefacts known from the site correspond to the El Bosque and La Selva archaeological phases [51, 64]. Only recently has a single, possible La Montaña-age artefact surfaced (from near the Machita Swamp); no La Cabaña artefacts are known from the reserve [64]. The maize pollen evidence from La Selva also highlights the need for modern ecological studies to consider the possible effect of prehistoric activity on current landscape patterns [55].

Further evidence of Pre-Columbian agriculture in the Central Atlantic Watershed subregion comes from the sediments of *Laguna María Aguilar* (10°18′09″ N, 84°11′24″W, 770 m elevation), located approximately 25 kilometers upslope from La Selva (see Figure 27-1). At both this site and nearby *Laguna Hule* (10°17′50″ N, 84°12′43″W, 740 m), pollen assemblages deposited between about 700 BC, when Laguna Hule formed, and about AD 1950

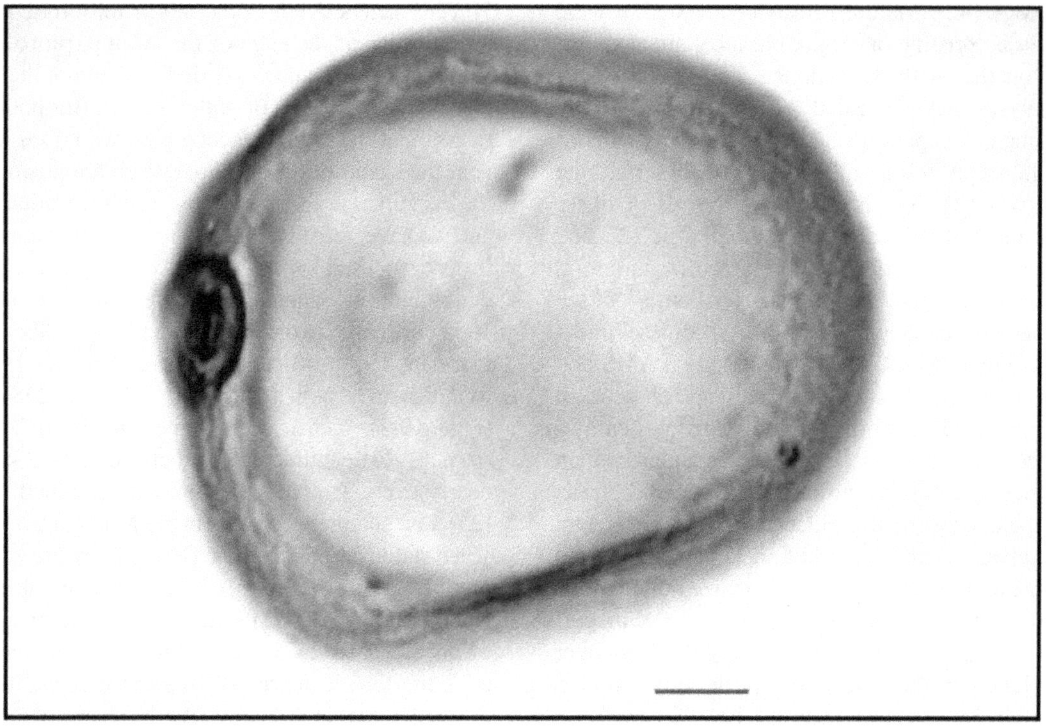

FIGURE 9-3 Maize pollen from the Machita swamp, La Selva Biological Station. Scale bar is 10 µm. Photograph by Karen Burhenn.

indicate relatively intact rainforest vegetation, with evidence of natural disturbances but few signs of agricultural activity [20, 26]. But older sediments from Laguna María Aguilar, buried by tephra from the eruption that formed the explosion crater occupied by Laguna Hule [22], contain abundant maize grains indicating local cultivation of the crop during the La Montaña phase of Period IV [26] (bracketing AMS dates of about 1030 BC and about 700 BC).

The earliest maize pollen evidence from Laguna Bonillita, Machita Swamp, and Laguna María Aguilar—all dating to the La Montaña phase of Period IV—constitutes the oldest botanical evidence of maize agriculture in the Central Atlantic Watershed subarea of the Central Highlands–Atlantic Watershed archaeological region. However, older evidence of maize cultivation exits in the Arenal–Tilaran subarea (see Figure 9-1). A maize macrofossil unearthed during the Arenal Prehistory Project (at the present site of the Arenal Reservoir at the foot of Volcán Arenal, see Figure 27-1) is the oldest evidence of maize in the archaeological region and the oldest maize macrofossil in the country. This carbonized maize kernel was dated indirectly (by charcoal in the same deposit) to 4450 ± 70 ^{14}C yr BP, or about 3140 BC [56]. This date falls within Period III and the subregional Tronadora phase. Archaeological excavations in the Arenal project also yielded maize pollen and phytoliths from this and later periods [7, 56]. In a sediment core we recovered from nearby *Lago Cote* (10°35′00″ N, 84°54′42″W, 650 m elevation), Martin Arford [2] found maize pollen directly above an interval of charcoal dated to 3630 ± 70 ^{14}C yr BP, or about 2000 BC.

The Guanacaste–Nicoya Archaeological Region

The Arenal project maize macrofossil constitutes the oldest maize macrofossil in Costa Rica—but it is not the earliest evidence of maize in the country. That title presently belongs to maize pollen in a sediment core from *Laguna Martínez,* one of six lakes that Martin Arford and I have cored on the lower Pacific slope of Volcán Miravalles in the Guanacaste–Nicoya archaeological region [3] (see Figure 9-1). Laguna Martínez (10°38′31″ N, 85°11′49″W, 340 m elevation) is a small (1.5 ha) lake that formed, as did others in the zone, over 8000 years ago as a result of lava and lahar flows from Volcán Miravalles. Our core from Laguna Martínez contains about 5 meters of lacustrine sediment and has a basal date of 7610 ± 50 ^{14}C yr BP (about 6470 BC). AMS radiocarbon dates on charcoal and dicotyledonous leaf fragments indicate rapid initial sedimentation in the basin followed by much slower sedimentation after 6060 ± 40 ^{14}C yr BP (about 4970 BC). The lowest maize pollen occurs in a sample with abundant tree pollen, but percentages of tree pollen begin to decline 5 centimeters above the lowest maize interval, and percentages of grass pollen and charcoal abundance both increase. Based on AMS dating of tiny charcoal fragments picked from the same sediment interval, this lowest occurrence of maize pollen dates to 4760 ± 40 ^{14}C yr BP (about 3550 BC). As a check on this date, we obtained a second AMS date on charcoal fragments 5 centimeters higher in the core, from an interval that also contains maize pollen along with increased grass pollen and charcoal. This sample dated to 4410 ± 40 ^{14}C yr BP (about 3060 BC). That these earliest maize grains occur just before and during times of declining tree pollen percentages, increases in other grasses, and abundant microscopic and macroscopic charcoal argues for local forest clearance and maize cultivation, likely right along the lake shore, during Period III. Pollen data from higher levels in the core indicate that once maize agriculture arrived at the site, it persisted: Almost every sample above the initial interval of prehistoric cultivation contains maize pollen, until the uppermost section corresponding to the historic period, in which maize occurs only sporadically.

Pollen is poorly preserved in sediment cores we have recovered from marshes at the Palo Verde Biological Station in the Tempisque River Valley (see Figure 9-1), but an archaeological site excavated by Sergio Chávez [11] of the University of Costa Rica yielded a carbonized maize kernel in a burial offering. The kernel, directly dated by AMS, yielded a radiocarbon age of 820 ± 40 ^{14}C yr BP (about AD 1210), which places it within the subregional Bebedero phase of Period VI. Blanco and Mora [7] listed maize macrofossils from Period VI at two other sites in the Guanacaste–Nicoya region and Period V macrofossils from one site. They also cited pollen evidence from the Nascoscolo archaeological site of maize cultivation during Period IV. Lynette Norr [45] described isotopic analyses of bone collagen from sites along the northern Pacific coastal region that suggest maize consumption between AD 300 and AD 1500 (later Period IV and Periods V and VI.) Our pollen results from six lakes on the lower slope of Volcán Miravalles are consistent with these archaeological finds [3].

The Diquís Archaeological Region

Four lakes in the Diquís archaeological region provide evidence of maize agriculture. Our longest record is from *Laguna Zoncho* (8°48′47″ N, 82°57′47″W, 1190 m), a small (0.75 ha) lake near the Panamanian border that occupies a basin that formed from mass wasting, faulting, or both [12]. AMS dates on charcoal and on leaf and wood fragments indicate that sediment began accumulating in the lake about 1330 BC, during the subregional Curré phase of Period III, and from the beginning of the record, included maize pollen (Figure 9-4) [7]. Percentages of herb and tree pollen fluctuate significantly upcore, suggesting varying levels of agricultural activity and human impact that are closely tracked by stable carbon isotope values [36]. The Zoncho record differs from the Bonillita record (see Figure 9-2) in showing strong forest recovery following the conquest, with steep increases in tree pollen and declines in grass pollen and charcoal abundance. The implication is a marked reduction in local human populations; however, low power scans reveal the presence of maize pollen in most samples from this zone, indicating that the site was not completely abandoned.

The maize pollen in basal sediments from Laguna Zoncho constitutes the oldest botanical evidence of maize agriculture in the Diquís region; the oldest maize macrofossils reported by Blanco and Mora [7] are from Periods V and VI. Pollen data from Laguna Santa Elena (8°55′50″N, 82°55′39″W, 1100 m elevation) [1], and pollen and stable isotope data from Laguna Vueltas (8°58′01″N, 83°10′41″W, 270 m elevation) [28, 69] provide additional evidence of maize cultivation during Periods V and VI (and the late Period IV at Santa Elena). Both of these records show evidence of significant forest clearance along with maize agriculture during Periods V and VI, corresponding to the subregional Aguas Buenas and Chiriquí phases, followed by postconquest regeneration of forests.

Nearby *Laguna Gamboa* (8°43′19″N; 82°59′17″W, 1460 m elevation) is the highest site in Costa Rica to have yet yielded evidence of maize agriculture. A maize pollen grain in a sample 1 centimeter above an AMS radiocarbon date on charcoal of 2450 ± 70 ^{14}C yr BP (about 590 BC) suggests local maize cultivation during Period IV.

CONCLUSION

Pollen, charcoal, and stable carbon isotope evidence from lacustrine sediments complements and extends archaeological evidence for Pre-Columbian maize cultivation in Costa Rica, as elsewhere in Central America [16, 23, see Chapter 26]. Maize microfossils in sediment cores predate macrofossils from archaeological sites in the Diquís and Guanacaste–Nicoya archaeological regions, and in the Central Atlantic Watershed subregion of the Central Highlands–Atlantic Watershed archaeological region. However, the earliest maize pollen grains yet discovered in Costa Rica are only slightly older than the earliest macrofossil, and neither date seems unreasonable when compared with directly dated macrofossils from Mexico. The two-sigma calibrated age ranges of both dates [3] overlap with the calibrated age range of the earliest AMS date on directly dated maize cobs from the San Marcos (Tehuacán) cave [19, 38]. However, the maize macrofossil and pollen grains from Costa Rican maize are about 700–1000 radiocarbon years younger than directly dated maize cobs from Guila Naquitz cave in Oaxaca, Mexico [3, 50], and their calibrated age ranges do not overlap with the calibrated age ranges of the Guila Naquitz cobs.

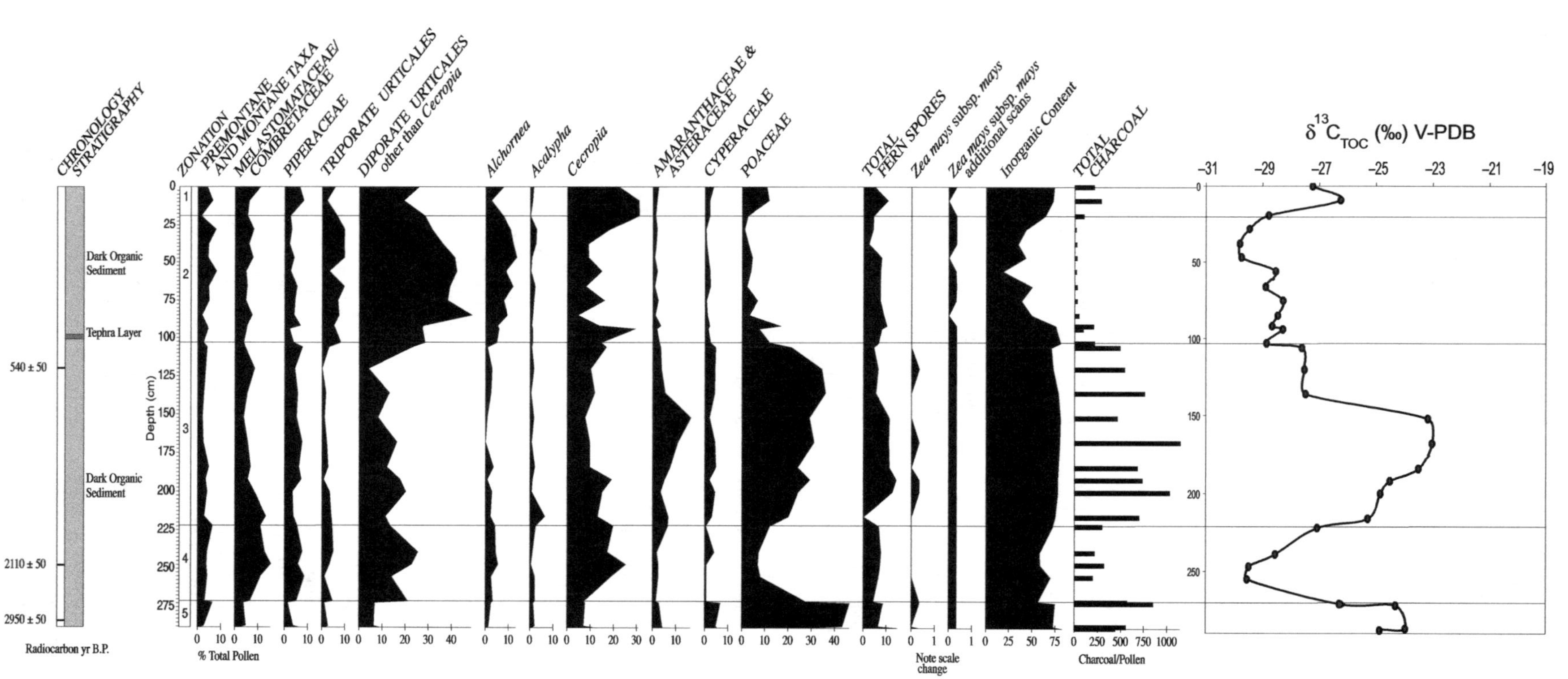

FIGURE 9-4 Sediment stratigraphy, pollen, microscopic charcoal, and $\delta^{13}C_{TOC}$ data for the upper lacustrine section of the Laguna Zoncho sediment core. Radiocarbon dates are uncalibrated. The Premontane-Montane category includes *Podocarpus, Quercus, Alnus, Myrica,* and *Weinmannia,* and the diporate and triporate Urticales groups include all two- and three-pored pollen in this order other than *Ficus* and *Cecropia.* Pollen percentages are calculated as a percent of total pollen (excluding indeterminates), whereas fern spore values are shown as the percent of total pollen plus fern spores. The additional scans curve for *Zea mays* subsp. *mays* indicates presence or absence of the pollen type as determined from complete scanning at low magnification of three slides prepared from each of 32 levels in the core. Pollen and charcoal curves are redrawn from Figure 4 in R. M. Clement, S. P. Horn. (2001). Pre-Columbian land-use history in Costa Rica: A 3000-year record of forest clearance, agriculture and fires from Laguna Zoncho. *The Holocene, 11,* 419–426, copyright 2001 Edward Arnold (Publishers) Ltd., used with permission of the publisher; isotope curve is from Figure 2 in C. S. Lane, S. P. Horn, C. I. Mora. (2004). Stable carbon isotope ratios in lake and swamp sediments as a proxy for prehistoric forest clearance and crop cultivation in the Neotropics. *Journal of Paleolimnology, 32,* 375–381; copyright 2004 Kluwer Academic Publishers, used with kind permission of Springer Science and Business Media.

The recent suggestion of Hugh Iltis [29] that maize may have been initially domesticated for its sugary pith or other edible parts rather than its grain led John Smalley and Michael Blake [58] to propose that the early and rapid spread of maize southward from Mexico was in part fueled by the importance of alcohol production from stalk sugar. If the focus of early maize production at sites in southern Central America and South America was corn stalks, rather than kernels, we might expect maize kernels and cobs to be less commonly preserved in hearths and to appear later than sedimentary maize pollen. The present discrepancy between microfossil and macrofossil ages for maize in Costa Rica seems not to require that we imagine early maize moving through Costa Rica with primary attention directed toward stalk sugar, rather than grain. Maize stalks may have been an important source of sugar for alcohol production in Pre-Columbian Costa Rica, but kernels were also important. The gaps in time that exist between the earliest maize microfossils in lake and swamp sediments and the earliest excavated maize macrofossils may be a consequence of the high preservation potential of maize pollen grains in sediments, and limited archaeological data [3]. Further archaeological research in Costa Rica may narrow these gaps, particularly if researchers can focus attention and resources on sites with paleoecological evidence of early maize, as our collaborators at the University of Costa Rica have been doing [54, 65]. Additional, direct AMS dating of archaeobotanical material (including already excavated materials in storage cabinets as well as new research finds) may also result in a narrowing of discrepancies, where they exist, between sedimentary and archaeological evidence of maize cultivation in archaeological regions of Costa Rica.

Acknowledgments

I am grateful to a large number of students and colleagues who have collaborated in the recovery, analysis, and interpretation of sedimentary records that have yielded evidence of maize cultivation in Costa Rica. Douglas Baird, Jr., Isabel Avendaño, Jairo Mora, Andrew Schiller, Rafael Arce, Lisa Northrop, Glenn Hyman, John Rodgers III, Edward Gray, Lisa Kennedy, Martin Arford, Rachel Clement, Kevin Anchukaitis, Duane Cozadd, Chad Lane, Zachary Taylor, Robert L. Sanford, Jr., Maureen Sánchez, Sergio Chávez, Ken Orvis, and Claudia Mora played key roles in developing the sediment records described here. Lynn Clark shared her knowledge of Costa Rican grasses, and herbarium material for making reference slides. I am pleased to also acknowledge the hospitality and logistical and other support provided in Costa Rica by Carolyn Hall, Luis Guillermo Brenes, Gilbert Vargas, and other faculty as well as students in the School of Geography at the University of Costa Rica; by faculty, students, and staff in the Laboratory of Archaeology at the University of Costa Rica; and by station directors and staff members of the Organization for Tropical Studies. The Costa Rican government and private landowners have graciously permitted and facilitated our field sampling. Funds for field and laboratory work were provided by the National Science Foundation (SES-9111588, BCS-0242286), the A.W. Mellon Foundation, The National Geographic Society, and the University of Tennessee. I thank Robert Dull and the editors for helpful reviews.

References Cited

1. K. J. Anchukaitis, S. P. Horn. (2005). A 2000-year reconstruction of forest disturbance from southern Pacific Costa Rica. *Palaeogeography, Palaeoecology, Palaeoclimatology, 221,* 35–54.
2. M. R. Arford. (2001). Late Holocene environmental history and tephrostratigraphy in northwestern Costa Rica: A 4000 year record from Lago Cote. M.S. thesis. Knoxville: University of Tennessee.
3. M. R. Arford, S. P. Horn. (2004). Pollen evidence of the earliest maize agriculture in Costa Rica. *Journal of Latin American Geography, 3,* 108–115.
4. M. M. Bender. (1971). Variations in the $^{13}C/^{12}C$ ratios of plants in relation to the pathway of photosynthetic carbon dioxide fixation. *Phytochemistry, 10,* 1239–1244.
5. J. Bennetzen, E. Buckler, V. Chandler, J. Doebley, J. Dorweiler, B. Gaut, M. Freeling, S. Hake, E. Kellogg, R. S. Poethig, V. Walbot, S. Wessler. (2001). Genetic evidence and the origin of maize. *Latin American Antiquity, 12,* 84–86.
6. B. E. Berglund. (1986). *Handbook of Holocene palaeoecology and palaeohydrology.* New York: John Wiley and Sons.
7. A. Blanco V., G. Mora S. (1994). Plantas silvestres y cultivadas según la evidencia arqueobotánica en Costa Rica. *Vínculos, 20,* 53–77.
8. M. Bowler, V. A. Hall. (1989). The use of sieving during standard pollen pre-treatment of samples of fossil deposits to enhance the concentration of large pollen grains. *New Phytologist, 111,* 511–515.
9. E. T. Burden, J. H. McAndrews, G. Norris. (1986). Palynology of Indian and European forest clearance and farming in lake sediment cores from Awenda Provincial Park, Ontario. *Canadian Journal of Earth Sciences, 23,* 43–54.
10. R. Byrne, S. P. Horn. (1989). Prehistoric agriculture and forest clearance in the Sierra de Los Tuxtlas, Veracruz, Mexico. *Palynology, 13,* 181–193.
11. S. Chávez Ch. (1999). Los pueblos antiguous y los humedales: un nexo vital. *Revista Informe Semestral, 35,* 53–67.
12. R. M. Clement, S. P. Horn. (2001). Pre-Columbian land-use history in Costa Rica: A 3000-year record of forest clearance, agriculture and fires from Laguna Zoncho. *The Holocene, 11,* 419–426.
13. P. Colinvaux, P. E. De Oliveira, P. J. E. Moreno. (1999). *Amazon pollen manual and atlas—Manual e Atlas Palinológico da Amazônia.* Amsterdam: Harwood Academic Publishers.
14. J. Doebley. (1990). Molecular evidence and the evolution of maize. *Economic Botany, 44*(3 suppl.), 6–27.
15. J. Doebley. (2004). The genetics of maize evolution. *Annual Review of Genetics, 38,* 37–59.
16. R. A. Dull. (2004). An 8000-year record of vegetation, climate, and human disturbance from the Sierra de Apaneca, El Salvador. *Quaternary Research, 61,* 159–167.
17. M. W. Eubanks. (2001). An interdisciplinary perspective on the origin of maize. *Latin American Antiquity, 12,* 91–98.
18. M. L. Fearn, K-B Liu. (1995). Maize pollen of 3500 B.P. from southern Alabama. *American Antiquity, 60,* 109–117.
19. G. J. Fritz. (1994). Are the first American farmers getting younger? *Current Anthropology, 35,* 305–309.
20. E. G. Gray, S. P. Horn. (1994). A Late Holocene history of a Costa Rican rainforest. Paper presented at the annual meeting of the Association of American Geographers, San Francisco, California, March 29–April 2, 1994.
21. B. E. Hammel, M. H. Grayum, C. Herrera, N. Zamora, (Eds.). (2003). *Manual de Plantas de Costa Rica, Vol. III Monocotiledóneas (Orchidiaceae-Zingiberaceae).* St. Louis: Missouri Botanical Garden.
22. S. P. Horn. (2001). The age of the Hule explosion crater, Costa Rica, and the timing of subsequent tephra eruptions: Evidence from lake sediments. *Revista Geológica de America Central, 24,* 57–66.
23. S. P. Horn. (In press). Late Quaternary climate and environmental history from lake and swamp sediments in Central America. In:

J. Bundschuh, G. E. Alvarado I., (Eds.), Central America: Geology, resources, hazards. Rotterdam: Balkema Publishers.

24. S. P. Horn, K. A. Haberyan. (1993). Costa Rican lakes: Physical and chemical properties. *National Geographic Research and Exploration, 9,* 86–103.
25. S. P. Horn, L. M. Kennedy. (2001). Pollen evidence of maize cultivation 2700 B.P. at La Selva Biological Station, Costa Rica. *Biotropica, 33,* 191–196.
26. S. P. Horn, J. C. Rodgers III. (1996). Environmental history in Costa Rica: Sediment records from Laguna María Aguilar. Paper presented at the annual meeting of the Association of American Geographers, Charlotte, North Carolina, April 9–13, 1996.
27. S. P. Horn, R. L. Sanford Jr. (1992). Holocene fires in Costa Rica. *Biotropica, 24:* 354–361.
28. S. P. Horn, M. Sánchez, D. Cozadd, K. H. Orvis, K. Anchukaitis. (2004). Paleoecological and archaeological investigations at Laguna Vueltas, Costa Rica. Paper presented at the annual meeting of the Association of American Geographers, Philadelphia, Pennsylvania, March 14–19, 2004.
29. H. H. Iltis. (2000). Homeotic sexual translocations and the origin of maize *(Zea mays*, Poaceae): A new look at an old problem. *Economic Botany, 54,* 7–42.
30. H. H. Iltis, B. F. Benz. (2000). *Zea nicaraguensis* (Poaceae), a new teosinte from Pacific coastal Nicaragua. *Novon, 10,* 382–390.
31. G. A. Islebe, H. Hooghiemstra, M. Brenner, J. H. Curtis, D. A. Hodell. (1996). A Holocene vegetation history from lowland Guatemala. *The Holocene, 6,* 265–271.
32. L. M. Kennedy. (1998). Prehistoric agriculture, fires, and droughts at the La Selva biological station, Costa Rica: Paleoecological evidence from the Cantarrana swamp. M.S. thesis. Knoxville: University of Tennessee.
33. L. M. Kennedy, S. P. Horn. (1997). Prehistoric maize cultivation at the La Selva Biological Station, Costa Rica. *Biotropica, 29,* 368–370.
34. L. M. Kennedy, S. P. Horn. (in prep). A 3200-year record of forest disturbance at La Selva Biological Station, Costa Rica.
35. W. Kennedy. (1968). Archaeological investigations in the Reventazón river drainage area, Costa Rica. Ph.D. thesis. New Orleans: Tulane University.
36. C. S. Lane, S. P. Horn, C. I. Mora. (2004). Stable carbon isotope ratios in lake and swamp sediments as a proxy for prehistoric forest clearance and crop cultivation in the Neotropics. *Journal of Paleolimnology, 32,* 375–381.
37. C. S. Lane, Z. Taylor, C. I. Mora, S. P. Horn. (2004). Reassessment of prehistoric human impacts at Laguna Bonillita, Costa Rica, using stable carbon isotopes. Paper presented at the annual meeting of the Association of American Geographers, Philadelphia, Pennsylvania, March 14–19, 2004.
38. A. Long, B. Benz, J. Donahue, A. Jull, A., L. Toolin. (1989). First direct AMS dates on early maize from Tehuacán, Mexico. *Radiocarbon, 31,* 1035–1040.
39. B. Ludlow-Wiechers, J. L. Alvarado, M. Aliphat. (1983). El polen de *Zea* (maíz y teosinte): Perspectivas para conocer el origen de maíz. *Biotica, 8,* 235–258.
40. R. S. MacNeish, M. E. Eubanks. (2000). Comparative analysis of the Río Balsas and Tehuacán models for the origin of maize. *Latin American Antiquity, 11,* 3–20.
41. Y. Matsuoka, Y. Vigouroux, M. M. Goodman, J. Sanchez G., E. Buckler, J. Doebley. (2002). A single domestication for maize shown by multilocus microsatellite genotyping. *Proceedings of the National Academy of Sciences of the United States of America, 99,* 6080–6084.
42. L. A. McDade, G. S. Hartshorn. (1994). La Selva biological station. In: L. A. McDade, K. S. Bawa, H. Hespenheide, G. S. Hartshorn, (Eds.), *La Selva: Ecology and natural history of a neotropical rain forest.* Chicago: University of Chicago Press. pp. 6–14.
43. P. D. Moore, J. A. Webb, M. E. Collinson. (1991). Pollen analysis, 2nd ed. Oxford: Blackwell Scientific Publications.
44. L. C. Norr. (1991). Nutritional consequences of prehistoric subsistence strategies in lower Central America. Ph.D. thesis. Urbana-Champaign: University of Illinois at Urbana-Champaign.
45. L. Norr. (1996). Prehistoric coastal subsistence in northwestern Costa Rica: Geographical diversity and chronological trends. In: F. W. Lange, (Ed.), *Paths to Central American prehistory.* Niwot: University Press of Colorado. pp. 253–269.
46. L. A. Northrop, S. P. Horn. (1996). Pre-Columbian agriculture and forest disturbance in Costa Rica: Paleoecological evidence from two lowland rainforest lakes. *The Holocene, 6,* 289–299.
47. M. H. O'Leary. (1981). Carbon isotope fractionation in plants. *Phytochemistry, 20,* 553–567.
48. D. R. Piperno. (1988). *Phytolith analysis: An archaeological and geological perspective.* Orlando, FL: Academic Press.
49. D. R. Piperno. (1994). Phytolith and charcoal evidence for prehistoric slash and burn agriculture in the Darien rain forest of Panama. *The Holocene, 4,* 321–325.
50. D. R. Piperno, K. V. Flannery. (2001). The earliest archaeological maize *(Zea mays* L.) from highland Mexico: New accelerator mass spectrometry dates and their implications. *Proceedings of the National Academy of Sciences of the United States of America, 98,* 2101–2103.
51. I. Quintanilla. (1991). Ocupaciones precolombinas en el bosque tropical lluvioso: evaluación arqueológico de la estación biológica La Selva. Unpublished report. San José, Costa Rica: Organization for Tropical Studies.
52. G. S. Raynor, E. C. Ogden, K. V. Hayes. (1972). Dispersion and deposition of corn pollen from experimental sources. *Agronomy Journal, 64,* 420–427.
53. P. J. Reimer, M. G. L. Baillie, E. Bard, A. Bayliss, J. W. Beck, C. J. H. Betrand, P. G. Blackwell, C. E. Buck, G. S Burr, K. B. Cutler, P. E. Damon, R. L. Edwards, R. G. Fairbanks, M. Friedrich, T. P. Guilderson, K. A. Hughen, B. Kromer, F. G. McCormac, S. Manning, C. Bronk Ramsey, R. W. Reimer, S. Remmele, J. R. Southon, M. Stuiver, S. Talamo, F. W. Taylor, J. van der Plicht, C. E. Weyhenmeyer. (2004). IntCalL04 terrestrial radiocarbon age calibration, 0–26 cal kyr BP. *Radiocarbon, 46,* 1029–1058.
54. M. Sánchez P., P. Rojas H. (2002). Asentamientos humanos antiguos en las tierras intermedias del canton de Coto Brus (avance de investigación). *Cuadernos de Antropología, 12,* 87–106.
55. R. L. Sanford Jr., S. P. Horn. (2000). Holocene rain-forest wilderness: A neotropical perspective on humans as an exotic, invasive species. In: S. F. McCool, D. N. Cole, W. T. Bourie, J. O'Loughlin. (Eds.), *Wilderness science in a time of change Volume 3: Wilderness as a place for scientific inquiry.* Ogden, UT: U.S. Department of Agriculture Forest Service, Rocky Mountain Research Station. pp. 168–173.
56. P. D. Sheets, B. R. McKee, (Eds.). (1994). *Archaeology, volcanism, and remote sensing in the Arenal Region, Costa Rica.* Austin: University of Texas Press.
57. A. Sluyter. (1997). Analysis of maize *(Zea mays subsp. mays)* pollen: Normalizing the effects of microscope-slide mounting media on diameter determinations. *Palynology, 21,* 35–39.
58. J. Smalley, M. Blake. (2003). Sweet beginnings: Stalk sugar and the domestication of maize. *Current Anthropology, 44,* 675–703.
59. B. D. Smith. (1994–1995). The origins of agriculture in the Americas. *Evolutionary Anthropology, 3,* 174–184.
60. B. D. Smith. (2001). Documenting plant domestication: the consilience of biological and archaeological approaches. *Proceedings of the National Academy of Sciences of the United States of America, 98,* 1324–1326.
61. M. J. Snarskis. (1975). Excavaciones estratigráficas en la vertiente Atlántica de Costa Rica. *Vínculos, 1,* 2–17.

62. M. J. Snarskis. (1981). The archaeology of Costa Rica. In: E. P. Benson, (Ed.), *Between continents/between seas: Precolumbian art of Costa Rica.* New York: Harry N. Abrams Publishers. pp. 15–84.
63. M. J. Snarskis. (1984). Central America: The lower Caribbean. In: F. W. Lange, D. S. Stone, (Eds.), *The archaeology of lower Central America.* Albuquerque: University of New Mexico Press. pp. 195–232.
64. R. F. Sol C. (2000). Asentamientos prehispánicos en la reserva biológica La Selva, Sarapiquí, Costa Rica: Sistemas de explotación de recursos naturales en un bosque tropical lluvioso. Licenciatura thesis. San José: University of Costa Rica.
65. K. Soto S., L. Gómez B. (2002). Sitio arqueológico el Zoncho (CAT. U.C.R. No.168): Una manifestación de los agricultores especializados en las tierras intermedias de San Vito, Cantón de Coto Brus, Puntarenas. Licenciatura thesis. San José: University of Costa Rica.
66. J. E. Staller, R. G. Thompson. (2002). A multidisciplinary approach to understanding the initial introduction of maize into coastal Ecuador. *Journal of Archaeological Science, 29,* 33–50.
67. D. Stone. (1956). Date of maize in Talamanca, Costa Rica: A hypothesis. *Journal de la Société des Américanistes, 45,* 189–194.
68. M. Stuiver, P. J. Reimer. (1993). Extended C-14 database and revised CALIB 3.0 C-14 age calibration program. *Radiocarbon, 35,* 215–230.
69. Z. P. Taylor, S. P. Horn, C. I. Mora, D. Cozadd, D., K. H. Orvis, M. Sánchez, K. Anchukaitis. (2004). Stable carbon isotope ratios at Laguna Vueltas, Costa Rica, as indicators of land-use history and paleoclimate. Paper presented at the annual meeting of the Geological Society of America, Denver, Colorado, November 7–10, 2004.
70. R. J. Telford, E. Heegaard, H. J. B. Birks. (2004). The intercept is a poor estimate of a calibrated radiocarbon age. *The Holocene, 14,* 296–298.
71. A. Traverse. (1965). Preparation of modern pollen and spores for palynological reference collections. In: B. Kummel, D. Raup, (Eds.), *Handbook of paleontological techniques.* San Francisco: W. H. Freeman and Company. pp. 598–613.
72. N. J. van der Merwe, E. Medina. (1991).The canopy effect, carbon isotope ratios and foodwebs in Amazonia. *Journal of Archaeological Science, 18,* 249–259.
73. D. R. Whitehead, E. J. Langham. (1965). Measurement as a means of identifying fossil maize pollen. *Bulletin of the Torrey Botanical Club, 92,* 7–20.
74. D. R. Whitehead, M. C. Sheehan. (1971). Measurement as a means of identifying fossil maize pollen, II: The effect of slide thickness. *Bulletin of the Torrey Botanical Club, 98,* 268–271.
75. H. E. Wright Jr., D. H. Mann, P. H. Glasner. (1984). Piston corers for peat and lake sediments. *Ecology, 65,* 657–659.

CHAPTER

10

Caribbean Maize

First Farmers to Columbus

LEE A. NEWSOM

Department of Anthropology, The Pennsylvania State University, University Park, Pennsylvania

Glossary

Archaic Age (also Lithic Age) General designation for original period of human inhabitation of the Caribbean Islands, beginning approximately 6000 years before present.

Mangrove Maritime coastal tropical–subtropical forest association.

Ostionoid Collective term for later Ceramic Age human occupations of the Caribbean, with emphasis on the Greater Antilles subregion.

Phytolith Plant silica originally deposited in gel form, which then occurs either as an incrustation or impregnation of the cell wall or in the form of translucent deposits (silica-bodies) in the cell lumina; some are crystalline in form, but more common are opaline structures.

Saladoid (also Huecoid) Collective term for early Ceramic Age human occupations of the Caribbean Islands.

Evidence for the Pre-Columbian presence of *Zea mays* in the Caribbean Islands has been recovered from a few locations in the central portion of the archipelago. Thus far, maize macroremains—cob and kernel specimens—are associated with cultural materials of a style characteristic of relatively late (ca. 800 AD and later) human occupations of the region. However, recent identifications of microremains—particularly starch grains and **phytoliths**—provide additional evidence for the presence of the plant and suggest a lengthier time range. Nevertheless, a central question remains as to the role and importance of maize among the indigenous people of the islands, given that in general evidence for the plant is scant. The apparent scarcity of archaeological maize, together with ethnohistoric descriptions suggestive of low intensity forms of use, provides a basis to hypothesize that this particular exotic plant introduction never achieved the status of a staple in the islands. A secondary role for maize in the Caribbean may have been a consequence of a stronger previous commitment to tropical root crop production, an economic focus that likely predated human colonization of the region, coupled with a relatively stable, productive marine-resource protein base. Alternative or additional social factors that may partially account for the patterns observed in the archaeological record include the perception and role of maize as a plant food reserved for elites that was used in ways apart from a common edible resource.

INTRODUCTION

The Caribbean Islands have a human history that spans several millennia before the fateful arrival of Europeans late in the fifteenth century, with indigenous people whose roots can be traced back to at least two continental homelands. Some, if not all, of the ancient human settlers in the Caribbean were plant cultivators. By the time of Columbus, Caribbean Indians managed a diverse array of native and introduced plants, including maize, in their gardens and fields, and they had adapted and developed agricultural practices entailing various technological improvements well suited to the island setting.

The cultural significance and long history of maize use in other regions of the Americas is well documented, but corresponding information on the antiquity and use of the plant in Caribbean has until recently been limited. Archaeological evidence for a Caribbean presence of maize has been recovered from one subregion of the archipelago. In this chapter I outline this evidence, which is composed of cob fragments, kernels, starch, phytoliths, and pollen from several sites. These identifications, although limited to a single subregion, shed important new light on the relative antiquity, types, and use of maize in the Caribbean, and provide a benchmark by which to compare ethnohistoric records of maize use in the region. Beyond this, I consider maize's status among a suite of edible plant resources that were cultivated prehistorically in the islands, and explore the possibility that maize served a role in Caribbean Indian life apart from being strictly a food item.

I begin with a brief synopsis of Caribbean biogeography and physical geography to touch on some of the constraints and potential to successfully grow crops in the region, particularly maize. I then briefly outline some relevant details of the culture history, as it is currently understood, followed by the archaeological records for maize. Finally, I discuss some of the ethnohistoric records concerning maize and Taíno Indian plant cultivation practices, and explore related issues to try to rectify that information with the archaeological data, as well as to suggest some explanations for the patterns observed archaeologically.

CARIBBEAN BIOGEOGRAPHY AND PHYSICAL GEOGRAPHY IN BRIEF

The Caribbean Islands stretch in an arc from northern South America to southern Florida (Figure 10-1), with the Caribbean Sea to the interior, western side, and bordered on the east by the Atlantic Ocean. The several hundred islands comprising the archipelago vary in size and conformation from those, which are quite large and mountainous, with sizable rivers and permanent streams, to others that are relatively small, with low relief and few of the resources necessary to sustain human life, especially freshwater. Offshore, the presence and extent of **mangroves** and shallow marine shelves supporting reef ecosystems also varies from one

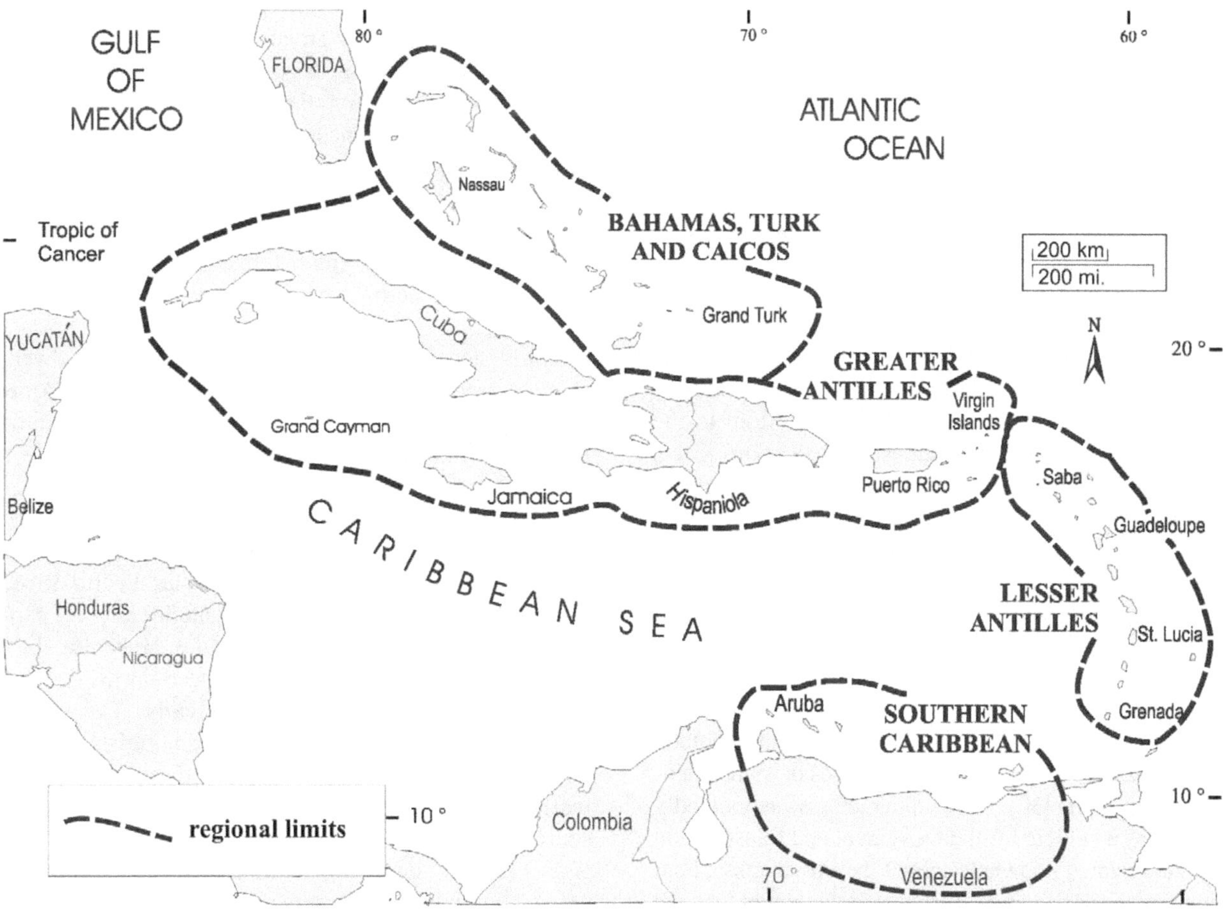

FIGURE 10-1 Caribbean Islands showing four major subregions (redrawn from original produced by F. E. Sergile).

island to the next. Moreover, while some of the islands are relatively isolated, others are clustered in small groups, and some smaller islands are situated relatively close to a large island with more diverse habitats or an adjacent mainland.

Aside from the basic details described earlier, different combinations of prevailing wind, rainfall pattern, latitude, altitudinal gradient, aspect, soil, and hydrology result in a number of distinct life zones and edaphic situations in the Caribbean. These, in turn, display diverse vegetation formations, any and several of which may be found on individual islands depending on the local circumstances [30]. Some islands and areas receive abundant tropical rainfall, enough to sustain wet tropical forest associations, while others are arid and desert-like, having either relief too low and unvaried to influence rainfall from the incoming moisture-laden trade winds, or by virtue of being positioned in the rain shadow of mountain ranges, and support dry forest communities. In fact, fully half of the vegetation in the Caribbean is classified within the dry forest life zones, areas that regularly experience a moisture deficit ranging from mild to relatively severe [30]. Furthermore, some of the islands are volcanic in origin and have deep fertile soils, while others were formed of limestone and coralline rock, the result of sea level fluctuations, and have little or no arable soil. Thus, while we can generally specify that the Caribbean is a region characterized by tropical to subtropical climatic conditions, it is important to recognize that different constraints on floristic patterns and biotic resources operate at the subregional, island, and local levels.

The inherent environmental diversity of the Caribbean is an important consideration with regard to human survival in the past. The individual islands, separately and as articulating groups, differed in the abundance and quality of the natural resources available for human use as indicated earlier. Consequently, they represented different opportunities for sustained human settlement, of course depending on available technology and social organization. More to the point, such a diverse ecosystem translates to various levels of potential and constraints on food resources and the production of crops such as maize [10]. This would likely have been a significant factor in the development of gardening practices in the region. It is important to note that where island clusters or mainland proximity are concerned, Caribbean archaeologists infer that ancient human inhabitants likely developed and maintained social and economic resource interaction spheres that allowed them to inhabitant less resource-rich islands and take full advantage of patchy island resources by traveling between islands or to the mainland in watercraft [30]. Such interactions undoubtedly served also as a vehicle for the movement and transmission of crops and related knowledge about their production, both within the region and between islands and mainland source regions.

SYNOPSIS OF THE HISTORY OF HUMAN SETTLEMENT AND CULTIVATION PRACTICES

Archaeologists, geographers, and others recognize four major subregions in the Caribbean archipelago (see Figure 23-1), defined in part by the biogeography outlined earlier, but also with some reference or relevance to internal histories of prehistoric human settlement. The earliest human colonists to venture into the islands came separately from two directions. Among the first were people from the Yucatan region of Central America, based on evidence from Cuba and Hispañiola, beginning around 3000–4000 BC [5, 13, 51, 54]. As early as 5500 BC there is evidence of the beginnings of a separate migration of people from South America who probably entered the archipelago from the south, based on evidence from Trinidad, and the southern Windward Islands. Little is known of the interactions between these groups of migrants from Central and South America, respectively, if and when they came into contact, perhaps in the vicinity of Puerto Rico or the northern Leeward Islands. In general, these first settlers in the region, especially during the so-called **Archaic Age** [13], are characterized as fisher–foragers who may have been residentially or logistically mobile to varying degrees. I would anticipate that some of these early groups became essentially or fully sedentary, much as has been inferred for coastal fisher–foragers living in south Florida at about the same time period, and who also made abundant use of coastal marine and subtropical forest resources [23]. A number of us working in the region have suggested that the Caribbean's Archaic Age inhabitants included at least some groups that were low level plant cultivators, who probably managed useful trees and herbs in home gardens or repeatedly visited settings [6, 30, 38].

During the middle first millennium BC, village-dwelling, ceramic-producing people from lowland South America (coastal Guiana and the Orinoco river valley) initiated migrations into the islands, or at least by this time they became more visible in terms of the archaeological record. One or more such groups entered the Caribbean on a relatively large scale and settled widely [15]. Their migrations through the Lesser Antilles to as far as Puerto Rico are readily traced by their distinctive material culture, particularly the ceramics tradition(s) (**Saladoid, Huecoid;** absolute relationships between the two are uncertain [31]). There is abundant evidence to suggest these early "Ceramic Age" [43, 44] colonists were extensive horticulturists who established settlements in places situated to take advantage of marine resources, arable soils, and reliable freshwater sources. Caribbeanists generally agree that early Ceramic Age villagers depended considerably on tropical root crops—familiar cultigens transferred from their homeland—and adapted their cultivation practices and ideas about root crop production to the island setting. The primary evidence for such an emphasis on root crops is the widespread presence of ceramic griddles at the relevant sites, and in many cases also microlithics, coral abraders, and other items, all of which are inferred to have been used to process bitter, potentially toxic tubers, particularly manioc (*Manihot esculenta,* Euphorbiaceae), into an edible form.

In the Greater Antilles and on some of the other islands, population growth dynamics and related social developments later in the so-called Ceramic Age culminated in regionally integrated, kinship-based polities—often described as chiefdoms, some more complex than others—which were present at the time of European contact [14–16, 52]. Strongly stratified societies emerged and thrived on some islands during this third overall phase of migration and human settlement, especially the Taíno Indians, who were centered in the Greater Antilles. European chroniclers detailed various aspects of Caribbean Indian (especially Taíno) social dynamics in the early historic era, including descriptions of complex multifaceted interrelations and interactions between elites. Hereditary leaders, or *caciques,* governed large villages and territories, and undoubtedly controlled many aspects of life within the community and their broader social networks. The separate segments of Taíno society included those with privileged social standing, or *nitainos,* as well as commoners and servants, or *naborías* [16, 52]. Within the elite group were specialists known as *behiques,* who seem to have been shamans responsible for a range of demonstrative and transformative religious practices including divination and curing rituals. There is strong evidence from the archaeology and ethnohistory of this latest period that contact between island dwellers and the mainland occurred or continued from earlier periods, including trade and the maintenance of myriad other social relationships [see various chapters in 53].

Based on a number of ethnohistoric observations and zooarchaeological research, it is clear that the Taíno and other late prehistoric groups of the Caribbean relied extensively on aquatic, especially marine, faunal resources, and they may have domesticated the muscovy duck and a native cat-sized rodent [30]. Guinea pigs were among the captive fauna introduced and maintained in the islands by at least the late Ceramic Age, and there is evidence that other small mammals were earlier introduced by people migrating and settling in the islands.

Spanish colonial records from the early historic period mention some of the plants that were cultivated and esteemed by the Taíno and other Caribbean indigenes. Specifically concerning maize, notable is a clear description of the plant dating to the second voyage of Columbus in 1494: "It is a grain of very high yield, of the size of the lupine, of the roundness of the chick-pea, and yields a meal ground to a very fine powder; it is ground as is wheat and yields a bread of very good taste; many chew the seeds when in need of nourishment" (Coma, quoted in [47, p. 55]). Some of the chroniclers indicate that two types or races of maize were grown in the islands, possibly with slightly different maturation periods. The following excerpt from Layfield [21, p. 154] is an example:

> Además del casabe cuentan con el maíz, del cual se hace un pan muy fino, que usan mucho. Hay *dos clases de maíz, el más pequeño no se diferencia mucho del arroz;* en proporción, tamaño y gusto: éste nunca le ví en plantación o crudo, pero lo he visto en la fuente, y al principio lo tome por arroz, exceptuando que yo me creí que estaba un poco inflado. Los que lo comían decían que sabía a arroz. *La otra clase la he visto en plantíos y es la misma o se parece mucho al grano que nosotros llamamos trigo;* crece con un tallo nudoso como una caña con grandes hojas esparcidas; crece hasta la altura de braza y media por lo menos, y en la misma punta brota la mazorca." (My emphasis added previously in italics; translation: Beyond cassava they depend on corn, from which a fine bread is made, which they make much use of. There are *two classes of corn; the smallest does not differ much from rice* in proportion, size, and flavor. I never saw this [type] in crop or raw, but I have seen it in the bowl, and at first I took it for rice, except that I believed it was a little inflated. Those that ate it [said] it tasted like rice. *I have seen the other class in fields, and it is the same, or appears very similar, to the grain that we call wheat.* It grows with a knotted stalk like a cane with large scattered leaves. It grows up to a height of at least a fathom and a half, and the ear grows from the very tip.)

The colonial documents also describe the preparation of agricultural plots by cutting and burning vegetation to clear ground: "The Indians first cut down the cane and trees where they wish to plant it [maize]. . . . After the trees and cane have been felled and the field grubbed, the land is burned over and the ashes are left as dressing for the soil, and this is much better than if the land were fertilized" [36, pp. 13–14]. This initial preparation was followed in some areas by the construction of small earthen mounds as growing platforms, described as having been about 1 foot high and 3 to 4 feet in diameter. Ultimately, the fields, or *conucos,* consisted of a series of the small circular earthen mounds, on which a variety of plants were said to have been grown [20, pp. 110; 47, pp. 51–54]. In some locations, field preparation and management was more labor intensive, including extensive terraforming to construct irrigation ditches (Hispaniola [19, p. 488]) and the construction of curved growing platforms or terraces on mountain slopes (Puerto Rico [33, 35]). Manioc, sweet potato (*Ipomoea batatas*, Convovulaceae; another South American native), other root crops, maize, and various other plants were reportedly grown in these fields. In addition, separate home gardens may have been the venues where particularly esteemed fruit trees, utilitarian plants such as gourds, cotton, and calabash tree (*Cresentia* sp., Bignoniaceae), specialty plants such as tobacco and *cojóbana* (*Anandenanthera peregrina,* Fabaceae; the source of a narcotic snuff), and perhaps some types of maize, were cultivated for comestibles, useful products, and for medicinal and ritual purposes [25, 27, 30].

THE EVIDENCE FOR MAIZE: ARCHAEOLOGICAL RESEARCH

Archaeobotany

Archaeobotanical research in the Caribbean began with the seminal work of Deborah M. Pearsall on the Krum Bay site in St. Thomas, Virgin Islands, which was excavated by Emily Lundberg, and described in a 1983 report. Pearsall's work was pivotal in demonstrating the wider potential of recovering plant remains from Caribbean sites, and the depth of insights that might derive from excavated plant materials. Since that time a number of sites on different islands have been analyzed archaeobotanically: including sites on three of the southern Caribbean Islands (see Figure 23-1); a number of others in the Lesser Antilles stretching from Grenada and Barbados in the south, to Nevis, St. Martin, and other islands in the north; several sites in the Virgin Islands have been analyzed, as well as others in the Greater Antilles, including on Puerto Rico and nearby Vieques, the Dominican Republic, Haiti, Cuba, and various islands in the Bahamas group [30]. At this juncture, archaeobotanical analyses in one form or another have been conducted nearly throughout the region, and to varying degrees are representative of the archaic, and early and later Ceramic Age human occupations.

Appropriate techniques for locating, recovering, and processing plant macroremains and microremains from the region have been developed over the course of the past two decades. Some sites have produced large volumes of carbonized material that have been extensively analyzed, in comparison to others. Because work with microremains is more recent, particularly starch grain analysis, the number of sites and samples that have been studied are necessarily fewer. Thus, regionally the data are somewhat uneven, but nevertheless preservation is generally good and collectively quite a bit of information is now available concerning subsistence and settlement dynamics in the region from a plant-use, economic, and landscape perspective, including a growing body of data on the presence of crop species [29, 30, 38].

Currently, we can specify from carbonized wood assemblages that many different woods were used for fuel, including some evidently preferred because of their hardness or other properties. These and other tree species may also have been used in construction or for wooden implements, or for their nontimber forest products. The last category potentially encompassed a wide range of uses: edible fruits, beverages, dyes, fish poisons, narcotic snuff, and other extracts from their bark, fruit, or seeds. The actual seed or fruit remains of a number of different edible tropical tree fruits have been identified (e.g., palm and members of the sapodilla family), as well as those of assorted herbs and vines, including several that may have served medicinal or ritual purposes [30]. A number of these plants may have been cultivated in home gardens, as suggested earlier, and several denote ancient or continued mainland contacts: for example, wild avocado *(Persea americana),* yellow sapote *(Pouteria campechiana),* níspero or chicle *(Manilkara bidentata),* and pepper *(Capsicum* sp., e.g., chili pepper, *C. annuum),* from Central America; papaya *(Carica papaya),* achiote *(Bixa orellana),* and jagua *(Genipa americana)* from South America, among others [30]. Carbonized tuber fragments recovered from the En Bas Saline site on Haiti, and smaller bits of carbonized starchy tissues from other sites provide direct evidence of root crops, at least manioc, if not also sweet potato. Production of both manic and sweet potato, as well as some additional tropical root crops from South America, is suggested by plant microremains, based primarily on research from Puerto Rico as described later.

Specifically with regard to maize, thus far macroremains of the plant are known from only two sites in the Caribbean. It was once thought that there was a third site, based on specimens from Sugar Factory on St. Kitts, but the identification proved erroneous [25, 28]. Definitive maize kernels were identified by Pearsall [40] from the Tutu site on St. Thomas (U.S. Virgin Islands), excavated by Elisabeth Righter. Tutu demonstrates a lengthy occupation history with earlier and later ceramic age components, but the maize remains come exclusively from Chican **Ostionoid** deposits (AD 1140–1350), thus, the later period of occupation [29, 40]. At Tutu, maize remains were found to be generally sparse and infrequent, amounting to fewer than a dozen individual specimens. Due to excessive fragmentation, it was not possible to specify any details of kernel morphology.

Maize macroremains have also been recovered from En Bas Saline, Haiti, excavated by Kathleen Deagan of the Florida Museum of Natural History, in cooperation with the Bureau National de'Ethnologie de'Haiti and the Musee de Guahaba in Limbe, Haiti [28]. En Bas Saline is a late prehistoric to contact era site. Like Tutu, the earliest maize remains came from well sealed contexts that are associated with the Chican Ostionoid tradition—specifically with Carrier variant cultural materials—which is relatively late, and is consistent with radiometric dating that places the occupation from about AD 1150 continuously into the early historic era. One sample with maize remains is attributed to the historic period by virtue of its stratigraphic position.

En Bas Saline was a Taíno settlement on the northeastern coast that encompassed an area of nearly 100,000 square meters. It is one of the largest and most elaborate of Taíno village sites reported for Haiti and is thought to have been either a principle settlement under the jurisdiction of or the actual residence of the Taíno *cacique* Guacanacaric (Guacanagarí), who interacted with Columbus during both his first and second voyages [7]. Thus, En Bas Saline was the exact location or was near where Columbus established his fortification of La Navidad in 1492. Significantly, maize is mentioned in the diary of Columbus's first voyage, and is thought to have been among the first plants brought from the Caribbean back to Europe. During that first voyage the explorers several times witnessed a grain crop in their travels through the Greater Antilles that they compared with their familiar barley and millet [8, pp. 89, 139, 209].

A prominent feature of the town, En Bas Saline, was a large central plaza that included a slightly elevated area on which stood a substantial wooden structure (actually a pair of rebuilt structures). By all ethnohistoric accounts, such structures conform to elite residential areas. Two large debris filled pits were located in the immediate vicinity of the central structure in the plaza. A raised earthen embankment and a band of concentrated midden debris form the northern and southern perimeters of the site, respectively. Samples of plant remains were recovered from each of these locations and cultural contexts.

The combined maize assemblage from En Bas Saline is small, but includes 9 whole or nearly complete kernels, 1 cob fragment, 28 individual cupules, and 10 additional kernel and cupule fragments. The sole cob fragment from En Bas Saline consists only of a single rank, with one nearly complete and two partially intact alicoles. The glumes are incompletely preserved, generally only their bases are present, and virtually no rachis tissue has survived; evidence of any pubescence is lacking. Conversely, some of the kernels are relatively intact and well preserved. These clearly represent two races of maize, which recalls the ethnohistoric mention of two forms noted earlier. Peg-shaped kernels (Figure 10-2) suggest a type of popcorn, possibly the race "Chandelle" or something similar, and crescent-shaped specimens conform to races with floury endosperm, perhaps the maize known as Early Caribbean [1, 2, 3, 4, 28]. Interestingly, Early Caribbean is thought to be a old indigenous race, a "remnant of one of the earliest types introduced into Europe" [2, p. 161]. It is not closely related to any known maize in either Central or South America, which would seem to support developments in relative isolation (i.e., a landrace that evolved in the islands); however, this does not clarify when it may have first been introduced into the Caribbean [1]. The maize kernel heights range from 4.20 to 7.90 millimeters, and widths at midsection from 5.40 to 7.40 millimeters; the measurement D1 (distance from the base to the widest part of the kernel, following King [18]) demonstrates a range from as low as 1.9 to 5.35. Additional morphometric data for the kernels, loose cupules, and the cob fragment from En Bas Saline are reported in [28].

The overall (i.e., sitewide) ubiquity of maize at En Bas Saline is 34%. However, maize, along with seeds of pepper (*Capsicum* sp., e.g., chili, *Capsicum annuum,* Solanaceae) and evening primrose (*Oenothera* sp., Onagraceae) are primarily associated with what presumably were high status contexts at the site, specifically the elite structure(s) and the two large pit features in the central plaza area. In fact, more than half of the maize specimens came from one of these deposits, Feature 11, a large prehistoric pit measuring 2

FIGURE 10-2 Maize kernels from En Bas Saline, Haiti, field sample 6316. Specimen on left conforms to popcorns; kernel dimensions are 6.90 millimeters height, 5.40 millimeters wide, and 5.10 millimeters thick at midsection, D1 = 5.35. Crescent-shaped specimen on right conforms with floury maize; kernel dimensions are 4.20 millimeters height, 7.25 millimeters wide, and 5.70 millimeters thick at midsection, D1 = 1.90. (Reprinted from *On Land and Sea: Native American Uses of Biological Resources in the West Indies,* Newsom, L.A., 2004, with permission from the University of Alabama Press.) See color insert.

meters deep and 1.5 meters square. The feature was densely filled with ash, shells, animal bone, carbonized manioc tuber fragments, clay griddle fragments, and pottery sherds from more than 140 ceramic vessels. We believe that Feature 11 functioned in communal–ritual feasting activities, and some of the broken ceramics may represent vessels used for the ceremonial consumption of beer [28]. All but three of the remaining samples with maize remains came from floor and wall deposits in one of the large central structures, which, to reiterate, was perhaps the chiefly residence. All of this suggests that at En Bas Saline maize was among a subset of edible plants that were reserved for individuals with higher social standing, or otherwise was among plants that were valued apart from simply supplying food. That is, some special social significance may have been attached to maize and some of the other plant taxa, either as special foods reserved for elites and/or that were deemed integral to feasting and communal–ritual activities at the site [26, 28].

Hints of a broader and perhaps earlier presence for maize have been suggested from several analyses of plant microremains in the region. Maize pollen possibly dated to around AD 1020 is reported for El Jobito, in the Dominican Republic [9, p. 36], and similarly, from sediment cores from Lake Miragoane, Haiti (dating between ca. AD 1000–1500 [12]). M. Sanoja [45, p. 532] mentions an additional pair of archaeological sites in the Dominican Republic that have produced maize pollen, in both cases said to date to around 1450 BC. This particular date may seem early, nevertheless, the dates for early maize in Central and northern South America indicate that the plant was potentially available to people migrating into or communicating with Caribbean people well before this date. More recently, Pearsall [29] has reported maize cob phytoliths from a sediment core from a pond adjacent to the Maisabel site in northern Puerto Rico, beginning in levels dated by extrapolation to around the middle second millennium BC. Finally, and most recently, Jaime R. Pagán-Jiménez's [38] analyses of starch residues on stone tools recovered from two Archaic Age sites in southern Puerto Rico (Maruca and Puerto Ferro; ca. third to first millennia BC) has revealed the presence of grains suggestive of maize, along with manioc and other native and introduced root crops. Because of the cultural associations, this evidence potentially provides important insights not only into vital plant introductions, including maize, but also lends further credence to suppositions about plant cultivation during this early period. Pagán-Jiménez [37] has also identified maize starch on artifacts from UTU-27, a late Ceramic Age site in the central mountains of Puerto Rico.

Human Bone Chemistry

Another relatively recent development in Caribbean archaeology is the use of isotopic analysis of human bone to detect dietary patterns. Among the most recent and exhaustive of these studies to date is Anne Stokes's [49] doctoral research, based on isotopic data derived from 102 human skeletons from different islands and belonging variously to the Ceramic Age. The results demonstrate that diet in terms of the contributions of marine and terrestrial foods did not change significantly between the early and later ceramic periods; however, people who lived on the larger, more diverse, and less isolated islands had a greater contribution of terrestrial foods in their diets, whereas those who lived on the smaller, more remote, low limestone islands with poor soils and impoverished land faunas relied more heavily on marine foods. In general, where plant foods are concerned, the isotopic signatures seem to indicate an emphasis on C_3 plants and cultigens such as manioc as staple foods, with maize evidently adopted late in prehistory (if at all), and then only in some areas and in some low intensity form of use [49]. The material culture remains in the form of the artifacts mentioned in the preceding section—to the extent most are accurately interpreted as having been exclusively or primarily associated with tropical root crops—and the plant remains described earlier, corroborate this broad conclusion of a low profile and a late emphasis on C4 plants (i.e., maize).

DISCUSSION

Maize remains have been recovered archaeologically from sites in the Caribbean in several forms: cob fragments, kernels, pollen grains, and it appears now also, phytoliths and starch grains. Both of the sites that have produced maize macroremains—Tutu and En Bas Saline—are relatively late in the overall sequence of human occupation in the region, and at each the remains were minimally present relative to other plant remains from the two sites. Moreover, at En Bas Saline the spatial distribution of maize may suggest a rather restricted pattern of use. Bearing in mind that research with microremains is comparatively recent in the Caribbean, that line of evidence suggests an earlier presence for maize, but in general seems to mirror the overall impression stemming from the work with plant macroremains from throughout the region, specifically that maize's prehistoric presence is quite sparse and spotty.

Why Such a Low Signal?

At this juncture, it is appropriate to consider in greater depth the general paucity of evidence for prehistoric maize in the Caribbean. Between the mentions of maize in the ethnohistoric documents, and now also with actual specimens from archaeological sites, we know that the plant was part of Caribbean subsistence, at least in the Greater Antilles and Virgin Islands. We now have some indication that this plant appeared relatively early in the sequence of human occupation in the region, perhaps even from the outset of colonization, thanks to the efforts of specialists working with plant microremains. But how do we explain the fact that there is no evidence of the plant in the form of macroremains until quite late in the history of human occupation? And even then, why is there so little evidence for maize—just two sites with few specimens that are quite limited in their distributions—given the number of sites and quantities of samples that have now been studied [29, 30]? Presently, there are no definitive answers to these questions, but I suggest that the explanations may be found in how maize was used in the Caribbean, including its real and perceived role among the suite of plant taxa exploited by Caribbean people at various times and places.

One possibility is that the pattern observed archaeologically may signify that in the Caribbean maize was originally grown for its vegetative tissues and/or immature reproductive organs, perhaps somewhat casually in home gardens, and at least in part for the sugary growth, as has been suggested for other areas (see Chapters 3 and 4). A focus on these earlier developmental or vegetative tissues would leave fewer traces of the plant archaeologically. Only later, at least by the time we see two races of maize—a flour type and popcorn—in late prehistory (i.e., at En Bas Saline), and we have ethnohistoric mention of its presence among *conuco* crops, as well as a discernable signal in the human bone chemistry, was the plant relegated to a more robust subsistence role in the realm of staple crops grown for the mature grains. However, in this regard note again my mention earlier of the limited distribution overall and the relatively restricted cultural contexts of maize at En Bas Saline, as a possible indication that it was not a staple food in the Caribbean at all.

Certainly from an archaeobotanical perspective, there is no evidence to suggest that maize was ever used as intensively in the islands as on the mainland. I reason that basically, given the volume of research and number of paleoethnobotanical studies that have now been conducted or are underway in the Caribbean, it is reasonable to expect that maize remains should be turning up more frequently and abundantly if it had ever achieved the level of importance it held prehistorically in other areas of the Americas. The archaeological evidence (or lack thereof) suggests it was not grown and stored as quantities of dried grain (i.e., as a food surplus), or it would be more evident or conspicuous in the archaeobotanical record. By all accounts and avenues of study, maize seems never to have achieved the status of a storable staple in the West Indies, at least not until late, if at all.

The Development of a Uniquely Caribbean Cuisine

The seemingly low profile of maize in the prehistoric Caribbean is in fact consistent with what is indicated in the ethnohistoric documents, that is, that maize was a secondary or minor crop, not a pivotal staple. The predominant forms in which maize seems to have been consumed in the islands are telling, generally implying consumption in the green state, that is, immature, or as roasted kernels [47, 50]. For example, "When the ears are tender they are eaten almost like milk" [36, pp. 14–15], and recall the quote earlier about eating kernels directly. These descriptions conform to low intensity use, more as a vegetable (e.g., creamed corn), supplement, or "curiosity" [39], than as a staple carbohydrate source. Harvesting the cobs and grain before fully ripened to a state with hardened sclerified tissues implies a low level of human reliance, and incidentally, one that would necessarily leave fewer traces in the archaeological record. This may mean that our current understanding of the timing and areas where maize may have been adopted in the Caribbean is potentially biased. Nevertheless, this assessment agrees with the fact that we are not seeing evidence archaeobotanically for maize in quantities or forms suggestive of an important staple: no abundance of cobs used for fuel or otherwise discarded mature cobs from which the kernels have been removed, nor of stocks of removed kernels, all of which imply mature, dried maize stored as a surplus for later preparation of ground meal and flour.

Perhaps this pattern of use and perceived role of maize in the Caribbean reflects a strong northern South American influence where crop production and food preferences are concerned, especially considering the ceramic age migrations, which reflect direct ties with that region and the primary direction from which most of island colonization was derived. The Spanish colonial records, as I indicated, mention maize, along with at least seven types of root crop that were cultivated by the Taíno at the time of European contact. Sweet potato and particularly manioc are singled out in the chronicles as the most important plant staples in the islands. The premier position of manioc, as Petersen [41, p. 128] and others have pointed out, was reflected in the fact that one of the two supreme Taíno deities was *Jocahu* or *Yucahu,* the "lord of cassava [manioc] and the sea" (in the Spanish-speaking islands of the Caribbean manioc is still known as *yuca,* a name possibly recalling the Taíno epithet). Sugary sweet potatoes called *batata* were distinguished from less sweet cultivars that were known as *age* [47, p. 54]. Aside from manioc and sweet potatoes, other root crops mentioned in the historic chronicles include *yautía* (*Xanthosoma* sp.), *araru* or arrowroot *(Maranta arundinacea), ileren* (*Calathea sp.*), *yampee (Dioscorea trifida),* and an edible-rooted form of *Canna* sp. [36, 47, 50]. All of the previously mentioned crops, including manioc and sweet potato, with the possible exception of *yampee* (that may actually represent an early historic introduction from Africa [e.g., 11]), and depending on the particular species, are native to northern South America [22, pp. 250, 259, 262]. The several South American tuber introductions make more plausible inferences that the maize grown in the Caribbean also originated from races developed or long cultivated in lowland South America [28, 32], although this remains to be verified and certainly maize from Mesoamerica remains a possibility as well. Nevertheless, it seems clear that a commitment to familiar South American tuber crops persisted among the groups who settled in the Caribbean, and these remained an integral part of the cultivation practices and cuisines that developed in the islands.

It is worth noting also that in northern South American prehistory, maize does not appear to have become a staple until well after the first Ceramic Age migrants entered the Caribbean islands, and then perhaps only in certain areas [24, 32, 39, 42, 46, 55]. This delayed or deflected emphasis on maize may relate at least in part to the availability of productive alternative food sources (e.g., the diverse array of edible tubers mentioned earlier), as well as aquatic protein sources (e.g., drawing on Keeley's [17] research among neotropical hunter–gatherers and low level horticulturists in which he has clarified the primacy and pivotal role of aquatic protein sources when available). In the Caribbean, marine fauna were abundant except in the interior-most areas of larger islands, and the zooarchaeological data demonstrate concerted reliance on those resources during virtually the entire span of human occupation in the region [30]. Moreover, at least at the scale of individual islands as described earlier, the Caribbean environment may have been less conducive to growing maize than that of the source regions. Maize is not widely or intensively grown in the islands today [10]; without the benefit of modern fertilizers and adequate irrigation it may have been problematic for ancient producers. We can speculate that perhaps the early introduction of maize into the region entailed a period of adjustment and experimentation, even rejection on the path to becoming incorporated into gardening practices or as a plant staple. Effectively to this point, Keegan (personal communication, 1993) has suggested that maize would have represented an expensive source of protein and, therefore, would never have been emphasized in the Caribbean as long as alternative protein sources were available at a lower economic cost. Again, Keeley comes to mind as to the pivotal role of aquatic food resources, which in this case would be the abundant, relatively sustainable marine fauna (but cf. [30] regarding local extirpations of marine faunal resources by Caribbean Indians).

It is also possible that maize was simply overshadowed in the Caribbean by the primary system of root cropping that seems to have been fully a part of Ceramic Age subsistence from the beginning [43, 44], if not also that of the earlier Archaic Age fisher–forager–gardeners [6, 30, 38]. This would parallel Pearsall's [41] explanation for the delayed spread of maize into northeastern South America. Finally, the rather restricted nature of maize presence at En Bas Saline may be significant. Deagan and I [28] have suggested that maize's strong association with the centrally positioned elite area of the site provides a basis for hypothesizing that restricted elite access to maize may have operated in the prehistoric Caribbean. Restricted access to maize by privileged social segments has been suggested as a factor in the early stages of the introduction of maize into other geographic regions, for example, as Bruce Smith [48] has suggested for southeastern North America. Because maize seems to have been primarily associated with the high status, or perhaps better described as the public area, of the site at En Bas Saline, Haiti (i.e., the central area where presumably the *cacique*'s residence was situated), this would seem to denote a status apart from more mundane food items and a strong association with power and status, and/or feasting and communal activities. This possibility currently is being further explored through analyses of additional archaeobotanical samples from En Bas Saline, and further testing of the ceramic vessels hypothesized to have been integral to beer consumption.

CONCLUSION

Plant remains and related evidence from the Caribbean Islands illustrate flexibility in the use of plant resources, but at the same time a clear focus on root-crop horticulture derived in large part from stock and planting systems originally developed in lowland tropical South America. There is archaeological evidence from at least two islands for the development of terrace or irrigation systems that we may presume served as means to intensify production of plant staples, and certainly the social infrastructure and wherewithal was in place during at least the later periods of occupation to facilitate and mobilize such heightened production. But in the Caribbean this is likely to have concerned primarily the tropical root crops, rather than maize, or before maize assumed any level of importance, based on all the lines of evidence and reasoning presented earlier. Such a crucial subsistence role for maize in the Caribbean, if it was ever to have come to pass, was unfortunately truncated by the presence of Europeans after 1492. Instead maize seems to have been among a suite of plant taxa, at least in the Greater Antilles, that were grown as vegetables or supplements, adding diversity to the diet, or that were restricted in use either to certain individuals within the society or perhaps as a plant important to the ritual and communal life of the group, possibly consumed as a fermented beverage during feasting or other ceremonial occasions.

Acknowledgments

Many thanks to Michael Blake, John Hart, and John Staller for their vision and monumental efforts in pulling together the interesting and successful maize symposia that were held at the SAA meetings in Montreal.

Likewise, I am grateful that Staller, Robert Tykot, and Bruce Benz have devoted their time and expertise to bring this publication to reality, and that they have chosen to include the Caribbean in this endeavor. I thank Jaime Pagán-Jiménez for his willingness to share data on Carribbean sites. Finally, without the superb archaeological field research conducted by individuals such as Kathy Deagan, L. A. Chanlatte Baik, Yvonne Narganes Storde, Jose Oliver, Holly Righter, Miguel Rodríguez López, Peter Siegel, and others, we would probably still have little or no idea of the role and presence of maize in the region. The research with En Bas Saline maize was supported in part by the National Science Foundation (BNS 8706697) and the University of Florida.

References Cited

1. P. K. Bretting, M. M. Goodman, C. W. Stuber. (1987). Karyological and isozyme variation in West Indian and allied American mainland races of maize. *American Journal of Botany, 74,* 1601–1613.
2. W. L. Brown. (1953). Maize of the West Indies. *Tropical Agriculture, 30,* 141–170.
3. W. L. Brown. (1960). *Races of maize in the West Indies.* National Academy of Sciences, National Research Council Publication 792. Washington, D.C.: Government Printing Office.
4. W. L. Brown, M. M. Goodman. (1977). Races of corn. In: G. F. Sprague, (Ed.), *Corn and corn improvement.* Madison, WI: American Society of Agronomy. pp. 49–88.
5. R. Dacal Moure, M. Rivero de la Calle. (1996). *Art and archaeology of Pre-Columbian Cuba.* Pittsburgh: The University of Pittsburgh Press.
6. D. D. Davis. (1988). Coastal biogeography and human subsistence: Examples from the West Indies. *Archaeology of Eastern North America, 16,* 177–185.
7. K. Deagan. (2004). Reconsidering social dynamics after Spanish conquest: Gender and class in culture contact studies, *American Antiquity, 69,* 597–626.
8. O. Dunn, J. E. Kelley, Jr. (1989). *The diario of Christopher Columbus's first voyage to America, 1492–1493: Abstracted by Fray Bartolome de las Casas.* Norman: University of Oklahoma Press.
9. M. A. García Arévalo, J. Tavares. (1978). Presentation (arqueología de Sanate). *Boletin del Museo del Hombre Dominicano, 10,* 31–44.
10. F. Gumbs. (1981). Agriculture in the wider Caribbean. *Ambio, 10,* 335–339.
11. R. L. Hall. (1991). Savoring Africa in the New World. In: H. J. Viola, C. Margolis, (Eds.), *Seeds of change: Five hundred years since Columbus.* Washington, D.C.: Smithsonian Institution Press. pp. 160–171.
12. A. Higuera-Gundy. (1991). *Antillean vegetational history and paleoclimate reconstructed from the paleolimnological record of Lake Miragoane, Haiti.* Ph.D. Dissertation. Gainesville: Botany Department, University of Florida.
13. W. F. Keegan. (1994). West Indian archaeology 1. Overview and foragers. *Journal of Archaeological Research, 2,* 255–284.
14. W. F. Keegan. (1996). West Indian archaeology 2. After Columbus. *Journal of Archaeological Research, 4,* 265–294.
15. W. F. Keegan. (2000). West Indian archaeology 3. Ceramic age. *Journal of Archaeological Research, 8,* 135–167.
16. W. F. Keegan, M. MacLachlan, B. Byrne. (1998). Social foundations of *caciques.* In: E. M. Redmond, (Ed.), *Chiefdoms and chieftancy in the Americas.* Gainesville: University of Florida Press. pp. 217–244.
17. L. H. Keeley. (1995). Protoagricultural practices among hunter-gatherers: A cross-cultural survey. In: T. D. Price, A. B. Gebauer, (Eds.), *Last hunters, first farmers.* Santa Fe, NM: School of American Research Press. pp. 243–272.
18. F. B. King. (1987). *Prehistoric maize in Eastern North America: An evolutionary evaluation.* Ph.D. Dissertation. Urbana-Champaign: Department of Agronomy, University of Illinois.
19. H. W. Krieger. (1930). *The aborigines of the ancient island of Hispaniola.* Annual Report of the Board of Regents of the Smithsonian Institution for 1929. Washington, D.C.: Government Printing Office. pp. 473–506.
20. B. de Las Casas. (A. Collard, Trans. and Ed.) (1971) [originally 1527–1565]. *History of the Indies.* New York: Harper and Row Publishers.
21. J. Layfield. (1995) [originally 1598]. Relación del viaje a Puerto Rico de la expedición de Sir George Clifford, tercer conde de Cumberland, escrita por el Reverendo Doctor John Layfield, capellán de la expedición. (Fragmentos) Año 1598. In: E. Fernandez Mendez (compiled and edited), *Cronicas de Puerto Rico: Desde la Conquista Hasta Nuestros Días (1493–1955).* San Juan, Puerto Rico: Editorial Universitaria, San Juan, Puerto Rico.
22. H. A. Liogier, L. F. Martorell. (2000). *Flora of Puerto Rico and adjacent islands: A systematic synopsis.* 2nd ed. Rio Piedras, Puerto Rico: Editorial de la Universidad de Puerto Rico.
23. W. H. Marquardt. (2001). The emergence and demise of the Calusa. In: D. S. Brose, C. W. Cowan, R. C. Mainfort, Jr., (Eds.), *Societies in eclipse: Archaeology of the Eastern Woodlands Indians, AD 1400–1700.* Washington, D.C.: Smithsonian Institution Press. pp. 157–171.
24. N. J. van der Merwe, A. C. Roosevelt, J. C. Vogel. (1981). Isotopic evidence for prehistoric subsistence change at Parmana, Venezuela. *Nature, 292,* 536–538.
25. L. A. Newsom. (1993). *Native West Indian plant use.* Ph.D. Dissertation. Gainesville: Department of Anthropology, University of Florida.
26. L. A. Newsom. (1998). Mangroves and root crops: Archaeobotanical record from En Bas Saline, Haiti. In: G. Richard, (Ed.), *Proceedings of the 16th International Congress for Caribbean Archaeology.* Basse Terre: Conséil Regional de la Guadeloupe. pp. 52–66.
27. L. A. Newsom. (2005). Caribbean gardens and gardeners (unpublished manuscript).
28. L. A. Newsom, K. A. Deagan. (1994). *Zea mays* in the West Indies: The archaeological and early historic record. In: S. Johannessen, C. A. Hastorf, (Eds.), *Corn and culture in the prehistoric New World.* Boulder, CO: Westview Press. pp. 203–217.
29. L. A. Newsom, D. M. Pearsall. (2003). Trends in Caribbean island Archaeobotany. In: P. E. Minnis, (Ed.), *People and plants in Ancient Eastern North America.* Washington, D.C.: Smithsonian Institution. pp. 347–412.
30. L. A. Newsom, E. S. Wing. (2004). *Land and sea: Native American uses of biological resources in the Caribbean.* Tuscaloosa: University of Alabama Press.
31. J. Oliver. (1999). The "La Hueca problem" in Puerto Rico and the Caribbean: Old problems, new perspectives, possible solutions. In: C. L. Hofman, M. L. P. Hoogland, (Eds.), *Archaeological investigations on St. Martin (Lesser Antilles): The sites of normal estate, Anse des Pères, and Hope Estate, with a contribution to the "La Hueca Problem."* Archaeological Studies Leiden University. Leiden, The Netherlands: Faculty of Archaeology, Leiden University. pp. 253–297.
32. J. Oliver. (2001). The archaeology of forest foraging and agricultural production in Amazonia. In: C. McEwan, C. Barreto, E. Neves, (Eds.), *Unknown Amazon.* London: The British Museum Press. pp. 50–85.
33. J. Oliver, J. A. Rivera Fontán, L. A. Newsom. (1999). Arqueología del Barrio Caguana: resultados preliminares de las temporadas 1996–1997. In: J. A. Rivera Fontán, (Ed.), *Trabajos de investigación arqueológica en Puerto Rico.* Puerto Rico: Publicaión ocasional de la División de Arqueologia, Instituto de Cultura Puertorriqueña. pp. 7–26.
34. E. Ortega, J. Guerrero. (1981). *Cuatro nuevos sitios paleoarchaicos en las Isla de Santo Domingo.* Santa Domingo: Ediciones Museo del Hombre Dominicano.
35. J. J. Ortíz Agilú, J. Rivera Meléndez, A. Principe Jácome, M. Mélendez Maiz, M. Lavergne Colberg. (1991). Intensive agriculture in

pre-Columbian West Indies: The case for terraces. In: A. Cummins, P. King, (Eds.), *Proceedings of the XIV Congress of the International Association for Caribbean Archaeology*. St. Ann's Garrison: Barbados Museum and Historical Society. pp. 278–285.

36. G. F. de. Oviedo y Valdes (S. A. Stoudemire [Trans. and Ed.]). (1959) [originally 1526]. *Natural history of the West Indies (1526)*. University of North Carolina Studies in the Romance Languages and Literature. No. 32. Chapel Hill: University of North Carolina Press.
37. J. R. Pagán-Jiménez. (2003). Reporte de progreso: estudio de almidones en artifactos líticos de los sitios arqueológicos SR-1 y Utuado-27, Proyecto Utuado-Caguana, Puerto Rico. Report submitted to J. Oliver (University College, London) and L. Newsom (The Pennsylvania State University).
38. J. R. Pagán-Jiménez, M. A. Rodríriguez López, L. A. Chanlatte Baik, Y. Narganes Storde. (2005). La temprana introducción y uso de algunas plantas domésticas, silvestres y cultivos en Las Antillas precolombinas. *Diálogo Antropológico, 3,* 7–33.
39. D. M. Pearsall. (1994). Issues in the analysis and interpretation of archaeological maize in South America. In: S. Johannessen, C. A. Hastorf, (Eds.), *Corn and culture in the Prehistoric New World*. Boulder, CO: Westview Press. pp. 245–272.
40. D. M. Pearsall. (2002). Analysis of charred botanical remains from the Tutu site, U.S. Virgin Islands. In: E. Righter, (Ed.), *The Tutu archaeological village site: A multidisciplinary case study in human adaptation*. London: Routledge. pp. 109–134.
41. J. B. Petersen. (1977). Taíno, Island Carib, and prehistoric Amerindian economies in the West Indies: Tropical forest adaptations to island environments. In: S. M. Wilson, (Ed.), *The indigenous people of the Caribbean*. Gainesville: University Press of Florida. pp. 118–130.
42. A. C. Roosevelt. (1980). *Parmana: Prehistoric maize and manioc subsistence along the Amazon and Orinoco*. New York: Academic Press.
43. I. Rouse. (1986). *Migrations in prehistory: Inferring population movements from cultural remains*. New Haven, CT: Yale University Press.
44. I. Rouse. (1996). *The Tainos: Rise and decline of the people who greeted Columbus*. New Haven, CT: Yale University Press.
45. M. Sanoja. (1989). From foraging to food production in northeastern Venezuela and the Caribbean. In: D. R. Harris, G. C. Hillman, (Eds.), *Foraging and farming: The evolution of plant exploitation*. London: Unwin Hyman. pp. 523–537.
46. M. Sanoja, I. Vargas. (1983). New light on the prehistory of eastern Venezuela. *Advances in World Archaeology, 2,* 205–244.
47. C. O. Sauer. (1966). *The early Spanish main*. Berkeley: The University of California Press.
48. B. D. Smith. (1990). *The Mississippian Emergence*. Washington, D.C.: Smithsonian Institution Press.
49. A. V. Stokes. (1998). *A biogeographic survey of prehistoric human diet in the West Indies using stable isotopes*. Ph.D. dissertation. Gainesville: Department of Anthropology, University of Florida.
50. W. C. Sturtevant. (1961). Taino agriculture. In: J. Wilbert, (Ed.), *The evolution of horticultural systems in Native South America: Causes and consequences*. Anthropologica Supplement 2. pp. 68–73.
51. M. Veloz Maggiolo. (1993). *La Isla de Santo Domingo antes de Colon*. Santa Domingo, Dominican Republic: Banco Central de la República Dominicana.
52. S. M. Wilson. (1990). *Hispaniola: Caribbean chiefdoms in the age of Columbus*. Tuscaloosa: University of Alabama Press.
53. S. M. Wilson, (Ed.). (1997). *The indigenous people of the Caribbean*. Gainesville: University of Florida Press.
54. S. M. Wilson, H. B. Iceland, T. R. Hester. (1998). Preceramic connections between Yucatan and the Caribbean. *Latin American Antiquity, 9,* 342–352.
55. A. Zucchi, K. Tarble, E. Vaz. (1984). The ceramic sequence and new TL and C-14 dates for the Aguerito Site of the Middle Orinoco, Venezuela. *Journal of Field Archaeology, 11,* 155–180.

PART III

Histories of Maize:

Isotope Analysis and Human Diet in Ancient Mesoamerica

C H A P T E R

11

Isotope Analyses and the Histories of Maize

ROBERT H. TYKOT

Glossary

collagen organic portion of bone mainly reflecting dietary protein

apatite (bone, tooth enamel) inorganic portion reflecting whole diet

isotopes (stable, unstable) elements with different numbers of neutrons in the atomic nucleus

mass spectrometer instrument used to produce precise measurements if isotope ratios

fractionation the change in isotope ratios due to chemical processes such as photosynthesis and metabolism

PDB international standard for C and O isotope analysis, based on Belemnitella americana from the Peedee formation in South Carolina

carbon, nitrogen, and oxygen isotopes stable isotopes representing the main dietary components of skeletal tissues

strontium isotopes heavy stable isotope representing geological area of food acquisition

The contribution of isotope analyses to archaeology has significantly increased ever since the introduction of radiocarbon dating by Willard Libby in 1949 [60], with applications of isotope analyses developing ever since. Today, stable isotope analysis is regularly applied to address questions concerning human diets around the world, and this is illustrated in this volume with the large number of studies on the origins, importance, and spread of maize throughout the Americas. In this introductory chapter, an overview of the history and principles of isotope analyses is provided, while the individual chapters in this section provide specific details on the samples and procedures used in each study, and how their results have contributed to our understanding the Histories of Maize. In particular, "we are what we eat" and isotope analyses provide quantitative data that complement floral, faunal, ethnohistoric, and other information about dietary practices. The details to follow include the contributions of quite recent methodological and interpretive isotope studies, but like some of the previous synthetic publications on isotopic analysis applied to archaeology, focus on relevance to maize studies [111, 107, 23, 1, 97, 44, 93, 112, 2, 8, 94, 99, 46, 54, 65, 81, 114, 45, 6, 108].

ISOTOPE DEFINITIONS

Isotopes are defined as one of the two or more forms of an element (e.g. carbon), which have the same number of protons in the nucleus (known as the atomic number) of an atom, but different numbers of neutrons in the nucleus, thus resulting in different atomic weights. Radioactive isotopes (e.g. carbon-14) decay over time, while stable isotopes (e.g. carbon-12, carbon-13) do not. Carbon occurs in three isotopic forms: 12C, 13C, and 14C, where the number to the upper left is the sum of the protons (6) and neutrons (6, 7, 8) in the carbon nucleus. Carbon dioxide (CO2) in the atmosphere consists of about 98.9% 12CO2 and 1.1% 13CO2 (with radioactive 14CO2 a tiny fraction of about 1 x 10-10%). The isotopic composition of atmospheric CO2 is established by equilibration with the much more abundant pool of dissolved inorganic carbon (mainly bicarbonate, HCO3-) in sea water. There is also considerable carbon in calcium carbonate (CaCO3), the main component of limestone and marble. Atmospheric carbon dioxide is photosynthesized by plants and metabolized into complex molecular compounds that we categorize as carbohydrates, proteins, and lipids, while many plants (wild, and now domesticated) are consumed by organisms which then appropriate and/or convert these compounds into their body tissues. In biosynthetic chemical reactions, especially photosynthesis [76], the lighter weight isotopic compounds will react faster, using less energy. If, as in the case of photosynthesis, the reaction is incomplete, the product (e.g. maize) will be enriched in 12C relative to the substrate (e.g. atmospheric CO2), resulting in changes in the 13C/12C ratio. This is called isotopic fractionation. When plants are consumed by herbivorous animals, the metabolic processes involved reverse the direction of fractionation, increasing the proportion of the heavier carbon (and nitrogen) isotope in the body tissues. Laboratory-based and other isotope studies have specifically determined that bone collagen is produced mainly from dietary protein, while bone apatite and tooth enamel represent the whole diet [7, 106]. In addition, it is also clear that turnover rates for bone are slow, so that the isotope values obtained represent at least the last several years of an individual's life.

Because of the small difference in $^{13}C/^{12}C$ ratios between all samples, these and other isotope abundances, instead of being reported as simple ratios (e.g. $^{13}C/^{12}C$) they are converted to ratios relative to international standards using the delta notation:

$$\delta^{13}C \text{ (in ‰ or per mil)}$$

$$= [\{(\text{sample } ^{13}C/^{12}C)/(\text{standard } ^{13}C/^{12}C)\} - 1] \times 1000$$

where the standard reference material was originally derived from a Cretaceous marine fossil sample, *Belemnitella americana*, recovered from the Peedee formation in South Carolina [19, 20]. Since the original material is no longer available, sample carbon isotope ratios are now reported relative to VPDB (Vienna Peedee belemnite), with raw values typically calibrated using the NBS (National Bureau of Standards) 19 reference sample so that results from different laboratories may be reliably compared [18].

The same mathematical formula is used for reporting nitrogen ($^{15}N/^{14}N$) and oxygen ($^{18}O/^{16}O$) isotope ratios, with the standard reference materials being AIR for nitrogen, and VPDB for oxygen. Since the $^{13}C/^{12}C$ ratio of all photosynthesized carbon is lower than that of the reference VPDB material, the ^{13}C values for plants and animals are mostly negative (see eq. above).

HISTORY OF ISOTOPE STUDIES

Variation in the relative proportions of stable carbon isotopes was first measured in 1939 by A.O. Nier and E.A. Gulbransen [71], and the general distribution of carbon isotopes in nature was explored by H. Craig in 1953 [19]. It was less than 20 years later then that the archaeological significance of such analyses was recognized, both for studying prehistoric human diets [110], and for sourcing marble [21, 41]. In between, important advances had been made by M. Calvin and A. Benson [16] in the understanding of the chemical pathway for carbon during photosynthesis, followed by the realization that there were multiple photosynthetic pathways, commonly referred to as C_3, C_4, and CAM (crassulacean acid metabolism), by H. Hatch and R. Slack [38] and S.L. Ransom and M. Thomas [85]. At about the same time that it was determined that maize followed the C_4 photosynthetic pathway [39], it was realized by archaeologists that radiocarbon dates on preserved carbonized maize samples were consistently about 200 years younger than expected [36], with the result that stable isotope measurements were made and correction factors were applied to radiocarbon dates being done on maize and other non-wood charcoal samples [11]. A conference held in Australia in 1970 led to the explicit realization that plants following the C_3 photosynthetic pathway had different stable carbon isotope ratios (ave. $\delta^{13}C = -26.5‰$) than those following the C_4 pathway (ave. $\delta^{13}C = -12.5‰$), which include maize, sugarcane, millet, sorghum, and certain species of amaranth and chenopodium [101]. These now domesticated C_4 plants, as well as a range of wild C_4 grasses, are originally native to areas of hot and dry climates, while CAM plants switch between C_3 and C_4 depending on their actual location and environmental circumstances.

The idea that carbon isotope values could be used for dietary information was developed at least by 1964, when P.L. Parker [77] published an article on carbon isotope analysis of marine plants and animals, but it was not until 1971 when a dietary study was done on humans. N.J. van der Merwe and J.C. Vogel specifically tested an Iron Age Khoi skeleton from the Transvaal of South Africa, whose bone collagen (protein made of multiple amino acids) carbon isotope value ($\delta^{13}C = -10.4$) was interpreted then as indicating dependence on sorghum (or other C_4 plants) in the Transvaal Lowveld [111]. It should also be noted that while the earliest radiocarbon dating of bone was done on whole bone samples, by 1970 demineralization of bone and extraction of humic and fulvic acids had been developed to produce much more accurate dating results on bone collagen [51, 62], and this was the specific sample material tested by van der Merwe and Vogel.

Shortly afterwards, discussions in van der Merwe's 1973 seminar at SUNY-Binghamton led to the isotopic analysis of human skeletal remains from sites in New York to address the question of when maize agriculture was introduced to that region [110]. The results were first published in *American Antiquity* in 1977 [116]. This was followed by a larger study of Archaic, Woodland and Mississippian period sites in North America, which achieved greater scientific recognition with a publication in *Nature* in 1978 [115]. Both studies clearly showed that the introduction of maize agriculture in North America dramatically changed the isotope values on human bone collagen in measurable, quantitative ways, demonstrating that isotope analyses could strongly complement botanical and other evidence for dietary interpretations. These early studies also demonstrated that for collagen $\Delta^{13}C$ values, 'you are what you eat, plus 5‰'. The percentage of C_4 foods in the diet could also be estimated by simple interpolation of the adjusted collagen value between the C_3 and C_4 endpoints (e.g. a collagen result of -14‰ suggests about 50% C_4 in the diet) This initial work was quickly followed by other studies by several scholars, with M.J. DeNiro and S. Epstein, who had already studied isotopic variation among animals [24], reporting results for early maize consumption at Tehuacan in highland Mexico [25], at the same time that a study was done on maize in the Hopewell period of the mid-western United States, by M.M. Bender, D.A. Baerreis and R.L. Steventon [12], while yet another study was done by van der Merwe and Vogel, with Anna C. Roosevelt, on maize in the rain forests of the Orinoco Valley of Venezuela [113]. This particular project emphasized the importance when interpreting isotopic results for humans of establishing baseline values for the plants and animals in a given region, which in this case were especially negative because of the recycling of CO_2 under a dense forest canopy [66]. Another early study specifically addressed maize in the diet of domestic dogs in Peru and Ecuador [15].

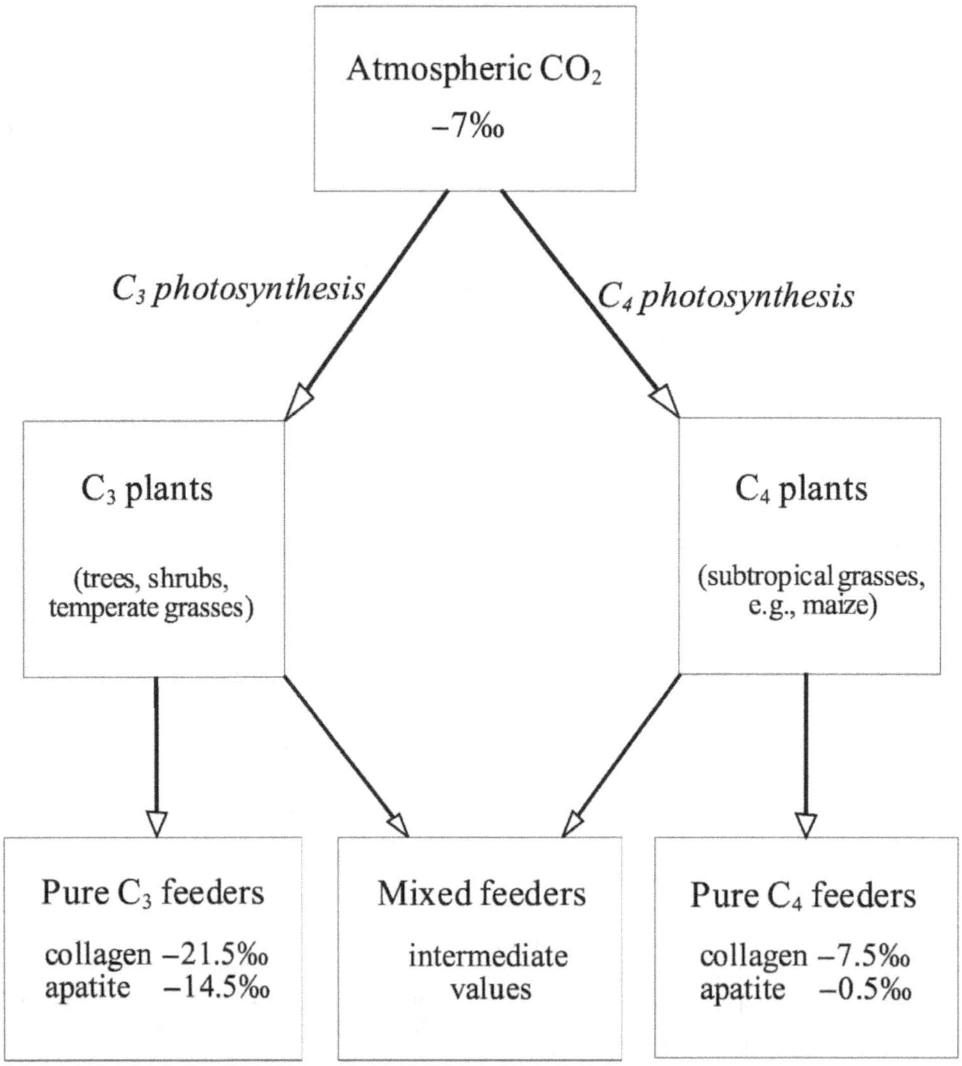

FIGURE 11-1 C_3 and C_4 pathways, showing isotopic differences passed on to skeletal tissues

Stable isotope analysis of nitrogen also developed in the early 1980s, with clear indications of differences due to trophic level effects, especially in marine systems [25, 26, 90, 91, 96, 5, 100]. More recent studies have added significantly to our understanding and interpretation of nitrogen isotope ratios, in particular the effects of climate and environment on both plant and animal values, and trophic level increases in both terrestrial and marine ecosystems [3, 14]. Nitrogen isotope analysis has also been demonstrated to be very useful for investigation of ancient weaning practices [95, 126, 123]. In general, nitrogen isotope ratios increase 2-3‰ with each trophic level, with values for terrestrial plants and animals generally much lower than for fish and mammals in most freshwater or marine ecosystems (Figure 11-2).

Other major advances to bone chemistry studies came in the 1980s regarding the preparation and carbon isotope analysis of inorganic bone apatite (calcium hydroxyphosphate, $Ca_5[PO_4]_3OH$, with some carbonate substitution). Because this mineral, which accounts for more than 75% of whole bone, is inorganic, it is much more susceptible to weathering and chemical reactions with soil deposits and thus alteration (diagenesis) of its original isotopic signature. This was quite clear from early radiocarbon dating of whole bone samples, which produced unreliable results [50], so that when stable isotope analysis of bone apatite for dietary purposes was first proposed by C.H. Sullivan and H.W. Krueger in 1981 [105, 53], it was not accepted by other researchers [89]. Since then, however, bone apatite sample pretreatment procedures have matured [58, 52, 48], while analysis of non-porous tooth enamel has been shown to be reliable going back millions of years [55, 59, 49, 102, 56]. Increased understanding of fossilization pathways is also now leading to better assessment of the isotopic integrity of bone apatite samples [117, 57], yet there does seem to be some systematic offsets in isotope values produced using different sample preparation methods [34].

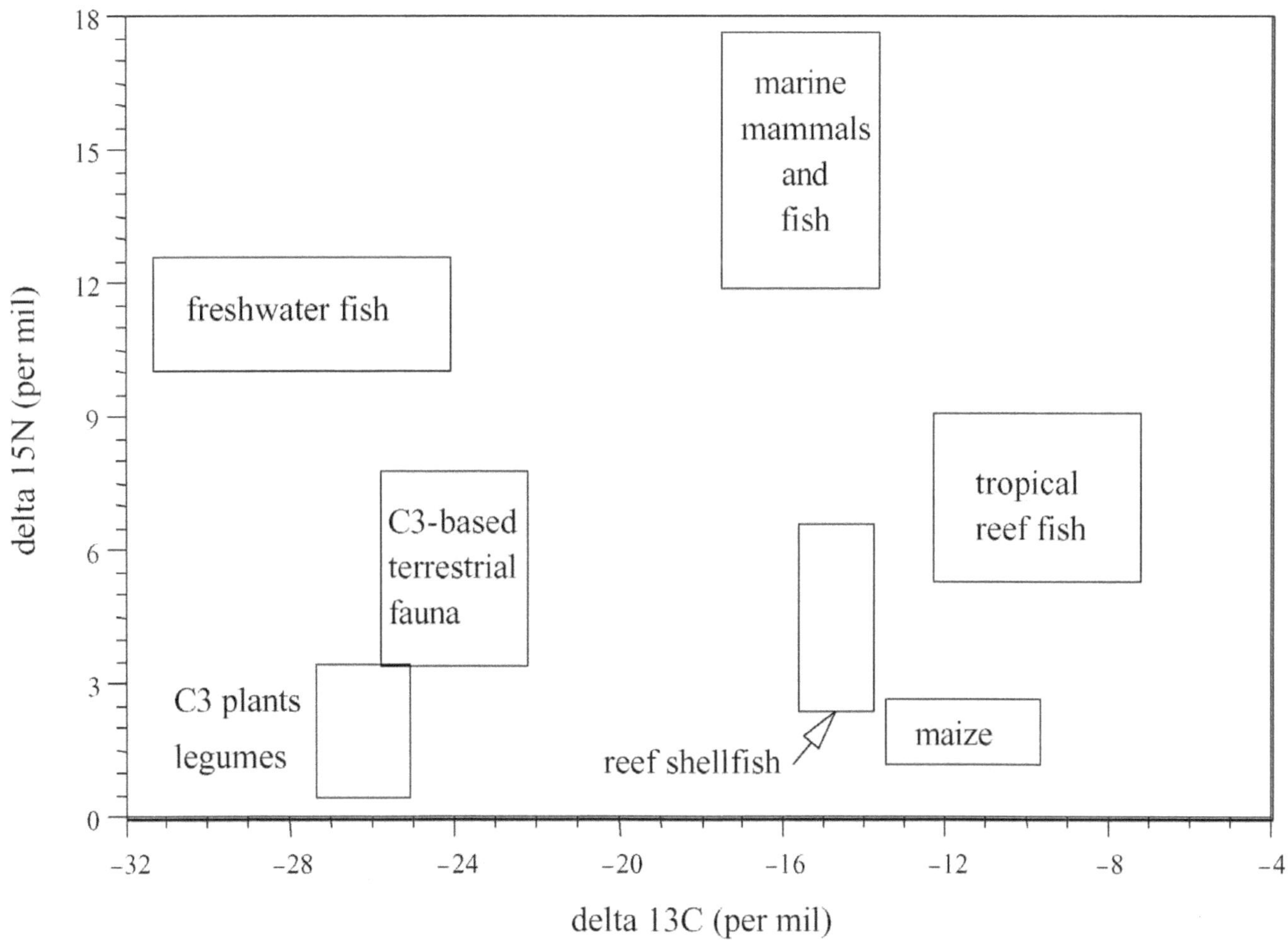

FIGURE 11-2 A Gulf Coast example of stable carbon versus nitrogen isotope ratios for plant and animal groups. Faunalbone collagen carbon corrected –2% to simulate flesh; modern samples corrected +1.5% for industrial effect. Althoughvalues for plants and terrestrial consumers are similar in other regions, values for aquatic resources may be different.

Overall, isotopic analysis of archaeological materials has increased tremendously since the early studies in the 1970s, when carbon isotope analysis of bone collagen were regularly done by radiocarbon laboratories to properly correct C14 dates, with both carbon and nitrogen isotope analysis now done in many other laboratories to address dietary questions. Until recently, I have personally maintained a detailed database of all published isotope studies in the New World, but the feasibility of this is steadily decreasing as the number of studies exponentially increase, not to mention more studies being done by graduate students, or for CRM firms, without widespread publication of the results.

SAMPLE PREPARATION AND ISOTOPIC ANALYSIS

Archaeological isotope studies, in contrast to most forensic studies of the recently deceased, must deal first with issues of preservation and contamination, in order to produce reliable scientific results with which to address ancient dietary questions. Over time, slightly different laboratory preparation procedures have been developed for preparing bone collagen and bone apatite samples. In most cases, a single 1 gram sample of reasonably preserved bone is sufficient for the preparation of both collagen and apatite (Figure 11-3); the actual amount of sample analyzed by most mass spectrometers is no more than a few milligrams, and as little as several hundred micrograms. For preparation of collagen, most techniques use an acidic solution to demineralize the bone, and other chemicals to remove contaminants and residual lipids. For preparation of bone apatite samples, sodium hypochlorite (bleach) is typically used to dissolve any preserved organic components, while a weak acid solution is employed to remove any non-biogenic carbonates. The specific procedures used are discussed in each of the chapters on isotope studies, as well as the particular mass spectrometer used and the precision of the measurements obtained. The precision of carbon and nitrogen isotopic results is almost always ≤ 0.3‰, and in many cases about 0.1‰, so that actual dietary isotope differences of just a few percent are noticeable.

Even more important is assessment of the reliability of the isotope results produced. While mass spectrometers have significantly changed over time, from manually operated systems in which solid samples were converted to CO_2 and N_2 gas off-line (Figures 11-4a, 11-4b), to modern, largely computer-operated instruments with automated in-line sample introduction systems (CHN analyzers for organic samples, common or individual acid baths for apatite/enamel samples) (Figures 11-5a, 11-5b), the users are always responsible for the type, quality, and purity of the sample being tested. For bone collagen, this is typically dealt with by measuring the percentage of

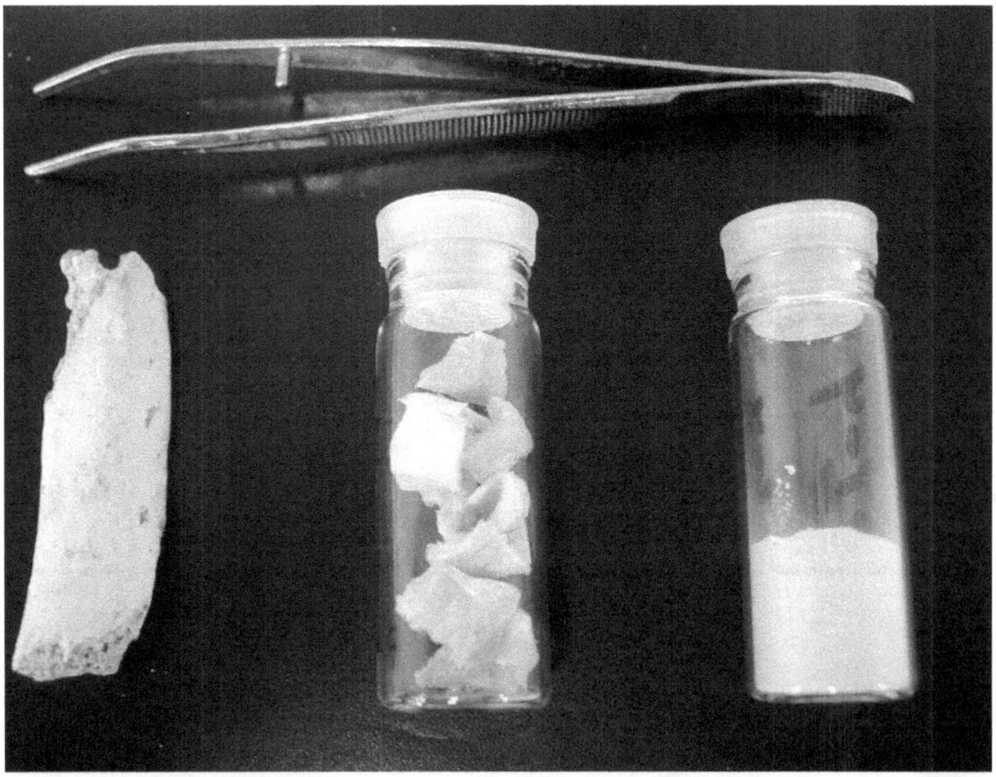

FIGURE 11-3 Typical bone sample, and collagen and apatite produced in Tykot's Laboratory for ArchaeologicalScience at the University of South Florida. (Courtesy of Robert H. Tykot)

FIGURE 11-4a (at left) Off-line gas sample preparation system for stable isotope samples, with Nikolaas J. van der Merwe.University of Cape Town, Stable Light Isotope Laboratory. Photo by R. H. Tykot. (Courtesy of Nikolaas J. van derMerwe, Ian Newton, Mark Pollard)

FIGURE 11-4b (below) A venerable stable isotope ratio mass spectrometer (VG602E), which requires manual input of CO2and N2gas samples. University of Cape Town, Stable Light Isotope Laboratory. Photo by R. H.Tykot.

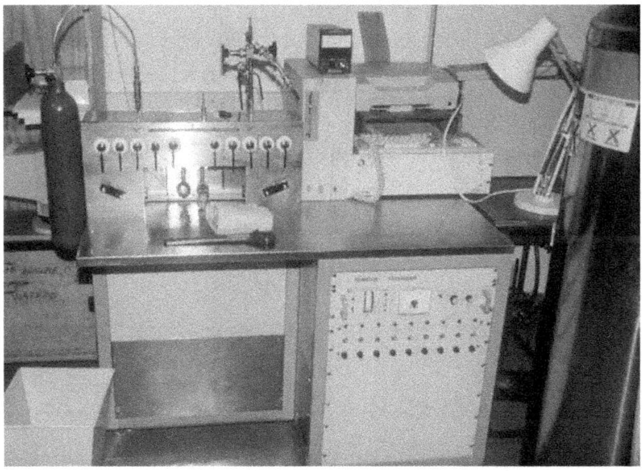

collagen produced from the original whole bone sample (<1% is usually considered unreliable, since bone originally was more than 20% collagen); the amount of C and N measured by the mass spectrometer, relative to the size of the sample introduced; and the actual C:N ratio (which should be the same as in living organisms). Problematic results mainly occur when preservation has not been so poor that no collagen remains (and thus no isotope data produced!), but where degradation has resulted in unequal breakdown and loss of the different amino acids, which individually would have different isotope values because of the different chemical reactions involved in their initial production. For bone apatite and tooth enamel samples, other than measures of sample loss during the preparation procedures, and CO_2 yield during the mass spectrometry analysis, more complex tests of sample reliability, using Fourier transform infrared spectroscopy (FTIR), have been developed [125, 48, 69, 70, 57, 34]. One logistical approach to producing reliable isotope results, in addition to measures of sample preparation and analysis, is the need for performing repeat isotope analyses for "outliers" when testing a group of individuals thought to have had similar diets.

INTERPRETATION AND SIGNIFICANCE OF CARBON AND NITROGEN ISOTOPE DATA

From the time of the earliest isotope studies in the 1970s, it was realized that bone collagen has a very slow turnover rate, at least for adults, so that the isotope results obtained would represent the average diet over at least the last several years of an individual's life. While the exact rate of turnover is still unknown, it is estimated to be at least 5-7 years if not more, and may vary depending on the density of each bone. In any case it has become evident that isotope studies generally may be done on any available part of skeletal remains, although micro-sampling of interior and exterior parts of long bones could produce different values if diet isotopically changed significantly during the time of their recent production. While the same lengthy turnover time applies to bone apatite as well, it is clear that tooth enamel, tooth roots (dentin), hair, finger nails, and flesh are very different. Teeth form only once, and thus represent at most a few years of diet, at the age when the tooth formed, with third molars the latest to form, at about ages 10-12 for the enamel and by young adulthood for the tooth dentin (roots). Analysis of other permanent teeth can reveal the age of weaning, at which point there is no longer a trophic level increase in isotope values resulting from breast-feeding [126]. Analysis of sequential samples in tooth dentin may be used to investigate seasonal isotope variation in diet, as can micro-sampling of tooth enamel layers [9, 10, 122]. When preserved,

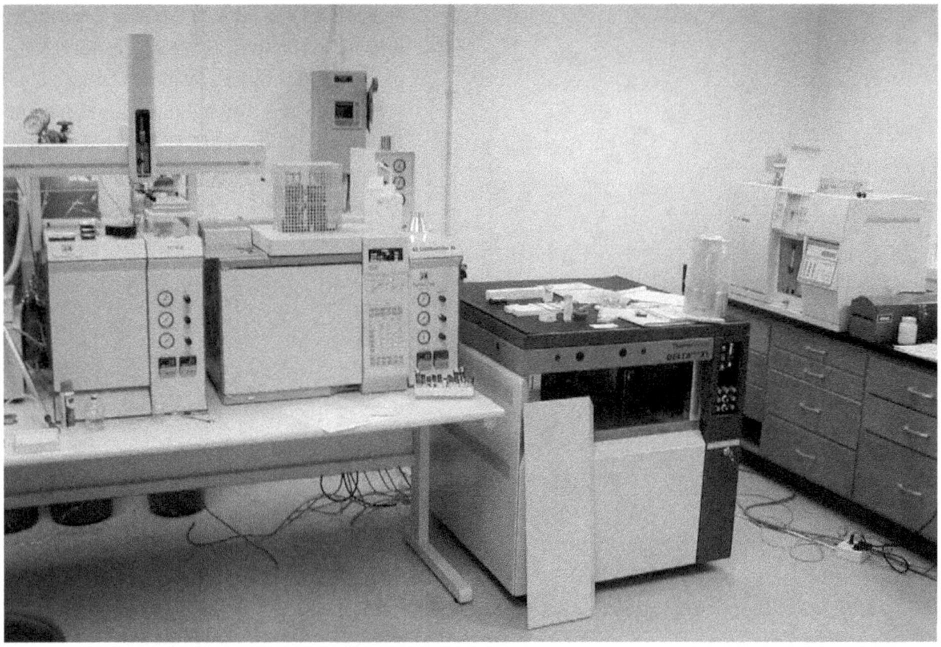

FIGURE 11-5a Example of a modern stable isotope ratio mass spectrometer (Finnigan MATDelta Plus XL), runby computer (far right), and connected with several input devices including a Carlo Erba CHN analyzer (upper rightcorner) that converts organic samples into $CO_2$and N_2gas. University of South Florida, Paleolab. Photo by R. H. Tykot.

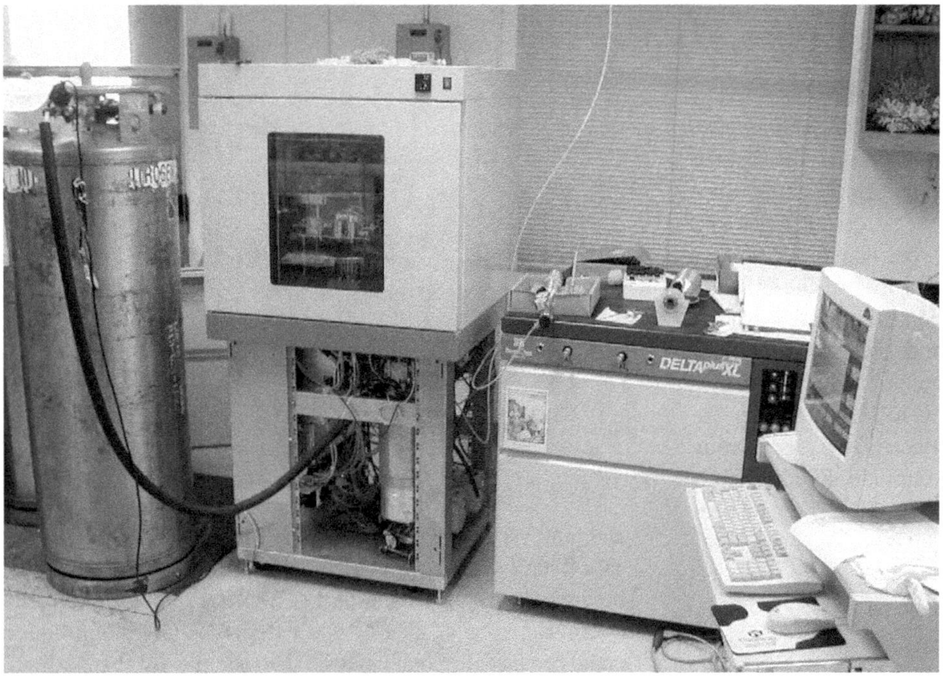

FIGURE 11-5b Another stable isotope ratio mass spectrometer, connected to an automated Kiel III individual acid-bath system (left) that converts inorganic carbonates into CO_2. University of South Florida, Paleolab. Photo by R. H. Tykot.

sequential hair and finger nail samples may also be used to assess short-term dietary change [88, 64, 73, 74, 75, 32, 98].

As for specific aspects of diet that are revealed through these different isotope studies, it became clear from studies of animals raised on controlled diets that bone collagen is produced primarily from dietary protein, while bone apatite and tooth enamel are products of the whole diet [7, 106]. For collagen, it appears that essential amino acids are transferred directly from appropriate foods, while non-essential amino acids may be transferred or produced independently, and thus represent the whole diet, including maize [37, 33, 43]. Isotopic analysis of individual amino acids would be especially useful for studies of the importance of maize in coastal environments.

Controlled diet studies have also provided a more precise assessment of the difference in $\delta^{13}C$ values between dietary resources and bone apatite, with variation in fractionation seemingly attributable to differences in body size, digestive physiology, and amount of methanogenesis [58, 17, 9, 43, 78]. For humans (not directly tested with controlled diets!), the fractionation between diet and bone apatite is estimated to be at least 9.5‰ [7] and as much as 11-12‰ [40]. While it is currently possible to quantitatively *compare* human individuals and groups of individuals, further studies are necessary if we are to have better precision on the *quantity* of maize and other C_4 plants in the diet.

Overall, interpretations are currently as follows. For herbivores, where the entire diet is based on carbohydrate dominated plant foods, any mixtures of C_3, C_4, and CAM plants would contribute equally to both collagen and apatite, unless there were significant differences in the percentage of protein in those plants (e.g. beans with about 20% protein, vs. maize with about 10% protein). For omnivores (including humans), animal food - whether wild or domestic, terrestrial or aquatic - would make a much greater contribution to bone collagen. So for a hypothetical New World inland society with maize agriculture, and dogs the only domesticated animals, it would be expected to have isotope values for both collagen and apatite that indicate maize consumption, with the percentage represented at least somewhat higher for bone apatite than for bone collagen, unless the dogs (or wild hunted animals) ate more maize than the humans (one reason for testing the isotope values of animals that were consumed). Differences between individuals in the percentage of maize vs. C_3 plants in the diet would be far more apparent in apatite than in collagen, thus making analysis of bone and tooth apatite much more effective than collagen for addressing questions about variation within a population or at a single site, comparison of individuals based on sex, status, and/or specialization, and changes over time [e.g. 109, 35, 4].

For studies where C_4 (or CAM) plants other than maize existed, or where seafood may have been consumed, it is strongly recommended to test archaeological faunal remains to learn whether they may have contributed more positive carbon isotope values to human consumers, as well as to use other archaeological and ethnohistoric information to interpret the isotope results. Marine plants, and their consumers, have more positive $\delta^{13}C$ values since their original sources of carbon include dissolved CO_2 and carbonic acid, as well as detritus from local terrestrial plants [93]. Because of the many trophic levels for marine fauna, there are continuous increases in both carbon and nitrogen isotope ratios from level to level, resulting in many cases with large fish and marine mammals with $\delta^{13}C$ values similar to C_4 plants such as maize - but also very enriched $\delta^{15}N$ values. Analysis of nitrogen isotopes in bone collagen is thus necessary for looking at diets in coastal regions, and in conjunction with bone carbonate carbon isotope analysis, may be used to assess the relative contributions of maize and marine foods in the whole diet and in protein in particular. Most freshwater aquatic fish, from inland lakes and rivers, also have much higher $\delta^{15}N$ values than terrestrial foods, but $\delta^{13}C$ values similar to C_3 plants, so that consumption of maize or other C_4 plants is readily apparent. It should be noted that in recent years some scholars have developed and used mathematical mixing models to interpret isotope results [61, 79, 80, 68].

OXYGEN AND STRONTIUM ISOTOPES

Even more significant in recent years has been the expansion of isotope analyses, from carbon and nitrogen in bone collagen and apatite, to charred ceramic residues [e.g. 67], carbonates and humic matter in soils [72, 118], and to carbon isotope analysis of cholesterol preserved in bones and teeth, of animal fats, and of lipid residues in ceramics [104, 29, 30, 86, 87]. Since the turnover time of cholesterol in bone samples is much greater than that of collagen and apatite, cholesterol analyses may be specifically used to address dietary change during the time more immediately preceding death, while analyses of residues and soils provide information on cultural and spatial landscape practices.

Oxygen and strontium isotope analyses have also been added to the isotope repertoire applied to ancient dietary issues. Oxygen isotope values relate directly to local climate, temperature and humidity [50, 63], and thus have been used not only for determining the seasonality of shells, and presumably their consumers [22], but also for climate studies [103, 92], and mobility [120, 119, 121], with appropriate changes in dietary patterns. Isotope ratios of strontium, which does not isotopically fractionate like biological C, N, and O, directly represent the geographic area of food production/acquisition, and thus the mobility of dietary resources and/or their consumers [27, 28, 82]. Strontium isotope analysis has thus been applied to significant migration periods in Europe [83, 13], immigration to Mesoamerican sites such as Teotihuacan and Tikal [84, 42, 124], and migration and mortuary patterns in South America [47].

ISOTOPE STUDIES IN THIS VOLUME

The following three chapters represent the latest stable isotope studies done in Mesoamerica. While a number of Maya sites in Belize, and in the Peten region, have been tested, these papers add significantly to the Pacific side of Mesoamerica, the Yucatan, and the overall sociocultural issues that relate to maize in the diet. Isotope studies such as these will continue to make significant contributions to our understanding of the *Histories of Maize.*

Bibliography

1. S.H. Ambrose, Preparation and characterization of bone and tooth collagen for isotopic analysis, *Journal of Archaeological Science* 17 (1990) 431-451.
2. S.H. Ambrose, Isotopic analysis of paleodiets: methodological and interpretive considerations, *Investigations of Ancient Human Tissue: Chemical Analysis in Anthropology*, M.K. Sandford (Ed.), Langhorne, PA, Gordon and Breach Science Publishers. 1993 pp. 59-129.

3. S.H. Ambrose, Controlled diet and climate experiments on nitrogen isotope ratios of rats, S.H. Ambrose and M.A. Katzenberg (Eds.), *Biogeochemical Approaches to Paleodietary Analysis*, New York, Plenum. 2000 pp. 243-67.
4. S.H. Ambrose, J. Buikstra and H. Krueger, Status and gender difference in diet at Mound 72, Cahokia, revealed by isotopic analysis of bone, *Journal of Anthropological Archaeology* 22 (2003) 217-26.
5. S.H. Ambrose and M.J. DeNiro, Reconstruction of African human diet using bone collagen carbon and nitrogen isotope ratios, *Nature* 319 (1986) 321-324.
6. S.H. Ambrose and J. Krigbaum, Bone chemistry and bioarchaeology, *Journal of Anthropological Archaeology* 22 (2003) 193-199.
7. S.H. Ambrose and L. Norr, Experimental evidence for the relationship of the carbon isotope ratios of whole diet and dietary protein to those of bone collagen and carbonate, J. B. Lambert and G. Grupe (Eds.), *Prehistoric Human Bone: Archaeology at the Molecular Level*, New York, Springer-Verlag.1993 pp. 1-37.
8. G.J. Armelagos, You are what you eat, K.D. Sobolik (Ed.), *Paleonutrition: The Diet and Health of Prehistoric Americans*, Southern Illinois University. 1994 pp. 235-244.
9. M. Balasse, Reconstructing dietary and environmental hsitory from enamel isotopic analysis: time resolution of intra-tooth sequential sampling, *International Journal of Osteoarchaeology* 12 (2002) 155-165.
10. M. Balasse, Potential biases in sampling design and interpretation of intra-tooth Isotope analysis, *International Journal of Osteoarchaeology* 13 (2003) 3-10.
11. M.M. Bender, Mass spectrometric studies of carbon 13 variations in corn and other grasses, *Radiocarbon* 10 (1968) 468-472.
12. M.M. Bender, D.A. Baerreis and R.L. Steventon, Further light on carbon isotopes and Hopewell agriculture, *American Antiquity* 46 (1981) 346-353.
13. A. Bentley, T.D. Price, J. Luning, D. Gronenborn, J. Wahl and P. Fullagar, Prehistoric migration in Europe: strontium isotope analysis of Early Neolithic skeletons, *Current Anthropology* 43 (2002) 799-804.
14. H. Bocherens and D. Drucker, Trophic level isotopic enrichment of carbon and nitrogen in bone collagen: case studies from recent and ancient terrestrial ecosystems, *International Journal of Osteoarchaeology* 13 (2003) 46-53.
15. R. Burleigh and D. Brothwell, Studies on American dogs, 1. Carbon isotopes in relation to maize in the diet of domestic dogs from early Peru and Ecuador, *Journal of Archaeological Science* 5 (1978) 355-362.
16. M. Calvin and A.A. Benson, The path of carbon in photosynthesis, *Science* 107 (1948) 476-480.
17. T.E. Cerling and J.M. Harris, Carbon isotope fractionation between diet and bioapatite in ungulate mammals and implications for ecological and paleoecological studies, *Oecologia* 120 (1999) 247-363.
18. T.B. Coplen, Reporting of stable hydrogen, carbon, and oxygen isotopic abundances, *Pure & Applied Chemistry* 66 (1994) 273-276.
19. H. Craig, The geochemistry of the stable carbon isotopes, *Geochimica et Cosmochimica Acta* 3 (1953) 53-92.
20. H. Craig, Isotope standards for carbon and oxygen and correction factors for mass-spectrometric analyses of carbon dioxide, *Geochimica et Cosmochimica Acta* 12 (1957) 133-149.
21. H. Craig and V. Craig, Greek marbles: determination of provenance by isotopic analysis, *Science* 176 (1972) 401-403.
22. M. Deith, Shell Seasonality: An Appraisal of the Oxygen Isotope Technique, R.E. Webb (Ed.), *Recent Developments in Environmental Analysis in Old and New World Archaeology*, Oxford, BAR International Series 416 (1988) 37-49.
23. M.J. DeNiro, Stable isotopy and archaeology, *American Scientist* 75 (1987) 182-191.
24. M.J. DeNiro and S. Epstein, Influence of diet on the distribution of carbon isotopes in animals, *Geochimica et Cosmochimica Acta* 42 (1978) 495-506.
25. M.J. DeNiro and S. Epstein, Influence of diet on the distribution of nitrogen isotopes in animals, *Geochimica et Cosmochimica Acta* 45 (1981) 341-351.
26. M.J. DeNiro and M.J. Schoeninger, Stable carbon and nitrogen isotope ratios of bone collagen: variations within individuals, between sexes, and within populations raised on monotonous diets, *Journal of Archaeological Science* 10 (1983) 199-203.
27. J.E. Ericson, Strontium isotope characterization in the study of prehistoric human ecology, *Journal of Human Evolution* 14 (1985) 503-14.
28. J.E. Ericson, Some problems and potentials of strontium isotope analysis for human and animal ecology, P.W. Rundel, J.R. Ehleringer and K.A. Nagy (Eds.), *Stable Isotopes in Ecological Research*, New York, Springer-Verlag. 1989 pp. 252-259.
29. R.P. Evershed, C. Heron, S. Charters and L.J. Goad, The survival of food residues: new methods of analysis, interpretation and application, A.M. Pollard (Ed.), *New Developments in Archaeological Science*, New York, Oxford University Press. 1992 pp. 187-208.
30. R.P. Evershed, S.N. Dudd, M.S. Copley, R. Berstan, A.W. Stott, H. Mottram, S.A. Buckley and Z. Crossman, Chemistry of archaeological animal fats, *Accounts of Chemical Research* 35 (2002) 660-668.
31. P. Farnsworth, J.E. Brady, M. DeNiro and R.S. MacNeish, A re-evaluation of the isotopic and archaeological reconstructions of diet in the Tehuacan Valley, *American Antiquity* 50 (1985) 102-116.
32. J.N. Fenner, *Putting a Hair in the Soup: An Investigation into the Archaeological Analysis of Single Hairs*, Master's Thesis, Eastern New Mexico University. 2002.
33. M.L. Fogel and N. Tuross. Extending the limits of paleodietary studies of humans with compound specific carbon isotope analysis of amino acids, *Journal of Archaeological Science* 30 (2003) 535-545.
34. S. Garvie-Lok, T. Varney and M. Katzenberg, Preparation of bone carbonate for stable isotope analysis: the effects of treatment time and acid concentration, *Journal of Archaeological Science* 31 (2004) 763-776.
35. J.P. Gerry, Bone Isotope Ratios and Their Bearing on Elite Privilege Among the Classic Maya, *Geoarchaeology* 12 (1997) 41-69.
36. R.A. Hall, Those late corn dates: isotopic fractionation as a source of error in carbon-14 dates, *Michigan Archaeologist* 13 (1967) 171-180.

37. P.E. Hare, M.L. Fogel, T.W. Stafford Jr., A.D. Mitchell and T.C. Hoering, The isotopic composition of carbon and nitrogen in individual amino acids isolated from modern and fossil proteins, *Journal of Archaeological Science* 18 (1991) 277-292.
38. M.D. Hatch and C.R. Slack, 1966, Photosynthesis by sugarcane leaves. A new carboxylation reaction and the pathway of sugar formation, *Biochem. J.* 101 (1966) 103-111.
39. M.D. Hatch, C.R. Slack and H.S. Johnson, Further studies on a new pathway of photosynthetic carbon dioxide fixation in sugarcane, and its occurrence in other species, *Biochem. J.* 102 (1967) 417-422.
40. R.E.M. Hedges, On bone collagen-apatite carbonate isotopic relationships, *International Journal of Osteoarchaeology* 13 (2003) 66-79.
41. N. Herz and D.B. Wenner, Assembly of Greek marble inscriptions by isotopic methods, *Science* 199 (1978) 1070-1072.
42. D.A. Hodell, R.L. Quinn, M. Brenner and G. Kamenov, Spatial variation of strontium isotopes (87Sr/86Sr) in the Maya region: a tool for tracking ancient human migration, *Journal of Archaeological Science* 31 (2004) 585-601.
43. M.A. Katzenberg, Advances in stable isotope analysis of prehistoric bones, S.R. Saunders and M.A. Katzenberg (Eds.), *Skeletal Biology of Past Peoples: Research Methods*, New York, Wiley-Liss. 1992 pp. 105-120.
44. M.A. Katzenberg, Stable isotope analysis: a tool for studying past diet, demography, and life history, in M.A. Katzenberg and S.R. Saunders (Eds.), *Biological Anthropology of the Human Skeleton*. Wiley-Liss, New York. 2000 pp. 305-328.
45. M.A. Katzenberg and R.G. Harrison, What's in a bone? Recent advances in archaeological bone chemistry, *Journal of Archaeological Research* 5 (1997) 265-293.
46. K.J. Knudson, T.D. Price, J.E. Buikstra and D.E. Blom, The use of strontium isotope analysis to investigate Tiwanaku migration and mortuary ritual in Bolivia and Peru, *Archaeometry* 46 (2004) 5-18.
47. P.L. Koch, N. Tuross and M. L. Fogel, The effects of sample treatment and diagenesis on the isotopic integrity of carbonate in biogenic hyroxylapatite, *Journal of Archaeological Science* 24 (1997) 417-429.
48. M. Kohn, M.J. Schoeninger and W.W. Barker, Altered states: effects of diagenesis on fossil tooth chemistry, *Geochimmica et Cosmochimica Acta* 63 (1999) 2737-2747.
49. Y. Kolodny, B. Luz and O. Navon, Oxygen isotope variations in phosphate of biogenic apatites, *Earth and Planetary Science Letters* 64 (1983) 398-404.
50. H.W. Krueger, The preservation and dating of collagen in ancient bones, R.M. Chatters and E.A. Olson (Eds.), *Proceedings of the 6th International Conference on Radiocarbon and Tritium Dating*, National Bureau of Standards, Washington, DC, 1965 pp. 332-337.
51. H.W. Krueger, Exchange of carbon with biological apatite, *Journal of Archaeological Science* 18 (1991) 355-361.
52. H.W. Krueger and C.H. Sullivan, Models for carbon isotope fractionation between diet and bone, in J. Turnlund and P.E. Johnson (Eds.), *Stable Isotopes in Nutrition*, American Chemical Society, Washington, DC. ACS Symposium Series 258 (1984) 205-222.
53. C.S. Larsen, *Bioarchaeology: Interpreting Behavior from the Human Skeleton,* Cambridge, Cambridge University Press. 1997.
54. J.A. Lee-Thorp, *Stable carbon isotopes in deep time: the diets of fossil fauna and hominids*, Ph.D. thesis, University of Cape Town, 1989.
55. J.A. Lee-Thorp, Preservation of biogenic carbon isotopic signals in Plio-Pleistocene bone and tooth mineral, S.H. Ambrose and M.A. Katzenberg (Eds.), *Biogeochemical Approaches to Paleodietary Analysis*, Plenum, New York, 2000 pp. 89-116.
56. J.A. Lee-Thorp and M. Sponheimer, Three case studies used to reassess the reliability of fossil bone and enamel isotope signals for paleodietary studies, *Journal of Anthropological Archaeology* 22 (2003) 208-216.
57. J.A. Lee-Thorp and N.J. van der Merwe, Carbon isotope analysis of fossil bone apatite, *South African Journal of Science* 83 (1987) 712-15.
58. J.A. Lee-Thorp and N.J. van der Merwe, Aspects of the chemistry of modern and fossil biological apatites, *Journal of Archaeological Science* 18 (1991) 343-354.
59. W.F. Libby, *Radiocarbon Dating*. University of Chicago Press. 1955.
60. J.D.C. Little and E.A. Little, Analysing Prehistoric Diets by Linear Programming, *Journal of Archaeological Science* 24 (1997) 741-747.
61. R. Longin, New method of collagen extraction for radiocarbon dating, *Nature* 230 (1970) 241-242.
62. B. Luz, Y. Kolodny and M. Horowitz, Fractionation of oxygen isotopes between mammalian bone-phosphate and environmental drinking water, *Geochimica et Cosmochimica Acta* 48 (1984) 1689-1693.
63. S.A. Macko, M.H. Engel, V. Andrusevich, G. Lubec, T.C. O'Connell and R.E.M. Hedges, Documenting the diet in ancient human populations through stable isotope analysis of hair, *Philosophical Transactions of the Royal Society of London* B 354 (1999) 65-76.
64. S. Mays, *The Archaeology of Human Bones*, London and New York, Routledge.1998.
65. E. Medina and P. Minchin, Stratification of □^{13}C values of leaves in Amazonian rain forests, *Oecologia* 45 (1980) 377-378.
66. J. Morton and H.P Schwarcz, Paleodietary implications from isotopic analysis of food residues on prehistoric Ontario ceramics, *Journal of Archaeological Science* 31 (2004) 503-517.
67. S.D. Newsome, D.L. Phillips, B.J. Culleton, T.P. Guilerson and P.L. Koch, Dietary reconstruction of an early to middle Holocene human population from the central California coast: insights from advanced stable isotope mixing models, *Journal of Archaeological Science* 31 (2004) 1101-1115.
68. C.M. Nielsen-Marsh and R.E.M. Hedges, Patterns of diagenesis in bone I: the effects of site environments, *Journal of Archaeological Science* 27 (2000) 1139-1150.
69. C.M. Nielsen-Marsh and R.E.M. Hedges, Patterns of diagenesis in bone II: the effects of acetic acid treatment and the removal of diagenetic CO_3^{2-}, *Journal of Archaeological Science* 27 (2000) 1151-1159.
70. A.O. Nier and E.A. Gulbransen, Variations in the relative abundance of the carbon isotopes, *Journal of the American Chemical Society* 61 (1939) 697-698.

71. L.C. Nordt, Stable carbon and oxygen isotopes in soils, P. Goldberg, V.T. Holliday, and C.R. Ferring (Eds.), *Earth Sciences and Archaeology*, New York, Kluwer. 2001 pp. 419-448.
72. T.C. O'Connell and R.E.M. Hedges, Investigations into the effect of diet on modern human hair isotopic values, *American Journal of Physical Anthropology* 108 (1999) 409-425.
73. T.C. O'Connell and R.E.M. Hedges, Isotopic comparison of hair and bone: archaeological analyses, *Journal of Archaeological Sciences* 26 (1999) 661-665.
74. T.C. O'Connell, R.E.M. Hedges, M.A. Healey and A.H.R.W. Simpson, Isotopic comparison of hair, nail and bone: modern Analyses, *Journal of Archaeological Science* 28 (2001) 1247-55.
75. R. Park and S. Epstein, Carbon isotope fractionation during photosynthesis, *Geochimica et Cosmochimica Acta* 21 (1960) 110-126.
76. P.L. Parker, The biogeochemistry of the stable isotopes of carbon in a marine bay, *Geochimica et Cosmochimica Acta* 28 (1964) 1155-1164.
77. B.H. Passey, T.F. Robinson, L. K. Ayliffe, T. E. Cerling, M. Sponheimer, M.D. Dearing, B.L. Roeder and J.R. Ehleringer, Carbon isotope fractionation between diet, breath CO2, and bioapatite in different mammals, *Journal of Archaeological Science* 32 (2005) 1459-1470.
78. D.L. Phillips and P.L. Koch, Incorporating concentration dependence in stable isotope mixing models, *Oecologia* 130 (2002) 114-125.
79. D.L. Phillips and J.W. Gregg, Source partitioning using stable isotopes: coping with too many sources, *Oecologia* 136 (2003) 261-269.
80. A.M. Pollard, Archaeological reconstruction using stable isotopes, H. Griffiths (Ed.), *Stable Isotopes: Integration of Biological, Ecological and Geochemical Processes*, Oxford, Bios Scientific Publishers. 1998 pp. 285-302.
81. T.D. Price, J.H. Burton and R.A. Bentley, The characterization of biologically available strontium isotope ratios for the study of prehistoric migration, *Archaeometry* 44 (2002) 117-35.
82. T.D. Price, G. Grupe and P. Schröter, Reconstruction of migration patterns in the Bell Beaker period by stable strontium isotope analysis, *Applied Geochemistry* 9 (1994) 413-417.
83. T.D. Price, L. Manzanilla and W.D. Middleton, Immigration and the ancient city of Teotihuacan in Mexico: a study using strontium isotope ratios in human bone and teeth, *Journal of Archaeological Science* 27 (2000) 903-913.
84. S.L. Ransom and M. Thomas, Crassulacean acid metabolism, *Annual Review of Plant Physiology* 11 (1960) 81-110.
85. E.A. Reber and R.P. Evershed, How did Mississippians prepare maize? The application of compound-specific carbon isotope analysis to absorbed pottery residues from several Mississippi Valley sites, *Archaeometry* 46 (2004) 19-33.
86. E.A. Reber and R.P. Evershed, Identification of maize in absorbed organic residues: a cautionary tale, *Journal of Archaeological Science* 31 (2004) 399-410.
87. M.K. Sandford and G.E. Kissling, Chemical analyses of human hair: anthropological applications, M.K. Sandford (Ed.), *Investigations of Ancient Human Tissue: Chemical Analyses in Anthropology,* Gordon and Breach Science Publishers. 1993 pp. 131-67.
88. M.J. Schoeninger and M.J. DeNiro, Carbon isotope ratios of apatite from bone cannot be used to reconstruct diets of animals, *Nature* 197 (1982) 577-578.
89. M.J. Schoeninger and M.J. DeNiro, Stable nitrogen isotope ratios of bone collagen reflect marine and terrestrial components of prehistoric human diet, *Science* 220 (1983) 1381-1383.
90. M.J. Schoeninger and M.J. DeNiro, Nitrogen and carbon isotopic composition of bone collagen from marine and terrestrial animals, *Geochimica et Cosmochimica Acta* 48 (1984) 625-639.
91. M.J. Schoeninger, M.J. Kohn and J.W. Valley, Tooth oxygen isotope ratios as paleoclimate monitors in arid ecosystems, S.H. Ambrose and M.A. Katzenberg (Eds.), Biogeochemical Approaches to Paleodietary Analysis, New York, Plenum. 2000 pp. 117-40.
92. M.J. Schoeninger and K. Moore, Bone stable isotope studies in archaeology, *Journal of World Prehistory* 6 (1992) 247-296.
93. M.J. Schoeninger and M.R. Schurr, Interpreting carbon stable isotope ratios, S. Johannessen and C.A. Hastorf (Eds.), *Corn and Culture in the Prehistoric New World*, Boulder, Colorado, Westview Press. 1994 pp. 55-66.
94. M.R. Schurr, Using stable nitrogen-isotopes to study weaning behavior in past populations, *World Archaeology* 30 (1998) 327-42.
95. H.P. Schwarcz, J. Melbye, M.A. Katzenberg and M. Knyf, Stable isotopes in human skeletons of southern Ontario: reconstructing paleodiet, *Journal of Archaeological Science* 12 (1985) 187-206.
96. H.P. Schwarcz and M. J. Schoeninger, Stable isotope analysis in human nutritional ecology, *Yearbook of Physical Anthropology* 34 (1991) 283-321.
97. H.P. Schwarcz and C.D. White, The grasshopper or the ant?: cultigen-use strategies in ancient Nubia from C-13 analyses of human hair, *Journal of Archaeological Science* 31 (2004) 753-62.
98. J. Sealy, R. Armstrong and C. Schrire, Beyond lifetime averages: tracing life histories through isotopic analysis of different calcified tissues from archaeological human skeletons, *Antiquity* 69 (1995) 290-300.
99. J.C. Sealy, N.J. van der Merwe, J.A. Lee-Thorp and J. Lanham, Nitrogen isotopic ecology in southern Africa: implications for environmental and dietary tracing, *Geochimica et Cosmochimica Acta* 51 (1987) 2707-17.
100. B.N. Smith and S. Epstein, Two categories of $^{13}C/^{12}C$ for higher plants, *Plant Physiology* 47 (1970) 380-384.
101. M. Sponheimer and J. Lee-Thorp, Isotopic Evidence for Diet of an Early Hominid, *Australopihecus africanus*, *Science* 283 (1999) 368-370.
102. E. Stephan, Oxygen isotope analysis of animal bone phosphate: method refinement, influence of consolidants, and reconstruction of palaeotemperatures for Holocene sites, *Journal of Archaeological Science* 27 (2000) 523-35.
103. A.W. Stott and R.P. Evershed, □^{13}C analysis of cholesterol preserved in archaeological bones and teeth, *Analytical Chemistry* 68 (1996) 4402-4408.
104. C.H. Sullivan and H.W. Krueger, Carbon isotope analysis of separate chemical phases in modern and fossil bone, *Nature* 292 (1981) 333-335.

105. L.L. Tieszen and T. Fagre, Effect of diet quality and composition on the isotopic composition of respiratory CO2, bone collagen, bioapatite, and soft tissues, J.B. Lambert and G. Grupe (Eds.), *Prehistoric Human Bone: Archaeology at the Molecular Level*, New York, Springer-Verlag. 1993 pp. 121-155.
106. J.R. Turnlund and P.E. Johnson (Eds.), *Stable Isotopes in Nutrition,* American Chemical Society. 1984.
107. R.H. Tykot, Stable isotopes and diet: you are what you eat, M. Martini, M. Milazzo and M. Piacentini (Eds.), *Physics Methods in Archaeometry. Proceedings of the International School of Physics "Enrico Fermi",* Bologna, Italy, Società Italiana di Fisica. 2004 pp. 433-444.
108. D.H. Ubelaker, M.A. Katzenberg and L.G. Doyon, Status and diet in precontact highland Ecuador, *American Journal of Physical Anthropology* 97 (1995) 403-411.
109. N.J. van der Merwe, New wrinkles in radiocarbon, paper presented at the Plains Anthropology Conference, Columbia, MO, 1973.
110. N.J. van der Merwe, Carbon isotopes, photosynthesis, and archaeology, *American Scientist* 70 (1982) 596-606.
111. N.J. van der Merwe, Light stable isotopes and the reconstruction of prehistoric diets, *Proceedings of the British Academy* 77 (1992) 247-264.
112. N.J. van der Merwe, A.C. Roosevelt and J.C. Vogel, Isotopic evidence for prehistoric subsistence change at Parmana, Venezuela, *Nature* 292 (1981) 536-38.
113. N.J. van der Merwe and H. Tschauner, C4 Plants and the Development of Human Societies, R.F. Sage and R.K. Monson (Eds.), *C4 Plant Biology*, New York, Academic Press. 1999 pp. 509-549.
114. N.J. van der Merwe and J.C. Vogel, 13C content of human collagen as a measure of prehistoric diet in woodland North America, *Nature* 276 (1978) 815-16.
115. J.C. Vogel and N.J. van der Merwe, Isotopic evidence for early maize cultivation in New York State, *American Antiquity* 42 (1977) 238-242.
116. Y. Wang and T.E. Cerling, A model of fossil tooth and bone diagnesis: implications for paleodiet reconstruction from stable isotopes, *Palaeogeography, Paleoclimatology, Palaeoecology* 107 (1994) 281-289.
117. E.A. Webb, H.P. Schwarcz and P.F. Healy, Detection of ancient maize agriculture in the Maya lowlands using the stable carbon isotope compositions of soil organic matter: Evidence from Caracol, Belize, *Journal of Archaeological Science* 31 (2004) 1039-1052.
118. C.D. White, F.J. Longstaffe and K.R. Law, Exploring the effects of environment, physiology and diet on oxygen isotope ratios in ancient Nubian bones and teeth, *Journal of Archaeological Science* 31 (2004) 233-50.
119. C.D. White, M.W. Spence, F.J. Longstaffe and K.R. Law, Testing the nature of Teotihuacán imperialism at Kaminaljuyú using phosphate oxygen-isotope ratios, *Journal of Anthropological Research* 56 (2000) 535-58.
120. C.D. White, M. Spence, F.J. Longstaffe and K.R. Law, Demography and ethnic continuity in the Tlailotlacan enclave of Teotihuacan: the evidence from stable oxygen isotopes, *Journal of Anthropological Archaeology* 23 (2004) 385-403.
121. J.W. Wilson, R.H. Tykot and D.H. Ubelaker, Intradental variation in stable carbon isotope ratios in human tooth enamel, *American Journal of Physical Anthropology* 114 (2001) 166.
122. L.E. Wright, Correspondence between stable carbon, oxygen and nitrogen isotopes in human tooth enamel and dentine: infant diets at Kaminaljuyu, *Journal of Archaeological Science* 26 (1999) 1159-70.
123. L.E. Wright, Identifying immigrants to Tikal, Guatemala: defining local variability in strontium isotope ratios of human tooth enamel, *Journal of Archaeological Science* 32 (2005) 555-566.
124. L.E. Wright and H.P. Schwarcz, Infrared and isotopic evidence for diagenesis of bone apatite at Dos Pilas, Guatemala: palaeodietary implications, *Journal of Archaeological Science* 23 (1996) 933-944.
125. L.E. Wright and H.P. Schwarcz, Stable carbon and oxygen isotopes in human tooth enamel: identifying breastfeeding and weaning in prehistory, *American Journal of Physical Anthropology* 106 (1998) 1-18.

Figures

Figure 1. C_3 and C_4 pathways, showing isotopic differences passed on to skeletal tissues.

Figure 2. A Gulf Coast example of stable carbon vs. nitrogen isotope ratios for plant and animal groups. Faunal bone collagen carbon corrected -2‰ to simulate flesh; modern samples corrected +1.5‰ for industrial effect. While values for plants and terrestrial consumers are similar in other regions, values for aquatic resources may be very different.

Figure 3. Typical bone sample, and collagen and apatite produced in Tykot's Laboratory for Archaeological Science at the University of South Florida.

Figure 4a. Off-line gas sample preparation system for stable isotope samples, with Nikolaas J. van der Merwe. University of Cape Town, Stable Light Isotope Laboratory. Photo by R.H. Tykot.

Figure 4b. A venerable stable isotope ratio mass spectrometer (VG602E), which requires manual input of CO_2 and N_2 gas samples. University of Cape Town, Stable Light Isotope Laboratory. Photo by R.H. Tykot.

Figure 5a. Example of a modern stable isotope ratio mass spectrometer (Finnigan MAT Delta Plus XL), run by computer (far right), and connected with several input devices including a Carlo Erba CHN analyzer (upper right corner) that converts organic samples into CO_2 and N_2 gas. University of South Florida, Paleolab. Photo by R.H. Tykot.

Figure 5b. Another stable isotope ratio mass spectrometer, connected with an automated Kiel III individual acid bath system (left side) that converts inorganic carbonates into CO_2. University of South Florida, Paleolab. Photo by R.H. Tykot.

C H A P T E R

12

Social Directions in the Isotopic Anthropology of Maize in the Maya Region

CHRISTINE D. WHITE*, FRED J. LONGSTAFFE†, AND HENRY P. SCHWARCZ‡

*Department of Anthropology, The University of Western Ontario, London, Ontario, Canada
†Department of Earth Sciences, The University of Western Ontario, London, Ontario, Canada
‡School of Geography and Earth Sciences, McMaster University, Hamilton, Ontario, Canada

Glossary

Apatite The mineral portion of bone, of which carbonate is a part.

Ascribed status Social status that is inherited.

Classic period collapse A phenomenon that occurred in the Late Classic period, which was manifest in the depopulation of ceremonial centers mostly in the lowlands.

Collagen The dominant component of the organic portion of bone.

Ethnohistory The record of events and cultural descriptions from sources provided by early observers of a people, in this case, Spanish chroniclers.

Gender The social rather than biological meaning of sex.

Mesoamerica The archaeological definition of Central America that includes most of modern Mexico, Guatemala, Belize, and Honduras.

Postprocessual archaeology A school of thought that uses the perspectives that theory and data are related, context is important, individuals are actors, and interpretations of the past are influenced by the political present.

Stelae Upright stone monuments found in Maya ceremonial centers. Histories of rulers and events are carved into them.

Trophic level A position in the food chain.

Isotopic data from human skeletal remains have been used to reconstruct the rise of maize agriculture in **Mesoamerica** and its outward spread, test the theory that ecological degradation caused the collapse of the Maya during the Classic period, and understand the relationship between diet and nutritional pathology. In keeping with theoretical approaches in Mesoamerican archaeology, isotopic research involving maize consumption is now addressing issues involved in understanding ideology, social structure, gender, economic relationships, and intergroup political relationships. Examples of these are provided from a variety of Maya sites (Altun Ha, Cahal Pech, Colha, Copán, Lagartero, Lamanai, Marco Gonzalez, Pacbitun, San Pedro, Tikal).

A BRIEF HISTORY OF ISOTOPIC ANTHROPOLOGY IN MESOAMERICA

The simple fact that maize is a C_4 plant has made it a focus of isotopic research in Mesoamerica, an area that includes modern Mexico, Guatemala, Belize, Honduras, and El Salvador. Iconography, pollen profiles, macroscopic remains, and ethnohistoric documents [10, 16, 26, 40, 45, 46, 66, 73] all provide evidence that maize was the most important plant food in Mesoamerica since its presumed domestication in Tehuacan Valley, Mexico [37]. As the primary cultigen in this area, maize is arguably the only major source of ancient plant food that followed the C_4 pathway (see Chapter 10). Before the Spanish conquest the only other possible plant sources that could confound isotopic diet reconstructions were crassulacean acid metabolism (CAM) plants (e.g., the nopal cactus *[Opuntia]*). These were undoubtedly consumed, particularly in the form of alcoholic beverages in Mexico, but there is no evidence that they were food staples. Other foods contributing to the appearance of a C_4 signature in food systems are animals that scavenge or are fed maize, and marine–reef resources. The second largest reef system in the world is located off the coast of Belize. This would have been a rich source of food for coastal Maya sites and may also have been accessed by inland sites through trade. The confounding dietary interpretations created by these resources can only be clarified through the combined use of $\delta^{13}C_{col}$ (to determine the protein component of the diet as it is derived from plant sources), $\delta^{13}C_{ap}$ (to determine the "whole" diet, i.e., protein, carbohydrates, lipids), $\delta^{15}N$ (to determine protein source and **trophic level**), and $\Delta^{13}C_{ap\text{-}col}$ (to determine the degree of carnivory versus herbivory) (see Chapter 14). Similarly, it is unlikely that the "routing" model that views $\delta^{13}C_{col}$ values as representations of dietary proteins can be completely supported with Maya data. Therefore, both $\delta^{15}N$ and $\Delta^{13}C_{ap\text{-}col}$ values are needed to determine relative quantities of maize-consuming animals versus other protein sources.

The use of human tissues for isotopic analyses meant that for the first time diet could be directly measured from consumers, rather than making assumptions from floral and faunal remains. About 10 years after the discovery that maize is a C_4 plant [69], the first anthropological applications appeared. These began with attempts to track the process of maize domestication indirectly [12, 15] and to determine the timing of its diffusion and adoption as a staple throughout North America [80, 2, 31] and South America [4, 14, 65, 78, 80]. Such investigations continue [75].

Maize had profound social and economic meaning throughout Mesoamerica. The Maya believed that they had been created from maize, adding ideological meaning to the chemical reality that we are quite literally "what we eat." For the Maya, maize was also perceived to be the foundation of their civilization. The intensive archaeological focus on maize production as an indirect measure of social and economic health led, in large part, to what is now known as the "ecological model" of the Classic period Maya collapse [67]. Demographic analysis at Tikal [22, 23] had shown that ancient population densities were much higher than originally thought, indeed even higher than those today [22, 23]. It was postulated that the widespread practice of slash and burn agriculture needed to support such large numbers of people could not have been ecologically sustainable and ultimately led to the depopulation of many Maya ceremonial centers [95]. The possibility that this great civilization was sent spinning into a decline because it may have outstripped its resources resonated with contemporary sensibilities about the potential effects of human-made environmental degradation, particularly in North America.

However, the relationship between cultural and environmental change during the period of collapse has been difficult to disentangle. Temporal differences in the consumption of maize have been found in isotopic analyses at most sites, but are not patterned uniformly [9, 27, 39, 48, 74, 76, 87–89, 96, 98, 99]. Although a full review of the isotopic evidence is beyond the scope of this chapter, the hypothesis that production was limited during the collapse is not consistently supported. This issue will not be resolved by simply measuring the maize content of the diet. Social factors such as trade, war, and political activity, as well as natural factors such as climatic change, also affect agricultural productivity and diet. Social heterogeneity in diet and small sample sizes are additional confounding factors. Therefore, archaeological, paleoenvironmental, and ethnohistoric lines of evidence should be integrated to reach an understanding of the relative contributions of culture and environment to dietary and social change.

Although reconstructions of Maya diet began as an attempt to identify temporal differences as they related to the collapse, they logically led to an understanding of regional diversity in food consumption [17, 85, 101]. These results challenged the stereotypic perception of a pan-Mesoamerican diet. Although maize was a major dietary component almost everywhere in the Maya area, extremely heterogeneous environments and different time periods caused significant variation in quantities of maize consumed.

Great effort has been made to understand the relationship between maize-dependent diets and pathology. Nutritional disease caused by heavy maize dependency was first suggested as a factor in the collapse by Ernest Hooton [28] and then later incorporated in the ecological model. High frequencies of iron deficiency anemia have been found in Maya populations [7, 28, 41, 68, 84, 94, 98], presumably a result of the iron-chelating properties of maize and the low iron content of a maize-dependent diet [3, 30]. Isotopic data have been used to test the relationship between maize consumption and the occurrence of anemia, but the results have been ambiguous. Pathology does not appear to have increased in general during the period of collapse [101]. By contrast, paleodietary and pathology data at Lamanai, Belize, suggest that anemia was associated with increasing maize consumption until the Post-Classic period. In the following Historic period, however, anemia was more likely related to an increase in parasitic or other infections introduced by the Spanish [84].

The relationship among other nutrient deficiencies and infectious and parasitic conditions now needs to be revisited with a more holistic epidemiological approach. Heavy consumption of maize may not be the primary factor in the manifestation of skeletal lesions traditionally interpreted as indicators of anemia (i.e., porotic hyperostosis and cribra orbitalia). Such lesions have since been understood to reflect a suite of conditions and differential diagnoses have changed [50, 82]. For example, the manifestation of vitamin C deficiency in the cranium can mimic anemia and has recently been found to co-occur with it [43]. Thus, we are in need of re-examining diagnoses of anemia made previously.

Until recently, understanding changing relationships among biology, culture, and environment has been the main emphasis in Mesoamerican isotopic anthropology. Thus, much of this research has been framed, at least indirectly, in terms of processual archaeology. The rest of this chapter illustrates how the ideological and social value of maize and the use of isotopic paleodietary data can move us into the realm of **postprocessual archaeology** and reconstructions of the social, political, and economic dynamics of Maya society.

IDEOLOGY

We must begin with the relationship between maize and belief. According to the mythology in the Popul Vuh, a Quiché Maya creation text, there were three unsuccessful attempts to create humans out of various materials, one of which failed because humans (then made of wood) maltreated and did not properly feed their dogs. Humans only survived as a species after they had been molded from maize and more generously fed their own food to their dogs. Artistic and ethnohistoric evidence suggests that dogs, as well as deer, were associated with political rituals and regeneration, and they were sacrificed in yearly renewal ceremonies as proxies for humans [73]. Dogs were domesticated long before the rise of the Maya, but it has been suggested that deer were husbanded during Maya times [73].

Because maize carries such strong symbolic currency, the assumption that it was a socially valued food among all Maya groups underlies all research involving the use of diet to reconstruct social organization. Because of this, we expect to observe differential access to it based on social rank. To test the degree of belief in creation symbolism (and, hence, the assumption of the social value of maize), the diets of dogs and deer from a variety of sites and contexts were reconstructed to see if they were purposefully fed maize [89, 90].

At the site of Colha, Belize (Figure 12-1). bone **collagen** of Pre-Classic dogs was analyzed from both midden and cache contexts [91]. Across the sequence from Early Middle to Terminal Late Pre-Classic periods, the midden dogs exhibited an increase in the consumption of C_4 foods and a decrease in the trophic level from which their protein sources were derived (Figure 12-2a). This decrease is consistent with increased reliance on maize. It appears that the diets of midden dogs were probably tracking those of their masters (although this remains to be confirmed using human samples) and reflects the intensification of maize agriculture toward the end of the Pre-Classic period. It is also strong evidence for the domesticity of dogs. Most of the cache dogs, however, had distinct $\delta^{13}C_{col}$ values that indicated pure maize diets. Because the bone collagen in these dogs would have represented diet for their entire lifespan, they must have been raised exclusively on maize from the beginning of their lives. In addition, they must have been physically restrained to prevent them from scavenging other food. In essence, these were dogs "made of maize" whose special context suggests they were probably sacrificed in some sort of ritual. Thus, the Colha cache dogs were likely created to be proxies in sacrificial rituals for humans, who were ideologically made of maize. It is ironic that the midden dogs may now be taken as proxies for humans by us.

The bone collagen of deer was also analyzed at Colha, as well as Lagartero, Copan, and Tikal (see Figure 12-1). [90, 91]. The majority of deer from these sites consumed C_3 foods, suggesting that they were wild and, therefore, most likely captured by hunting (see Figure 12-2b). There is a great deal of evidence for the hunting of deer in Maya art and **ethnohistory** [5, 53, 59]. Some of the deer at Copan and Lagartero had consumed significant quantities of C_4 plants. These deer were probably reflecting their natural tendency to feed opportunistically from the edges of maize fields [57]. With the exception of Lagartero, they all came from ordinary middens.

By contrast, the deposit at Lagartero represents a feasting event that was most likely tied to a ritual. It contains a deer that, like the cache dogs, was fed a pure maize diet all its life while being physically restrained (see Figure 11-2b). The physical restraint and use of maize to purposefully feed both deer and dogs (presumably to create human proxies) reflects the ideological importance of both maize and the animals themselves and their use in accessing the supernatural for humans. Such use of isotopic measures in anthropology, therefore, has great potential for further understanding human–animal interactions and their social implications.

SOCIAL STRUCTURE

The Maya deer and dogs demonstrate the strong social and ritual value of maize in Maya society. If we accept the premise that the high social value of maize can also be inferred from the association of C_4 foods with individuals who have wealth and power, then we can assume that high-ranking individuals were able to control the degree and social organization of labor, tribute, redistribution, and trade of foods or control of the technology, land, and water needed to produce or procure foods, or both [19, 89]. Such a "food system" would be characteristic of George Gumerman's [21] model for regional economies such as the Maya had. Elite individuals or groups should, therefore, have had the greatest choice of foodstuffs and may have used the consumption of ideologically important, or rare and hard-to-obtain, foods as a metaphor for their superiority. The definition of an elite food would naturally have been conditioned by what was

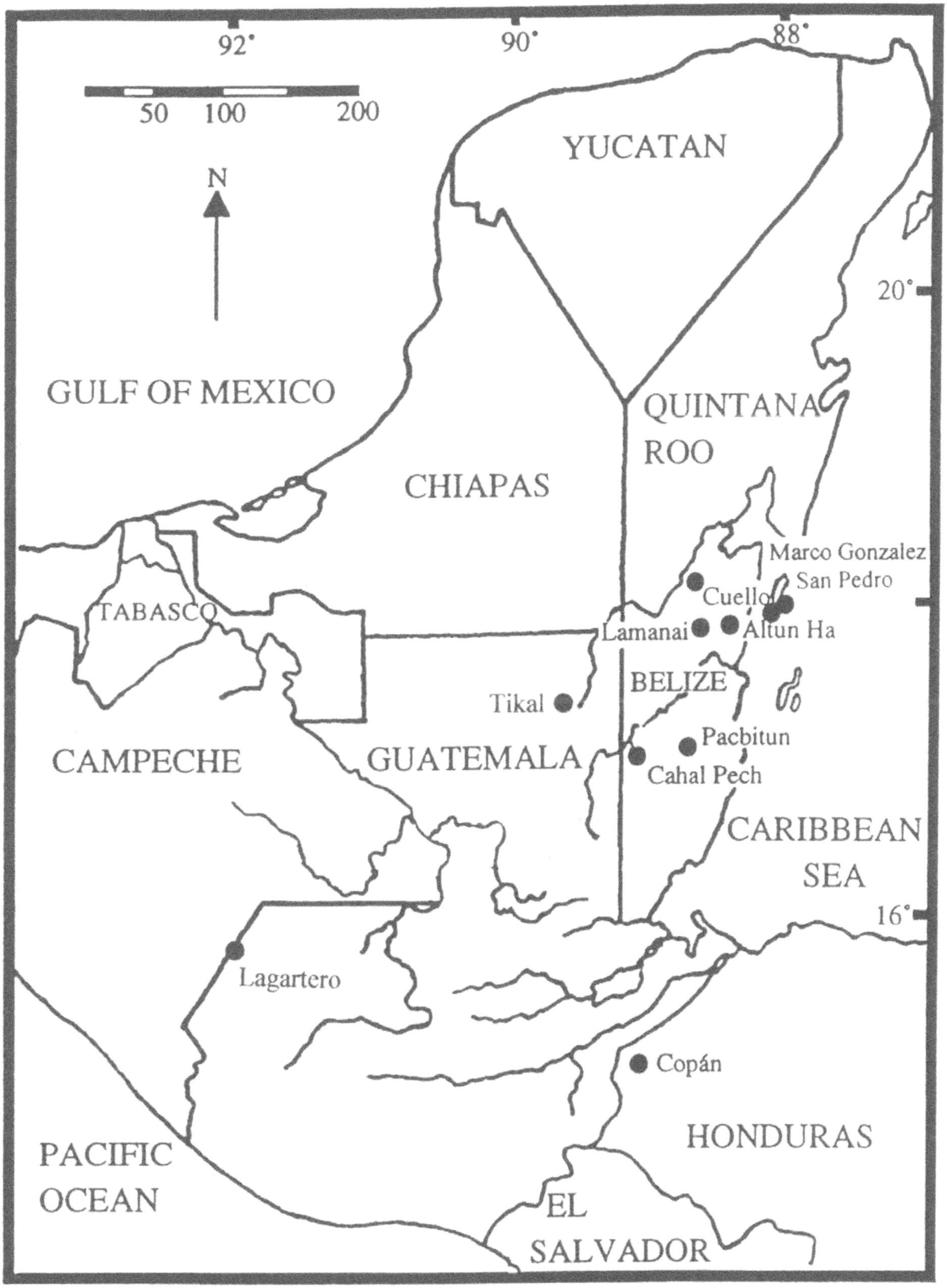

FIGURE 12-1 Map of Mesoamerica showing location of sites mentioned in the text.

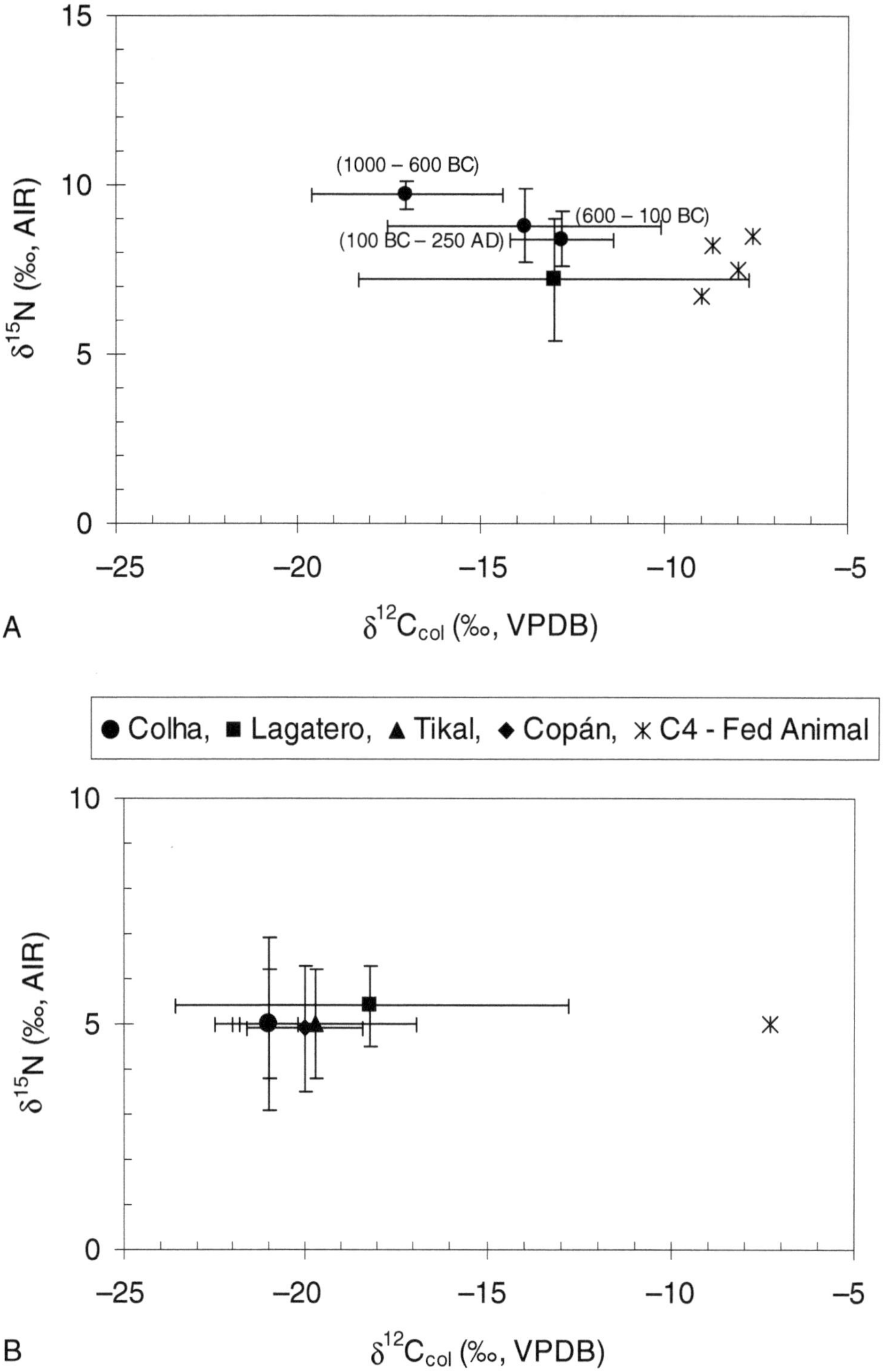

FIGURE 12-2 Isotopic values of bone collagen from Colha and Lagartero. **A,** dogs (*denotes dogs from special contexts); **B,** deer.

common and rare in the local environment. In addition, differential quantity was probably also as important as the kinds of foods chosen. Evidence that elites likely ate *more* can be found in the many artistic representations of corpulent high ranking individuals on pottery and **stelae.** Conspicuous consumption in quantity and quality was probably expressed through feasting.

Rise of Social Differentiation

The role of maize in the rise of Maya civilization, the relationship between maize production and population density, and the shift from egalitarianism to social ranking have not been well studied with human remains because of poor bone preservation and a variety of archaeological and political factors [24, 25]. In addition to the dogs who acted as proxies for humans at Colha, a number of sites (Altar de Sacrificios, Seibal, Lamanai, Altun Ha, see lFigure 12-1) have yielded small samples of Pre-Classic human remains that have been analyzed isotopically [87, 89, 97]. The oldest and largest human sample, without doubt, has come from Cuello, Belize (see Figure 12-1) [76, 79]. Pre-Classic economies at all of these sites, with the exception of Seibal, were based on maize agriculture. However, none of these sites (Cuello included) have provided samples with contexts that allow the use of humans in testing the hypothesis that status differences included differential access to food [97].

Cahal Pech, Belize (see Figure 12-1), is one of the earliest sites in the central Maya lowlands, dating to 1200 BC [25] and has yielded samples that came from socially distinct contexts for which both bone collagen and **apatite** were analyzed [61, 92]. Pottery impressions of cobs indicate that maize was grown in the area from an extremely early date (2900 BP [34]). However, because faunal remains in middens indicate a broad spectrum of riverine, terrestrial, and coastal resources [61, 71], it has been speculated that maize was less important here than at other Pre-Classic sites [34, 82].

Unlike Cuello, which is believed to have been a fairly egalitarian agricultural community [76, 79], Cahal Pech was constituted of three socially distinct settlement groups. Individuals from the rural agricultural community (Tolok group) are particularly distinctive and consumed foods from the lowest position in the food chain [61]. Their lower $\delta^{15}N_{col}$ values and higher $\Delta^{13}C_{ap\text{-}col}$ values indicate protein sources that were composed of a mixture of terrestrial herbivores and marine–reef fish (which is consistent with the faunal sample in the Tolok midden) (Figure 12-3A, B) Their $\delta^{13}C_{col}$ values also suggest that they consumed the fewest C_4 foods among the three groups (see Figure 12-3A). Although they were probably the primary producers of maize, the Tolok group individuals apparently did not have equal access to its consumption. This kind of inverse relationship between production and consumption is not uncommon in ranked societies where socially valued foods are produced by one group for consumption by another [52]. The differential consumption between social groups at Cahal Pech could simply mean that the lower status group on the site periphery had a greater natural access to wild C_3 foods or that the production and distribution of maize during the Pre-Classic period was likely controlled by elites.

By contrast, the Cas Pek group, which probably represents artisans [35], consumed more animal and/or marine–reef protein and more C_4 foods than the Tolok group [92] (from higher $\delta^{15}N_{col}$ and $\delta^{13}C_{col}$ values, lower $\Delta^{13}C_{ap\text{-}col}$ values, see Figure 12-3A, B). The differentiation in C_4 signals could be a result of marine–reef foods or greater maize consumption, or both. The most extreme C_4 signatures, however, come from the highest ranking individual at the site (Zotz group). He was more carnivorous and consumed more reef fish than his contemporaries. His $\delta^{15}N$ value (in combination with the $\delta^{13}C_{col}$ and $\delta^{13}C_{ap}$ values) suggests that some of his protein may have come from C_4-fed terrestrial animals.

The high social value of maize can, thus, be inferred from the association of C_4 foods with the highest ranking individuals at the site. More importantly, the existence of dietary distinctions between social groups in the Pre-Classic period supports theories of an early rise of social rank and elite control over the production and distribution of food.

Socioeconomic Status

The past tendency for archaeologists to focus excavations on ceremonial centers has resulted in skeletal samples that are biased in socioeconomic status. This situation is exacerbated by the difficulty in determining social status from grave type or goods [6, 18]. Consequently, the elucidation of status from diet is not ensured. The ecological setting of the site may play a role in determining the relative social value of foods. Although maize was always an important food for elites, it was more likely to have been consumed in greater quantity in areas where it was more difficult to produce [85]. Maize was also consumed by high status individuals in the form of maize-fed animals [9, 88]. The identification of high status individuals is based on the assumption that maize, maize-fed animals, and imported foods would have been most accessible to elites and, as mentioned earlier, their association with archaeological markers of status.

The way in which the social value of maize *can* be reflected in socioeconomic status, is illustrated in Terminal Classic period burials at the site of Pacbitun. Located in west-central Belize (see Figure 12-1). Pacbitun was a medium-sized ceremonial center where the population peaked during the Late-Terminal Classic period when intensive agriculture in the form of hillside terracing was developed. In a study of temporal trends at the site, this time period also represents a reduction in the general

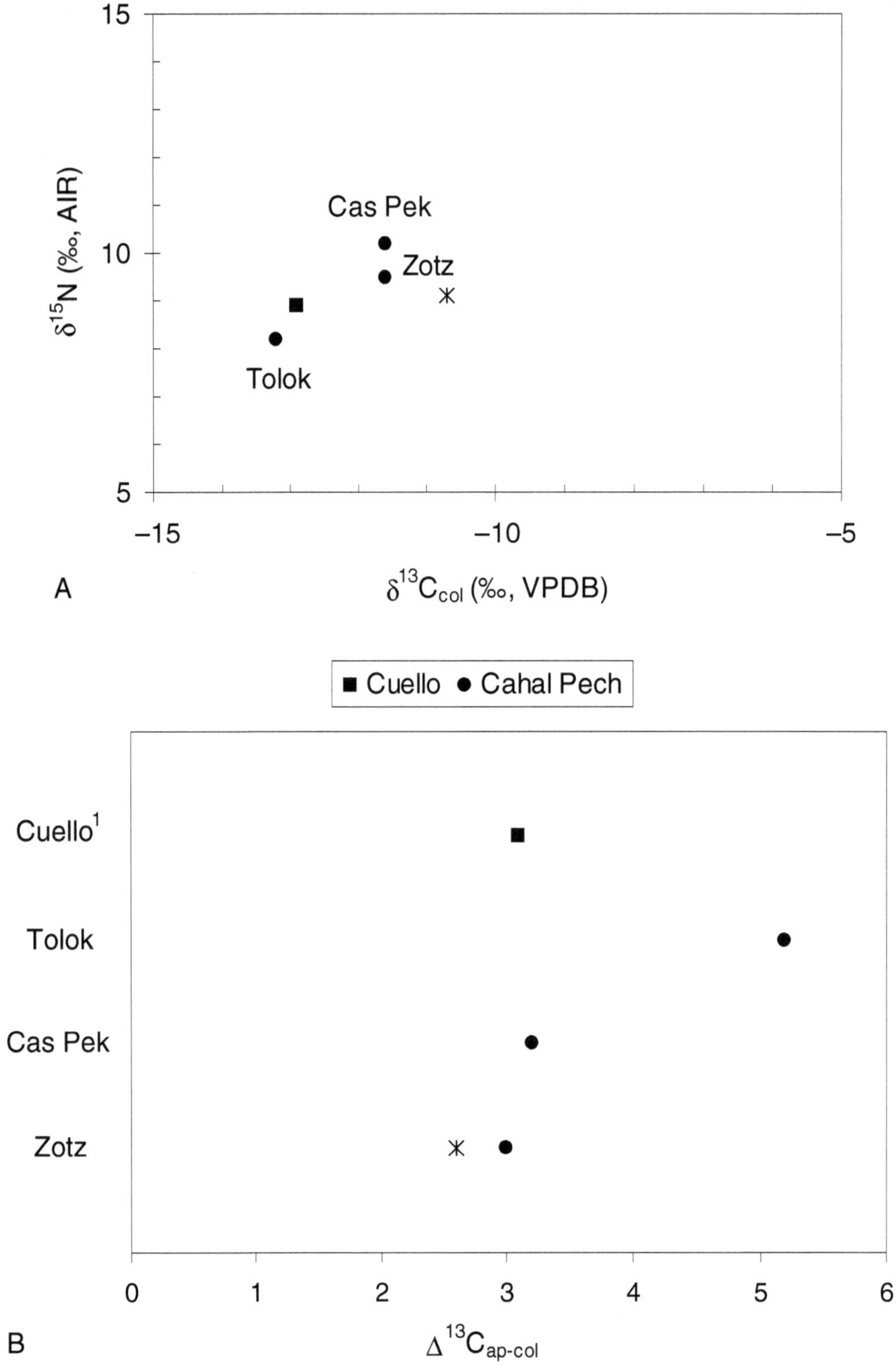

FIGURE 12-3 Isotopic values from human skeletons at Cahal Pech. (*denotes the high status individual from the Zotz group); [1]Cuello data from R. H. Tykot, N. J. van der Merwe, N. Hammond. (1996). Stable isotope analysis of bone collagen, bone apatite, and tooth enamel in the reconstruction of human diet: A case study for Cuello, Belize. In: M. V. Orna, (Ed.), *Archaeological chemistry V*. Washington, D.C.: American Chemical Society. pp. 355–365.) **A,** $\delta^{13}C_{col}$ and $\delta^{15}N$ values for collagen; **B,** $\Delta^{13}C_{ap\text{-}col}$.

consumption of maize at the site. The extremely labor-intensive activity of terracing was possibly a response to an inadequate supply of maize for a burgeoning population, but it appears that the increase in productivity was still not sufficient to meet the demand [88].

The skeletal sample includes individuals buried in both rural households and a variety of elaborate grave contexts in the ceremonial core [88]. Decreasing maize consumption (from $\delta^{13}C_{col}$ values) correlates significantly with increasing distance from the site (Figure 12-4A). In addition, grave type (crypt/cist, pit, urn) indicates that the higher the social rank, the greater access to maize (Figure 12-4B). Individuals with more elaborate grave goods also consumed more maize.

INTRAELITE DIFFERENTIATION

The majority of paleodiet reconstructions represent rank in fairly simplistic ways, in that high status individuals may be separated on the basis of grave type and goods or pooled for comparison with lower status groups, or both [as preceding, see also 8, 62, 87, 88]. The assumption underlying the archaeological distinctions is that meaningful social differences can be measured by the effort spent on construction of the interment facility and the economic and social value of the artifacts contained therein.

Tomb burials within ceremonial core samples often illustrate intraelite differences. One example comes from Lamanai, Belize (see Figure 12-1). Lamanai was a large ceremonial center with an ecologically heterogeneous environment that allowed extremely productive agriculture in the form of raised fields [32]. Although all the individuals analyzed at this site were probably elite, those found in the most elaborate tomb were dietarily distinct from other elites and from each other [9, 87]. Combined collagen and apatite carbon isotope analyses indicate that the female consumed the greatest quantity of maize at the site, but the male subsisted on a combination of maize-fed terrestrial animals and marine–reef resources. Chau Hiix, a nearby medium-sized ceremonial center, also yielded a male buried in a tomb who consumed far more maize–maize products than anyone else at the site, to the extent that his diet was almost pure C_4 [44].

An alternate way of examining social relations is to compare dietary data from different high status residential structures or locations, or both, within a site. Extensive excavation of different elite structures in the ceremonial core of Altun Ha, Belize, allowed a glimpse of how different elite groups may have presented themselves to each other and to the wider community. Altun Ha was a small ceremonial center in northern Belize located about 7 kilometers from the Caribbean coast (see Figure 12-1). Its occupation began in the Middle Pre-Classic period (ca. 800 BC), but it was largely abandoned by the beginning of the Early Post-Classic period [55]. Faunal remains indicate availability of a wide variety of terrestrial and marine resources [55]. The largest piece of worked jade in the Maya lowlands [55], and a postinterment tomb offering from the Mexican polity of Teotihuacan [54] indicate that, despite its relatively small size, Altun Ha held great symbolic significance for both the Maya and Teotihuacanos (perhaps because it is one of the easternmost sites in the Maya world).

Bone collagen and apatite were analyzed in 72 individuals from 9 residential zones or activity areas in the ceremonial core [89]. Diet at the site is generally characterized by a combination of maize as the agricultural staple and a higher degree of dependence on marine–reef resources than previously found at any Maya site. This is a reflection of its physical location and perhaps also its ideological association with the sea. There are temporal shifts in food consumption, but during each time period, dietary differences exist between structures. For brevity, the Late Classic period configuration is used for illustration (Figure 12-5A, B). Inhabitants of zone C, which is in the most central location, consumed significantly more C_4 foods than the peripheral zones H, J, and K (see Figure 12-5A), but were intermediate in terms of their carnivory (see Figure 12-5B). Zone C inhabitants were probably getting protein from both C_4-fed animals and the sea. Zone H inhabitants were getting protein from a higher trophic level (probably marine–reef resources) than any of the other zones (see $\delta^{15}N_{col}$ values in Figure 12-5A), and zone E and J inhabitants were the most strongly herbivorous according to $\Delta^{13}C_{ap\text{-}col}$ values (see Figure 12-5B), The elites at this site did not just use food as a means of expressing their general superiority over lesser beings who did not occupy the ceremonial core but also saw the necessity to differentiate themselves from each other. These data may provide evidence of the existence of lineages or political groupings who found different ways to demonstrate their identities and power, and who also may have competed for authority.

GENDER

Sex differences in diet are not investigated in all isotopic studies, often because of poor preservation that reduces sample size or does not allow identification of sex, or both. Nonetheless, there is strong archaeological and ethnohistorical evidence that women were the gatekeepers of food [1, 49, 63, 73]. In addition to the fact that women prepared food for both family and community events, the production and social distribution of food was probably an important source of female power. For example, women provisioned rituals with food and drink, which were used for offerings, as well as consumption. Raising domestic animals and birds was also the domain of women [60]. Although it is unclear how much control women had over the production of maize across the span of Maya history, the Spanish observed them

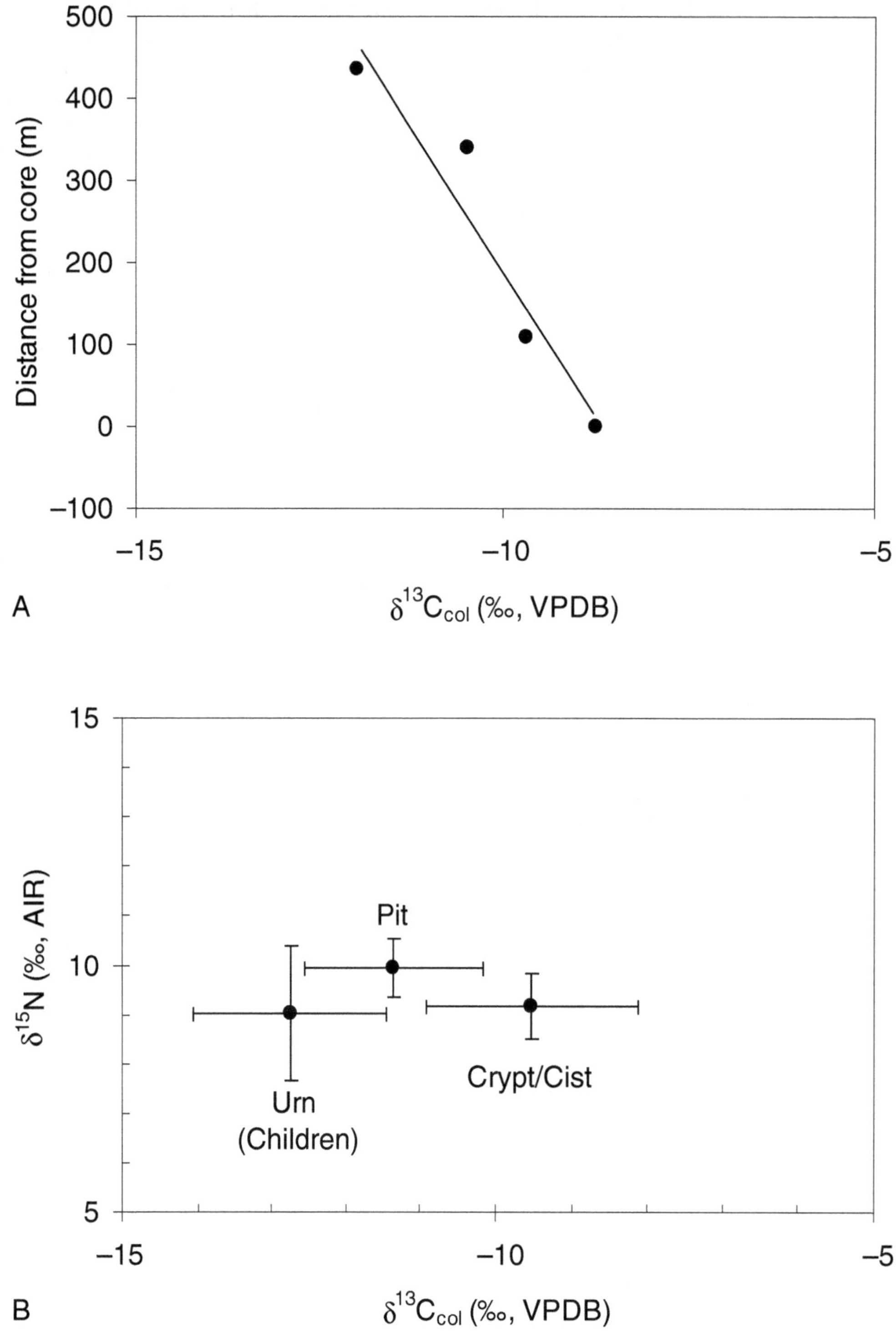

FIGURE 12-4 Isotopic variation at Pacbitun. **A,** Correlation of $\delta^{13}C$ values for collagen with distance from the ceremonial core; **B,** $\delta^{13}C$ and $\delta^{15}N$ values for collagen by burial type.

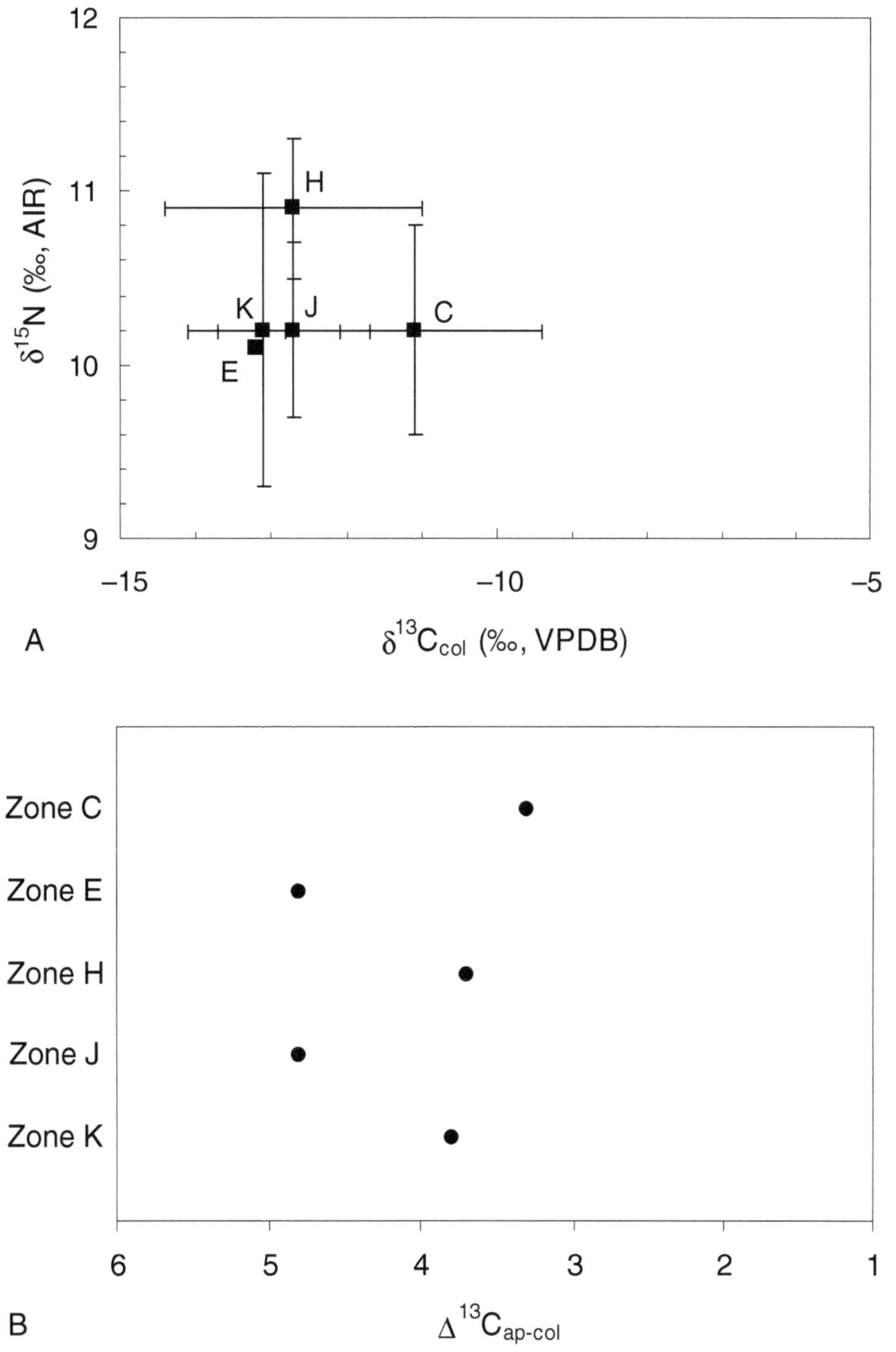

FIGURE 12-5 Isotopic values from human skeletons at Altun Ha. **A,** $\delta^{13}C_{col}$ and $\delta^{15}N$ values for collagen; **B,** $\Delta^{13}C_{ap\text{-}col}$.

working in the fields, harvesting crops where necessary, and marketing produce [73].

The relationship between the important role that women played in the subsistence economy of the Maya and their food consumption behavior is poorly studied. Access to the products of their own labor may have been limited. (Recall that the agricultural group at Cahal Pech consumed the least amount of maize despite that they were the primary producers.) Even though there is a **gender** complementarity to the Maya subsistence system as a whole, data from several sites suggest that women may have not have been able to consume as many ideologically valued or ritual foods.

At Altun Ha, for example, during the Late Classic period, higher $\delta^{13}C_{col}$ and $\Delta^{13}C_{ap\text{-}col}$ values indicate that women consumed significantly less maize and meat than men from the same locations (see Figure 12-6A, B). Although males and females were consuming protein from a similar trophic level (as deduced from the $\delta^{15}N$ values, see Figure 12-6A), it appears from the $\Delta^{13}C_{ap\text{-}col}$ values (see Figure 12-6B), and $\delta^{13}C_{ap}$ values that males had access to more C_4-fed animals [89]. This pattern may have been fairly widespread and deep-rooted. For example, markedly greater C_4 consumption by males was also found during the Classic period at Pacbitun (see Figure 12-6A, B). It is possible that diet was used to symbolically differentiate males and females. Alternately, maize, maize-fed animals, alcohol that may have been made from C_4 plants, and meat may all have been ideologically or socially valued foods that were used in ritual. Some combination of these items appears to denote status at most sites. Therefore, greater male access to these foods could also indicate that males either had a higher general status or were more regular participants in ritual, which may have physically situated them closer to the gods.

Much smaller sex differences were found at Pre-Classic Cuello and San Pedro, which is a small fishing village off the coast of Belize (see Figure 12-6A, B) [76, 96]. Although Robert Tykot and colleagues [76] have suggested that the slightly greater C_4-food consumption by males at Cuello might have been the result of heavier imbibement of maize-based alcohol, the greater similarity of male–female isotopic compositions at both sites might be a reflection of their lower general status or the likelihood that they saw relatively less public ritual activity, or both. Further research into the effects of status, time period, and region on gendered food behavior is needed. At present, however, female diets appear to reflect either some sort of symbolic differentiation or less "direct" access to the supernatural than male diets. The existing data indicate that gendered status and power were variable by time and place and that some form of gender hierarchy may have coexisted with gender complementarity in the domain of spiritual behavior [29].

TRADE

To identify the consumption of imported food, the import(s) must have isotopically distinct compositions and must be consumed in large quantities. For these foods to register their isotopic composition in bone, their consumption must also be sustained for prolonged periods of time because bone has a slow turnover rate, which means that it only represents diet over the last 10 to 15 years of an individual's life before their death [38, 51]. The imported foods that are easiest to identify are those found out of their environmental context (e.g., marine foods in a terrestrial environment or C_3 foods in a C_4 environment). Trade with coastal regions has long been implied by marine remains found in ritual contexts [47, 58, 72] or exotic artifacts that are composed of, or manufactured from, marine products [72]. The theory that marine resources were exported to interior sites as a source of food was first proposed in 1971 by Fred Lange [33]. Although it has been supported at some sites by faunal remains [e.g., 59, 92], it has not be well examined using isotopic data.

The earliest example of imported food comes from the Pre-Classic site of Cahal Pech [93]. The highest status individual at the site and the craft specialists who worked with shells consumed significant quantities of marine–reef resources relative to the agricultural population (identified with an asterisk in Figure 12-3A, B). Because Cahal Pech is an inland site, these resources must have been obtained through trade. Furthermore, they must have been consumed in quantity over a long period of time to be recorded in the isotopic composition of these individuals.

Trade in maize within a generalized maize-based economy is, however, almost impossible to identify and requires the analysis of a site where maize production is minimal or nonexistent. Marco Gonzalez, which is located at the tip of Ambergris Cay off the coast of Belize, is such a site. The cay would not have provided fertile ground for maize production, nor is there any evidence for maize agriculture. Maize in the diets of its inhabitants is, therefore, assumed to have been imported [93]. Marco Gonzalez was probably a gateway for trade that took place between the inland ceremonial center of Lamanai; the Yucatan to the north; and Guatemala or El Salvador, or both, to the south [56]. The contemporaneous, but smaller fishing village of San Pedro, slightly to the north and on the windward side of the cay, provides a social comparison [20].

Not surprisingly, the inhabitants of both sites exhibited a strong marine–reef component in their diets [96]. Marco Gonzalez, however, appears to have had greater access to maize and possibly to terrestrial animals. The differential was most likely a reflection of both site function and socioeconomic status. As a community whose main purpose was trading, Marco Gonzalez had greater opportunity to access mainland resources. Furthermore, because it was not engaged in significant production of food, trading for food also would have been a necessity.

IDENTIFICATION OF THE "OTHER" IN SACRIFICES

Intersite comparisons tell us that although maize was the staple throughout ancient Mesoamerica, it was consumed in different amounts and in different combinations with other foods between sites [86]. This regional variation has the potential to allow us to identify people who were not local.

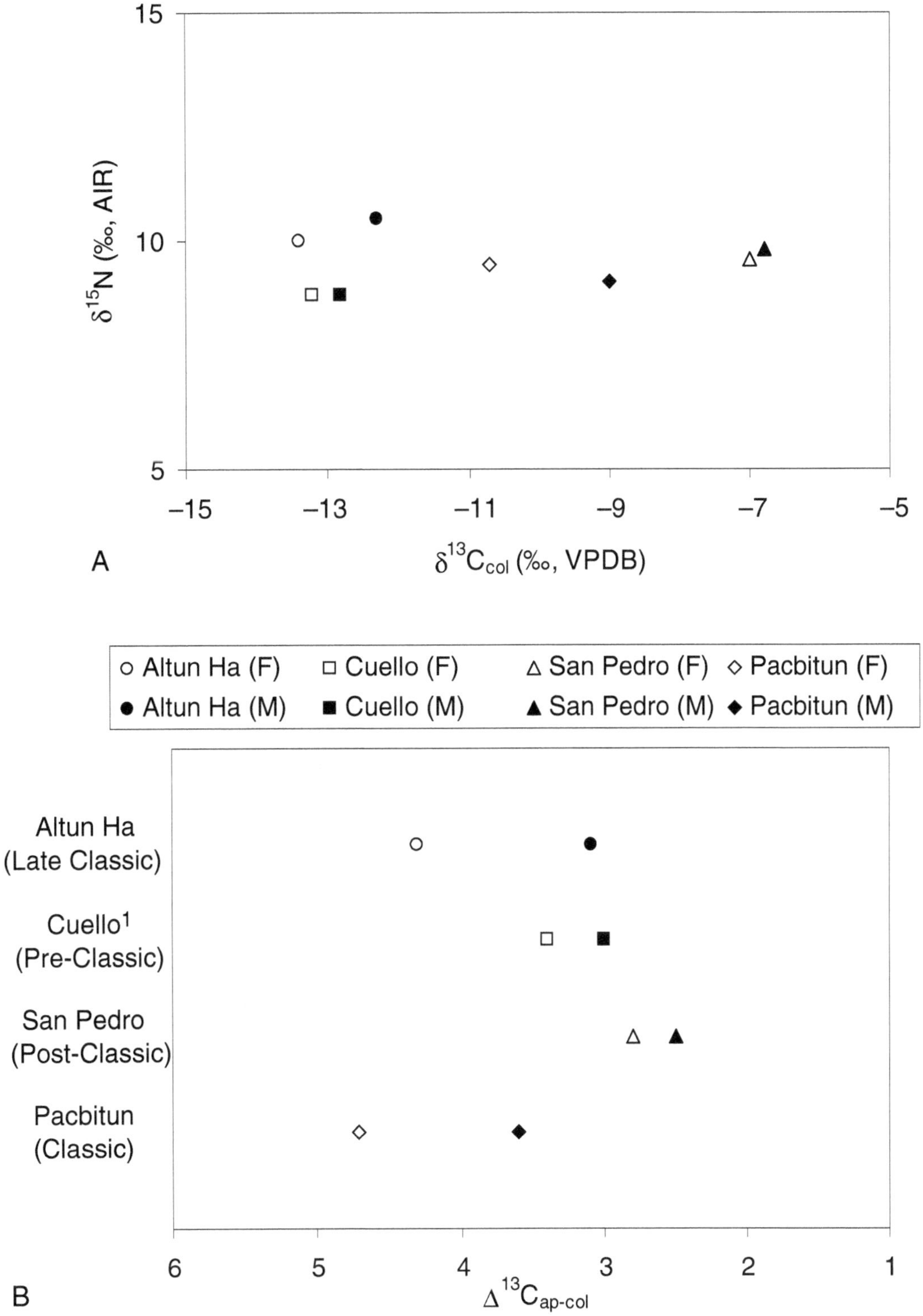

FIGURE 12-6 Variation in isotopic values from males and females at selected sites. ([1]Cuello data from R. H. Tykot, N. J. van der Merwe, N. Hammond. (1996). Stable isotope analysis of bone collagen, bone apatite, and tooth enamel in the reconstruction of human diet: A case study for Cuello, Belize, In: M. V. Orna, (Ed.), *Archaeological chemistry V.* Washington, D.C.: American Chemical Society. pp. 355–365.) **A,** $\delta^{13}C$ and $\delta^{15}N$ values for collagen; **B,** $\Delta^{13}C_{ap\text{-}col}$.

In recognition of individual food preferences and social status as a source of variation in diet, evidence for the presence of foreigners needs to be interpreted within intrasite variation where status and time period can be controlled.

Sacrifice was common among the Maya. Arthur Demarest [11] has suggested that the role of sacrifice was to justify economic and political power, and that the practice changed over time in "scale and intensity" (p. 237). It was used to justify the right to rule during the Early Classic Period and to legitimize competing states during the Post-Classic Period. Its most extreme expression in Mesoamerica was the justification of territorial expansionism by the Mexica [11].

Sacrifice involving humans ranged from blood letting, to autosacrifice of body parts (i.e., fingers, teeth), to purposeful killing [13, 42, 64, 70, 83]. It occurred on small to large scales (e.g., from single individuals to groups, such as war captives or ball teams). Sacrifice did not automatically denote the killing of foreigners. Some of the "victims" may have been chosen from lower social groups. An example of intragroup sacrifice might be found at Pacbitun. The urns at this site all contained child burials and were all found in high status locations in the ceremonial core or in crypts. If these were high ranking children of **ascribed status,** we would have expected them to share the diets of elite adults. Instead, they had diets consistent with those of the lowest status groups analyzed at the site (see Figure 12-3B). [88]. It is, therefore, possible that they may have been selected for sacrifice from the lower ranks of society, and the ritual resulted in exalted locations as places of burial.

Two examples of the possible mass sacrifice of foreigners come from Altun Ha and Cuello. Single structures at both sites have yielded Pre-Classic mass burials that have been interpreted as possible sacrificial victims. Because of poor preservation, interpretations of the groups at Altun Ha and Cuello have been based dominantly on mortuary pattern rather than trauma analysis, so their identification as sacrifices is still speculative. Both groups contained men, women, and children, so if they are foreigners they would most likely be captives of war or raiding rather than warriors or members of losing ball teams.

Individuals in the mass burial of Structure C-13 at Altun Ha were dietarily distinct (Figure 12-7A, B). and, therefore, may not have been local. They consumed the least amount of maize at the site (seeFigure 12-7A). and their $\Delta^{13}C_{ap\text{-}col}$ values suggest they were the least carnivorous. Protein was based on mainly terrestrial animals, rather than the marine–reef resources that dominated the diet of the rest of the site's inhabitants (see Figure 12-7B). Because there are no other Pre-Classic data to use for comparison within Altun Ha, it is possible that the distinction of this group is caused simply by a temporal difference, despite their different burial treatment. Oxygen- and strontium-isotope data are currently being used to determine if the individuals from Structure C-13 are local.

At Cuello, the mass burial individuals were also isotopically distinct (see Figure 12-7A, B). [76]. They consumed the most C_4 foods at the site (many of which came from marine–reef resources), and appeared to be more carnivorous than the rest of the site. Exaggerated carnivory is expected where marine–reef resources are the main source of protein [14, 36]. In contrast with the mass burial at Altun Ha, this group appears to have come from a more coastal location.

Both of these studies address the importance of warfare in the rise of state-level society among the Maya and the identification of general regions from which the "vanquished" came. Other studies of mass burials or suspected sacrifices are underway, including the comparison of bone-versus-enamel values to identify timing of relocations.

CONCLUSION

Carbon and nitrogen isotope analysis has provided a means of reconstructing ancient Maya diets that is particularly valuable given that the generally poor preservation of bone in the region does not allow many kinds of traditional osteological investigations. Lori Wright [100] notes that the Maya area is, in fact, one of the most intensively used for dietary stable isotope analysis. Under the challenge of the ecological model of the **Classic period collapse,** paleodiet researchers have gained a better understanding of the regional and temporal heterogeneity in Maya diets. Social heterogeneity became recognized during the process of reconstructing relationships among culture, biology, and environment, laying a good foundation for further work using new archaeological models and social theory to explore social interactions.

Because maize was so economically and ideologically central to Maya culture from its earliest expressions to modern times, it provides an ideal means of addressing questions relating to political, economic, and social dynamics. Meals become metaphors, whether they are interpreted from individuals (e.g., high status tomb burials, dogs in caches) or groups (e.g., by gender, age, structure, location, or time period). This chapter has emphasized the potential of Maya paleodiet studies to explore the meaning of these metaphors and to use the social heterogeneity that is often seen as a confounding variable for understanding the cultural complexity of these ancient people.

Acknowledgments

We thank Mary Pohl, Paul Healy, Hermann Helmuth, Jaime Awe, David Pendergast, and Elizabeth Graham for providing samples and for collaboration on the papers from which this synthesis was drawn. We also thank Kimberley Law and Martin Knyf for technical assistance and Matt Longstaffe and Karyn Olsen for assistance with the figures. Funding for the original work came from Natural Sciences and Engineering Research

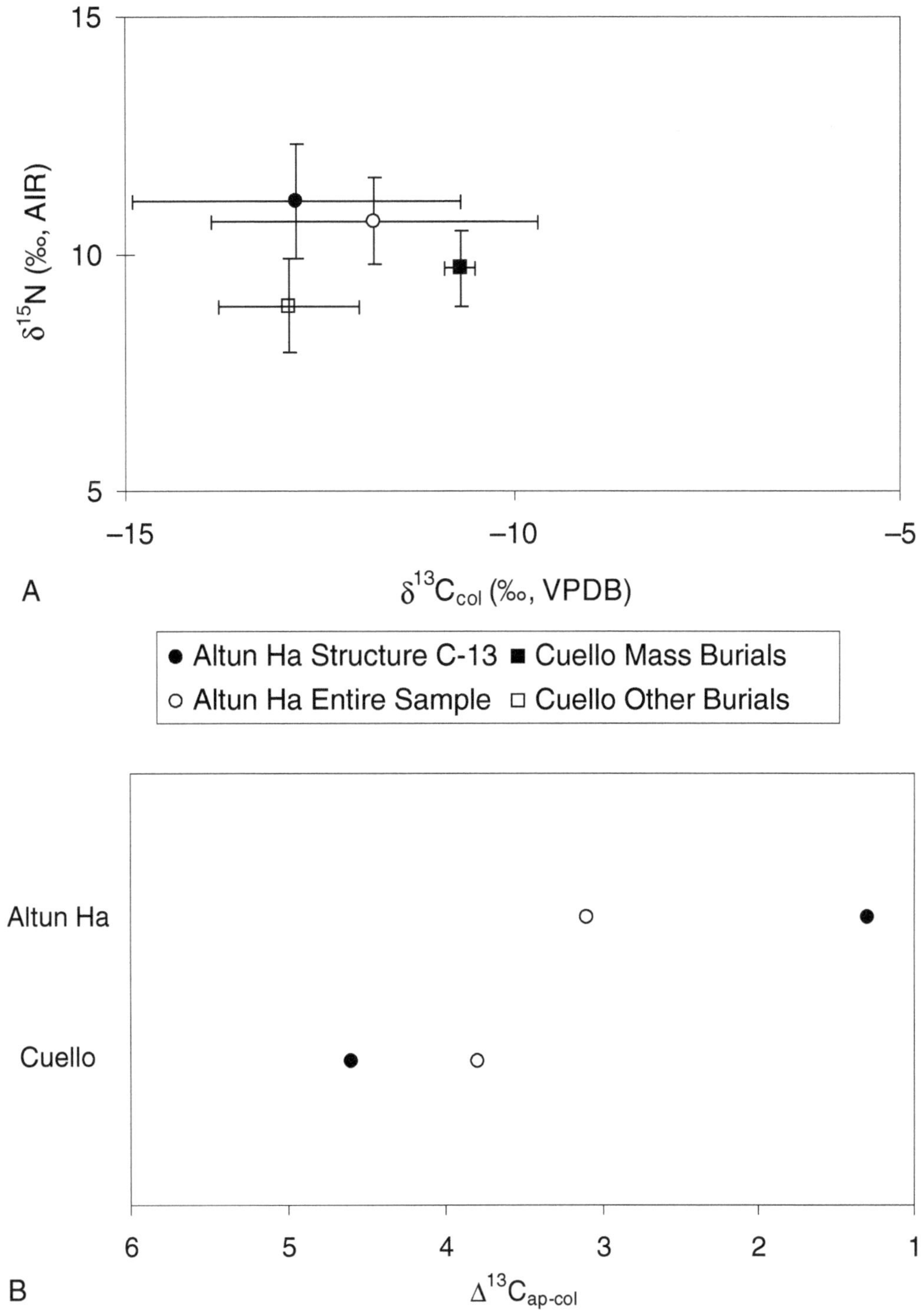

FIGURE 12-7 Isotopic values of mass burials compared with single burial averages ([1]Cuello data from R. H. Tykot, N. J. van der Merwe, N. Hammond. (1996). Stable isotope analysis of bone collagen, bone apatite, and tooth enamel in the reconstruction of human diet: A case study for Cuello, Belize, In: M. V. Orna, (Ed.), *Archaeological chemistry V.* Washington, D.C.: American Chemical Society. pp. 355–365.) **A,** $\delta^{13}C$ and $\delta^{15}N$ values for collagen; **B,** $\Delta^{13}C_{ap\text{-}col}$.

Council and Social Sciences and Humanities Research Council grants to Fred Longstaffe, Henry Schwarcz, and Christine White.

References Cited

1. T. Ardren. (2004). *Ancient Maya women.* Walnut Creek, CA: Altamira Press.
2. M. M. Bender, D. A. Baerreis, R. L. Steventon. (1981). Further light on carbon isotopes and Hopewell agriculture. *American Antiquity, 46,* 346–353.
3. R. R. Bressani, D. Paz, N. S. Scrimshaw. (1958). Corn nutrient losses: Chemical changes in corn during the preparation of tortillas. *Journal of Agricultural and Food Chemistry, 6,* 770–773.
4. R. L. Burger, N. J. van der Merwe. (1990). Maize and the origin of Chavín civilization: An isotopic perspective. *American Anthropologist, 92,* 85–95.
5. H. S. Carr. (1985). Subsistence and ceremony: Faunal utilization in a Late Preclassic community at Cerros, Belize, In: M. D. Pohl, (Ed.), *Prehistoric lowland Maya environment and subsistence economy.* Papers of the Peabody Museum of Archaeology and Ethnology, Vol. 77. Cambridge, MA: Harvard University. pp. 115–132.
6. A. Chase, D. Chase. (1992). *Mesoamerican elites: An archaeological assessment.* Norman: University of Oklahoma Press.
7. M. N. Cohen, K. O'Connor, M. Danforth, K. Jacobi, C. Armstrong. (1994). Health and death at Tipu. In: C. S. Larsen, G. Milner, (Eds.), *In the wake of contact: Biological responses to conquest.* New York: Wiley–Lis. pp. 121–133.
8. S. Coyston. (1995). An application of carbon isotope analysis of bone apatite to the study of Maya diets and subsistence at Pacbitun and Lamanai, Belize. Master's Thesis. Peterborough, Ontario: Department of Anthropology, Trent University.
9. S. Coyston, C. D. White, H. P. Schwarcz. (1999). Dietary carbonate analysis of bone and enamel for two sites in Belize. In: C. D. White, (Ed.), *Reconstructing ancient Maya diet.* Salt Lake City: University of Utah Press. pp. 199–220.
10. C. J. Crane. (1986). Late preclassic Maya agriculture, wild plant utilization and land-use practices. In: D. A. Freidel, (Ed.), *Archaeology at Cerros, Belize, Central America, Vol. 1: An Interim Report.* Dallas, TX: Southern Methodist University. pp. 147–151.
11. A. Demarest. (1984). Overview: Mesoamerican human sacrifice in evolutionary perspective. In: E. Boone, (Ed.), *Ritual human sacrifice in Mesoamerica.* Washington, D.C.: Dumbarton Oaks. pp. 227–243.
12. J. DeNiro, S. Epstein. (1981). Influence of diet on the distribution of nitrogen isotopes in animals. *Geochimica et Cosmochimica Acta, 45,* 341–351.
13. B. D. Dillon, L. Brunker, K. O. Pope. (1984). Ancient Maya autoamputation? A possible case from Salinas de los Nueve Cerros, Guatemala. *Journal of New World Archaeology, 5,* 24–38.
14. J. E. Ericson, M. West, C. H. Sullivan, H. W. Krueger. (1989). The development of maize agriculture in the Viru Valley, Peru. In: T. D. Price, (Ed.), *The chemistry of prehistoric bone.* Cambridge, England: Cambridge University Press. pp. 680–704.
15. P. Farnsworth, J. E. Brady, M. J. DeNiro, R. S. MacNeish. (1985). Re-evaluation of the isotopic and archaeological reconstruction of diet in the Tehuacan Valley. *American Antiquity, 50,* 102–116.
16. K. Flannery. (1982). *Maya subsistence: Studies in memory of Dennis Puleston.* New York: Academic Press.
17. J. P. Gerry, H. W. Krueger. (1997). Regional diversity in classic Maya diets. In: S. Whittington, D. Reed, (Eds.), *Bones of the Maya: Recent studies of ancient Maya skeletons.* Washington, D.C.: Smithsonian Institution Press. pp. 196–207.
18. S. D. Gillespie. (2001). Personhood, agency, and mortuary ritual: A case study for the ancient Maya. *Journal of Anthropological Archaeology, 20,* 73–112.
19. J. Goody. (1982). *Cooking, cuisine, and class: A study in comparative sociology.* Cambridge, UK: Cambridge University Press.
20. E. E. Graham, D. M. Pendergast. (1994). The sands of time: San Pedro, Ambergris Cay, 1993. *Royal Ontario Museum Archaeological Newsletter Series, II*(52), 1–4.
21. G. Gumerman IV. (1997). Food and complex cocieties. *Journal of Archaeological Method and Theory, 4,* 105–139.
22. W. A. Haviland. (1969). A new population estimate for Tikal, Guatemala. *American Antiquity, 34,* 429–433.
23. W. A. Haviland. (1972). Family size, prehistoric population estimates, and the ancient Maya. *American Antiquity, 37,* 135–139.
24. P. F. Healy, J. J. Awe, (Eds.). (1995a). *Belize Valley preclassic Maya project: Report on the 1994 Field Season.* Occasional Paper in Anthropology 10. Peterborough, Ontario: Trent University.
25. P. F. Healy, J. J. Awe. (1995b). Preclassic Maya of the Belize Valley: 1994–1995 Project Research Objectives. In: P. F. Healy, J. J. Awe, (Eds.), *Belize Valley preclassic Maya project: Report on the 1994 field season.* Occasional Paper in Anthropology 10. Peterborough, Ontario: Trent University. pp. 1–17.
26. N. Hellmuth. (1977). Cholti-Lacandon (Chiapas) and Peten-Itza Agriculture. In: N. Hammond, (Ed.), *Social process in Maya prehistory.* New York: Academic Press. pp. 421–428.
27. H. H. Henderson. (1998). The organization of staple crop production in Middle Formative, Late Formative, and Classic Period Farming Households at K'axob, Belize. Ph.D. Dissertation. Pittsburgh, PA: University of Pittsburgh, Department of Anthropology.
28. E. A. Hooton. (1940). Skeletons from the Cenote of sacrifice at Chichen Itza. In: C. Hay, R. L. Linton, S. K. Lothrop, H. L. Shapiro, G. C. Vaillant, (Eds.), *The Maya and their neighbors.* New York: Appleton-Century. pp. 272–280.
29. R. A. Joyce. (2000). *Gender and power in prehispanic Mesoamerica.* Austin: University of Texas Press.
30. S. H. Katz, M. L. Hediger, L. A. Valleroy. (1974). Traditional maize processing techniques in the New World. *Science, 184,* 765–763.
31. M. A. Katzenberg, H. P. Schwarcz. (1984). Paleonutrition in Southern Ontario: Evidence from strontium and stable isotopes. *Canadian Journal of Anthropology, 5,* 15–21.
32. J. D. Lambert, J. T. Arnason, A. H. Siemens. (1984). Ancient Maya drained field agriculture: Its possible application today in the New River flood plain, Belize, C.A. *Agriculture, Ecosystems and Environment, 11,* 67–84.
33. F. Lange. (1971). Marine resources: A viable subsistence alternative for the prehistoric Lowland Maya. *American Antiquity, 73,* 619–639.
34. E. J. Lawlor, A. J. Graham, S. L. Fedick. (1995). Preclassic floral remains from Cahal Pech, Belize, In: P. F. Healy, J. J. Awe, (Eds.), *Belize Valley preclassic Maya project: Report on the 1994 field season.* Occasional Paper in Anthropology 10. Peterborough, Ontario: Trent University. pp. 150–172.
35. D. Lee, J. Awe. (1995). Middle formative architecture, burials, and craft specialization: Report on the 1994 investigations at the Cas Pek group, Cahal Pech, Belize, In: P. F. Healy, J. J. Awe, (Eds.), *Occasional Paper in Anthropology 10.* Peterborough, Ontario: Trent University. pp. 95–115.
36. J. Lee-Thorp, J. C. Sealy, N. J. van der Merwe. (1989). Stable carbon isotope ratio differences between bone collagen and bone apatite, and their relationship to diet. *Journal of Archaeological Science, 16,* 585–599.
37. R. S. MacNeish. (1967). A summary of the subsistence. In: D. S. Beyers, (Ed.), *Environment and subsistence: The prehistory of the Tehuacan Valley.* Austin: University of Texas Press. pp. 290–309.
38. S. Manolagos. (2000). Birth and death of bone cells: Basic regulatory mechanisms and implications for the pathogenesis and treatment of osteoporosis. *Endocrine Reviews, 21,* 115–137.
39. E. B. Mansell, R. H. Tykot, D. Freidel, T. Ardren, B. Dahlin. (2002). Ancient Maya diet in the northern and southern lowlands: A stable

isotope study of diet from the Yucatán, Mexico, and La Milpa, Belize. Paper presented at the 67th Annual Meeting of the Society for American Archaeology, Denver, Colorado, April.

40. J. Marcus. (1982). The plant world of the sixteenth- and seventeenth-century lowland Maya. In: K. Flannery, (Ed.), *Maya subsistence: Studies in memory of Dennis Puleston*. New York: Academic Press. pp. 239–273.
41. L. Márquez Morfin, M. E. Peraza, J. Gamboa, T. Miranda. (1982). *Playa del Carmen: Una población de la costa oriental en el Postclásico (un estudio osteológico)*. Colección Científica. Antropología Física 119. Instituto Nacional de Antropología e Historia, México: Sección de Antropología Física, Centro Regional del Sureste.
42. V. Massey, D. G. Steele. (1997). A Maya skull pit from the terminal classic period, Colha, Belize. In: S. Whittington, D. Reed, (Eds.), *Bones of the Maya: Studies of ancient skeletons*. Washington, D.C.: Smithsonian Institution Press. pp. 62–77.
43. J. Maxwell. (2003). Evaluating morbidity and mortality in skeletal remains: An analysis of paleopathology at the Maya sites of Marco Gonzalez and San Pedro, Belize. Master's Thesis. The University of Western Ontario: Department of Anthropology.
44. J. Metcalfe, C. D. White, F. J. Longstaffe. (2004). Postclassic Maya socio-economic complexity and ideology at Chau-Hiix, Belize. Paper Presented at the Annual Meeting of the Canadian Association for Physical Anthropology, London, Ontario, November 2004.
45. C. H. Micksicek. (1983). Microfloral remains of the Pulltrouser Swamp area: Settlements and fields. In: B. L. Turner II, P. D. Harrison, (Eds.), *Pulltrouser swamp: Ancient Maya habitat, agriculture, and settlement in Northern Belize*. Austin: University of Texas Press. pp. 94–104.
46. C. H. Micksicek. (1991). The ecology and economy of Cuello, Belize. In: N. Hammond, (Ed.), *Cuello: An early Maya community in Belize*. Cambridge, MA: Harvard University Press. pp. 70–84.
47. H. Moholy-Nagy. (2004). Vertebrates in Tikal burials and caches. In: K. Emery, (Ed.), *Maya zooarchaeology: New directions in method and theory*. Los Angeles: Cotsen Institute of Archaeology, University of California. pp. 193–208.
48. C. R. Nance, S. L. Whittington, B. E. Borg. (2003). *Archaeology and ethnohistory of Iximche*. Gainesville: University Press of Florida.
49. L. S. Neff. (2004). Gender divisions of labor and lowland terrace agriculture. In: T. Arden, (Ed.), *Ancient Maya women*. Walnut Creek, CA: Altamira Press. pp. 31–51.
50. D. J. Ortner. (2003). *Identification of pathological conditions in human skeletal remains,* 2nd ed. San Diego, CA: Academic Press.
51. A. M. Parfitt. (1983). The physiologic and clinical significance of bone histomorphometric data. In: R. R. Becker, (Ed.), *Bone histomorphometry: Techniques and interpretation*. Boca Raton, FL: CRC Press. pp. 143–223.
52. G. Pelto. (1987). Social class and diet in contemporary Mexico. In: M. Harris, E. B. Ross, (Eds.), *Food and evolution: Towards a theory of human food habits*. Philadelphia, PA: Temple University Press. pp. 517–540.
53. D. M. Pendergast. (1969). The prehistory of Actun Balam, British Honduras. *Royal Ontario Museum Art and Archaeology Occasional Papers 16*. Toronto: Royal Ontario Museum.
54. D. M. Pendergast. (1971). Evidence of early Teotihuacan-lowland Maya contact at Altun Ha. *American Antiquity, 36,* 455–460.
55. D. M. Pendergast. (1979). *Excavations at Altun Ha, Belize, 1964–1970, Vol. 1*. Toronto: Royal Ontario Museum.
56. D. M. Pendergast, E. Graham. (1990). An island paradise (?): Marco Gonzalez 1990. *Royal Ontario Archaeological Newsletter, 41,* 1–4.
57. M. D. Pohl. (1977). Hunting in the Maya village of San Antonio Rio Hondo, Orange Walk District, Belize. *Journal of Belizean Affairs, 5,* 52–97.
58. M. D. Pohl. (1983). Maya ritual faunas: Vertebrate remains from burials, caches, caves, and cenotes in the Maya Lowlands. In: R. Leventhal, A. Kolata, (Eds.), *Civilization in the Ancient Americas*. Albuquerque: University of New Mexico Press. pp. 55–103.
59. M. D. Pohl. (1990). The ethnozoology of the Maya: Faunal remains from five sites in Peten, Guatemala. In: G. R. Willey, (Ed.), *Excavations at Seibal*. Peabody Museum Memoirs, Vol. 18, No. 3. Cambridge, MA: Harvard University Press. pp. 144–174.
60. M. D. Pohl, L. H. Feldman. (1982). The traditional role of women and animals in lowland Maya economy, In: K. Flannery, (Ed.), *Maya subsistence*. New York: Academic Press. pp. 295–312.
61. T. G. Powis, N. Stanchly, C. D. White, P. F. Healy, J. J. Awe, F. J. Longstaffe. (1999). A reconstruction of Middle Preclassic Maya subsistence economy at Cahal Pech, Belize. *Antiquity, 73,* 364–376.
62. D. M. Reed. (1994). Ancient Maya diet at Copán, Honduras, as determined through the analysis of stable carbon and nitrogen isotopes. In: K. Sobolik, (Ed.), *Paleonutrition: The diet and health of prehistoric Americans*. Occasional Papers Series, Center for Archaeological Investigations. Carbondale, IL: Southern Illinois University. pp. 210–221.
63. C. Robin. (2004). Gender and Maya farming. In: T. Ardren, (Ed.), *Ancient Maya women*. Walnut Creek, CA: Altamira Press. pp. 12–30.
64. F. Robicsek, D. M. Hales. (1984). Maya heart sacrifice: Cultural perspective and surgical technique. In: E. Boone, (Ed.), *Ritual human sacrifice in Mesoamerica*. Washington, D.C.: Dumbarton Oaks. pp. 49–87.
65. A. Roosevelt. (1980). *Parmana*. London: Academic Press.
66. D. Rue. (1987). Early agriculture and early postclassic Maya occupation in Western Honduras. *Nature, 326,* 285–286.
67. R. S. Santley, T. W. Killion, M. T. Lycett. (1986). On the Maya collapse. *Journal of Anthropological Research, 42,* 123–159.
68. F. P. Saul. (1972). *The human skeletal remains of altar de sacrificios: An osteobiographic analysis*. Papers of the Peabody Museum of Archaeology and Ethnology, Vol. 63, No. 2. Cambridge, MA: Harvard University Press.
69. B. N. Smith, S. Epstein. (1971). Two categories of 13C/12C ratios for higher plants. *Plant Physiology, 4,* 380–384.
70. L. Schele. (1984). Human sacrifice among the Classic Maya. In: E. Boone, (Ed.), *Ritual human sacrifice in Mesoamerica*. Washington, D.C.: Dumbarton Oaks. pp. 7–45.
71. N. Stanchly. (1995). Formative period Maya faunal utilisation at Cahal Pech, Belize: Preliminary analysis of the animal remains from the 1994 field season. In: P. F. Healy, J. J. Awe, (Eds.), *Belize Valley preclassic Maya project: Report on the 1994 field season*. Occasional Paper in Anthropology 10. Peterborough, Ontario: Trent University. pp. 124–149.
72. W. Teeter. (2004). Animal utilization in a growing city: Vertebrate exploitation at Caracol, Belize. In: K. Emery, (Ed.), *Maya zooarchaeology: New directions in method and theory*. Los Angeles: Cotsen Institute of Archaeology, University of California. pp. 177–192.
73. A. M. Tozzer (Ed. and Trans.). (1941). *Landa's Relación de las Cosas de Yucatán*. Papers of the Peabody Museum of Archaeology and Ethnology, Vol. 18. Cambridge, MA: Harvard University Press. p. 1566.
74. R. H. Tykot. (2002). Contribution of stable isotope analysis to understanding dietary variation among the Maya. *Archaeological Chemistry. Materials, Methods, and Meaning*. ACS Symposium Series 831. Washington, D.C.: American Chemical Society. pp. 214–230.
75. R. H. Tykot, J. E. Staller. (2002). The importance of maize agriculture in Coastal Ecuador: New data from La Emerenciana. *Current Anthropology, 43,* 666–677.
76. R. H. Tykot, N. J. van der Merwe, N. Hammond. (1996). Stable isotope analysis of bone collagen, bone apatite, and tooth enamel in the reconstruction of human diet: A case study for Cuello, Belize, In: M. V. Orna, (Ed.), *Archaeological chemistry V*. Washington, D.C.: American Chemical Society. pp. 355–365.

78. N. J. van der Merwe, J. C. Vogel. (1978). ^{13}C content of human collagen as a measure of prehistoric diet in Woodland North America. *Nature, 276,* 815–816.
79. N. J. van der Merwe, R. H. Tykot, N. Hammond. (2000). Diet and animal husbandry of the Preclassic Maya at Cuello, Belize: Isotopic and zooarchaeological evidence, In: S. Ambrose, M. A. Katzenberg, (Eds.), *Biochemical approaches to paleodietary analysis.* New York: Academic/Plenum Press. pp. 22–38.
80. N. J. van der Merwe, J. A. Lee-Thorp, J. S. Raymond. (1993). Light stable isotopes and the subsistence base of Formative cultures at Valdivia, Ecuador. In: J. B. Lambert, G. Grupe, (Eds.), *Prehistoric human bone: Archaeology at the molecular level.* New York: Springer-Verlag. pp. 63–97.
81. U. Wapler, E. Crubézy, M. Schultz. (2004). Is cribra orbitalia synonymous with anemia? Analysis and interpretation of cranial pathology in Sudan. *American Journal of Physical Anthropology, 123,* 333–339.
82. A. Weisen, D. Lentz. (1997). *Floral remains at Cahal Pech and surrounding sites: Summary report.* New York: New York Botanical Garden.
83. B. Welsh. (1988). *An analysis of classic lowland Maya burials.* British Archaeological Reports, International Series 409. Oxford, UK: BAR.
84. C. D. White. (1994). Dietary dental pathology and cultural change in the Maya. In: L. Chan, A. Herring, (Eds.), *Strength and diversity.* Toronto: Canadian Scholar's Press. pp. 279–302.
85. C. D. White. (1997). Ancient diet at Lamanai and Pacbitun: Implications for the ecological model of collapse. In: S. Whittington, D. Reed, (Eds.), *Bones of the Maya: Studies of ancient skeletons.* Washington, D.C.: Smithsonian Institution Press. pp. 171–180.
86. C. D. White. (1999). *Reconstructing ancient Maya diet.* Salt Lake City: University of Utah Press.
87. C. D. White, H. P. Schwarcz. (1989). Ancient Maya diet as inferred from isotopic and elemental analysis of bone. *Journal of Archaeological Science, 16,* 451–474.
88. C. D. White, P. F. Healy, H. P. Schwarcz, Intensive agriculture, social status and Maya diet at Pacbitun, Belize. *Journal of Anthropological Research, 49,* 347–375.
89. C. D. White, D. M. Pendergast, F. J. Longstaffe, K. R. Law. (2001a). Social complexity and food systems at Altun Ha, Belize: The isotopic evidence. *Latin American Antiquity, 12,* 371–393.
90. C. D. White, M. D. Pohl, H. P. Schwarcz, F. J. Longstaffe. (2001b). Isotopic evidence for Maya patterns of deer and dog use at Preclassic Colha. *Journal of Archaeological Science, 28,* 89–107.
91. C. D. White, M. D. Pohl, H. P. Schwarcz, F. J. Longstaffe. (2004). Feast, field and forest: Deer and dog diets at Lagartero, Tikal and Copán, In: K. Emery, (Ed.), *Maya zooarchaeology: New directions in method and theory.* Monograph 51. Los Angeles: Cotsen Institute of Archaeology, University of California. pp. 141–158.
92. C. D. White, L. E. Wright, D. M. Pendergast. (1994). Biological disruption in the northern Maya Lowlands, In: C. Spencer-Larsen, G. Milner, (Eds.), *In the wake of contact: Biological responses to conquest.* New York: Wiley-Liss. pp. 135–145.
93. C. D. White, F. J. Longstaffe, K. R. Law, R. J. Song. (In prep). Preclassic Maya diet at Cahal Pech, Belize: Isotopic evidence, In: J. J. Awe, P. F. Healy, (Eds.). *The Preclassic Maya of the Belize Valley.*
94. S. L. Whittington. (1989). *Characteristics of demography and disease in low status Maya from classic period Copán, Honduras.* University Park, PA: Pennsylvania State University, Department of Anthropology.
95. G. R. Willey, D. B. Shimkin. (1973). The Maya collapse: A summary view, In: T. P. Culbert, (Ed.), *The Classic Maya collapse.* Albuquerque: University of New Mexico Press. pp. 457–502.
96. J. S. Williams, C. D. White, F. J. Longstaffe. (In press). Maya marine subsistence: Isotopic evidence from Marco Gonzalez and San Pedro, Belize. *Latin American Antiquity.*
97. E. Wing, S. J. Scudder. (1991). The exploitation of animals. In: N. Hammond, (Ed.), *Cuello: An early Maya community.* New York: Cambridge University Press. pp. 84–97.
98. L. E. Wright. (1994). Sacrifice of the earth? Diet, health, and inequality in the Pasión Maya. Ph.D. Dissertation, Chicago: Department of Anthropology, University of Chicago.
99. L. E. Wright. (1997). Biological perspectives on the collapse of the Pasion Maya. *Ancient Mesoamerica, 8,* 267–273.
100. L. E. Wright. (2004). Osteological investigations of ancient Maya lives, In: C. W. Golden, G. Borgstede, (Eds.), *Continuities and changes in Maya archaeology: Perspectives at the millenium.* New York: Routledge. pp. 201–215.
101. L. E. Wright, C. D. White. (1996). Human biology in the Classic Maya collapse: Evidence from paleopathology and paleodiet. *Journal of World Prehistory, 10,* 147–198.

CHAPTER

13

Early to Terminal Classic Maya Diet in the Northern Lowlands of the Yucatán (Mexico)

EUGENIA BROWN MANSELL[*], ROBERT H. TYKOT[†], DAVID A. FREIDEL[‡], BRUCE H. DAHLIN[§], AND TRACI ARDREN[¶]

[*]*Department of Anthropology, University of South Florida, Tampa, Florida*

[†]*Department of Anthropology, University of South Florida, Tampa, Florida*

[‡]*Department of Anthropology, Southern Methodist University, Dallas, Texas*

[§]*Department of Sociology and Anthropology, Howard University, Washington, D.C.*

[¶]*Department of Anthropology, University of Miami, Coral Gables, Florida*

Glossary

Bone apatite The mineral component of bone that reflects the overall diet.

Bone collagen The fibrous protein component of bone that reflects the protein in the diet.

Chunchucmil Maya site located in the western part of the Yucatán peninsula.

Classic Maya Cultural period ca. AD 250–900. Typically divided into Early Classic (ca. AD 250–550) and Late Classic (ca. AD 550–900).

Stable carbon and nitrogen isotope analysis Used to divide consumed foods into discrete, isotopic groups (i.e., ^{13}C to ^{12}C and ^{15}N to ^{14}N), leading to the understanding of the diet.

Tooth enamel The outer surface of teeth, which reflects the diet at the time of tooth formation.

Yucatán peninsula Peninsula in southeast Mexico that extends into the Gulf of Mexico.

Yaxuná Maya site located in the north central part of the Yucatán peninsula.

For more than a quarter century, **stable isotope analysis** of human skeletal remains has been used to determine the diet of ancient people. For the ancient Maya, the main questions have been focused on the reliance on maize and how it changed over time. Although many areas of the Maya highlands and southern lowlands in Belize, Honduras, and Guatemala have been the subject of isotopic studies, recently the northern lowlands, in particular the **Yucatán peninsula** of Mexico, have been the subject of such research. Twenty-two individuals from **Yaxuná,** in the interior, and five from **Chunchucmil,** on the coastal plain, were specifically selected to provide some data for the Yucatán. Bone and tooth samples were prepared using well-established procedures to ensure integrity and reliability, especially considering the poor preservation of many skeletal remains from this region. Stable carbon and **nitrogen isotope** ratios were then measured for **bone collagen,** and carbon isotope ratios were measured for **bone apatite** and **tooth enamel.** The results suggest significant differences between the two **Classic Maya** sites, with the residents of Yaxuná consistently the most dependent on maize. There was also greater dietary variation among the individuals at Chunchucmil, probably because of the availability of diverse resources and differences related directly to social status.

INTRODUCTION

In 1990, Vogt [20] wrote that maize composes up to 70% of the modern Maya diet, but ethnographic information does not necessarily translate into the diet of the ancient Maya. Although current research tells us that maize was a vital component of their diet, how vital was it? Were there differences in consumption among social groups, between geographic locales, or between genders? This stable isotope study has given us one of our first opportunities to evaluate the importance of maize in the Classic-Period Maya (AD

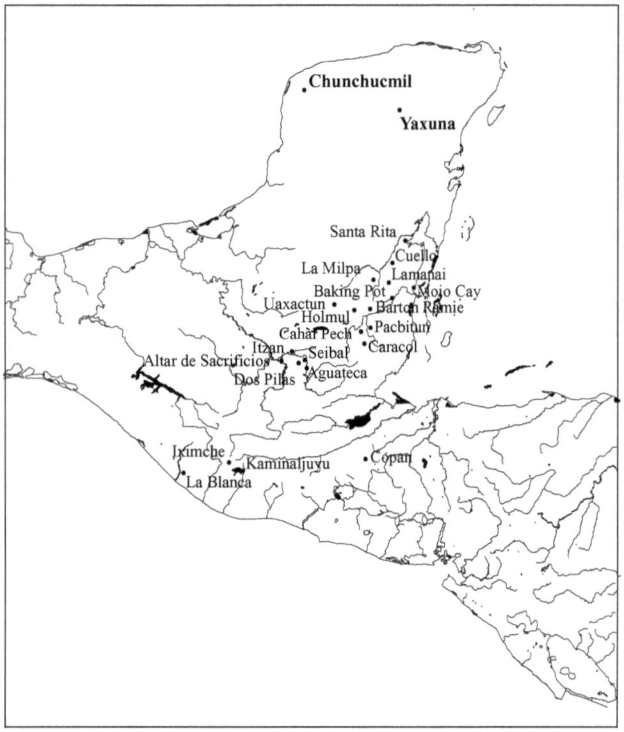

FIGURE 13-1 Map showing sites in the Yucatán and adjacent regions. (Courtesy of Eugenia Brown Mansell)

250 to AD 1050) of the northern Yucatán peninsula and to study differences among the diets of differing status groups at Yaxuná and Chunchucmil, including elites at Yaxuná and a possible middle class at Chunchucmil (Figure 13-1). The two assemblages also have given us an opportunity to further assess differences in diet among the different geographical areas of Yucatán.

Thanks to the previous stable isotope studies by White and Schwarcz [22], Powis and colleagues [13], Gerry [7], Wright and Schwarz [25], Tykot and colleagues [19], and others we have an increasingly complete picture of the diet of the Southern Lowland Maya in ancient times [17]. However, our knowledge of the Northern Lowland Maya has been lacking. This chapter presents what we hope is the first of many other isotopic studies in the Northern Lowlands. Our study looks specifically at the diet of Early to Late–Terminal Classic people who lived at the sites of Yaxuná and Chunchucmil, both located in the state of Yucatán, Mexico. The purpose of this research is to assess dietary trends and how they reflect social and economic status, gender (at Yaxuná), and access to nourishment within the communities both chronologically and spatially.

METHODS

Because of typical Maya Lowland environmental factors (i.e., humidity and limestone underlying thin soil), the human skeletal material found at both sites was poorly preserved. Collagen was extracted following established procedures [18], including using a dilute hydrochloric acid solution to demineralize the bones, followed by sodium hydroxide to neutralize humic acids, and a defatting solution with methanol and chloroform to remove residual lipids. Powdered bone apatite and tooth enamel samples were treated in sodium hypochlorite to remove organics, and then in an acetic acetate buffer to remove nonbiogenic carbonates. Most samples yielded no collagen at all, with only three individuals each from Yaxuná and from Chunchucmil yielding 1% or more of sample. However, apatite was extracted from all bone samples and most of the teeth available. All samples were run on Finnigan MAT stable isotope ratio mass spectrometers at the University of South Florida, one using a CHN analyzer for collagen samples and the other using a Kiel III device for bone apatite and tooth enamel samples. Collagen yields and C : N ratios were used to confirm the integrity of the collagen isotope results, which are reported in Table 13-1. The analysis of the stable carbon and nitrogen isotopes for collagen yields the percentage of dietary protein in the individual tested, whereas the analysis of the stable carbon isotopes for apatite yields knowledge of the overall diet in the test subject. The analysis of tooth enamel provides information on diet at the age of tooth formation.

ISOTOPIC STUDIES OF THE MAYA

The Maya adapted themselves to a range of ecological zones, including highlands, lowlands, and the coastal plains. In Belize most isotopic studies have been performed at sites on the wide, flat coastal plains, including Lamanai [22], Baking Pot and Barton Ramie [7], Pacbitun [21], Cuello [19], Cahal Pech [13], Colha [23], Altun Ha [23], and on the coast at Mojo Cay [12]. One study has been reported from the west central mountains at Caracol [4]. Elsewhere in Mesoamerica, isotope studies also have been published from the Peten [25]. Research has been done on the Peten area of Belize with a study at La Milpa [11]; this chapter presents a new study that adds the Northern Lowlands to the growing volume of knowledge.

Yaxuná

The site of Yaxuná is located approximately 20 kilometers southwest of Chichén Itzá (Figure 13-1). Its placement in the central northern third of the peninsular has several advantages for subsistence. There is a greater rainfall than the coastal plain for agriculture, low scrub jungle rather than the high tropical rain forest of the Southern Lowlands for ease of seeing prey and clearing land for milpa (maize fields), and a good trade location at a crossroads linking the site

TABLE 13-1 Results of Stable Isotope Analysis at Yaxuná and Chunchucmil

Yaxuná	Sex/Age	$\delta^{13}C_{co}$	$\delta^{15}N_{co}$	$\delta^{13}C_{ap}$	Tooth	$\delta^{13}C_{en}$
Burial 6	F 25–35			–3.3	I	–6.2
Burial 7	F 25–35			–3.6	I	–2.3
Burial 8	F 30–40	–11.8	6.6	–4.1	I2	–0.1
Burial 9	F ?			–2.9	M	–0.8
Burial 11	F A			–3.3		
Burial 13B	M YA			–4.1	M1	–0.9
Burial 13C	M 40+			–3.4		
Burial 15A	M 40+			–2.9		
Burial 16	M 30–35			–4.2	M	–2.6
Burial 17	M 25–40	–12.9	7.3	–1.2	M1	–1.0
Burial 23	M 40–50			–2.6	M3	–3.7
Burial 24-1	A 22–28			–1.4	M1	–2.8
Burial 24-2	J 12–15			–3.0	PM1	–2.8
Burial 24-3	J 7–9			–2.8	I	–3.4
Burial 24-4	J 10–12			–3.1	I	–3.4
Burial 24-5	F 34–45			–2.3	C	–2.1
Burial 24-6	F 20–25			–3.0	M2	–1.7
Burial 24-7	M 55+			–2.3	PM	–4.0
Burial 24-10	F 20–25			–2.5	PM	–2.4
Burial 24-11	M 50+			–2.5	PM	–2.4
Burial 24-12	J 0–0.5	–12.3	7.3	–8.1		
Burial 24-13	F 40+			–3.5		
ave.		–12.3	7.1	–3.2		–2.5
std. dev.		0.4	0.3	1.4		1.5
Chunchucmil	**Sex/Age**	$\delta^{13}C_{co}$	$\delta^{15}N_{co}$	$\delta^{13}C_{ap}$	**Tooth**	$\delta^{13}C_{en}$
Kaab' 9D1.4 1999	? J/YA	–14.7	6.4	–8.2		
'Aak 9C1.8 2000	? 12+	–13.4	7.6	–4.5	M2	–6.3
'Aak 9C1.9 1999				–5.2	M1	–7.7
Kaab' 9D1.2 2000	? A			–5.6	M1	–5.8
Muuch 10A22.9.1 2001	? YA?	–16.0	7.0	–3.4		
ave.		–14.7	7.0	–5.4		–6.6
std. dev.		1.1	0.5	1.6		0.8

with the east coast of the peninsular via a 100-kilometer sacbe (raised causeway) to Cobá. From 1986 to 1996, the Yaxuná Archaeological Project under the direction of David Freidel mapped 1.5 square kilometers of the site center and other portions of the site, and excavated and consolidated several large structures (Figure 13-2). We examined 22 individuals from the site, described in detail by S. Bennett [1–3]. We sampled the entire contents of two royal tombs, dating to the Early Classic (AD 250–550 or 600), and also 10 individuals (five females, five males) recovered from nonelite contexts around the site. These burials were found primarily within crypts in modest house mounds that dated, based on ceramics, to the Late–Terminal Classic period (AD 550/600–1000), probably between AD 700 and 900 (Figure 13-3).

The North Acropolis at Yaxuná is a large centrally located platform 11 meters high with an approximately 300-square-meter plaza. On this plaza are three large buildings, two of which contained tombs. In 1993, excavations were begun on the 16.5-meter building on the north side of the plaza. It was during these excavations that the first undisturbed royal tomb in Yucatán was uncovered. The floor of the tomb was

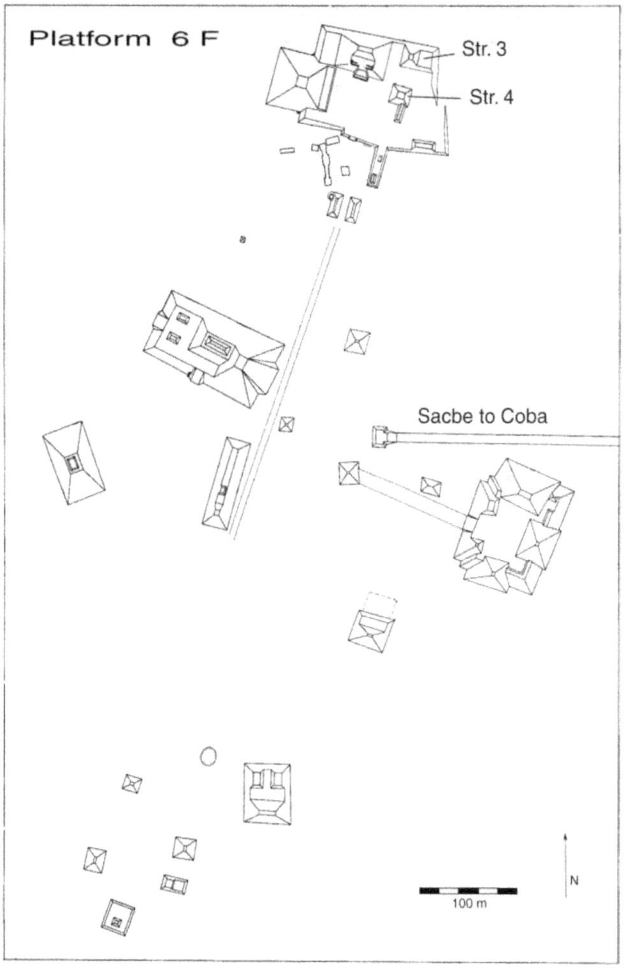

FIGURE 13-2 Map of the central portion of Yaxuná. (Courtesy of Eugenia Brown Mansell)

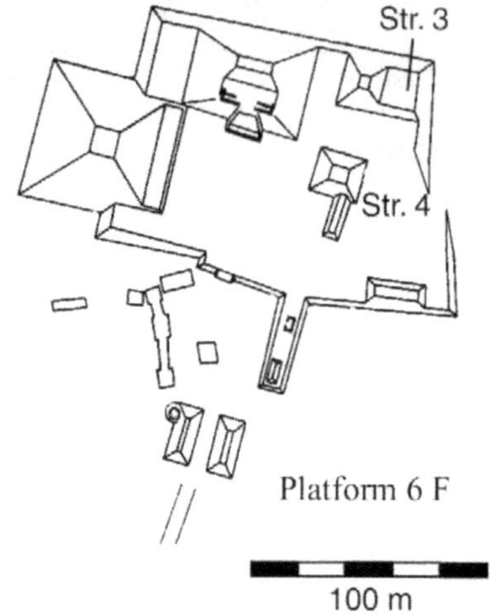

FIGURE 13-3 Map of Platform 6F, showing Structures 3 and 4. (Courtesy of Eugenia Brown Mansell)

fine, white marl, and the walls were covered in coarse stucco. There were multiple ceramic vessels placed in the tomb, such as four sealed miniature vessels, a polychrome shallow bowl, black bowl with a spout, two cups, and a jar. Also included in the grave goods were a turtle carapace, carved deer bones, three carved jade jewels, a shell pendant carved in the shape of a turtle or frog, and a death head carved from a spondylus shell. The body of one individual (Burial 23) was that of a 40–50-year-old male, who until death had apparently enjoyed good health, with no serious infections or trauma, although there was evidence of arthritis. It was determined by the Yaxuná-Project archaeologists that this individual had been a ruler at Yaxuná [16]. There was no collagen present for this individual; however the $\delta^{13}C$ values obtained for apatite (–2.6‰) and for third molar tooth enamel (–3.7‰) suggest a high dependence on maize or maize-fed animals, both as a youngster and as an adult.

The following year a tomb in an 8-meter structure on the eastern side of the plaza was discovered. It contained the remains of 11 individuals: one male, four females, one individual of undetermined sex, and four children placed around a decapitated adult male who had been centrally placed in the tomb. The excavators believe that the royal people had been terminated, ending their lineage. They were entombed in a terminated elite ceremonial building, which had another structure built over it. As in the previous tomb, there were elite grave goods: ceramics, carved shell, greenstone, and obsidian blades. A description of each individual follows.

A. Burial 24-7 was a centrally placed, older male (55 years old or older), who had been decapitated, and his skull placed in another part of the tomb; all physical data from the Yaxuná burials were taken from the field reports of Sharon Bennett [1–3]. He had an apatite $\delta^{13}C$ value of –2.3‰ and tooth-enamel value of –4.0‰ from a premolar, which were similar values to the older male in Burial 23.

B. Burial 24-1 was an adult between the age of 22 and 28 years; a definite sex assignment was not possible. The apatite signature was –1.4‰, and the tooth-enamel value was –2.8‰ for a first molar.

C. Burial 24-2 was at least 12 years old, but no more than 15 years old, based on dental evidence, and no sex was assigned. The skull had been crushed. The individual was interred with a necklace of jade and shell. The $\delta^{13}C$ value for apatite was –3.0‰, and the value for tooth enamel from a first premolar was –2.8‰.

D. Burial 24-3 was a small child, with the dental and epiphyseal evidence indicating that the child was between 7 and 9 years old. Apatite $\delta^{13}C$ was –2.8‰ whereas the enamel from an incisor yielded –3.6‰.

E. Burial 24-4 was also a child between the age of 10 and 12 with an apatite signature of –3.1‰ and enamel value of –3.4‰, also from an incisor. Both burial 24-3 and 24-4, although they were children, contained a rich array of grave goods.

F. Burial 24-5 was an older female who was between the age of 34 to 45 years, with an apatite $\delta^{13}C$ reading of –2.3‰ and enamel from a bicuspid reading –2.1‰.

G. Burial 24-6 was a female, 20 to 25 years old, with cut marks on the left tibia and the right femur. This woman was interred with many grave goods of jade, shell, and so on, and a unique polychrome ceramic figurine. The apatite $\delta^{13}C$ value was –3.0‰, and the enamel value from a second molar was –1.7‰.

H. Burial 24-10 was a female, 20 to 25 years old. The apatite value was –2.5‰, and tooth-enamel value from a premolar was –2.4‰.

I. Burial 24-11 was a male more than 50 years old. The apatite $\delta^{13}C$ value was –2.5‰, and the tooth-enamel value from a premolar was –2.4‰.

J. Burial 24-12 was an infant, approximately newborn to 6 months. Despite the young age, this was the only individual in the tomb to produce collagen data ($\delta^{13}C$ = –12.3‰; $\delta^{13}C$ = 7.3) along with bone apatite results ($\delta^{13}C$ = –8.1).

K. Burial 24-13 was a decapitated female who was, in all probability, more than 40 years old. Her apatite signature was –3.5‰. There were no teeth associated with this individual.

The centrally placed male and the young female with the figurine were carefully placed in the tomb. All of the other individuals were literally tossed into the crypt, but the richness of the grave goods indicates that they were interred with all of the trappings of royalty. The single collagen result for this tomb, from the infant, is in line with the two other collagen values produced for the Late Classic nonelite individuals. This value most likely represents less dependence on C_4 resources; however, there is a trophic level effect in breastfeeding. The apatite $\delta^{13}C$ value of –8.1‰ for this infant was more negative than any of the other individuals tested, suggesting far less direct dependence on maize and maize products for children of such a young age.

The other 10 individuals tested, five females and five males, were recovered from nonelite contexts around the site. The burials were found primarily within crypts in modest house mounds that date, based on ceramics, to the Late–Terminal Classic Period (AD 550 or 600–1000), probably between AD 700–900.

A. Burial 6 was a 25–35-year-old female recovered from a crypt. She had a Cehpech sphere (Late–Terminal Classic) vessel over her head and another below the femur. The apatite $\delta^{13}C$ value was –3.3‰, and the tooth enamel $\delta^{13}C$ value from an incisor was –6.2‰.

B. Burial 7 was a 25–35-year-old female with a Cehpech sphere vessel and probable deer bones. The apatite reading was –3.6‰, and the tooth enamel $\delta^{13}C$ value from an incisor was –2.3‰.

C. Burial 8 was a 30–40-year-old female with an apatite signature of –4.1‰ and tooth enamel value of –0.1‰ from an incisor. Burial 8 also had a collagen value of –11.8‰. She was interred with Cehpech sphere vessel over the skull.

D. Burial 9 was a female of indeterminate adult age, interred with shell pendants and deer bones. The apatite $\delta^{13}C$ value was –2.9‰, and the tooth-enamel value for a molar was –0.8‰.

E. Burial 11 was an adult female with an apatite $\delta^{13}C$ reading of –3.3‰. There were no teeth associated with this burial. She had three Cehpech sphere vessels interred with her.

F. Burial 13C was a robust male older than 40 years, who was probably a warrior because the decapitated skull of a young adult male (Burial 13B) was interred with him, and he also had multiple grave goods. The ceramics are late Cehpech, suggesting Terminal Classic. The apatite $\delta^{13}C$ for the young adult male (13B) was –4.1‰ and the tooth enamel value was –0.9‰ for a first molar. The warrior (13C) has an apatite signature of –3.4‰, and there were no teeth from this individual.

G. Burial 15A was a male more than 40 years old with a Cehpech vessel over his head. The apatite $\delta^{13}C$ value was –2.9‰. There were no teeth.

H. Burial 16 was a 30–35-year-old male interred in a crypt. His apatite $\delta^{13}C$ signature was –4.2‰, and his tooth-enamel value from a molar yielded a reading of –2.6‰.

I. Burial 17 was a male who was 25–40 years old, and his crypt was disturbed by the construction of the crypt for Burial 16. He had a collagen signature of –12.9‰, an apatite value of –1.2‰, and the results for tooth enamel from a first molar was –1.0‰.

Overall there is a high degree of consistency in stable isotope values for all of these individuals, indicating a great dependence—about 60–70% on average—on maize and maize products for the whole diet (as visible from the apatite). The collagen data, however, suggest that only 50% of dietary protein was represented by C4-based plants (e.g., maize), indicating that the diet was supplemented by C_3 plants (e.g., beans that have a higher protein percentages than maize) and wild animals (e.g., deer).

Chunchucmil

Chunchucmil is located 27 kilometers east of the Gulf of Mexico (Figure 13-4). It is an area of the Yucatán peninsula that has low rainfall and poor soil with little depth. Current

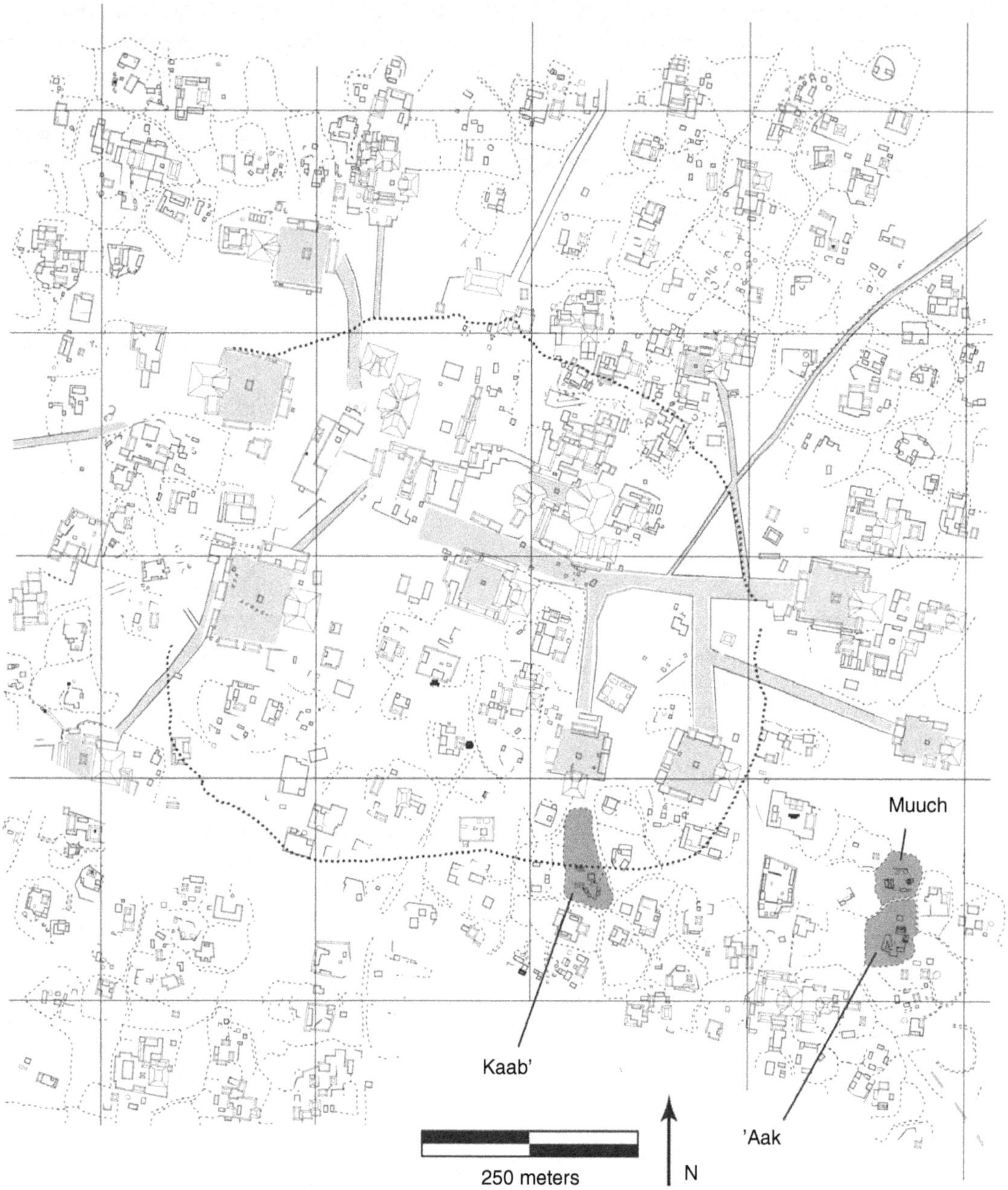

Map of Central Chunchucmil, with quadrangle groups and sacbes highlighted in gray and the 'Aak, Muuch, and Kaab' groups highlighted in darker gray.

FIGURE 13-4 Map of the site of Chunchucmil (Yucatán) showing residential groups with analyzed human skeletal materials. (Courtesy of Eugenia Brown Mansell)

estimates state that at its peak during the late Early Classic (ca. AD 400–550 or 600) the population of the site could have reached 29,680 to 46,648 [6]. The estimate is based on the unusually large number of house mounds found at the site. Chunchucmil, with its aridity and thin, impoverished soil, could not have grown enough maize to sustain a population of this size. Chunchucmil residents would have been importing foodstuffs: maize from areas to the east with richer farm lands, marine resources from the nearby coast, and various floral and faunal foodstuffs available in the nearby savanna area.

The five individuals included in this study were found in modest house mounds, in nonelite contexts. At least one individual (Burial 2), however, had grave goods generally associated with higher-status individuals [8]. All residential groups of this period at Chunchucmil are characterized

FIGURE 13-5 Map of Kaab' residential group, Chunchucmil. (Courtesy of Eugenia Brown Mansell)

by structures arranged around a patio and enclosed by a stone wall [9]. Osteological analyses have been done by Stanley Serafin [14].

Burial 1 was recovered from the Kaab' Group, a residential unit of three medium-sized mounds arranged on three sides of a common patio and several smaller structures arranged around a second southeastern patio (Figure 13-5). The group is located 300 meters from the site center. Burial 1 was a dedicatory burial at the base of the only vaulted structure at the group, Structure 23 (Figure 13-6). It was contained in a cist in the construction fill of the structure under a red painted plaster floor. The skeletal material was in a flexed position. A Teabo-Red vessel was placed over the skull, and a Dzibical vessel was to the east of the long bones [10]. Both vessels date to the Late Classic (AD 550 or 600–800). A complete skeleton was not recovered, but the parts appear to be those of a juvenile or a small adult. The collagen $\delta^{13}C$ value of this individual was −14.7‰, and the apatite value was −8.2‰. There were no teeth.

Burial 5 also was recovered in Structure 32 of the Kaab' Group. It was found on top of the floor in the southeast corner of Room 2. Again, only a partial skeleton was recovered, the skull and long bones of an adult [10]. There was no collagen preserved, although the carbon isotope reading was −5.6‰ for apatite and −5.8‰ for tooth enamel from a first molar.

Burials 2 and 3 were recovered from the 'Aak Group, another residential group of six structures located 600

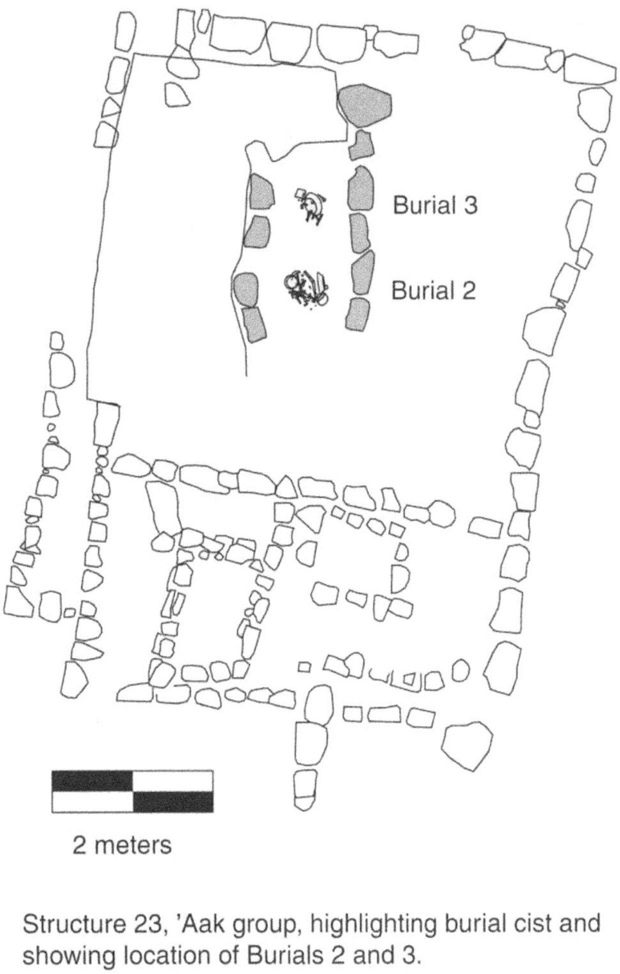

FIGURE 13-6 Map of Structure 23 showing burials, 'Aak residential group, Chunchucmil. (Courtesy of Eugenia Brown Mansell)

meters from the site center (Figure 13-7). Structure 23, the tallest (2 m) structure at the group, yielded two burials under large stone slabs at a depth of 130 centimeters. Burial 2 consisted of disarticulated bones surrounded by three complete vessels: one was an elaborately carved Acu bowl. In addition, there were numerous jade and greenstone objects, a celt, pendant, and multiple round and tubular beads (Figure 13-8). The individual was at least 12 years of age based on teeth recovered but probably older. The collagen $\delta^{13}C$ value was −13.4‰, the apatite value was −4.5‰, and the tooth-enamel value for a second molar was −6.3‰.

Burial 3, on the northern edge of the unit, had fewer remains than Burial 2, only the long bones and a scapula, gathered and perhaps bundled, were recovered under a tripod shallow bowl [8]. There was no collagen recovered. The apatite $\delta^{13}C$ value was −5.2‰ and the enamel value for a first molar was −7.7‰. The vessels in both burials date to the late Early Classic (AD 400–550 or 600).

Burial 6, possibly a young adult, was recovered under construction fill while excavating a unit in Structure 16 of the Muuch Group, a small residential albarrada group, just north of 'Aak (Figure 13-9). Burial 6 was a secondary burial; the remains, consisting of a portion of the zygomatic arch, several long bone fragments, and three premolars, had been placed in a large Maxcanu cazuela (basin) (Figure 13-10). Dental evidence indicates that the individual was a young adult. The collagen $\delta^{13}C$ value was −16.0‰, and the apatite value was −3.4‰. There was no enamel reading recovered. There were two identical Kanachen dishes and two nearly identical Hunabchen vessels dating to the Early Classic in association with the burial; all were crushed under construction fill [15]. As with Burial 3, these vessels dated to the late Early Classic period; all physical data from the Chunchucmil burials were taken from the field report of Serafin [14].

The burials from Chunchucmil, although not elite, had a large quantity of grave goods. This could represent a "middle class" with a market economy based on trade and exchange, because Chunchucmil probably controlled the second largest salt flats in Mesoamerica and the first salt flats reached when traveling from central Mexico. Even though the sample size is small, it is clear from the stable isotope data that neither maize nor maize-fed animals were quite the critical contributors to the diet as at Yaxuná, and supports the hypothesis that low rainfall and poor soil limited maize production. At the same time, there is no isotopic indication that maize was substituted for by any marine resources, because they likely would have increased $\delta^{13}C$ and $\delta^{15}N$ values in collagen and $\delta^{13}C$ values in bone apatite and tooth enamel. Instead, it appears that there was increased reliance on C3 plants (e.g., squash and beans), whether locally produced or imported, and their consumers.

DISCUSSION AND CONCLUSION

Comparison of the data from Yaxuná and Chunchucmil with published results for Pre-Classic, Classic, and Post-Classic sites across the Maya area indicates that the importance of maize to dietary protein of these two sites from the northern lowlands was similar to many sites in Belize (Figure 13-11) but clearly less important than for sites in the Peten and Guatemala (Figure 13-12). The collagen carbon isotope data for Chunchucmil clearly suggest that maize was far less important in the diet there, when compared with the Belize sites and Yaxuná. The bone apatite carbon isotope data, which measures the whole diet, illustrate the importance of doing isotope analysis on more than collagen samples. Apatite data are especially important when the research is meant to address the significance of plant foods such as maize, which have low protein content. The bone apatite and collagen data produced for Yaxuná strongly suggest that maize (and/or other C4/CAM plants that were directly consumed by humans) was of significance to the

FIGURE 13-7 Map of 'Aak residential group, Chunchucmil. (Courtesy of Eugenia Brown Mansell)

overall Maya diet in the central portion of the Yucatán peninsula, similar to Belize or other adjacent regions (Figure 13-13). In contrast, in Chunchucmil, which was located in a poor agricultural region and close to a variety of ecological regions and trading routes, people relied less on maize. At the same time, a greater percentage of dietary protein at Yaxuná and Chunchucmil, compared to Belize and elsewhere, must have ultimately come from foods with C3-like carbon isotope values. Although the lower nitrogen isotope ratios for both Chunchucmil and Yaxuná most likely indicate that their more negative collagen carbon isotope values did not result from greater consumption of typical terrestrial or riverine fauna, which are mostly protein and thus have a much greater effect on human collagen values than do plants, fauna in this region need to be isotopically tested, because they could have lower average nitrogen isotope values compared to those from Belize. Another interpretation would be that beans and other legumes were of greater importance in the Yucatán than in Belize.

The isotope data clearly indicate that other sources of information, perhaps faunal isotopic studies, about diet in the Yucatán must be examined to test the hypotheses and potential interpretations presented here. When compared with Chunchucmil, it appears that more rain and deeper soil allowed the people of Yaxuná to grow maize according to their needs; indeed, the people of modern Yaxuná rely on their milpa more than the contemporary people of Chunchucmil (Mansell, personal observation). The skeletal remains of ancient Yaxuná do not evidence a lack of sustenance; the population appears to have been in good health. There appears to be no evidence of gender and status differences in diet or a temporal difference. Their relatively

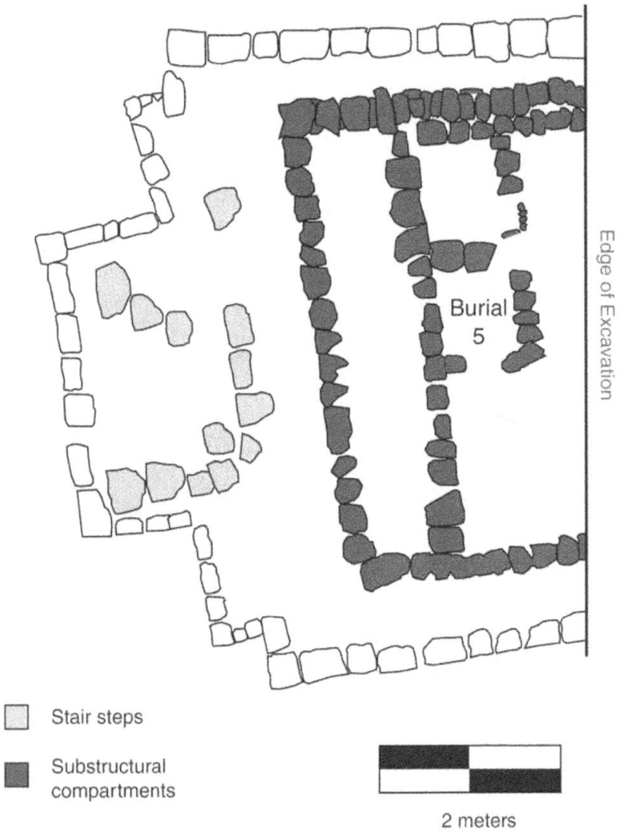

FIGURE 13-8 Map of Structure 22 showing burials, 'Aak residential group, Chunchucmil. (Courtesy of Eugenia Brown Mansell)

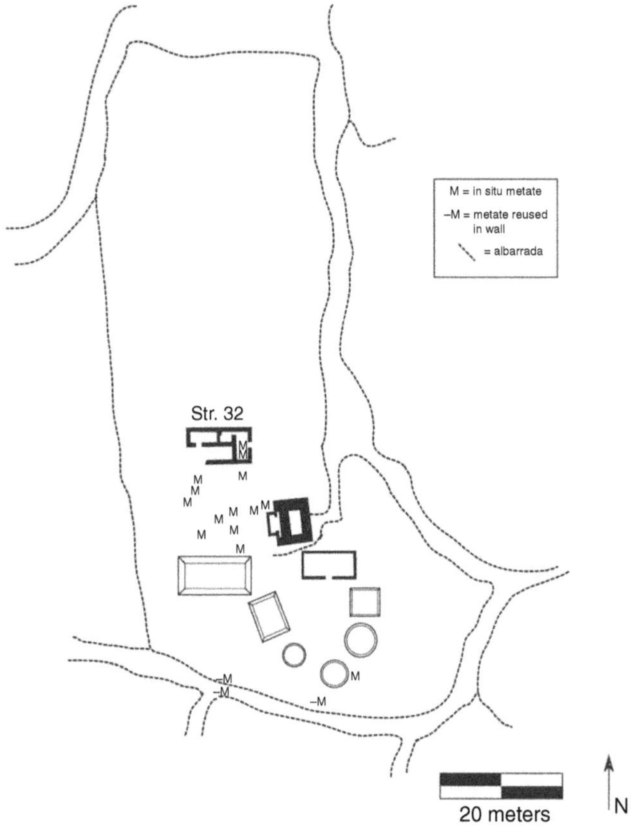

FIGURE 13-9 Map of Muuch residential group, Chunchucmil. (Courtesy of Eugenia Brown Mansell)

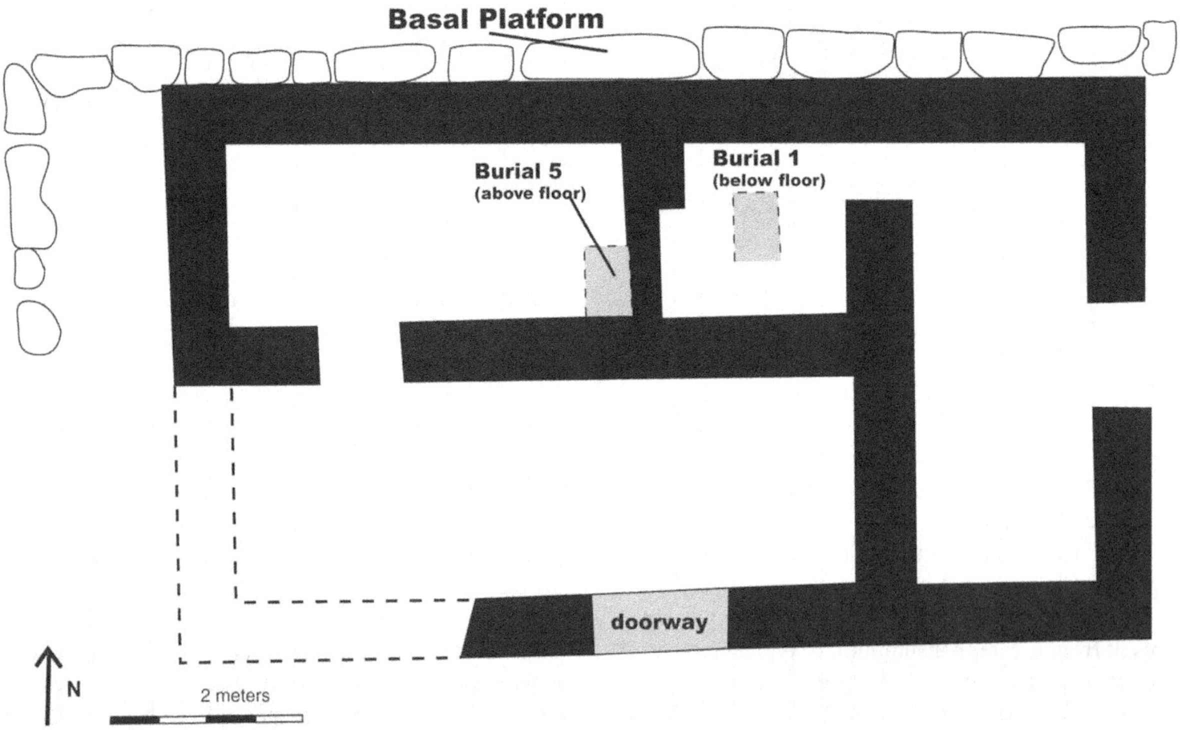

FIGURE 13-10 Map of Structure 16 showing burial, Muuch residential group, Chunchucmil. (Courtesy of Eugenia Brown Mansell)

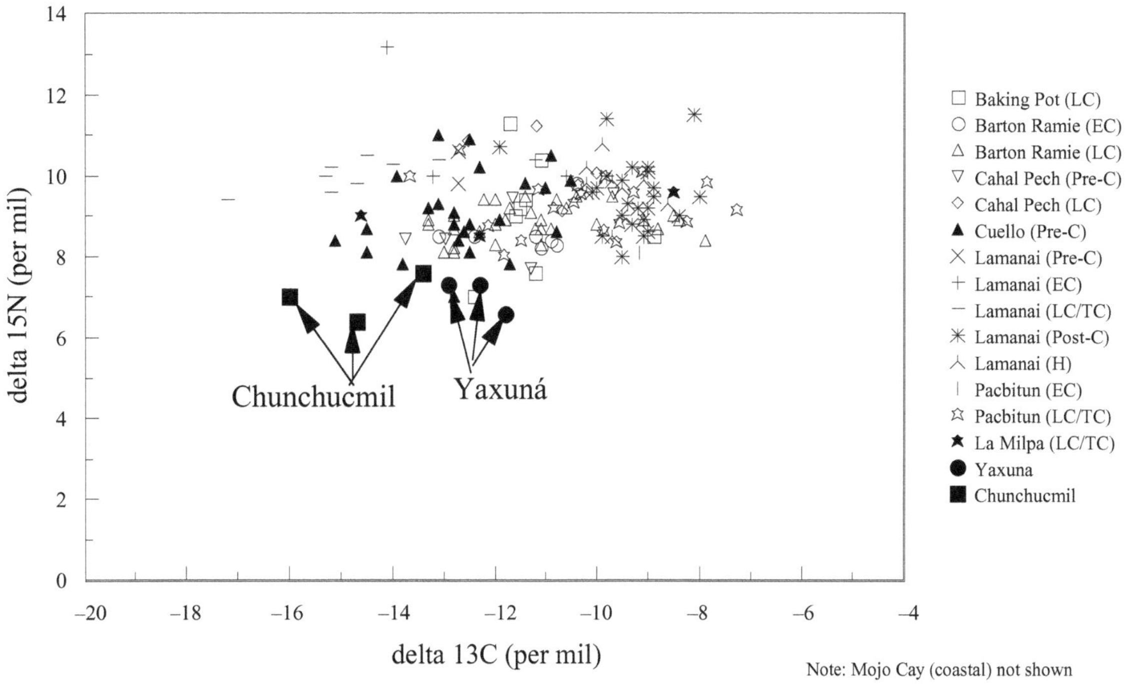

FIGURE 13-11 Collagen stable isotope data for Yucatán sites of Chunchucmil asnd Yaxuná compared with data for sites in Belize. (Courtesy of Eugenia Brown Mansell)

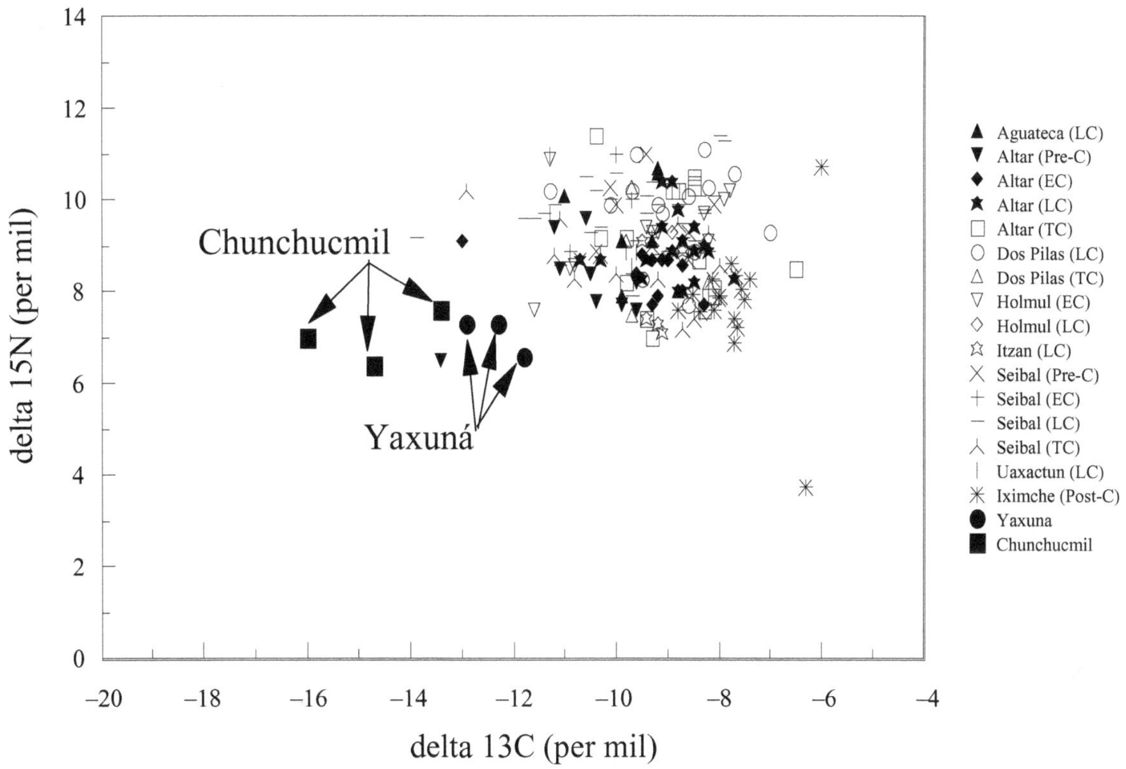

FIGURE 13-12 Collagen data for Yucatán sites of Chunchucmil and Yaxuná compared with data for sites in the Peten and Guatemala. (Courtesy of Eugenia Brown Mansell)

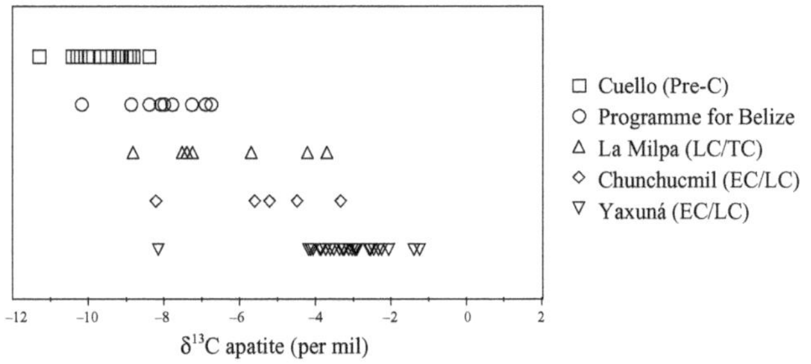

FIGURE 13-13 Bone apatite data for Chunchucmil and Yaxuná compared with selected sites in Belize. (Courtesy of Eugenia Brown Mansell)

greater reliance on maize, however, may have been directly related to their environment and the lack of diversity in its resources. In contrast, the people of Chunchucmil, unable to grow enough maize to feed their large population on the regionally marginal agricultural resources that the area provided, were probably fortunate that the site occupied a strategic location along a well-established maritime trade route through its port, Punta Canbalam, enabling the site to receive a diversity of subsistence goods. Also, the fact that they abutted an environmentally rich savanna and wetlands greatly improved their ability to provide a variety of nutrients for their population. Dietary gender differences cannot be studied with the Chunchucmil sample because the poorly preserved skeletal material cannot be sexed in most cases. Moreover, because all the burials were from nonelite contexts, we cannot evaluate dietary differences based on social status. Future investigations may possibly reveal whether the Chunchucmil elite and nonelite populations had a similar diet as those from Yaxuná.

From these two small studies, it appears likely that environmental factors are a great determinate to a population's reliance on maize in Northern Yucatán. We await other studies to confirm this conclusion. We feel confident that in the Northern Lowlands, the location of a site and the availability of local and foreign resources was a determining factor in the overall diet of the resident population.

Acknowledgments

We would like to thank Aline Magnoni, Scott Hutson, and Traci Ardren for their help with the illustrations, suggestions, and patience during the writing of the chapter.

References Cited

1. S. Bennett. (1991). Burials at Yaxuná, Yucatán. The Selz Foundation Yaxuná Project. In: D. Freidel, C. Suhler, R. Cobos, (Eds.), *Final report of the 1991 field season.* México: INAH, Mérida, Yucatán.
2. S. Bennett. (1992). Burials from Yaxuná, Yucatán. The Selz Foundation Yaxuná Project. In: C. Suhler, D. Freidel, (Eds.), *Final report of the 1992 field season.* México: INAH, Mérida, Yucatán.
3. S. Bennett. (1993). The burial excavations at Yaxuná in 1993. The Selz Foundation Yaxuná Project. *Final Report of the 1993 Field Season.* México: INAH, Mérida, Yucatán.
4. D. Z. Chase, A. F. Chase, C. D. White, W. Giddens. (1998). Human skeletal remains in archaeological context: Status, diet, and household at Caracol, Belize. Paper presented at the 14th International Congress of Anthropological and Ethnological Sciences, Williamsburg, Virginia, July 27, 1998.
5. B. H. Dahlin, T. Ardren. (2002). Modes of exchange and their effects on regional and urban patterns at Chunchucmil, Yucatán, México. In: M. A. Masson, D. Freidel, (Eds.), *Ancient Maya political economies.* Walnut Creek, CA: Altamira Press. pp. 249–284.
6. B. H. Dahlin, T. Beach, S. Lazzader-Beach, A. Magnoni, S. R. Hutson, E. B. Mansell, D. Mazeau. (In press). Reconstructing agricultural self-sufficiency at Chunchucmil, Yucatán, México. *Ancient Mesoamerica.*
7. J. P. Gerry. (1993). Diet and status among the Classic Maya: An isotopic perspective. Ph.D dissertation. Cambridge, MA: Harvard University.
8. S. Hutson. (2000). Excavations in the 'Aak Group. In: W. Stanton, (Ed.), *Pakbeh regional economy program: Report of the 2000 field season.* México: INAH, Mérida, Yucatán.
9. S. Hutson, A. Magnoni, T. Stanton. (2004). House Rules? The practice of social organization in Classic Period Chunchucmil, Yucatán, Mexico. *Ancient Mesoamerica, 15,* 73–90.
10. A. Magnoni. (2001). Excavations at the Kaab' Group. In: B. H. Dahlin, D. Mazeau, (Eds.), *Pakbeh regional economy program: Report of the 2001 field season.* México: INAH, Mérida, Yucatán.
11. E. B. Mansell, R. H. Tykot, F. Valdes, N. Hammond. (2000). Stable isotope analysis of human remains at La Milpa, Belize. Paper presented at the 65th Annual Meeting of the Society for American Archaeology, Philadelphia, Pennsylvania, April 5–9, 2000.
12. L. C. Norr. (1991). Nutritional consequences of prehistoric subsistence strategies in lower central America. Ph.D. dissertation. Urbana, IL: University of Illinois.
13. T. G. Powis, N. Stanchly, C. D. White, P. F. Healy, J. J. Awe, F. Longstaffe. (1999). A reconstruction of Middle Preclassic Maya subsistence economy at Cahal Pech, Belize. *Antiquity, 73,* 364–376.
14. S. Serafin. (2004). Osteological analyses of the human skeletal remains. In: S. R. Hutson, (Ed.), *Pakbeh regional economy program: Report of the 2004 field season.* México: INAH, Mérida, Yucatán.
15. T. W. Stanton. (2001). Horizontal excavations at the Muuch Group. In: B. H. Dahlin, D. Mazeau, (Eds.), *Pakbeh regional economy program: Report of the 2001 field season.* México: INAH, Mérida, Yucatán.
16. C. K. Suhler. (1996). Excavations at the North Acropolis, Yaxuná, Yucatán, México. Ph.D. dissertation. Dallas, TX: Southern Methodist University.
17. R. H. Tykot. (2002). Contribution of stable isotope analysis to understanding dietary variation among the Maya. In: K. Jakes, (Ed.), *Archaeological chemistry. Materials, methods, and meaning.* ACS Symposium Series 831. Washington, D.C.: American Chemical Society. pp. 214–230.
18. R. H. Tykot. (2004). Stable isotopes and diet: You are what you eat. In: M. Martini, M. Milazzo, M. Piacentini, (Eds.), *Physics methods in archaeometry.* Bologna, Italy: Società Italiana di Fisica. pp. 433–444.
19. R. H. Tykot, N. J. van der Merwe, N. Hammond. (1996). Stable isotope analysis of bone collagen, bone apatite, and tooth enamel in the reconstruction of human diet: A case study from Cuello, Belize. In: M. V. Orna, (Ed.), *Archaeological chemistry: Organic, inorganic, and biochemical analysis.* ACS Symposium Series 625. Washington, D.C.: American Chemical Society. pp. 355–365.
20. E. Z. Vogt. (1990). *The Zinacantecos of Mexico: A modern Maya way of life.* New York: Harcourt Brace Jovanovich.
21. C. D. White, P. F. Healy, H. P. Schwarcz. (1993). Intensive agriculture, social status, and Maya diet at Pacbitun, Belize. *Journal of Anthropological Research, 49,* 347–375.

22. C. D. White, H. P. Schwarcz. (1989). Ancient Maya diet: As inferred from isotopic and elemental analysis of human bone. *Journal of Archaeological Science, 16,* 451–474.
23. C. D. White, D. M. Pendergast, F. J. Longstaffe, K. R. Law. (2001). Social complexity and food systems at Altun Ha, Belize: The isotopic evidence. *Latin American Antiquity, 12,* 371–393.
24. C. D. White, M. E. D. Pohl, H. P. Schwarcz, F. J. Longstaffe. (2001). Isotopic evidence for Maya patterns of deer and dog use at Preclassic Colha. *Journal of Archaeological Science, 28,* 89–107.
25. L. E. Wright, H. P. Schwarcz. (1996). Infrared and isotopic evidence for diagenesis of bone apatite at Dos Pilas, Guatemala: Paleodietary implications. *Journal of Archaeological Science, 23,* 933–944.

C H A P T E R

14

Diet in Prehistoric Soconusco

BRIAN CHISHOLM* AND MICHAEL BLAKE†

*Department of Anthropology and Sociology, University of British Columbia, Vancouver, British Columbia, Canada

†Department of Anthropology and Sociology, University of British Columbia, Vancouver, British Columbia, Canada

Glossary

C_3 plants Those plants that follow the Calvin–Benson photosynthetic cycle and generate molecules with three carbon atoms as the first product of photosynthesis. They include trees, shrubs, flowering plants, and most grasses.

C_4 plants Those plants that follow the Hatch–Slack photosynthetic cycle and generate molecules with four carbon atoms as the first product of photosynthesis. They are mostly grasses that grow in high light intensity environments.

Fertilizer effect If the nitrogen isotope ratios in the fertilizer are different than the local values, there will be a noticeable effect seen in the local food chains; it is because of the addition of nitrogen-rich fertilizers to local fields and gardens.

Lipids The fats and oils found in biological materials. They yield different carbon isotope ratios than do proteins and carbohydrates and can lead to erroneous results if not removed from food species samples in studies such as this one.

Stable isotopes Those forms of an element that do not decay radioactively. They differ from other isotopes of an element in the number of neutrons in their nuclei and therefore have different masses. The ratio of the stable isotopes of an element often differs from one individual to another due to diet and environmental or geographic factors.

We have analyzed a number of plant and animal species from the Soconusco region of Mexico to provide baseline data on stable carbon and nitrogen isotope ratios of potential foods for the prehistoric inhabitants of the region. Human bone samples were analyzed to determine the importance and time of appearance of maize in prehistoric diets. Our results suggest that the Archaic inhabitants of the region obtained much of their protein from either maize or marine species such as shrimp. People from later times, in the Locona through Cuadros phases, seem to have obtained most of their protein from C_3 plant species or their herbivore consumers, or both. Following the Cuadros phase, **C_4 plants,** specifically maize, became dominant in the local diet.

INTRODUCTION

Ancient peoples living along the Pacific Coast of southeastern Mexico inhabited a lush environment with abundant and diverse plant and animal species. Thanks to the work of Barbara Voorhies [12], we have an increasingly clear picture of environment and cultures spanning the period from about 6000 years ago to the beginning of village life. Her research has shown that Archaic people lived seasonally on islands in lagoons and along river channels, fishing, hunting, and collecting plants and animals and other resources. One of the questions arising from this research is, What is the extent that these people may have been engaged in agriculture in general and maize cultivation in particular?

In 1992, we made an initial attempt to survey the isotopic evidence for the changing importance of maize agriculture in the Soconusco region of southeastern Mexico by looking at samples of human bone from burials dating from the Chantuto B phase (beginning ca. 5000 years ago) and every phase thereafter until the Late Post-Classic period [1, 2, 7].

We found that two Chantuto B phase individuals from the site of Tlacuachero had stable carbon isotope ratios substantially higher and nitrogen isotope ratios substantially lower than Formative period villagers in the neighboring Mazatan region, indicating that the Archaic Tlacuacheros may have been eating more C_4 plants, possibly maize, than the later villagers.

This observation, coupled with the lack of evidence for maize microremains or macroremains in Archaic period sites and the abundance of maize macrobotanical remains in Early Formative villages, created a paradox that is certainly puzzling and does not fit the expectations for the emergence and spread of agriculture. On the one hand, why and how would small groups of mobile fisher, hunter–gatherers who had access to the rich marine and terrestrial resources of the Pacific coastal lowlands become so reliant on the consumption of C_4 plants that their stable carbon isotopic signatures come to resemble those of Late Post-Classic people in the region who were known to have relied heavily on maize agriculture? On the other hand, why would Early Formative villagers, whose most ubiquitous plant remains left behind in their refuse deposits was maize, not show isotopic signatures representing high maize consumption similar to those of later peoples in the region [8]?

In an attempt to answer these questions we decided to expand our study to encompass an isotopic analysis of both modern and archaeological plant and animal samples representing the types of foods that were and are consumed in the region. We also wanted to remeasure some of the previously analyzed human bone samples and add to the sample of archaeological human bones. Here we report on these analyses and show that there is a possible alternative interpretation of the isotope measures of the Archaic period individuals from Tlacuachero: It is possible that high consumption of certain marine species, particularly shrimp, could also produce isotopic signatures similar to high consumption of C_4 plants. This pattern provides some confirmation of a proposal made by Voorhies, Michaels, and Riser [13] that many of the large island sites in the Chantuto lagoons, such as Tlacuachero, may have served as platforms for the preservation of shrimp. We also discuss our reanalysis of the Early Formative human bone samples and show that, indeed, the consumption of maize was not as important as in later times and that maize was a small but increasing component of the diet in this coastal environment. Finally, we present appendices showing the measurements of our samples so that they can be incorporated into existing databases and checked by other researchers.

SAMPLE SELECTION

The first step in diet reconstruction in any area should be the collection and analysis of local food species. Although this is not always easy, particularly in light of poor preservation conditions that severely limited the number of archaeological specimens available for analysis in this study, it was possible to obtain modern specimens from local markets and during fieldwork. Species collected were considered representative of the various environments that provided food for local inhabitants. Plant samples were mostly modern, but a few archaeological samples were also available for analysis. Animal data were obtained from either muscle tissue of modern samples or extracted bone collagen–gelatin from archaeological samples.

Human samples were from sites excavated over a number of years in the region by Michael Blake, John Clark, and their associates. Although the original intent was to obtain samples from a broad range of sites, preservation conditions limited the samples to those sites reported here (Figure 14-1).

SAMPLE PREPARATION AND ANALYSIS

Stable isotope studies of prehistoric diet have been used for more than 25 years now and do not need to be described here. Analytical procedures used in this study are those outlined by Chisholm [4, 5]. Modern food species were prepared for analysis by homogenization with acetone to remove soluble lipids, and then the residue was dried. **Lipids** were removed because they may differ considerably in isotope ratios from proteins and carbohydrates [7, 11]. Archaeological samples did not contain any soluble lipids and thus were not treated with acetone. Bone samples were prepared using HCl and NaOH treatments [4].

In 1992, we reported human data from the Soconusco region, however there were a number of samples that gave questionable results [1, 2, 6]. A few of the original problematic samples were prepared again and analyzed at Kyushu University, and the Tlacuachero samples were dated using accelerator mass spectometry (AMS) at the Tandem Accelerator Laboratory of Nagoya University. Additional samples were analyzed in the Department of Earth and Ocean Sciences at the University of British Columbia. As a result of these analyses, we have been able to confirm some of our earlier data and add further reliable data to the set.

PLANT RESULTS

The plant-food values (Appendix 1) form two clusters with no intermediary values, showing a clear distinction between C_3 and C_4 species. For the modern samples, the mean values for carbon fall almost exactly where literature values for similar species usually fall, with no observable outliers or unusual results (Table 12-1). For the archaeological samples, the carbon means are higher than those of the modern samples by 2.8‰ for C_3 species and 1.7‰ for C_4

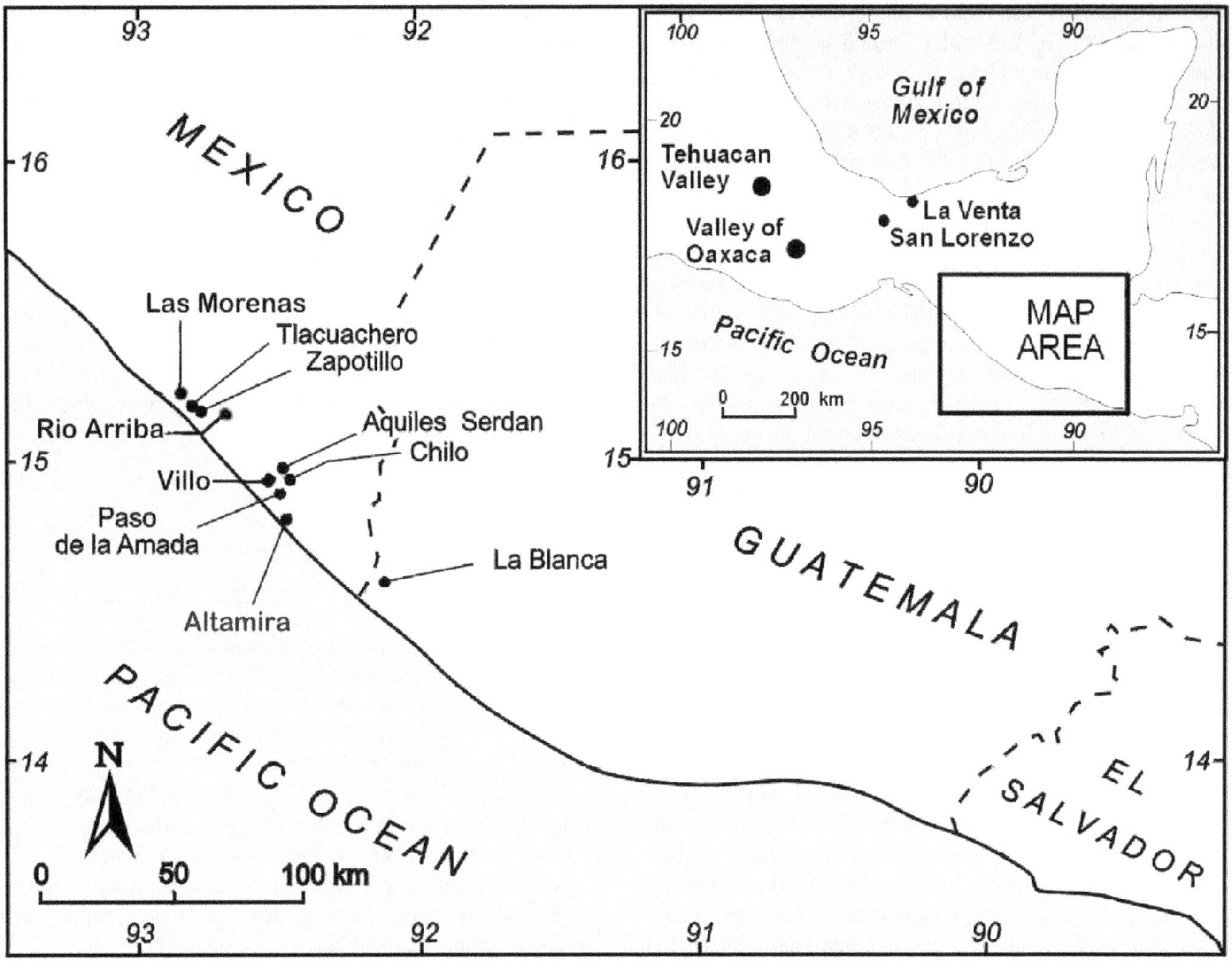

FIGURE 14-1 Map of the Soconusco region showing sites providing samples for this study.

species. Such differences could result from fossil fuel effects, forest canopy microenvironmental effects, changes in the local floral environments, or from diagenetic effects. However, the degree of difference does not obscure the distinction between C_3 and C_4 species in either period.

Although the archaeological samples sets are too small to make meaningful interpretations, it does appear that the mean values for nitrogen for both modern and archaeological samples fall within the normally observed range for such samples; the archaeological results are slightly but not significantly lower than those of the modern samples. The small degree of nitrogen displacement would tend to eliminate diagenesis as a possible cause for the displacement of the carbon results, favoring differences in reservoir values as an explanation.

Results for the modern samples show a larger standard deviation, particularly in nitrogen, indicating that the local microenvironments from which they were sampled were themselves varied, and therefore we must be careful in making interpretations on the basis of regional samples. It is possible that we are seeing a **fertilizer effect** in the nitrogen results, which could cause variation in the nitrogen results for these samples.

ANIMAL RESULTS

A total of 135 animal samples were prepared and analyzed (Appendix 2). Results for archaeological samples were obtained from collagen that was extracted from bones recovered during site excavations in the region. In some cases the collagen yield was too low to obtain both carbon and nitrogen results. Modern sample analyses were from muscle tissue.

The animal results are widely distributed (see Table 14-1), ranging over about 20‰ for both C and N, and at first glance difficult to explain. Mean values for the estuarine and terrestrial habitat species are around 4‰ higher for modern carbon and essentially similar for modern nitrogen. When temporal changes in the carbon reservoir because of fossil

TABLE 14-1 Mean Isotope Ratios for Food Species from the Soconusco Region, Mexico

Family	Environment	Arky/Modern	Mean	∂^{13}C(‰) St Dev	n =	Mean	∂^{15}N(‰) St Dev	n =
C_3 Plants	Terrestrial	A	−23.6	0.9	4	2.6	0.65	4
C_3 Plants	Terrestrial	M	−26.4	2.13	20	3.7	3.22	20
C_4 Plants	Terrestrial	A	−8.8		2	2.3		2
C_4 Plants	Terrestrial	M	−10.5	0.78	3	4.2	3.52	3
Birds	?	A	−23.0		2			
Birds	Estuary	M	−21.3		1	9.3		1
Birds	?	M	−19.0		2	7.8		2
Fish	Estuary	A	−20.8	1.98	16	2.9	2.81	7
Fish	Estuary	M	−19.0	5.59	13	10.9	4.15	13
Fish	Marine	M	−14.6	1.31	24	12.9	2.83	22
Fish	River	M	−29.4		1	12.3		1
Mammals	Terrestrial	A	−21.1	1.45	14	2.3	3.92	4
Mammals[a]	Terrestrial	M	−22.4	2.18	5	6.7	2.89	5
dogs	*Terrestrial*	*M*	*−19.7*		2	*9.3*		2
sheep	*Terrestrial*	*M*	*−16.6*		2	*7.4*		2
Mammals	Estuary	M	−26.3		2	9.4		2
Reptiles	Terrestrial	A	−22.4	0.64	3	−0.3		1
Reptiles	Terrestrial	M	−21.9	2.86	6	8.1	4.49	6
Reptiles	Estuary	A	−20.3	0.77	13	2.7	3.81	7
Reptiles	Estuary	M	−23.0	3.93	11	6.4	2.35	11
Reptiles	Marine	M	−14.7		1	12.6		1
Reptiles	River	M	−26.4		1	9.9		1
Shellfish	Estuary	A	−20.8		1	−1.8		1
Shellfish	Estuary	M	−23.5	2.33	5	5.7	0.57	5
Shellfish	Marine	M	−14.8	2.30	5	7.1	4.72	5
Shellfish	River	M	−16.5		2	6.9		2

[a]Modern terrestrial mammals group excludes dogs and sheep.

fuel effects, usually around 1‰ [3, 10], and the usual differences in an individual of about 3‰ between meat (modern samples) and collagen (archaeological samples) are accounted for, there are still unexplained differences between the archaeological and modern values. A few reasons that may explain these discrepancies are 1) the samples are not appropriately paired; 2) the sample numbers are small; or 3) the environments and reservoirs have changed.

Standard deviations for all of these means range from less than 1‰ to more than 5‰, and whether they are collagen or meat show considerable variation in values for the samples. This indicates that the animals sampled were selecting their foods from a number of isotopically varied sources; they subsisted on a variety of diets. This scatter is seen elsewhere—in Japan, on the Northwest Coast of North America, and so on—and most likely reflects opportunistic foraging of animal consumers, as well as the fact that they are mobile and pass through numerous microenvironments that vary in carbon and nitrogen reservoir values. The variation seen in these results does make it difficult to determine precise proportions of various dietary groups used by the humans in this region. However, patterns of change and general tendencies should be interpretable. We may not be able to say that 35 ± 10% of the diet was composed of a particular food group, but we should be able to see the introduction of new foods that differ in isotopic signature or whether there were noticeable changes in prehistoric diets.

When the species are grouped according to the habitat they occupy, the dispersion is minimized for most groupings (Table 14-2).. The greatest degree of variation is found in estuarine species, whereas results for species from the terrestrial and marine environments are less varied. Because estuaries can have considerable fresh water input, and are often depositional environments for terrestrial runoff materials, this variation comes as no surprise. Both the marine and terrestrial environments would be expected to show less

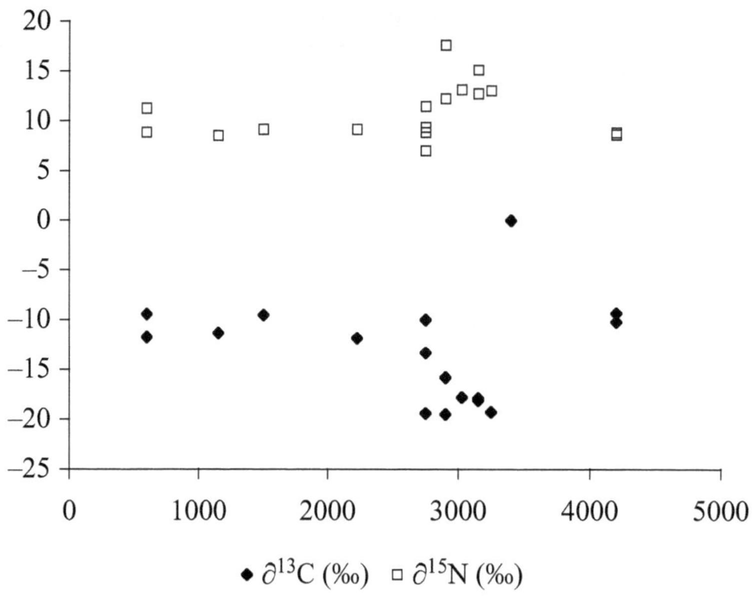

FIGURE 14-2 Temporal changes in stable isotope ratios for humans from the Soconusco region, Mexico.

TABLE 14-2 Animal Results According to Environment in the Soconusco Region, Mexico

Environment	Arky/Modern	Mean	$\partial^{13}C$(‰) St Dev	$n =$	Mean	$\partial^{15}N$(‰) St Dev	$n =$
Estuary	A	−20.5	1.55	29	2.5	3.32	15
	M	−21.6	4.90	32	8.4	3.72	32
Terrestrial	A	−21.4	1.41	17	1.7	3.58	5
	M	−22.2	2.46	11	7.5	3.73	11
Marine	M	−14.8	1.68	30	12.6	2.73	27
River	M	−16.5		2	6.9		2
	M	−27.9		2	14.0		2

mixing and hence less variation in isotope ratios for individuals that habituate them.

Comparing the food species results according to their habitat (see Table 14-2) defines three food groups with different average isotopic signatures. The estuarine and terrestrial species overlap virtually completely whereas the **C_3 plants** are close to them in carbon values. It would appear that the estuarine species obtained most of their carbon from terrestrial sources, possibly via runoff in the rainy periods of the year. Nitrogen values for the estuarine fish differ from the rest of this group by about 7–8‰, reflecting an aquatic habitat. A second group is made up of the marine species, which show little variation, but do have four outliers that yielded results more than two standard deviations from the mean for this group. We do not know the reasons for these outliers, but incomplete removal of lipids is one possibility. Local microenvironmental effects could also have influenced their results. The last food group is the C_4 plants (see Table 14-1), including maize, which show little variation in values although their sample set is small.

HUMAN RESULTS

Out of 78 sample preparations for 54 individuals, useful results were obtained from only 19 individuals (Table 14-3). In most cases the bone preservation was poor to begin with, and in some cases the use of Ducco as a preservative contaminated the bone. In three cases there was a strange-smelling orange residue recovered with the collagen portion that was clearly a contaminant. In all of the 19 cases acceptable carbon : nitrogen (C : N) values were obtained from collagen samples that looked clean and normal.

The "reconstructed diet" data (Table 14-3) were obtained by applying a diet to consumer collagen increment of 4.5‰ for carbon and 3‰ for nitrogen to the measured

TABLE 14-3 Stable Isotope Analysis Results for Humans from the Soconusco Region, Mexico

Period	Site	Measured		Reconstructed diet	
		$\delta^{13}C$	$\delta^{15}N$	$\delta^{13}C$	$\delta^{15}N$
Late Post-Classic (AD 1200–1524)	Zapotillo	−9.4	8.8	−13.9	5.8
	Zapotillo CS8[a]	−9.0	8.4	−13.5	5.4
	Altamira	−11.7	11.2	−16.2	8.2
Late Classic (AD 600–1000)	Las Morenas	−11.3	8.5	−15.8	5.5
Middle Classic (AD 300–600)	Rio Arriba	−9.5	9.1	−14.0	6.1
Late Formative (650 BC–AD 100)	Tlacuachero	−11.8	9.1	−16.3	6.1
Mid Formative:					
Conchas (850–750 BC)	La Blanca	−19.4	11.4	−23.9	8.4
		−13.3	7.0	−17.8	4.0
		−13.3	8.8	−17.8	5.8
		−10.0	9.3	−14.5	6.3
	Paso de la Amada	−14.8	8.3	−19.3	5.3
Cuadros (1000–900 BC)	Aquiles Serdan	−15.8	17.6	−20.3	4.6
	Villo	−19.5	12.2	−24.0	9.0
Cherla (1150–1000 BC)	Aquiles Serdan	−17.8	13.1	−22.3	10.1
Ocos (1250 –1150 BC)	Paso de la Amada	−17.9	15.1	−22.4	12.1
		−18.1	12.7	−22.6	9.7
Locona (1350–1250 BC)	Chilo	−19.3	13.0	−23.8	10.0
Barra (1800–1350 BC)					
Archaic:					
Chantuto B (2700–1800 BC)	Tlacuachero	−9.3	8.6	−13.8	5.6
		−10.2	8.8	−14.7	5.8

[a]Sample provided by Barbara Voorhies, see [12].

results and are representative of the long-term average diet of the individuals [4]. This allows easier comparison of the human results with the various diet alternatives in the region.

The data in Table 14-3 indicate the major dietary biases, whereas Figure 12-2 shows the changes over time. There are a few oddities here—one individual from La Blanca (−19.4; 11.4) does not fit with the others from that site. This could be a visitor, someone who ate a different diet, or a contaminated sample that still yielded a "good" C:N result. The bone sample from this individual appeared as well preserved as any of the others. Also note that there were no good results from individuals of the Barra period. The two Archaic, or Chantuto B phase, individuals are difficult to interpret and are discussed later.

The low carbon values and high nitrogen values for individuals from the Locona through Cuadros phases, before about 2800 BP are suggestive of a mix of C_3 plants and C_3-eating herbivores as the peoples' food sources. In particular the carbon values indicate a high level of C_3 protein in their diets. Their higher nitrogen values suggest that meat was an important dietary component, that possibly plant foods were not as important as they later became, and that marine foods were not a major component of the diet—although they were likely present. (Marine foods would typically lead to higher nitrogen results.) Freshwater species might confuse the issue somewhat, because they could be affected by terrestrial runoff of both carbon and nitrogen from the surrounding lands. Although maize remains have been found in the Locona through Cuadros deposits, the isotopic signature of the humans does not reflect its presence in their diets. This discrepancy may be explained by the use of maize for making beer and not for food, as suggested by Smalley and Blake [9].

After about 2800 BP the higher carbon values and lower nitrogen values for people of the Conchas through Late Post-Classic phases indicate a diet that is much heavier in C_4 plants. The most obvious explanation for this shift is that people had developed a heavy reliance on maize. The high carbon values are typical of heavy maize eaters, and the low nitrogen values suggest that most protein in those individuals' diets came from plants and not from animals.

The two Archaic-Period individuals from Tlacuachero show both carbon and nitrogen values that are similar to those of the maize eaters from the Mid-Formative times to the present; however, people from the intervening Locona to Cuadros phases seem to have not eaten much maize. It seems unlikely that there was a shift from large amounts of maize in the Archaic-Period diet to low-maize intake in the Locona to Cuadros times, followed by a shift back to maize in later periods. It is possible that the Archaic-Period people were eating some other C_4 species, but not likely obtaining too much meat from C_4-eating herbivores, because their nitrogen values are low. It is possible that there were some other food species in the region that gave similar values to maize and that were significant contributors to local diets. It was also considered possible that the Tlacuachero samples were intrusive. However, an AMS date of 4680 (NUTA 2230) confirmed that they were from the Archaic Period. We also confirmed our original isotope results for these two individuals by having them reanalyzed in a second laboratory, obtaining the same values.

There are three groups of food species in the region that have given similar isotope ratio values and that could explain the Archaic-Period humans' results: marine crabs and shrimp (from Paredon), maize, and Opuntia. Opuntia can likely be discounted, because it is not known to be eaten in the high quantities that would be necessary to attain the human results measured for the Tlacuachero individuals. Although shrimp and crab values do not represent all shrimp and crab in the coastal zone, they could represent a large portion of the available foods in estuaries and coastal lagoon areas and could be an important food alternative as Voorhies and colleagues [13] have suggested. Both maize and shrimp could have provided the human results seen here.

CONCLUSIONS

Although there is a great deal of variation in the food-species values presented herein, thus making it difficult to provide precise estimates of the relative importance of various foods in the diet of prehistoric Soconusco people, it is still possible to construct a model of changing diet in the Soconusco Region from the Archaic Period through the Late Post-Classic Period. In the Archaic Period, around 4800 BP, people subsisted on a diet that was high in either maize or marine species such as shrimp. Shrimp was the likely option because there was little maize evident in the archaeological deposits from that time. In the following Locona through Cuadros phases, the local diet likely consisted of a mix of C_3 plants, herbivores consuming C_3 plants, and estuarine species. Maize, which is evident in the archaeological deposits, may have been used for making beer and not used as food. Following the Cuadros phase, C_4 species, specifically maize, became important in the local diets, whereas C_3 plants and their herbivore consumers became much less important.

It is also clear that we need much more information on the food-species values throughout the region if we wish to attain any level of precision in diet reconstructions. Finally, it is clear that larger human data sets are necessary to reach a clear understanding of prehistoric diet in the Soconusco region.

Acknowledgments

Samples were analyzed by Bente Neilsen in the Department of Earth and Ocean Sciences, University of British Columbia and by Sho Mihara in the laboratory of Dr. Hiroko Koike in the Graduate School of Social and Cultural Studies of Nagoya University. AMS dating was provided by Toshio Nakamura of the Tandem Accelerator Laboratory of Nagoya University.

References Cited

1. M. Blake, B. Chisholm, J. E. Clark, B. Voorhies, M. W. Love. (1992). Prehistoric subsistence in the Soconusco region. *Current Anthropology, 33,* 83–94.
2. M. Blake, B. Chisholm, J. E. Clark, K. Mudar. (1992). Non-agricultural staples and agricultural supplements: Early Formative subsistence in the Soconusco, Mexico. In: T. D. Price, A. B. Gebauer, (Eds.), *Transitions to agriculture.* Madison, WI: Prehistory Press. pp. 133–151.
3. W. S. Broecker, T.-H. Peng. (1982). *Tracers in the sea.* New York: Columbia University.
4. B. Chisholm. (1989). Variation in diet reconstructions based on stable carbon isotopic evidence. In: T. D. Price, (Ed), *The chemistry of prehistoric human bone.* A School of American Research volume. Cambridge, MA: Cambridge University Press. pp. 10–37.
5. B. Chisholm. (1986). Reconstruction of prehistoric diet in British Columbia using stable-carbon isotope analysis. Ph.D. dissertation. Burnaby, Canada: Simon Fraser University.
6. B. Chisholm, M. Blake, M. W. Love. (1993). More on prehistoric subsistence in the Soconusco region: Reply to Ambrose and Norr. *Current Anthropology, 34,* 432–434.
7. M. J. DeNiro. (1977). I. Carbon isotope distribution in food chains, II. Mechanism of carbon isotope fractionation associated with lipid synthesis. Ph.D. dissertation. Pasadena: California Institute of Technology.
8. V. L. Feddema. (1993). Early formative subsistence and agriculture in southeastern Mesoamerica. Master's thesis. Vancouver, British Columbia: Department of Anthropology and Sociology, University of British Columbia.
9. J. Smalley, M. Blake. (2003). Sweet beginnings: Stalk sugar and the domestication of maize. *Current Anthropology, 44,* 675–703.
10. M. Stuiver. (1978). Atmospheric carbon dioxide and carbon reservoir changes. *Science, 199,* 253–258.
11. J. Vogel. (1978). Isotopic assessment of the dietary habits of ungulates. *South African Journal of Science, 74,* 298–301.
12. B. Voorhies. (1976). *The Chantuto people: An archaic period society of the Chiapas Littoral, Mexico.* Papers of the New World Archaeological Foundation No. 41. Provo, UT: Brigham Young University.
13. B. Voorhies, G. H. Michaels, G. M. Riser. (1991). Ancient shrimp fishery. *National Geographic Research and Exploration, 7,* 20–35.

APPENDIX 1 Plant Data for the Soconusco Region

Spanish common name	English common name	Scientific name	Sample provenience	∂^{13}C(‰)	∂^{15}N(‰)	Modern/archaeological
aguacate	avocado					archaeological
aguacate	avocado	*Persia americana*		–24.6	2.3	archaeological
frijole	bean	*Persia americana*		–22.5	2.3	archaeological
frijole	bean	*Phaseolus vulgaris*		–23.2	3.6	archaeological
maiz	maize	*Phaseolus vulgaris*		–23.9	2.3	archaeological
maiz	maize	*Zea mays*		–9.5	2.7	archaeological
aguacate	avocado	*Zea mays*		–8.0	1.9	modern
aguacate	avocado	*Persia americana*		–29.4	3.6	modern
cacao	chocolate	*Persia americana*	Buenos Aires	–25.9	1.4	modern
cacao	chocolate	*Theobroma cacao*		–30.3	4.6	modern
cacao	chocolate	*Theobroma cacao*	Soconusco	–26.6	3.9	modern
calabaza	squash	*Theobroma cacao*	Buenos Aires	–31.8		modern
calabaza	squash	*Cucurbita* sp.	San Cristobal	–25.3	6.7	modern
camote	sweet potato	*Cucurbita* sp.	Mazatan	–25.4	1.9	modern
camote (techcamote)	tuber	*Ipomoea batatas*	Tlapachula	–25.5		modern
camote morado	sweet potato	*Ipomoea batatas*	Tlapachula	–27.3		modern
caspirol	legume bean pod	*Ipomoea batatas*	Tlapachula	–26.8		modern
chicozapote	sapodilla plum	*Inga preussi*	Mazatan	–22.6	8.4	modern
chicozapote	sapodilla plum	*Achras sapota*	San Lorenzo	–23.9	–2.8	modern
chicozapote	sapodilla plum	*Achras sapota*	San Lorenzo	–24.2	0.1	modern
chicozapote	sapodilla plum	*Achras sapota*	Gulf Coast	–24.8	6.9	modern
chile	chili pepper	*Achras sapota*	Mazatan	–30.0	0.3	modern
chile	chile pepper	*Capsicum annuum*	San Lorenzo	–26.6	4.8	modern
frijol blanco	bean, white	*Capsicum annuum*	Mazatan	–28.8	9.6	modern
frijol negro	bean, black	*Phaseolus vulgaris*	Tlapachula	–26.6	0.5	modern
frijole	bean	*Phaseolus vulgaris*	Tlapachula	–25.2	5.8	modern
frijole	bean	*Phaseolus vulgaris*		–25.5		modern
frijole	bean, black	*Phaseolus vulgaris*		–25.2	0.5	modern
maiz	maize	*Phaseolus vulgaris*	San Lorenzo	–26.0	3.0	modern
maiz	maize	*Zea mays*		–10.1		modern
maiz	maize	*Zea mays*		–10.1	5.3	modern
maiz	maize	*Zea mays*	San Lorenzo	–10.2	0.3	modern
nopal	prickly pear	*Zea mays*	San Carlos	–10.3		modern
palma corozo (nuez)	macaw palm (nut)	*Opuntia* sp.	San Carlos	–11.9	7.1	modern
palma coyol (nuez)	coyol palm (nut)	*Orbignya cohune*	Mazatan	–24.6	4.6	modern
palma coyol (nuez)	coyol palm (nut)	*Acrocomia aculeata*	San Lorenzo	–27.3	7.4	modern
palma coyol	coyol palm (nut)	*Acrocomia aculeata*	Buenos Aires	–27.4	6.4	modern
yucca	manioc	*Acrocomia aculeata*	San Lorenzo	–27.3	7.4	modern
		Manihot esculenta	Mazatan	–26.8		modern

APPENDIX 2 Animal Data for the Soconusco Region

Family	Spanish common name	English common name	Scientific name	Sample provenience	∂^{13}C(‰)	>∂^{15}N(‰)	Modern/ archaeological	Habitat
bird	pato aguja	anhinga, snakebird	*Anhinga anhinga*	Aquiles Serdan, 1B/N8	−23.1		archaeological	mixed
bird		unknown	unidentified		−22.8		archaeological	unknown
bird	cigüeña, tántalo americano	wood stork	*Mycteria americana*	Pampa Cantilena	−21.3	9.3	modern	estuary
bird	gallareta	american coot	*Fulica americana*	Pampa Cantilena	−18.7	7.0	modern	mixed
bird	hocofaisán	great curassow	*Crax rubra*	Tapachula	−19.3	8.5	modern	mixed
fish	bagre	catfish	unidentified	Aquiles Serdan, 1B/10	−22.9		archaeological	estuary
fish	bagre	catfish	unidentified	Aquiles Serdan, 1B/6	−18.7	6.9	archaeological	estuary
fish	mojarra	mojarra	*Cichlasoma* sp.	Aquiles Serdan, 1B/6	−22.4		archaeological	estuary
fish	mojarra	mojarra	*Cichlasoma* sp.	Aquiles Serdan, 1B/6	−23.6		archaeological	estuary
fish	mojarra	mojarra	*Cichlasoma* sp.		−22.4		archaeological	estuary
fish	peje lagarto	alligator gar	*Atractosteus spatula*	Aquiles Serdan, 1B/6	−17.9		archaeological	estuary
fish	peje lagarto	alligator gar	*Atractosteus spatula*	Aquiles Serdan, 1B/6	−22.7		archaeological	estuary
fish	peje lagarto	alligator gar	*Atractosteus spatula*	Aquiles Serdan, 1B/6	−23.3		archaeological	estuary
fish	peje lagarto	alligator gar	*Atractosteus spatula*	Aquiles Serdan, 1B/6	−17.7	6.0	archaeological	estuary
fish	peje lagarto	alligator gar	*Atractosteus spatula*	Aquiles Serdan, 1B/6	−20.1		archaeological	estuary
fish	peje lagarto	alligator gar	*Atractosteus spatula*	Aquiles Serdan, 1B/N8	−19.3	3.3	archaeological	estuary
fish	peje lagarto	alligator gar	*Atractosteus spatula*		−19.4	−1.0	archaeological	estuary
fish	peje lagarto	alligator gar (scales)	*Atractosteus spatula*	Aquiles Serdan, 1B/6	−20.9	2.8	archaeological	estuary
fish	peje lagarto	alligator gar (vertebrae)	*Atractosteus spatula*	Aquiles Serdan, 1B/6	−20.8	0.7	archaeological	estuary
fish	pescado	fish	unidentified	Aquiles Serdan, 1B/6	−21.6		archaeological	estuary
fish	pescado	fish	unidentified		−18.8	1.6	archaeological	estuary
fish	bagre chijun	marine catfish	*Ariopsis* sp.	Barra San Jose	−25.5	5.9	modern	estuary
fish	bagre del mar	marine catfish	*Ariopsis* sp.	Puerto Madero	−14.2	12.9	modern	estuary
fish	bagre del mar	marine catfish	*Ariopsis* sp.	Barra San Jose	−24.4	6.6	modern	estuary
fish	bagre/tacazonte	marine catfish	*Ariopsis* sp.	Pampa Cantilena	−14.1	17.7	modern	estuary
fish	bagre/tacazonte	marine catfish	*Ariopsis* sp.	Barra San Jose	−13.4	15.7	modern	estuary
fish	barbudo	yellow bobo, threadfin	*Polydactylus opercularis*	Barra San Jose	−14.3	12.9	modern	estuary
fish	besugo	elongate grunt	*Pomadasys elongatus*	Barra San Jose	−18.8	10.5	modern	estuary
fish	camiseta	Panama spadefish	*Parapsettus panamensis*	Barra San Jose	−15.0	15.4	modern	estuary
fish	mojarra	mojarra	*Cichlasoma* sp.	Barra San Jose	−14.0	13.5	modern	estuary
fish	mojarra blanca	Peruvian mojarra	*Diapterus peruvianus*	Barra San Jose	−16.4	8.8	modern	estuary

(Continued)

APPENDIX 2 *(continued)*

Family	Spanish common name	English common name	Scientific name	Sample provenience	$\partial^{13}C$(‰)	>$\partial^{15}N$(‰)	Modern/ archaeological	Habitat
fish	mojarra negra	mojarra	*Chiclasoma* sp.	Barra San Jose	–26.7	6.7	modern	estuary
fish	peje gallo robato??	unidentified	unidentified	Barra San Jose	–22.3	10.4	modern	estuary
fish	vieja	spotted sleeper	*Eleotris picta*	Puerto Madero	–28.2	4.9	modern	estuary
fish	calamar dardo	dart squid	*Loliolopsis diomedeae*	Puerto Madero	–16.6	11.9	modern	marine
fish	calandria	steeplined drum	*Larimus acclivis*	Puerto Madero	–15.7		modern	marine
fish	chacalin, camaroncillo	Pacific seabob	*Xiphopenaeus riveti*	Puerto Madero	–13.9	10.2	modern	marine
fish	chivo	goatfish	*Pseudupeneus squamipinnis*	Barra San Jose	–14.1	16.0	modern	marine
fish	cocinero	bigeye scad	*Selar crumenophthalmus*	Barra San Jose	–16.1	12.0	modern	marine
fish	corvina	silver weakfish	*Isopisthus remifer*	Barra San Jose	–14.5	14.9	modern	marine
fish	huachinango	Pacific red snapper	*Lutjanus peru*	Puerto Madero	–28.7	8.7	modern	marine
fish	jurel	Pacific crevalle jack	*Caranx caninus*	Barra San Jose	–14.8	15.2	modern	marine
fish	liza	white mullet	*Mugil curema*	Paredon	–12.4	10.3	modern	marine
fish	liza	white mullet	*Mugil curema*	Barra San Jose	–20.9	4.9	modern	marine
fish	mojarra	mojarra	*Cichlasoma* sp.	Rio Angostura??	–13.3	22.7	modern	marine
fish	pampano (jurel verde)	green jack	*Caranx caballus*	Barra San Jose	–16.3	11.8	modern	marine
fish	peje caite	unidentified	unidentified	Barra San Jose	–14.6	13.4	modern	marine
fish	peje gallo	elephantfish	*Callorhinchus callorynchus*	Paredon	–14.2	16.7	modern	marine
fish	pelona	Leatherjacket	*Oligoplites saurus*	Paredon	–13.0	10.0	modern	marine
fish	pelona	Leatherjacket	*Oligoplites saurus*	Barra San Jose	–13.5	14.8	modern	marine
fish	pez vela	sailfish	*Istiophorus platypterus*	Puerto Madero	–15.4	13.2	modern	marine
fish	pulpo	octopus	*Octopus* sp.	Rio Coatan	–13.7	9.7	modern	marine
fish	raton del mar	Goode croaker	*Paralonchurus goodei*	Barra San Jose	–14.2	15.5	modern	marine
fish	raya gavilán	spotted eagle ray	*Aetobatus narinari*	Paredon	–12.0	18.5	modern	marine
fish	raya gavilán	spotted eagle ray	*Aetobatus narinari*	Barra San Jose	–15.8	13.0	modern	marine
fish	robalo	black snook	*Centropomus nigrescens*	Gulf Coast	–24.9	12.3	modern	marine
fish	sambuco de mar	inshore sand perch	*Diplectrum cf. pacificum*	Barra San Jose	–15.0	13.7	modern	marine
fish	bagre	river catfish	unidentified	Rio Angostura	–29.4	18.0	modern	river
fish	chacalin, camaroncillo	Pacific seabob	*Xiphopenaeus riveti*	Rio Coatan	–16.8	8.7	modern	river
fish	sardina	Pacific piquitinga	*Lile stolifera*	Rio Coatan	–19.2	6.8	modern	river
fish water	chato	Pacific smalleye croaker	*Nebris occidentalis*	Barra San Jose	–14.0	13.9	modern	salt
fish water	chavelita	yellowfin jack	*Hemicaranx leucurus*	Barra San Jose	–14.6	13.9	modern	salt
mammal	ardillón	pocket gopher	*Orthogeomys cuniculus*	Aquiles Serdan, 1B/6	–21.1		archaeological	terrestrial
mammal	ardillón	pocket gopher	*Orthogeomys cuniculus*	Aquiles Serdan, 1B/6	–23.9		archaeological	terrestrial
mammal	ardillón	pocket gopher	*Orthogeomys cuniculus*	Aquiles Serdan, 1B/N8	–23.1		archaeological	terrestrial

mammal	armadillo	armadillo	*Dasypus novencinctus*	Aquiles Serdan, 1B/10	–21.6		archaeological	terrestrial
mammal	armadillo	armadillo	*Dasypus novencinctus*	Aquiles Serdan, 1B/6	–20.7		archaeological	terrestrial
mammal	armadillo	armadillo	*Dasypus novencinctus*	Aquiles Serdan, 1B/N8	–22.6		archaeological	terrestrial
mammal	conejo	cottontail rabbit	*Sylvilagus* sp.	Aquiles Serdan, 1B/6	–20.0		archaeological	terrestrial
mammal	conejo	cottontail rabbit	*Sylvilagus* sp.	Aquiles Serdan, 1B/N8	–22.3		archaeological	terrestrial
mammal	jabali	collared peccary	*Tayassu tajacu*	Aquiles Serdan, 1B/6	–21.9		archaeological	terrestrial
mammal	perro	dog	*Canis familiaris*	Aquiles Serdan, 1B/6	–19.6	3.9	archaeological	terrestrial
mammal	perro	dog	*Canis familiaris*	Aquiles Serdan, 1B/N8	–19.7		archaeological	terrestrial
mammal	tlacuache	opossum	*Didelphis marsupialis*	Aquiles Serdan, 1B/6	–20.8	–2.8	archaeological	terrestrial
mammal	venado	white tail deer	*Odocoileus virginianus*	Aquiles Serdan, 1B/6	–20.1	6.4	archaeological	terrestrial
mammal	venado	white tail deer	*Odocoileus virginianus*	Aquiles Serdan, 1B/N8	–19.0	1.5	archaeological	terrestrial
mammal	nutria	nutria	*Myocastor coypus*	Pampa Cantilena	–25.9	8.0	modern	estuary
mammal	nutria	nutria	*Myocastor coypus*	Pampa Cantilena	–26.7	10.7	modern	estuary
mammal	borrego	sheep	*Ovis aries*	Tapachula	–16.4	5.9	modern	terrestrial
mammal	borrego	sheep	*Ovis aries*	Tapachula	–16.8	8.8	modern	terrestrial
mammal	conejo	cottontail rabbit	*Sylvilagus* sp.	Paso de la Amada	–20.8	4.3	modern	terrestrial
mammal	perro	dog	*Canis familiaris*	Paredon	–11.7	10.1	modern	terrestrial
mammal	perro	dog	*Canis familiaris*	Buenos Aires	–14.0	8.5	modern	terrestrial
mammal	tlacuache	opossum	*Didelphis marsupialis*	Buenos Aires	–19.9	8.9	modern	terrestrial
mammal	venado	white tail deer	*Odocoileus virginianus*	Pampa Cantilena	–25.1	5.3	modern	terrestrial
mammal	venado	white tail deer	*Odocoileus virginianus*	Pampa Cantilena	–24.1	10.7	modern	terrestrial
mammal	venado	white tail deer	*Odocoileus virginianus*	Buenos Aires	–22.2	4.5	modern	terrestrial
reptile	cocodrilo	crocodile	*Crocodylus acutus*	Aquiles Serdan, 1B/6	–18.8	6.8	archaeological	estuary
reptile	cocodrilo	crocodile	*Crocodylus acutus*	Aquiles Serdan, 1B/6	–20.1	5.4	archaeological	estuary
reptile	cocodrilo	crocodile	*Crocodylus acutus*	Aquiles Serdan, 1B/6	–20.5	–3.4	archaeological	estuary
reptile	cocodrilo	crocodile	*Crocodylus acutus*		–20.0	0.1	archaeological	estuary
reptile	masacuate	boa snake	*Boa constrictor*	Aquiles Serdan	–19.7		archaeological	estuary
reptile	masacuate	boa snake	*Boa constrictor*	Aquiles Serdan, 1B/6	–21.2		archaeological	estuary
reptile	masacuate	boa snake	*Boa constrictor*	Aquiles Serdan, 1B/6	–21.6		archaeological	estuary
reptile	tortuga	pond slider turtle	*Trachemys* sp.	Aquiles Serdan, 1B/6	–20.8	1.4	archaeological	estuary
reptile	tortuga	pond slider turtle	*Trachemys* sp.	Aquiles Serdan, 1B/6	–19.5		archaeological	estuary
reptile	tortuga casquito	mud turtle	*Kinosternon* sp.	Aquiles Serdan, 1B/6	–19.8	6.8	archaeological	estuary
reptile	tortuga casquito	mud turtle	*Kinosternon* sp.		–20.2	2.0	archaeological	estuary
reptile	tortuga casquito	mud turtle	*Kinosternon* sp.		–19.9		archaeological	estuary
reptile	tortuga casquito	mud turtle	*Kinosternon* sp.		–21.1		archaeological	estuary
reptile	iguana	iguana	*Iguana iguana*	Aquiles Serdan, 1B/6	–23.1		archaeological	terrestrial
reptile	iguana	iguana	*Iguana iguana*		–22.1	–0.3	archaeological	terrestrial
reptile	vibora	unidentified	unidentified		–21.9		archaeological	terrestrial

(Continued)

APPENDIX 2 *(continued)*

Family	Spanish common name	English common name	Scientific name	Sample provenience	$\partial^{13}C$(‰)	>$\partial^{15}N$(‰)	Modern/ archaeological	Habitat
reptile	cocodrilo	crocodile	*Crocodylus acutus*	Pampa Cantilena	−22.9	4.2	modern	estuary
reptile	cocodrilo	crocodile	*Crocodylus acutus*	Vivero	−23.7	3.3	modern	estuary
reptile	cocodrilo	crocodile	*Crocodylus acutus*	Tapachula	−26.9	6.5	modern	estuary
reptile	masacuate	boa snake	*Boa constrictor*	Buenos Aires	−19.1	7.9	modern	estuary
reptile	masacuate	boa snake	*Boa constrictor*	Tapachula	−23.4	9.3	modern	estuary
reptile	masacuate	boa constrictor	*Boa constrictor*	Tapachula	−16.5	11.5	modern	estuary
reptile	tortuga	pond slider turtle	*Trachemys* sp.	Pampa Cantilena	−26.6	5.7	modern	estuary
reptile	tortuga	pond slider turtle	*Trachemys* sp.	Mazatan	−27.7	4.9	modern	estuary
reptile	tortuga casquito	turtle		Pampa Cantilena	−23.2	5.3	modern	estuary
reptile	tortuga casquito	mud turtle	*Kinosternon* sp.	San Lorenzo	−25.9	5.9	modern	estuary
reptile	tortuga casquito	mud turtle	*Kinosternon* sp.	Paredon	−16.9	6.4	modern	estuary
reptile	parlama	green sea turtle	*Chelonia mydas*	Barra	−14.7	12.6	modern	marine
reptile	tortuga	pond slider turtle	*Trachemys* sp.	San Lorenzo	−26.4	9.9	modern	river
reptile	cascabel	rattle snake	*Crotalus durissus* ssp.	Tapachula	−22.0	10.2	modern	terrestrial
reptile	cascabel	rattle snake	*Crotalus durissus* ssp.	Tapachula	−20.1	7.4	modern	terrestrial
reptile	iguana	iguana	*Iguana iguana*	Mazatan	−25.7	2.3	modern	terrestrial
reptile	iguana	iguana	*Iguana iguana*	Tapachula	−24.4	4.3	modern	terrestrial
reptile	iguana	iguana	*Iguana iguana*	Soconusco	−21.6	14.9	modern	terrestrial
reptile	vibora/sumbadora	unidentified	unidentified	Tapachula	−17.8	9.3	modern	terrestrial
shellfish	concha	clam shell	unidentified	Aquiles Serdan, 1B/N8	−20.8	−1.8	archaeological	estuary
shellfish	camarone	Shrimp	unidentified	Barra San Jose	−25.4	6.3	modern	estuary
shellfish	camarone	shrimp	unidentified	Barra San Jose	−20.3	5.9	modern	estuary
shellfish	camarone	shrimp	unidentified	Barra San Jose	−25.9	4.8	modern	estuary
shellfish	cangrejo	crab	unidentified	Barra San Jose	−22.1	5.6	modern	estuary
shellfish	cangrejo	crab	unidentified	Barra San Jose	−23.9	6.0	modern	estuary
shellfish	camarone	shrimp	unidentified	Paredon	−11.7	2.2	modern	marine
shellfish	camarone	shrimp	unidentified	Tonala	−17.9	2.1	modern	marine
shellfish	camarone grande	giant shrimp	unidentified	Puerto Madero	−14.4	10.7	modern	marine
shellfish	camarone tigre	tiger shrimp	unidentified	Puerto Madero	−14.2	12.2	modern	marine
shellfish	mejillón	mussel	*Mytelis* sp.	Puerto Madero	−16.0	8.2	modern	marine
shellfish	camarone	shrimp	unidentified	Rio Coatan	−15.7	10.5	modern	river
shellfish	cangrejo	crab	unidentified	Rio Angostura	−17.2	3.3	modern	river

CHAPTER

15

Stable Carbon Isotope Analysis and Human Diet in Mesoamerica: A Synthesis

HENRY P. SCHWARCZ

School of Geography and Earth Sciences, McMaster University, Hamilton, Ontario, Canada

INTRODUCTION

The isotope chapters presented in this section are all focused upon the use of stable isotope analyses to characterize and quantify the use of maize in human diet in the Americas. These chapters exploit the fact that maize utilizes the Hatch-Slack, or C_4 photosynthetic pathway, whereas the majority of wild plants utilized either as cultigens or consumed by herbivorous sources of protein utilize the Calvin-Benson, or C_3 pathway. This distinction results in a large difference in the $^{13}C/^{12}C$ ratio in the products of photosynthesis in these plants as well as all other biochemical species in them (including proteins). Specifically, the average $\delta^{13}C$ of C_3 plants is around -26‰ whereas C_4 plants today have $\delta^{13}C$ values averaging around -12‰. Both of these would have been shifted to values about 1.5‰ more positive in the pre-industrial era before the beginning of large-scale burning of fossil fuel. Modern mass spectrometry can easily detect differences in $\delta^{13}C$ of 0.1‰ although in most cases larger differences between populations (c. 1‰) would be needed to establish significant differences in use of maize in diet, etc.

As a result of these isotopic differences, specialists are able to identify the presence of maize in various contexts in the archaeological record. In most of these papers, isotopic analyses of human tissues and especially bone, are reported and comparisons made between corresponding analyses of humans who subsisted on C_3- based nutrients (plants and herbivore flesh). In addition, researchers have shown how ^{13}C-enrichment in residues extracted from ceramic sherds can be used to detect maize-based foods that were cooked or stored in the complete vessel. A third approach is the detection of C_4 -based humic matter in ancient soils indicative of the cultivation of maize [25, 26].

Isotopic analyses of human tissues where maize was a potential nutrient opens a number of vantage points from which to discuss these data:

a) Historical: As is discussed elsewhere in this volume, maize originated as a cultigen in one place and time, somewhere in the Valley of Mexico (Chapter 2). It was subsequently dispersed to virtually all the regions of the Americas where climate and soil conditions permitted its cultivation. A number of the papers focus on specific regions and use radiocarbon or other dates to calibrate the rate and extent of adoption of maize as a cultigen.
b) Geographic: complementary to the historic viewpoint, authors have also considered the geography of the dispersal process, observing its spread from Mexico throughout the hemisphere and, in some cases, on a more local scale.
c) Status: In a few of these papers authors have attempted to define the differential use of maize as a nutrient by humans according to demographic parameters: age, sex and social status. This addresses the general question whether maize was considered a luxury food, a staple, or a low-status item.

THEORETICAL BASIS OF THE USE OF ISOTOPES

The use of $\delta^{13}C$ as an indicator of the presence of maize in human diet is predicated on the basic assumption [8] that "you are what you eat" plus an isotopic offset (fractionation). In fact, human tissues as preserved in archaeological sites offer a number of different opportunities to glimpse the isotopic signature imparted from the diet. Ancient bone preserves at least two important molecular species whose isotopic composition can be used for this purpose: collagen, the most abundant protein in living bone; and the carbonate (CO_3) molecule which forms part of the bone mineral, hydroxyapatite. In addition, ancient bone contains lesser amounts of other molecules whose $\delta^{13}C$ values are inherited from the diet: a) cholesterol [12]; b) lipids and non-collagenous proteins such as osteocalcin [1]. No data for any of these minor species are presented in these papers.

Collagen, which is the bone component most widely used in paleodiet studies, is a protein whose component amino acids are all present in the diet. In 1984, Krueger and Sullivan [20] suggested that the amino acids from the diet were selectively "routed" to the sites of protein building in humans; therefore the $\delta^{13}C$ value of collagen ($\delta^{13}C_{col}$) should more closely resemble that of the protein component of the diet. Ambrose and Norr [3] confirmed this model, using rats fed C_3 -and C_4 - labeled protein and non-protein fodder constituents, finding that a large fraction of the amino acids (c. 80%) incorporated into body protein was derived from amino acids in the diet. This concept has been widely accepted by researchers using $\delta^{13}C$ data to reconstruct diet, and is frequently cited [13].

On the other hand, the $\delta^{13}C$ value of CO_3 in hydroxyapatite (HA) of bone ($\delta^{13}C_{ap}$) is believed to represent the total $\delta^{13}C$ of the diet, that is, all the C atoms that are consumed and contribute to caloric value [20].

The fractionation between $\delta^{13}C_{col}$ and diet, $\Delta_{c\text{-}d} = \delta^{13}C_{col} - \delta^{13}C_{diet}$ is estimated to be about 5‰ for animals and humans but is not well-known in humans due to the obvious difficulty of obtaining collagen samples from humans with known diets. The corresponding fractionation between apatite-CO_3 and diet, $\Delta_{ap\text{-}d}$, is less certain. Passey and colleagues [18] have recently shown that this fractionation varies widely amongst animals, roughly increasing with body mass. As noted, an accurate value for humans has yet to be determined, but probably lies in the range from 9 to 11‰. Knowledge of these fractionations is essential for the quantitative determination of the proportion of maize in diet, but relative comparisons can be made without knowledge of these Δ values. It is important, however, that authors state what values are being assumed in dietary reconstructions.

Many of these papers also make use of analyses of $\delta^{15}N$ of collagen, for comparison with, and cross-plotting of $\delta^{13}C_{col}$ data. $\delta^{15}N$, representing the $^{15}N/^{14}N$ ratio, varies principally as a function of the trophic level of the food source, increasing by approximately 3‰ between trophic levels. Use of $\delta^{15}N$ as an additional variable was especially important when analyzing data from sites where marine foods were also available [see Chapters 12, 13] because fish and marine mammals, like maize, are also enriched in ^{13}C. Though the enrichment is only half as great as for maize (c. 7‰ *vs* 12‰) the effect on $\delta^{13}C$ of collagen is stronger because fish, seals, etc., are sources of protein. We should, however, be able to recognize increases in $\delta^{13}C_{col}$ from this source because they would be correlated with increase in $\delta^{15}N$ [9. 16].

THE SIGNIFICANCE OF ISOTOPES IN RECONSTRUCTION OF PALEODIET IN THE AMERICAS

Some chapters in this volume have emphasized the special role that maize plays in Mesoamerican prehistory. As in the Old World, we see an initial population of hunters and gatherers gradually beginning to exploit diverse cultigens, starting at dates ranging from several thousand years before the present (in Mexico) to a few hundred years ago (in eastern N. America). Maize appears to have been the most widely adopted species, and is certainly distributed more broadly across the hemisphere than any other food plant. For archaeologists interested in human subsistence, maize and its $\delta^{13}C$ signal serves as a "signal light" that marks, if not the onset, then at least the expansion of agriculture. This is because essentially all other foods, and especially wild plants would have carried the low-$\delta^{13}C$ signal characteristic of C_3 plants. Maize was in many ways a less satisfactory crop than other previously used cultigens at sites where maize eventually moved in. Its poorer nutrient qualities and greater difficulty of cultivation were more than compensated for by the large yield and ease of storage.

It is interesting to consider what we might expect to find in a study of variation in $\delta^{13}C$ of collagen and apatite in historic populations across the Americas. If we are studying sites far removed in time and space from the source of maize agriculture, we would expect that, prior to some date, all skeletal materials would give $\delta^{13}C$ signals indicative of pre-agricultural hunter-gatherers exploiting C_3 wild plants and the flesh of C_3-based herbivores. In some hotter, drier areas we could expect to see some enrichment in $\delta^{13}C$ of collagen because herbivores are feeding on grasses that include many C_4 species. However, this would only lead to a limited increase in $\delta^{13}C_{col}$ (2-5‰). In open-air sites (as opposed to rock shelters and caves) it is likely that the occupants would take up agriculture after some date. If the onset of agriculture is rapid and is marked by the cultivation of maize, then we would expect to see an abrupt increase in $\delta^{13}C_{col}$, whereas it is also possible that maize arrives on the scene in gradual stages, slowly replacing either wild or cultivated C_3 plant-foods. In either case we would not expect to see an increase in $\delta^{13}C$ of the animal-derived protein. Dependence on protein from C_3-based animals has the effect of suppressing the rate of increase of $\delta^{13}C_{col}$ due to routing of dietary amino acids.

Note however [20] that about 80 % of the C atoms in collagen occur in non-essential amino acids (neAA's). The human body has the enzymes needed to synthesize these amino acids from almost any nutrient that we consume. Therefore, these internally synthesized AA's may have $\delta^{13}C$ values similar to that of the total diet, plus the characteristic fractionation between AA's and diet [7]. The question as to whether a particular neAA is or is not synthesized internally is determined by the level of that AA in the blood plasma and therefore by the amount of protein in the diet. Therefore we expect that when humans are living on a relatively protein-poor diet, a significant portion of their neAA's will be "made at home" from whatever else they are eating including carbohydrates and fats. These include, of course, the constituents of maize.

With the exception of coastal populations we expect that the principal sources of protein for Mesoamerican populations will be "flavored" C_3 (but see below for an important exception). This includes the flesh of herbivorous animals, aquatic fish, and beans. When their diet is relatively rich in these proteins, the $\delta^{13}C$ of collagen will be relatively low (strongly negative) even though maize is a component of the total diet. However, in populations in which protein is less readily available, the human body will tend to use all foods, including maize, in order to synthesize the non-essential amino acids needed for collagen. Therefore these people will have relatively high $\delta^{13}C_{col}$ values. We see such values commonly in many of the populations reported in this section.

Because $\delta^{13}C$ of bone and tooth apatite represents the totality of the calorie-contributing carbon atoms in the diet (that is, excluding fiber), this value will directly reflect the presence and amount of maize in the diet. There is, however, some uncertainty as to the relationship between $\delta^{13}C_{ap}$ and $\delta^{13}C$ of total diet since the fractionation between them ($\Delta_{ap\text{-}d}$) is known to vary greatly in animals [18] and may even vary in humans as a function of the quality of the diet. The authors in this section who present values for $\delta^{13}C_{ap}$ have generally assumed $\Delta_{ap\text{-}d} \approx$ 10‰.

The difference between $\delta^{13}C_{ap}$ and $\delta^{13}C_{col}$ ($\Delta_{ap\text{-}col}$) has been used in the isotopic paleodiet literature as a measure of trophic level [11] where it is found that $\Delta_{ap\text{-}col}$ increases with decreasing trophic level, that is, with decreasing proportion of animal protein in the diet. We see here, however, that exactly the opposite would be expected. As the proportion of (C_3 –based) protein-rich foods increases, there would be a decrease in the fraction of amino acids synthesized internally from the total diet including maize. Therefore $\delta^{13}C_{col}$ would become more negative. Consider two individuals who both consumed the same percentage of maize as a fraction of their total caloric intake, but one of whom consumed twice as much meat as the other. The person with the meat-rich diet would display a larger $\Delta_{ap\text{-}col}$ value. Data on $\Delta_{ap\text{-}col}$ are given in [given in Chapter 12] but are not interpreted as indicators of trophic level.

An interesting and possibly important qualification to this discussion is the possibility that some of the fauna being harvested as sources of protein could have themselves been feeding on maize. For example, deer living near maize fields could be eating crops. However, samples of wild fauna from archaeological sites generally do show evidence of maize consumption. For example, deer living near Maya sites in Belize showed no ^{13}C enrichment [27]. Although domestic animals as sources of protein were rare in Mesoamerica, the turkey (*Meleagris gallopavo* L.) was domesticated in the American Southwest and possibly in Mesoamerica, and may have been maize-fed. However, faunal samples from these sites are always dominated by wild species whose $\delta^{13}C$ values show them to be feeding exclusively on C_3 plants.

RATE OF SPREAD OF MAIZ AND AGRICULTURE

In many of the areas of Mesoamerica studied here see maize as arriving as a cultigen through the history of a site or region. This raises the possibility that we could study the rate at which this new cultigen was adopted, by tracking the rise through time in $\delta^{13}C$ of human tissues. In effect, this allows us to construct models for the rate of dispersal of an agricultural technology and, in some cases, the spread of agriculture itself. This is, of course, a question that has long interested archaeologists in the Old World [4] where, unfortunately, no C_4 plants (with the exception of millet) were important cultigens. The American continents on the other hand provide the possibility of tracing the rate of agricultural dispersal. Some papers in this volume address this issue, by showing plots of $\delta^{13}C_{col}$ *vs* radiocarbon age. With some exceptions, these show that dispersal did not follow a simple pattern of gradual outward diffusion from a source. For example, in the Ohio Valley, the extent of dependence on maize as a nutrient increases abruptly over a few generations, whereas in S. Ontario we appear to see a gradual rise in $\delta^{13}C_{col}$ over a period of about 600 years. The causes of these differences in rate of dispersal are poorly understood, but presumably include issues such as changes in population density, invention of new agricultural or food-processing technologies, and rate of genetic adaptation of maize to regions with shorter growing seasons.

MESOAMERICA

In contrast to most other regions of the Americas, ancient Mesoamerica is dominated by the ruins of complex societies. Anthropological studies of human remains from this region allow us to study in much more detail the relationship between diet and the position of the individual in society. Isotopic studies of this region began in Belize (with studies at Lamanai [28]) and have now extended to more than twenty sites in Belize, Guatemala, Mexico, Honduras and El Salvador.

Maize is already an important food source in early Preclassic Maya, as is demonstrated by isotopic studies at Cuello [23]. White *et al.* (see Chapter 12) note that, although maize is clearly an essential food for all the Maya, data from many sites show that maize preferentially consumed by elite individuals. In some cases, by males more than by females. The absolute levels of maize consumption are clearly very high, with $\delta^{13}C_{col}$ ranging up to -9‰ at Pacbitun and -7‰ at San Pedro (which may be influenced by consumption of marine foods). These high levels of maize consumption do not appear to have compromised the health of the Maya. High $\delta^{13}C_{col}$ values might in some cases be associated with consumption of C_4 -fed animals. While dogs have clearly been maize-fed (possibly for ritual purposes) other fauna appear to have fed only on C_3 plants. At near-coastal sites (Cahal Pech , etc.) and even at some inland sites consumption of reef-based fish and other marine foods are believed to contribute to higher $\delta^{13}C_{col}$ values, associated with higher $\delta^{15}N$.

While most Maya studies have concentrated on the highlands and southern lowlands, the many famous sites of the northern lowlands of Yucatan have remained largely uninvestigated isotopically. At two Yucatan sites, one coastal, one in the interior, Mansell *et al.* (Chapter 13) encounter very poor sample preservation: most of their samples lack analyzable collagen. Bone and tooth apatite is apparently unaffected by diagenesis. At the inland site $\delta^{13}C_{ap}$ values of -3 ± 1‰ correspond to diets of c. 75% maize. At the near-coast site of Chunchucmil, $\delta^{13}C_{ap}$ is slightly lower (-5 ± 2‰). The apatite-collagen difference is the same at both sites (9‰). Their figure 11 shows that both $\delta^{13}C_{col}$ and $\delta^{15}N$ values for both sites are strikingly lower than have been reported for most sites in Belize and Guatemala, whereas $\delta^{13}C_{ap}$ is very high. While lower $\delta^{13}C_{col}$ suggests higher levels of protein in the diet, the ^{13}C-enriched apatite shows that maize was an important source of calories. The significance of the lower $\delta^{15}N$ values observed at these sites is unclear. Values for the coastal site (Chunchucmil) are low compared to expectations for consumers of reef-based marine resources. This region is relatively arid, which might also be expected to lead to higher $\delta^{15}N$ values. The authors suggest the possible role of bean-consumption in this region to account for low $\delta^{15}N$.

One surprising aspect of the Maya agricultural economy was their ability to provide sufficient maize to support the populations of their large urban centers. For the modern Maya, maize cultivation is principally carried out by milpa agriculture in which small fields are established in a forested region, exploited for a few years until the soil is exhausted, and then abandoned. By contrast, the ancient Maya are known to have cultivated maize in large fields in bottom-lands or on terraced slopes of the highlands. Isotopic evidence for the existence of these fields and their geographic extent is beginning to emerge from studies of the humic component of their soils [25, 26]. Organic matter derived from the roots and residual plant materials left in the fields after harvesting of maize crops is enriched in ^{13}C relative to the native forest vegetation in these areas . Studies at sites in Belize and Guatemala show that humic matter enriched in ^{13}C occurs only at depths c. 60 cm in the soil. This is because, after Maya farmers abandoned the soils, wild C_3 vegetation invaded the fields, and bioturbation has presumably translocated this later ^{13}C-depleted humic component downwards. Regardless, the presence of humic substances that are enriched in ^{13}C can indicate that the soil was formerly used for maize agriculture, and can be used to map out the locations of maize-growing fields.

CONCLUSIONS

When it was initially demonstrated that the $^{13}C/^{12}C$ ratios of an animal's diet was reflected in the isotopic composition of body tissues [8. 24], they presented anthropologists and archaeologists with a tool that could be used in many ways to analyze human dietary behavior. The first application of this concept by van der Merwe and Vogel [24] was, in fact, a study of the introduction of maize into aboriginal eastern North America. Since then humans who ate maize to varying extent have been one of the prime targets of students of isotopic paleodiet, largely because of the huge difference in $\delta^{13}C$ between C_3 and C_4 foods. The papers in this issue are a good sample of the possible range of applications that could be made using relatively standard isotopic procedures.

Today, however, we begin to see prospects for finer-scale studies that also exploit these isotopic differences. Notable amongst these is the analysis of single amino acids extracted from whole collagen [7,10]. Through analysis of specific essential and non-essential amino acids, we can better test the extent to which an individual has depended on ingested proteins to supply the non-essential amino acids need in the synthesis of collagen.

Other biochemical species besides collagen are potential targets for analysis. For example, serial samples of human hair can show seasonal variation in proportion of C_3 and C_4 foods [21]. In subtropical areas where maize alternates with other C_3 cultigens, we might expect to see such effects. Besides collagen and keratin (hair, nails), some other molecules survive burial including cholesterol and some non-collagenous proteins in bone such as osteocalcin [1, 12]. All of these have the property that their turnover rate is probably faster than collagen, and they would therefore record diet closer to time of death.

In general, there are good prospects for expanded and more detailed analyses of carbon isotopes in human tissues, but these must be set against the increasing sensitivity of native groups in North America and possibly in other regions to the scientific analysis of the remains of their ancestors. This issue has already significantly impeded this line of research in North America, and has forced isotopic archaeologists to look to other types of samples. For example, dogs have been found to provide an excellent proxy to human diet at N. American and Mesoamerican sites [5, 13]. Likewise, the analysis of coprolites, and of food residues (both charred and adsorbed) on and in ceramics remains a possible avenue for wider investigation [6, 17].

References

1. H. O. Ajie, I. R. Kaplan, P. V. Hauschka, D. Kirner, P. J. Slota, R. E. Taylor. (1992). Radiocarbon dating of bone osteocalcin-isolating and characterizing a non-collagen protein. *Radiocarbon, 34,* 296–305.
2. S. H. Ambrose, J. Buikstra, H. W. Krueger. (2003). Status and gender differences in diet at Mound 72, Cahokia, revealed by isotopic analysis of bone. *Journal of Anthropological Archaeology, 22,* 217–226.
3. S. H. Ambrose, L. Norr. (1993). Experimental evidence for the relation of the carbon isotope ratios of whole diet and dietary protein to those of collagen and carbonate. In: J. B. Lambert, G. Grupe (Eds.), *Prehistoric human bone: Archaeology at the molecular level.* Berlin: Springer-Verlag. pp. 1–37.
4. P. Bellwood, C. Renfrew, (Eds.). (2002). *Examining the farming language dispersal hypothesis.* Cambridge, MA: McDonald Institute for Archaeological Research, University of Cambridge.
5. A. Cannon, H. P. Schwarcz, M. Knyf. (1998). Marine-based subsistence trends and the stable-isotope analysis of dog bones fromNamu, British Columbia. *Journal of Archaeological Science, 26,* 399–407.
6. M. S. Copley, R. Berstan, S. N. Dudd, R. P. Evershed. (2005). Dairying in antiquity. I. Evidence from absorbed lipid residues dating to the British Iron Age. *Journal of Archaeological Science, 32,* 485–503.
7. L. T. Corr, J. C. Sealy, M. C. Horton, R. P. Evershed. (2005). A novel marine dietary indicator utilising compound-specific bone collagen amino acid □13C values of ancient humans. *Journal of Archaeological Science, 32,* 321–330.
8. M. J. DeNiro, S. Epstein. (1978). Influence of diet on the distribution of carbon isotopes in animals. *Geochimica et Cosmochimica Acta, 42,* 495–506.
9. M. J. DeNiro, P. L. Walker. (1986). Stable nitrogen and carbon isotope ratios in bone collagen as indices of prehistoric dietary dependence on marine and terrestrial resources in southern California. *American Journal of Physical Anthropology, 71,* 51–61.
10. M. L. Fogel, N. Tuross. (2003). Extending the limits of paleodietary studies of humans with compound specific carbon isotope analysis of amino acids. *Journal of Archaeological Science, 30,* 535–545.
11. R. E. M. Hedges. (2003). On bone collagen-apatite-carbonate isotopic relationships. *International Journal of Osteoarchaeology, 13,* 66–79.
12. S. Jim, S. H. Ambrose, R. P. Evershed. (2004). Stable carbon isotopic evidence for differences in the dietary origin of bone cholesterol, collagen and apatite: Implications for their use in palaeodietary reconstruction. *Geochimica et Cosmochimica Acta, 68,* 61–72.
13. M. A. Katzenberg. (1989). Stable isotope analysis of archaeological faunal remains from southern Ontario. *Journal of Archaeological Science, 16,* 319–329.
14. H. W. Krueger, C. H. Sullivan. (1984). Models for carbon isotope fractionation between diet and bone. In: J. E. Turnlund, P. E. Johnson, (Eds.), *Stable isotopes in nutrition.* American Chemical Society Symposium Series 258. Washington, D.C.: American Chemical Society. pp.205–222.
15. C. S. Larsen, D. E. Harn. (1994). Health in transition: Disease and nutrition in the Georgia Bight. In: K. D. Sobolik, (Ed.), *Paleonutrition: The diet and health of prehistoric Americans.* Occasional Papers 22. Carbondale, IL: Center for Archaeological Investigations. pp. 222–234.
16. D. R. Lubell, M. Jackes, H. P. Schwarcz, M. Knyf. (1994). The Mesolithic-Neolithic transition in Portugal: Isotopic and dental evidence of diet. *Journal of Archaeological Science, 21,* 201–216.
17. J. Morton, H. P Schwarcz. (2004). Paleodietary implications from isotopic analysis of food residues on prehistoric Ontario ceramics. *Journal of Archaeological Science, 31,* 503–517.
18. B. H. Passey, T. F. Robinson, L. K. Ayliffe, T. E. Cerling, M. Sponheimer, M. D. Dearing, B. L. Roeder, J. R. Ehleringer. (2005). Carbon isotope fractionation between diet, breath CO2, and bioapatite in different mammals. *Journal of Archaeological Science, 32,* 1459–1470.
19. H. P. Schwarcz. (1991). Some theoretical aspects of isotope paleodiet studies. *Journal of Archaeological Science, 18,* 261–275.
20. H. P. Schwarcz. (2000). Some biochemical aspects of carbon isotopic paleodiet studies. In: S. Ambrose, M. A. Katzenberg, (Eds.), *Biogeochemical approaches to paleodietary analysis.* New York: Kluwer Academic Press. pp. 189–210.
21. H. P. Schwarcz, C. D. White. (2004). The grasshopper or the ant? Cultigen-use strategies in ancient Nubia from C-13 analyses of human hair. *Journal of Archaeological Science, 31,* 753–762.
22. B. D. Smith. (1995). *The emergence of agriculture.* New York: Scientific American Library.
23. R. H. Tykot, N. J. van der Merwe, N. Hammond. (1996). Stable isotope analysis of bone collagen, bone apatite, and tooth enamel in the reconstruction of human diet: Acase study for Cuello, Belize. In: M. V. Orna, (Ed.), *Archaeological Chemistry V.* Washington, D.C.: American Chemical Society. pp. 355–365.
24. J. C. Vogel, N. J. van der Merwe. (1977). Isotopic evidence for early maize cultivation in New York State. *American Antiquity, 42,* 238–242.
25. E. A. Webb, H. P. Schwarcz, P. F. Healy. (2004). Detection of ancient maize agriculture in the Maya lowlands using the stable carbon isotope compositions of soil organic matter: Evidence from Caracol, Belize. *Journal of Archaeological Science, 31,* 1039–1052.
26. E. A. Webb, H. P. Schwarcz, C. T. Jensen, R. E. Terry, M. Moriarty, K. Emery. (2005). Stable carbon isotope signature of ancient maize agriculture in the soils of Motul de San José, Guatemala. *Geoarchaeology.*
27. C. D. White, M. D. Pohl, H. P. Schwarcz, F. Longstaffe. (2004). Feast, field and forest: Deer and dog diets at Lagartero, Tikal and Copan. In: K. Emery, N. Stanchly, (Eds.), *Mesoamerican zooarcheology: New directions in method and theory.* Monograph 51. Los Angeles, CA: Cotsen Institute of Archaeology, UCLA Press. pp. 141–158.
28. C. D. White, H. P. Schwarcz. Ancient Maya diet as inferred from isotopic and chemical analyses of human bone. *Journal of Archaeological Science, 16,* 451–474.

PART IV

Histories of Maize:

The Language of Maize in Mesoamerica

CHAPTER

16

Thipaak and the Origins of Maize in Northern Mesoamerica

JANIS B. ALCORN*, BARBARA EDMONSON†, AND CÁNDIDO HERNÁNDEZ VIDALES‡

*The Garfield Foundation, Chevy Chase, Maryland

†Anthropology Department, Tulane University, New Orleans, Louisiana

‡Tamjajnec, S.L.P., Mexico

If we don't remember the history, the seeds will abandon us.

Pedro Kinim, 9 July 1979

Glossary

Ach Literally Grandmother or Elder Woman, refers to a particular Teenek female deity who also takes the form of an eagle—demanding human sacrifice.

Boliim A large (12- to 14-inch) ceremonial *tamale*.

Brosium alicastrum A tall tree with edible fruit, seed, and sap, known as *ramón* in Spanish.

Genetic drift Random changes in gene frequency that contribute to evolution, in addition to natural selection, mutation, and gene flow (migration); effects are greatest in small populations; basis of "founder effect."

Maam Literally Grandfather or Elder, but refers to Teenek male supreme beings, who are associated with lightning.

Mesoamerica A culturally defined zone that includes parts of Mexico and Central America covering the area of Pre-Hispanic stratified agricultural societies [17].

Milpa The Mesoamerican system of swidden agriculture that incorporates maize, beans, squash, and other cultigens; also refers to the maize field itself.

Nixtamal Maize kernels soaked in *warm* lye water to soften the outer shell for easy removal *prior to grinding for making tortillas*.

Nahua A Mexican indigenous people who speak a dialect of Nahuatl, the language spoken by the Aztec.

Ojox The Teenek word for *Brosimum alicastrum*.

Teenek The preferred name for the Maya living in the area of Mexico known as "the Huasteca," also refers to their language; the double vowel in Teenek words indicates that the sound is extended, not pronounced as double vowels in English.

Thipaak The Teenek culture hero who brought maize to people.

Maize is more than a grain to Mesoamericans—it is a powerful cultural icon that has held societies together around a shared set of beliefs, economy, annual ritual cycles, and work patterns while adapting over time from hunting and gathering transhumant condition up to the Conquest, and through subsequent years to modern times and globalization. Like scientific theories, indigenous theories about the origins of maize vary but share main themes. We focus on the maize origin theories of the **Teenek** (Huastec Maya) of San Luis Potosí and Veracruz, Mexico, and present unpublished field data from San Luis Potosí. In Pre-Hispanic times, Teenek lived in an ecologically diverse region that includes the Sierra Madre Oriental caves with the earliest known small maize ears with teosinte [21], as well as large temple complexes along the Gulf Coast. Teenek oral histories refer to the escapades of a culture hero, **Thipaak,** who brought maize to the people to challenge their earlier religion linked to gathering *Brosimum alicastrum*. The Teenek histories indicate maize evolution may have been influenced by maize's ritual value in the dynamic political and cultural context of diverse multiple strategy subsistence systems. Comparison with other Mesoamerican traditions supports this suggestion.

INTRODUCTION

Mesoamerica lies roughly between the Tropic of Cancer and southern Honduras. The fuzzy northern border of Mesoamerica passes through southern Tamaulipas across San Luis Potosí to the Pacific. Within this zone, maize cultivation has been adapted to an ecologically diverse region with Atlantic and Pacific coasts, deserts, rainforests, high plateaus, and rugged high mountain ranges.

Throughout Mesoamerica, maize continues to be respected as a symbol of cultural identity; even in downtown Mexico City one sees the occasional maize plant left to flower wherever it appears along a street. Maize-based foods are a central part of the diet, and maize plays a vital role in annual ritual cycles across Mesoamerica.[1]

In Pre-Hispanic times, Teenek lived along the Gulf Coast in the forest on the eastern side of the Sierra Madre Oriental from modern day Veracruz up through Tamaulipas and westward into the dry areas west of the Sierra Madre Oriental mountains. The Teenek have been separated from their southern Maya linguistic relatives for some 3000 years.[2] It is debated whether their language separation arose from Teenek movement northward, whether Teenek stayed behind when the others moved south, or whether there was a split from some middle area [11].

The earliest evidence of agriculture in the Teenek zone dates to 7000 BC [21, 22], although permanent villages with ceremonial centers do not appear until around 500 BC. At the southern end of the zone, along the Veracruz coast, pre-agricultural villages depended on sea, river, and forest resources from 3000 BC until 1700 BC when agriculture began to increase in importance through AD 900 when major urban centers arose throughout the region [32, 44]. At the time of the Spanish conquest, the Teenek population was organized in a stratified agricultural society of small warring chiefdoms [36, 37]. The conquest of the Teenek area was especially brutal; in 1554, Friar Nicolas de Witte wrote that a Teenek urban center near present-day Tampico was almost empty, although "it had been the most populated area under the sun" [23]. Teenek life was reoriented to survive on haciendas or retreat along rivers into the forest to avoid slave traders.

The current Teenek population exceeds 120,000 people. Their immediate neighbors include **Nahua** to the southwest, Totonac to the south, and Otomi to the west. In the mid 1400s, Teenek were conquered by the Nahuas (Aztecs) to whom they were required to give tribute,[3] and even today there is a strong sense of separation between Nahua and Teenek speakers [9]. Today, Teenek settlements are scattered in rural areas on the fringes of the mestizo populations of eastern San Luis Potosí, Hidalgo, Queretaro, southern Tamaulipas, and northern Veracruz states. The majority live in San Luis Potosí and Veracruz.

THIPAAK MAIZE HISTORIES

When Teenek sit outside their homes talking under the moon, they speak of many things, including Thipaak. In conversation or when asked, they mingle the past and present in their presentations of information about Thipaak and his compatriots. Elements of the histories are transposed by some.[4] Curers refer to pieces of the history in passing, assuming and renewing the listeners' knowledge of Thipaak. Out of respect and love, Teenek sometimes do not use the name Thipaak, but rather call him Little Boy *(Tsakam Kwitool, 'itool).*

We present summaries of representative, lengthy texts describing the history of Thipaak spoken in typical Mayan couplet, repetitive form, and information from direct queries and casual conversations between 1978 and 1983.[5] The logic and sequence of the speakers' presentations are noted, as well as the information in the texts, because both offer insights.

Pedro Kinim (A:PK): Well, this is what I will say, what I will tell. I didn't see it, this was heard. This that can be told, we take it into account. When we began, we didn't work. We passed our days with **ojox** *(Brosimum alicastrum),* with *ojox;* that's what was eaten, this *ojox.* The *ojox,* they say, was the one that was cared for. The birds

[1]European religion, based on a northern hemisphere cereal agriculture calendar, allowed post-conquest Mesoamerican maize agriculture religious traditions to be incorporated into Catholic ritual, [15] but left little space for incorporation of nature deities and rituals related to other aspects of natural resource management.

[2]Until recent times a closely related language, Chicomuceltec, which separated from Teenek about 900 years ago, was spoken in an area near the Mexico–Guatemalan border. It is now nearly extinct.

[3]Tribute sent to the Aztec conquerors included cotton and woven garments, as well as hot climate fruits, roots, deerskins, and forest products.

[4]A structural analysis of these histories, and their transposed and transmuted elements, is beyond the scope of this chapter.

[5]As part of an ethnobotanical investigation from 1978 to 1983, Janis Alcorn collected texts and information about Thipaak and maize cultivation in 24 Teenek communities across a wide area in Tanlajas, San Antonio, Tancanhuitz, Tampamalón, Ciudad Valles, Huehuetlán, Tanquian, and Aquismon municipios of San Luis Potosí, and Tantoyuca, Veracruz. Texts collected for linguistic study by Barbara Edmonson [10] from 1978 to 1983 also contain references to Thipaak. Cándido Hernández Vidales, a Teenek curer and independent, self-taught researcher, worked closely with Alcorn from 1978 to 1983 was taped and transcribed by CHV. He also provided texts for Edmonson from 1979 to 1983, and continued to tape, transcribe, and mail texts to Edmonson through 2002. Pedro Kinim, an elderly man from Tokoy, S. L. P. (A:PK) and Epimeño Flores, an elderly man from Tampaxal, S. L. P (A:EF) were recorded and transcribed by Janis Alcorn, and translated in collaboration with Cándido Hernández Vidales. Don Mosilino, an elderly curer from Tamjajnec, S. L. P. (E:MOS) was taped by Cándido Hernández Vidales, and corrected and translated by Barbara Edmonson. These three men, portions of whose narratives are presented herein, lived in different municipios and did not know each other.

wouldn't leave it alone; they ate it. Every morning the Old Woman went to care for the *ojox*. And then in the afternoon she returned. The Old Woman had a beautiful daughter of whom she was very possessive. She didn't let her go anywhere, she kept her closed up. The daughter got bored, because she wasn't free to go. She could only go bathe and return to the house. And when "the hour" arrived, then this Old Woman arose early to protect the fruit from the birds. So she wasn't there when the girl had the desire to go out. The mother of Thipaak went to the creek alone to bathe, and she came near to a little bird with a very happy song. As she returned from the creek, carrying back water, the bird spoke to her. He entered, they say, her mouth; it was the medicine of this *Maamlab*[6] (deity). It didn't take long, maybe fifteen days, when the bird spoke again, because he saw that the Old Woman was beating the girl. Why? She had gone alone and had sinned with this little bird. When this little bird spoke, he told the Old Woman not to mistreat her, because there would arrive a day, a year, when things here would change; times would change and new words would be born. But no, the Old Woman wouldn't hear it, and she continued beating her and maltreated her badly. The boy was born quickly after three months. Now when this child arrived, he was a little boy who came walking. No one comes walking like that. He was a Wise One. And he told the Grandmother to stop the maltreatment, that he didn't want to be beaten. Then the Grandmother was very angry, and she wanted to kill him with beating. It didn't take long for the child to grow, because this is the one who would change the food, the one who would stay forever. Then the Grandma went off to watch over the planted ones, the food, the *ojox*. Because this *ojox* was kept in a storehouse a *wiil te'*. This was the food, this *ojox*. And the boy stayed there, cared for by his mother, with his mother. One day arrived in which he thought what he would do. He sent his mother to go to the creek. He told her, "I will stay here. I will watch over the house." So he stayed. Because this *ojox* was cooked as we cook **nixtamal** today (i.e., soaked and simmered before crushing), made into *nixtamal* like maize. He said he would stay there and make the *nixtamal;* he would put on the *nixkon*. And his mother went on, because he was just a child. And he stayed there alone. When she returned, she said to the boy, "You have made the *nixtamal*. You have done it. I will see if it needs to be stirred." And she saw that it was different; it wasn't *ojox,* it was maize. That was what we call maize that we know now. Then the Grandma arrived, and began her beating, to fight with the mother. "This boy is bad, he would have us eat this dirty mess. This boy went and defecated in the pot; that's what he calls maize to make *nixtamal*." She tied him up and did all sorts of things to him. She was evil. But he endured a little longer, and he ended up teaching her more. Then she said that he was going to have to go, that he couldn't live there forever. He was tied up, burned, shut up inside a fence, burned. And because he was a clever one, when he was shut up, there he appeared, yelling and singing, dancing outside. This child. He couldn't be scared with anything, he didn't burn up. The Grandmother wanted to end him, because she was evil. He wasn't like a real boy of today. He was burned to scare him to think that he would be ended. But no, he didn't end. He came back, no one could see how he revived. Tired of this maltreatment, one day the Grandmother scattered amaranth seeds, broadcast them with *pepian* pumpkin seed, and told him, if he could pick them up, she wouldn't burn him. And then they say, this boy, because they say he was wise in his heart, he didn't leave this amaranth, this *pepian*. They were picked up by little birds who came to help him, they lifted them. There it began, and there it appeared. It seemed that he brought the seeds. When he was bored, he said to the Grandmother, If you can stand it, I will close you up and burn you. And if you can change, then we will know it is true that you are also true, everyone will know. To end this suffering. Then because she wasn't scared, he closed her up, this Grandma, and burned her too. The house burned, and the Grandmother was destroyed. He didn't let her go. That's how it is told. There was a pile of ashes that held the power given by the gods. This ash was picked up and put in a *huyul* (large two eared pot), because if it were left there like that it would disturb the people, it is an evil being to be born. So it was put in the *huyul,* and they went to drop it off in the ocean, the big lake. But it wasn't half way there when, midpath to arrive to the ocean, the dust came alive, it yelled, and there was heard a *thuyuyul* sound, a buzzing. The carrier wanted to go throw it out. And he couldn't stand it, this one, this one who carried it. He opened the mouth of the *huyul* to see what was there. And when the *huyul* was opened, they scattered out—mosquitoes. And then the carrier couldn't walk. They bunched up on him to eat him. The Grandmother had transformed herself into mosquitoes because of his stupidity. He went all the way to the ocean. Then he was lost too. They say he is the one we call "teller of the rain," a little frog, a little toad. He has bumps on his back where he was struck by the sucking ones. That's the Grandma of Thipaak. And Thipaak, when he did this, said there would be no more of this disorder, this sinning here. When it was time to go, then he said he was going. Then he said he was going, just one afternoon, he spoke words like water being scattered. He spoke words, he spoke different words, that he was going. He went there to the sierra, where he was. We call it *T'ithach*.[7] That's where he went to rest. And here remained the seed, in each house it remained, *one plant per house,* but very fruitful this plant. In every leaf there was an ear, every one had its ear, in this seed that he left. When they wanted to maltreat him, he left and he never returned. He stayed in the sierra. And that's why we make an offering, in regard to our work, that's the sort of thing that's been said, and it's true. [pause] It came and it carried them off, that's the Eagle, that's the old Grandmother. If one had a child, he cried when it was born, because the child would be gone soon. The man moves with the fiesta, there with the dancers, and when he comes home, the baby is gone. The "wind-carrying off one" came. That's why we take very careful care of our babies. We watch over them and never leave them alone so the wind won't carry them off with the Grandmother. And after a while the child grows to an adult and is eaten in the presence of other people, as if there was no Goodness, like today. But the child Thipaak gave another offering. He prepared it with a beautiful dressing, a decorated cloth. This was a deception. Thipaak was able to win, because when this Eagle That Carries Off came and grabbed the hot **boliim,** the Eagle left screaming. It burned its foot. It burned its talons. It never returned.

A longer version of the Thipaak history was told by Epimeño Flores Hernandéz (A:EF). His version contains the same basic beginning, narrative, and elements as PK's version

[6]While there are many Maams in Teenek religion, there is only one Ach-Eagle. Other female deities are referred to as Miims (Ladies). Maam used alone in a text (as in this case) is ambiguous but probably refers to the Muxi' Maam (the highest deity). We have translated Ach as Old Woman or Grandmother for clarity in the unfolding story of the Ach-Eagle's relationship to Thipaak. There is another Teenek adjective yehtsel, "old man," referring to ordinary aged men, is used in these texts to refer to old man who was the Ach's helper, and to the old man who discovered the pumpkinseed girl—individuals whose history is not explained beyond their brief appearance in the Thipaak history. The Ach-Eagle is also called "the Wounded One" in some contexts, in reference to her talons being burned by a hot boliim that was substituted for a sacrificial victim.

[7]The highest jagged mountain peak in the region.

earlier, but with several additions. He focuses on a long series of battles between the Grandmother (of the *ojox*) and the Boy (of maize). Some of the interesting additional narrative elements added by EF include the following.

[Element inserted after the girl had been impregnated by the bird] Evening came. The girl was shut up alone in her room. The Old Woman told her, "Be careful not to speak to anyone." But then, oooh, there were sounds of a fiesta, of merriment, of dance, music, dancing in her room. The rats were happy that it would soon come to pass as predicted. Maize would be born. The Old Woman was mad, "Why this noise?!" When she came and opened the door, she didn't find anyone. The rats ran off. The Old Woman left. Happy, happy, happy were the rats. The Goodness was coming. The days passed until arrived the day when the baby was born. The girl had a baby. The Old Woman became a Grandmother. When the child was born, the Grandmother said, "Now we aren't going to let this baby stay, because he will bring disorder, white demon, he will kill the house. We'll let the rats eat this baby." The girl didn't want to destroy him. The child was born at night, but it wasn't known. The poor mother collapsed on the earth, lost. The Grandmother returned happy. She returned bringing the girl. It was good to return so that it would seem that nothing had happened, that there was nothing. Night fell, a week passed. She went to visit the baby where it had been left. And she found a "Planted One"—just like that. This was all there was left of the baby. Angered by the Planted One, this maize we call it, she struck it, she hit it with a staff, she doubled it, she killed it, it was lost. And she returned. She told the old man she wanted to win, perhaps she had won. So she was happy. But it came back again. The next week she went to visit the place and it had gotten taller. It was growing beautifully. She gave it another blow until it dried out, this Planted One. She rested. Then she went and told the old man who was her helper, "Perhaps I will win." "You won't win. He is just resting his strength. He is big and he rests," said the girl. Another week passed, she went back to visit. Uuuh, and now the ears were golden, twenty two days had passed. Goodness.

She wouldn't be able to win below or above. Another week passed, then the boy arrived at the house, tired. "Grandma," the boy called. "When are you going? Where do you go? Sit down," said the Old Woman. "Sit down, sit down boy, sit down Little Boy." Tiny but he already walked. She didn't know if she would strike. And she was just angry, angry. And the Little Boy, he was light, he was light. He yelled, he screamed, he laughed, he climbed high in the trees, way up there the boy climbed. Up there he yelled, he laughed, he squealed, happy. The Grandmother was angry. The boy yelled. "I am not happy and what can I do? This boy does anything," she said to the old man. "What to do? He's bad, bad. He yells, he jumps, he dances, he yells." Thipaak wanted to see what the Old Woman would do. Another day broke. The Grandmother, held her head, it hurt. "Go make me a little bow and arrow, just a small one, and make me an arrowhead so it will work. And I will kill a lizard." What could she do for her head? What could she do? She made him the bow and arrow. "Go kill it." True, true. The Old Woman was made foolish. The Little Boy had gotten stronger and now he ordered the Grandmother. "Make me music Grandma. Make it so I can play music." She made a little guitar and a little harp. And she was made foolish again. He climbed high in the tree and made music. The Grandmother was angry. "Now you go to work tomorrow boy, here you are tiresome. I am not happy. So go to work. Over there, go far away to work deep in the forest." And so he went since he was ordered. He carried his arrow. He pulled back the bow-string and shot the arrow upward, and it went way over there, far away it stood. And he drew another and shot it way over there, it fell in the corner.[8] Another one, way over there. Another one over there it stood. Fine, his work was done and the boy returned. "What are you doing boy? Returned already? I sent you to work." "I've finished my work, Grandma. Tomorrow I will burn it. I will burn the *milpa*." "What work? You liar, it's not true." "Go see if it is true. Today I felled it." He felled it by magic. Great was his power/authority. He has the power this boy. He is growing, and he is getting used to it. She was beaten. The next day began with him saying, "I'm going to burn the slash, Grandma. Now you go get the seed and drop it off." "Where?" "Now don't you remember where it was left? Remember where the seed was left!" "That's a lie, it's not true, not true." "Well, go!"

She went back to the place and truthfully there was the seed, the sign. It was dry, like that. An armful. She was going to beat that boy. Maybe this was it. Smart one she was, but she had found the smart one who came back even after being buried under the ground. When did it come up, so small? Oh, Santo! She harvested the maize ears. She wouldn't let herself be beat. She came back with these ears and set to removing the grains. Rubbed together, rubbed together with tobacco, tobacco, tobacco, so she wouldn't be beat. And she threw them into the big water so the maize seed would be eaten by the fish. When it fell into the water, the maize spoke and told the fish not to eat it. It spoke to the water. It slid down, slid down, slid down, there it went below. When the **milpa** rose up, when it was ripe, the Sky became great, the world became great with this food. She did it to win, but she didn't win. No, when the Old Woman came back, she told the boy, "Now you listen to me. Go carry this net bag and fetch water." He brought water back in the net bag. "Grandma, I have brought you the water." He was happy. The Old Woman felt as if the boy had struck her in the head. Because she knew she couldn't bring water in a net bag. Thipaak said, "You bring water back in your *telega* (cloth bag). I brought it in my net bag, you bring it in the *talega*." Turn about. She sat there. She wanted to win. And she didn't win, but came close. She tried to fill the bag with water, but she just returned with her clothes all wet. She wasn't able to bring the water. Now the Old Woman had to really go to work. But the boy didn't work hard, because he knew the maize and the *milpa* grew by itself. The boy had done a beautiful job.

He beat the Old Woman and was happy. The Old Woman was furious and began to hunt for the boy. She wanted to eat the boy. She was evil. She lived on the *T'ithach* pillar to the sky. She was wise, very wise and powerful. She didn't want to give up being the queen of the sky but she wouldn't be able to destroy the boy. She was stupid because she forgot to be respectful. But the boy was lucky and respectful of all the gods.

EF then elaborates on a section of the narrative where Thipaak finds a girl crying in her house, because she is going to be eaten by the Old Woman who has gone to the creek to bathe. Thipaak counsels her to cook the maize *nixtamal*. He makes the *nixtamal* by sticking his leg into the boiling water. And he tells her how make a large *boliim* to trick the Old Woman. EF then proceeds to tell the story of the Grandmother–Eagle, similar to that given by PK above, where the Eagle is tricked by the *boliim* that Thipaak puts in place of the expected human sacrifice.

After elaborating on this sacrificial switch, EF continues by telling about an old bachelor who is working in his field where various crops came up by themselves.

[8]The corners refer to the four cardinal directions.

He discovers a tiny girl inside a *thakpen thukuk* (pumpkin whose seeds are eaten, *Cucurbita ficifolia*). He has no wife and no children, so he is very happy. The girl is called *Thakpen* K'aach (pumpkinseed girl). This is a short form of tsik'aach "girl". When she grows up, she is invited to stay in the house of the Old Woman, where Thipaak is living. When the Old Woman goes off to bathe in the creek, Thipaak tells the girl that the Old Woman is getting ready to kill her, to go see for herself. The girl goes to the creek and spies. She sees the Grandmother sharpening her fingernails. The girl comes back to Thipaak, frightened and wondering what to do. Thipaak tells her to take all the weaving equipment (warping pegs, backstrap loom, and spindles) with her and leave pieces of it, one by one, along her path as she runs away. But first, he tells her, she has to defecate in the four corners of the house and in the center. She did as she was told. When the Grandmother returns from the creek, she asks Thipaak for the girl. Thipaak says he doesn't know where she is, so the Old Woman calls her name and hears answers from inside the house—five times. But when she goes into the house, she discovers the girl's feces, and then angrily pursues the girl. Every place that the girl has left a weaving tool, the Old Woman encounters a barrier of dense forest. The Old Woman was detained by crossing various barriers, including tar pits, but each time she overcomes the barrier and gets closer and closer to the girl. Just before the Old Woman reaches her, the girl turns herself into a palm tree *(Sabal mexicana)*.

The Grandmother was even more angry and came back to the house looking for Thipaak. He ran from her, and as he ran, fruit trees sprung up and crops matured behind him. They both had their advisors and their words, but in the end Thipaak's words won. Thipaak had the advice of the **Maam**[9] whom he visited underground. But he played tricks on the old Maam too, asking him to try out his walking staff—a stinging nettle tree trunk *(Cnidoscolus megacanthus)* and beating him at a ball game.[10] Finally, seeking to destroy Thipaak, the Old Woman reached a forest where *K'ubi* birds were playing. Thipaak saw that the *K'ubi* were planting rocks, so Thipaak planted maize for them. When they heard the Grandmother coming, they told Thipaak to hide in the little *xomom* (bottle gourd for carrying tobacco, *Lagenaria siceraria*). She asked them if they had seen the boy, and they said no. She knew he was there, and she looked around carefully. She felt he was with the tobacco. She asked them for tobacco, and they gave her the tiny *xomom* with tobacco liquid in it. Thipaak was hiding there, but he had come out of the gourd and was clinging as a drop on the bottom, where she couldn't see him. She took a big drink and chewed the tobacco, but he had escaped again. She admired the clothes of the *K'ubi* and wanted to know how she could have such a pretty dress. The *K'ubi* told her that they used a big fire to make their clothes—that she should come back the next day if she wanted new clothing. First they would have to gather firewood. They cut a big field from the forest, and gathered the firewood. The next morning she returned, because she wanted to transform her clothing. They built a huge fire, and told her she would have to jump from one corner to the other, crossing the fire twice. She had magical powers, so she jumped across without harm. Thipaak came and told the *K'ubi* to grab her, tie her up and throw her into the center in order to terminate her "growing shoot" and burst her head open. They waited for her head to pop. When it popped, it was very loud because she was so clever. The boy knew that if she wasn't destroyed, she would destroy everything. He told them to carefully gather the ashes into the two eared pot and give it to the one who was to take it to the ocean.

EF continues with a version of how the ashes of the Old Woman transformed themselves into mosquitos en route to the ocean. He concludes by noting that after Thipaak walked all the way to the Nahua border region of Huehuetlán, he retreated to rest in *T'ithach* mountain peak. "*He left one plant of maize in each house*. This plant had fruit on its leaves and on all parts of the plant." EF then concludes his narrative by saying that if anyone behaves as though they are the king–queen and do not respect the authority of the gods, they will lose, perhaps a subtext supporting the hierarchical authority related to maize-based society–theocracy.[11]

The narratives are clear that Thipaak did not give out seeds but rather used his body and his magic to create *milpas* or *nixtamal,* thus maintaining control of the seeds himself. It is also interesting to note that both PK and EF's versions (as italicized) mention that only one plant was left in each house, an indication that maize was not cultivated in large fields initially.

The conclusion of the history of Thipaak's adventures lies in the eventual breaking of the *T'ithach* mountain to reveal the seeds stored there. Texts collected on this stage vary from a simple statement of fact that people got maize seed when the pillar to the sky was broken by a blast from the heavens leaving the jagged peak "*La Sillita*" that dominates the skyline to the west today, to short explanations focused on ants or demons that led to the mountain being split by lightning, as mentioned in a previous section, to a more elaborate version that identifies Thipaak as the source behind the rise to power of the current top Teenek deity—Muxi' Maam[12]—according to EF, as follows:

> Thipaak had worked hard and traveled far, and one day the Maam called to him. Thipaak had declared that the times were going to change, that he was going to say what must happen according to God. So the Big Maam was sent to break the big mountain *T'ithach.* It was very tall and reached to the sky. The Big Maam is a bachelor. At one time in the past, he was sent to get water from the ocean, but he frightened the people. When asked why he frightened them, he said because Thipaak told him to do it. So he had been beaten and went home. At the time when they wanted to break *T'ithach,*

[9]There are three rain-bringing Maams, one of whom is the oldest and most powerful—the Muxi' Maam—who was forced to live under the sea after a period in the underworld, after losing the ballgame to Thipaak and then failing to carry out Thipaak's command that he serve as a raingod without frightening people. Teenek also relate histories of the two other rain carrying Maams, not presented here. The Maams bring the rains from the ocean as forceful storms rushing west to caves in the sierra to consort with the female spirits of the sierra in an orgy, and then, their energies spent, drift back toward the ocean as low clouds.

[10]Ball games were ritually significant events for Mesoamericans. To beat the Muxi' Maam at a ball game would indicate that Thipaak had significant powers—political or magical.

[11]At the time of the Spanish conquest, Teneek lived in a class society, with nobles and slaves [36].

[12]Teenek deities are perceived as belonging to a society that to mirrors society on Earth. Curers achieve cures by arguing the patient's case as lawyers before the godpowers. The gods have political relations among themselves as well as with people.

because there were ants/demons climbing it, the lesser Maams found they couldn't break it alone, without the old Muxi' Maam. They called him back, but he declined, saying he feared they would mistreat him again. But Thipaak told him to go, that he would win. He didn't want to go, so Thipaak carried him. He wasn't aware of the distance traveled, but he arrived to the big meeting where they asked him to help. He said he couldn't do it, that maybe he needed an offering to eat and drink. He was still cautious, but again Thipaak assured him that he could do it, and then he would become famous and big. Thipaak tells him what tools to use, but Muxi' Maam says no, he will just use his staff. He asks everyone to sit down to see, and then with one blow of his staff, with one lightning bolt, the mountain fell into pieces. Everyone celebrated and said they wanted him to stay, that they would make him a house. But when they carried him there, he looked at the house with nineteen rooms, smiled, and destroyed it. So they made him a house in the ocean and took him there. It was made of tar. They asked him to rest there, but he tried to destroy it. But because it was made of tar, it just stretched—so they said, this is your house where we can find you. And the other Maams went back to their task of bringing rain to the Earth. Thipaak is the one who commanded the work, but he is gone now. The story is beautiful, about the sierra. Thipaak was the most wise one, and we have survived and passed our time with maize until now. In the old days we worshipped the Maams, and the **Ach** (Old Woman) with food and dance, but now, no. Now we don't respect Thipaak as before, and so the *milpa* does not yield much. Thipaak was the most powerful. The Maams couldn't die[13] and come back to life like Thipaak. He was burned, his body became *nixkon*. He was beaten, he died, and returned anew. The Old Woman tried to kill Thipaak. She didn't respect him, and so he won. And because the Maam of the East (Muxi' Maam) didn't try to kill him, but obeyed him, that's why Muxi' Maam is the greatest—he got his powers from working with Thipaak.

Ariel de Vidas [6] reports that among the Veracruzano Teenek, Dhipaak (Veracruzano Teenek pronunciation) is seen as the soul of maize, whereas the Thundergod (Muxi' Mam to Potosino Teenek) is responsible for the origin of maize. As the preceding texts illustrate, the relationships are circular. In none of the texts collected is there any mention of a father for Thipaak, beyond the role of the bird and alluding to the bird as a messenger from a higher authority. In Teenek *¢'i¢iin* (bird), when possessed *in ¢'i¢i:n* (his bird) means his soul or spirit. As a language, Teenek abounds in passive constructions—things happen more than they are done by an agent, which often leaves an interpretable ambivalence in relation to responsibility for actions.

Other versions of the Thipaak history exist. One version says that the girl went to the creek to wash *ojox* when the bird impregnated her by eating one of the *ojox* fruits–seeds.[14] In another, the Old Woman grandmother kills Thipaak by putting him into the boiling water in a *nixtamal* pot on the fire while his mother is out gathering *ojox* to cook. The mother returns surprised to find the boy missing and maize *nixtamal* in the pot, as the boy dances in, unhurt, through the door. In the common versions, Thipaak is the main actor. Minor details vary but the story essentials remain constant—maize replaces ojox through divine intervention that engendered conflict among the gods.

An alternative set of maize origin stories focuses on the splitting of the mountain, and highlights the roles of animals and nature. One Veracruzano Teenek story reported by Ariel de Vidas [6] concerns a woman who spied a worker ant carrying a grain of maize. The woman asks the ant where it got the maize, but the ant replies that it is forbidden to tell. The ant is tortured by tying a thread around its stomach. Each time the ant refuses to tell where it got the maize the thread is tightened, and finally the ant cannot stand further suffering and confesses that the maize is under a mountain. When the men of the village reach the mountain they call on the god of lightning to split the rock open. He does so, and the kernels of maize fly out. A variation on the link between the ants' and maize is found in a Guatemalan Mam Maya tradition [43], in which ants brought maize out of the mountain called Paxil. However in this case, humans, after learning from the ants, got their maize seeds from wildcat excrement, because wildcats also entered the mountain to eat maize.

In an alternate version of the splitting of *T'ithach* collected in San Luis Potosí, ants also play crucial roles. The *piich* bird, who today sings in the *milpa* while eating young maize plants, is the bird who sang at the base of *T'ithach* drawing the attention of passing merchants to the maize seeds being carried out of *T'ithach* by leafcutter ants. *Laak'* ants attract lightning. *Pothots*, the bird who was Thipaak's nursemaid, eats *laak'* ants.

This suggests that the competing stories regarding the splitting of *T'ithach*—an ant–lightning connection versus a Muxi'–Thipaak connection—possibly indicate the survival of older oral histories of hunter–gatherers with nature (ants and lightning) as the prime actor as remnant subtextual elements within the agricultural tradition stories where Thipaak and an alliance of personified gods are the prime actors.[15]

THIPAAK AND MAIZE IN CONVERSATION AND DAILY LIFE

Thipaak is at the heart of Teenek identity. He is "the heart of maize," "the spirit of maize," and "the soul of maize"—having the same three aspects that every human being has. The landscape itself, between the sierra to the west and the

[13]The principal Maams are beings who have died and now live in the underworld; the earth itself is held up by four drowned men who are replaced each new year by new ones. Thipaak had the special ability to move between the dead and the living, between deities and mortals.

[14]A fresh *Brosimum alicastrum* fruit has a thin layer of sweet flesh around a single seed and is about the size of a cherry fruit. It shrinks as it dries until it becomes a dark, hard round seed, approximately the size of a grain of maize, enveloped in a papery residue.

[15]In his introduction to Alan Sandstrom's monograph about Otomi rituals, Peter Gold [33] postulates that hunter–gatherer societies explain their world through the actions of nature and animals, while agricultural societies tend to personify the powers of nature.

ocean to the east, is called Thipaak's fallow–field, bounded by the four pieces of the *T'ithach* pillar from which the Old Woman once ruled. The songs of birds are ever present in the Teenek environment, and their songs recall Thipaak. The song of the *Ta'tam* (Thipaak's mother) and *Ts'ok* (the father–deliverer of sperm) recall Thipaak's conception. The song of the *Pothots'* recalls Thipaaks' nursemaid. The *K'ubi* recalls the Old Woman's death and the victory of Thipaak.

Although many Teenek plant names have been documented [5], few relate to Thipaak, perhaps due to the singular respect shown him. Pineapple, an introduced cultigen, is called *chabcham huitz* (dancing staff flower) because Thipaak planted his dancing staff to give rise to this plant.

In almost every conversation about Thipaak, people lamented the loss of ritual attention and respect to Thipaak. It is said that, during the All Saints Days of the Dead, at time when the souls of the dead and the gods visit the houses, Thipaak visits the houses of those who have not made offerings to him during the year, and, when he leaves that house, the household's food disappears.

In 1977, there were several cases of Thipaak "sightings"—one by a Nahua boy who saw a small seated figure at the base of a red, white and green striped (Mexican flag colors) maize plant who identified himself as Thipaak, and another by a Teenek boy who encountered a talking palm tree[16] who claimed to speak for Thipaak. In both cases, Thipaak was threatening to abandon the Huasteca if people did not remember and respect him. According to reports, thousands of people came to dance and pay respects to the plants. Women danced carrying "babies" made of maize tassels in their baby slings.

Thipaak serves as a crucial reference, calling on Teenek cultural identity and strength. Teenek shamans invoke Thipaak. A lengthy curing ceremony includes this typically enigmatic text. After invoking Christian deities at length in Spanish, the curer shifts to Teenek, speaking in the formal couplets of Teenek ritual speech, and invokes the aid of Thipaak (E:MOS).

> He does not want any suffering, this child.
> For this reason he testifies here,
> he asks a favor of the Mighty Lord,
> the Mighty Lady,
> and during this hour now
> there is placed the altar,
> there is placed the cloth,
> there are placed the offerings,
> there are placed the gifts,
> on account of that illness
> which wanders about upon the earth,
> in the sky.
> And for this reason
> these indeed remain as his substitute,
> because he does not want this suffering,
> this poor one.
>
> Let potency be given,
> because the one who began it when the earth began,
> when life began
> for the child,
> complete at birth,
> then he/she comes,
> when a week has passed.
> The Mighty Death comes
> to carry off that child.
> He/she does not allow him to grow older,
> he /she does not allow him to develop,
> the poor one.
>
> Then he/she pondered,
> the Thipaak,
> he left for him a word,
> so that the child might speak.
> Then that child said thus,
> when he was born,
> he said that he wanted to be made for him an offering,
> to be made a meal.
> They are going to leave for him/her his (chicken) heart,
> and then he is going to hide, the child.
> This is the Thipaak,
> child Thipaak,
> boy Thipaak,
> girl Thipaak,
> and it was this one who left that sign.
>
> When he/she came,
> the Mighty Death,
> the Mighty Lord,
> an altar was left for him/her,
> a cloth was placed on it.
> then that child
> hid beneath the altar
> when the mighty one came.
> He/she is going to eat.
> This he/she grabbed, the heart,
> Only this he/she carried off.
> Over there indeed he/she rested.
> He/she went over there,
> He/she carried off only the heart,
> with that it was sufficient.
> Indeed he/she never returned,
> indeed he/she nurtured the thought
> that he/she really did not want to touch that child.

Maize also has its medicinal properties, as can be seen, for example, in a curer's text about a man who had tuberculosis, was sick and coughing up fluid. (E:MOS) "Then he prepared some maize, the red kind. . . . He put it outside at twilight so that it could collect the dew, so that it could collect the heat of the earth. That dew was the dew of the moon, the dew of the great star. All the power of the star the dew then casts down. And with that power he got well. That's how that friend got cured."

Teenek believe that the maize plant is like a young boy, aging from age 7 to an adolescent. When the maize needs to be weeded, it is said to the "sad," and when it suffers from

[16]The talking palm tree recalls the part of the Thipaak history where Thakpen 'Kaach turns into a palm tree to escape the Ach-Eagle.

drought, it is said to be in pain *(yaynal)*. Caring for maize is seen to be like caring for a child. The sound of a laughing boy in the *milpa* is a sign of a good harvest; a crying boy sound indicates a bad harvest.

MAIZE CULTIVATION RITUALS

Milpa refers to the Mesoamerican system of slash-and-burn agriculture that incorporates maize, beans, squash, and many other cultigens into long fallow or short fallow cycles. *Milpa* also refers to the maize field itself. At a landscape level, *milpa*s do not exist alone in vast monotypic stands, but rather small fields are scattered in clusters according to ecological parameters—along seasonal riverbeds in dry areas for example, in sheltered areas at higher altitudes, and within forest in moister areas.[17]

All stages of *milpa* management incorporate rituals. The first set of offerings are made 2 weeks before a man clears the bush in the area to be planted, to ask the Earth's permission to poke holes in the ground, saying that he is not cutting the trees to maltreat them, but because he is hungry and has to plant to survive. The second offering is made at the time of planting, when teams of men move across the burned field poking holes into the earth with dibble sticks and dropping maize seeds into the holes with skilled motion. The planting ritual is considered the most important of all rituals. The day before the exchange labor team arrives to begin planting, a chicken heart is left in the middle of the burned field so that the Earth will care for the seed; it is called *pihchixtalaab,* the ritual feeding of the spirits of the animals that could damage the *milpa*. The field's owner also gives the Earth water and aguardiente and censes the spot with copal—all done at daybreak, before the planters arrive. The seed is placed on the household altar laden with the ritual offering of *boliims* and aguardiente, with a girl and boy crowned with flowers sitting next to it, and prayers are made to the Maams to ensure the crop's safety and good rains. After the planters finish their work, they eat the *boliim*. Nowadays the planters move in lines across the field, but they say that in ancient times the planters moved in a counterclockwise spiral to converge on the center of the field at the end. Three more ritual ceremonies are done successively when the tassels first open, when the first green ears are ready, and when the dry ears are ready to harvest.

After harvest, the dry ears were traditionally stored in the field in raised enclosed platforms *(wiil te')*; according to many people *ojox* was once stored in these *wiil te'* and that's what the Grandmother was guarding from the birds when Thipaak was conceived. Seed ears, selected during harvest,

FIGURE 16-1 A Teenek grandmother makes the final selection of maize seed ears while her daughter's husband rubs the seeds off into the planters' bags in the presence of a young boy—renewing memories of the Thipaak oral history of the origins of maize. (Courtesy of Janis Alcorn).

are taken back to the house and hung over the fire where the smoke will keep insects from damaging them. Teenek consider their seed seriously, and each farmer has his or her own ideas about how to create and care for the best seed. Some travel to distant areas in the sierra to look for unusual or famous seed to add to their mix; others look for variation as it arises within their own fields. Some seek conformity of ears; others prize variation as a main characteristic for selecting seed. At planting time, seeds are taken only from the central part of the cob, because it is believed that the seeds from either end are weaker (Figure 16-1). Care is taken not to "frighten" the seed as it is being collected for the planters' bags; the seed is treated with respect. Teenek are aware that planting different varieties in the same field can influence the maize that is produced.

As part of their annual New Year's offerings in caves, Teenek farmers request that their *milpa* be blessed in the coming year. They leave their offerings in the sacred cave, but they leave the cave carrying dry *ojox* as ritual "seed" for planting.[18]

RELATION TO OTHER MESOAMERCAN TRADITIONS

There has been, to our knowledge, no intensive investigation into *Brosimum alicastrum*'s presence in oral traditions or Mayan glyphs, or its relationship to maize evolution. Substitution of the cacao god for the maize god in some

[17]The incredibly diverse systems of maize cultivation in all their different habitats have been extensively documented by the late Efraím Hernández Xolocotzi and his students in Mexico.

[18]*Ojox* seed is found in caves, because bats carry *ojox* into the caves.

glyphic situations, and the fact that the Foliated Maize God emerges from the World Tree offer hints of a relationship between maize and tree crops [8]. Puleston and Puleston's [27] hypothesis was that *ojox,* not maize, was the staple food during Classic Maya times. Post-conquest Spanish documents mention the presence of *ojox* in *milpa* orchards.[19] The multiple nutritional values of the fruit and its sap, storage qualities, and other characteristics of *ojox* make it a potential staple. Yucatec curers have used *ojox* seeds in their diviner rattles [31], possibly a marker of *ojox*'s past ritual importance. A population genetics study [26] indicated that populations of *Brosiumum alicastrum* in the classic Maya sites had been domesticated, producing fruit twice a year instead of once a year. There are also some indications[20] that the Pacific populations of *Brosimum* were introduced by people bringing seeds from the Gulf Coast populations.

Transhumant, hunting and gathering Mesoamericans, such as those who may have depended on *ojox,* were likely active participants in the domestication of maize [20]. As they traveled seasonally, people harvested grains and carried them to the next habitat, subjecting them to a variety of conditions. They also likely selected seed from plants with non-shattering, condensed rachis because these would have been the grains left on the plant when they again returned to an area. But they did not depend on maize as a staple. Under such conditions, maize may have been cultivated in small amounts as a special item for ritual needs, rather than as a staple, among forests, which yielded *ojox* and game, from distant times up until the last century. In the first decades of twentieth century, *Brosimum* was still a major subsistence food for some Teenek.[21] Some Teenek populations maintained a hunting and gathering lifestyle in the sierra above Aquism6n in the ecological transition zone between the moist eastern slopes and the dry western slopes of the Sierra Madre Oriental, on the fringes of settled society, wearing little clothing, eating *ojox,* and wild game. They committed suicide[22] in mass numbers when outside society finally enclosed them in the early 1900s.

Edwin Braakhuis [8] has exhaustively explored main elements of the *Popol Vuh* and other Mesoamerican traditions in relation to the Tonsured Maize God of the Classic Maya in his study of what he terms "the Mayan civilizing hero," substantially demonstrating the existence of a coherent body of Mesoamerican oral and glyphic histories that incorporate maize as an element of a culture hero. Within Braakhuis's material, the cannabalistic grandmother, the mother at the creek, the visit to the old god in the underworld (who is identified directly as the maize hero's father by Braakhuis), the ocean, and the fish elements are repeated in other Mesoamerican oral histories. However, there are important inversions in the Teenek oral tradition, so that the benevolent grandmother who tries to protect the maize hero in Classic Maya and other traditions, becomes in the Teenek version the grandmother who repeatedly kills or tries to kill the maize hero. Law [19] reports a Nahua tale from Mecayapan, on the coast of southern Veracruz concerning a hero called Tamakasti, who, like Thipaak, is the son of a bird and a woman. Like Thipaak he escapes from several attempts by his grandmother to kill him.

The Thipaak conception story is somewhat similar to the *Popol Vuh,* where Blood Girl is impregnated by spittle from the skull of 1 Hunter, which was hung from the branches of a gourd tree [12]. She then gives birth to Hunter and Jaguar Deer—the hero twins of the legend. There are no twins in the Teenek version, but rather a possible overlay of the Sun hero and Thipaak. As detailed by Brian Stross (see Chapter 42), the *Popol Vuh* and the *Chilam Balam* of Chumayel contain references to maize and heroes that echo the Thipaak histories.

In the *Chilam Balam,* the thunder god opens the mountain in which maize is hidden [40]. Many cultures repeat a story about the problem of getting at maize that is hidden inside a mountain. Most tales involve ants[23] who are seen carrying grains of maize from a cleft in a rock. Eventually the rock or mountain, like *T'ithach,* is split open so that people can get at the maize. There is a curious similarity between the name of the Guatemalan peak split to release maize, Xepaxa–Paxil, mentioned as the place of the origin of maize by Mam, Quiche, and Cakchiquel and the Teenek area called Tampaxal, literally "place of Paxal," which lies near the base of *T'ithach* [24]. To the north of Tampaxal lies Tamapats (place of the palm tree leaves), giving geographical reference to the Thak'pen K'aach character in the Thipaak history.

Mountain peaks also figure in legends recorded by Alan Sandstrom [34] in a remote Nahua village in the tropical forest of northern Veracruz where he has worked for many years. One legend revolves around Tonantsij, a mother deity who rules from her mountain cave over the seed spirits who the villagers believe are her children. Cut paper figures of the seed spirits are carefully preserved in a chest and are prayed over during the year. Maize is considered to be the

[19]*Milpa* orchards take many forms, one being the te'lom of the Teenek [1–4].

[20]C. Peters, personal communication, 1983.

[21]*Brosimum* seed was available in sufficient amounts to be remembered as the main food that saved people from starvation during the Mexican Revolution. Today *ojox* is still gathered, stored, and eaten casually by most families. When farmers are asked why *ojox* trees are left standing in a *milpa,* the invariable reply is "because *ojox* is the maize of ancient times."

[22]This assertion mentioned in conversations in various parts of Aquism6n municipio during 1978–83, confirms the information published by Laughlin [18]. Several old informants in Tamapatz noted that rope was scarce at the turn of the last century, because so many people were using rope to hang themselves.

[23]Ant-related mountain-splitting stories have also been recorded from Chorti [13] and Mam [43] in Guatemala and from central Mexican Nahua [39].

physical and spiritual link between human beings and the sun. Sandstrom [33] has described similar cut paper figures among the Otomí.

When viewed in the larger Mesoamerican context, the Thipaak history could also conceivably be interpreted as documenting a history of multigenerational conflicts—with wins, losses, and revivals—between competing leadership lineages each with their associated deities. A famous stone statue of a youth with a child on his back discovered in the Teenek urban center of Tamuin, constructed around AD 1000, may represent Thipaak. The statue is commonly identified as the Aztec god Quezalocoatl in his Evening Star aspect, a deity who is thought to have his origins in the Huasteca because of his conical hat and other cultural characteristics, and whose incarnation in the life history of a Toltec prince Ce Acatl Topiltzin around AD 1000 includes some elements similar to that of Thipaak, including opposition to human sacrifice. There is a Bas relief of an eagle eating a heart, just as the Teenek Ach–Eagle ate hearts from sacrificial victims, on Pyramid B at Tula, the capital of the Toltec empire, ca. AD 900–1200. The Grandmother–Eagle has similarities to the Aztec goddess Tlazolteotl (also called Toci, "our grandmother," a weaver), who has Teenek origins as the goddess of illicit sex, and is sometimes depicted with eagle claws[24] [38]. In the Aztec pantheon, Centeotl, the god of maize, is the son of Tlazolteotl, an apparent transmutation of Grandmother to Mother.

DISCUSSION AND CONCLUDING OBSERVATIONS

Aspects of the Teenek narratives are shared with the Nahua, Otomi, and Totonac histories, published from the Huasteca region [16, 33–35], other Mesoamerican narratives, and glyph analyses, and some analyzed as a body by Edwin Braakhuis [8]. Some elements are shared with ancient Mayan texts and embedded in Mayan language (Chapter 42), indicating a depth of age that may conceivably reach back to the time of the domestication and spread of maize cultivation, possibly associated with a shift in religious practice, political authority, and social structure.

Many Mayan ethnographers have reiterated that for Maya to live is to make *milpa* [28] or "everything they did or said so concerned maize that they almost regarded it as a god" [40]. But, taking the long view, it is worth considering that the importance of maize as a survival food may have arisen as a salient aspect of culture in the post-conquest situation after indigenous societies were forced into concentrated areas, lost access to their forests, and often reduced to slave labor for their haciendas—a condition in which their lives literally depended on the small plots of maize that they were allowed to grow for their own survival.[25]

There is substantial evidence, moreover, that forests have long been a fundamental part of Mayan subsistence systems that incorporated intensive raised fields and gardens in addition to *milpa* [7, 41, 42]. Unfortunately we know little about these systems, because ethnographers have focused on *milpa* field cultivation and its associated ritual, creating abundant literature on the subject while ignoring the management of forests in relation to *milpa*s despite the fact that many *milpa*s are by nature slash-and-burn systems that incorporate forest regrowth, and despite evidence of tree stand management in modern Mayan systems in Yucatan as well as in the Huasteca.

An open-minded reading of the glyphs, oral histories, and early Spanish documents indicate that maize was one among many cultigens, that rivers, coasts, forests and root crops[26] were also important sources of sustenance, and the *milpa* was but one organizing principle for life. This is not to deny the importance of maize within Mayan religion, because maize clearly occupies a special niche of deep respect rare among crops,[27] but rather to suggest that maize scholars should reconsider whether the history and early evolution of maize are directly related to its central economic and survival importance, or to its importance for social, ritual, and political reasons. Evolution in small isolated fields or dooryards, with attention to individual plants, and seed shared in small amounts would increase the importance of **genetic drift** as well as selection for special qualities (i.e., a fundamentally different scenario for evolution than would occur in mass plantings for a dietary staple as is the case today).

Oral histories hold encoded empirical information about changes and events of the past. One reading of the Thipaak history could be that through the introduction of maize, an older *Brosimum*-based religion with a vital female deity demanding human sacrifice coexisted with and was eventually displaced by a hierarchical male-dominant theocracy[28] that required maize products in their rituals and economy.

[24]www.firstwrites.com/authors/Educational/Gloria_Bertonis/Stone_Age_Divas_Medicine_Chapter.pdf

[25]This "landless" condition of survival dependent on small maize plots slowly worsened between 1700 and 1900, and ultimately led to the Mexican revolution in the early 1900s. Survival on small maize plots has contributed to the desperate conflict in Chiapas today.

[26]Reina and Hill [29] cite sixteenth-century Spanish accounts documenting Mayan dependence on roots, wild foods, and game for one third of the year. Murra [25] has documented the Spanish myopic, cereal-biased observations, showing how they ignored the importance of the potato and other root crops in the Andes.

[27]With the exceptions of rice in Asia and *dioscorea* yams in New Guinea.

[28]The most important Teenek deities are the male raingod Maams, but the lesser deities are male–female pairs—Maams and Miims (Ladies) for dance, for music, and so on. Teenek curers—today's main figures in the practice of indigenous religion—likewise work in male–female pairs, with male or female taking the lead, although some curers are bisexual–androgynous.

References Cited

1. J. B. Alcorn. (1981). Huastec non-crop resource management: Implications for prehhistoric rain forest management. *Human Ecology, 9,* 395–417.
2. J. B. Alcorn. (1983). El te'lom Huasteco: Presente, pasado y futuro de un sistema de silvicultura indígena. *Biótica, 8,* 315–331.
3. J. B. Alcorn. (1984). *Huastec Mayan ethnobotany.* Austin: University of Texas Press.
4. J. B. Alcorn. (1990). Indigenous agroforestry systems in the Latin American tropics. In: M. Altieri, S. Hecht, (Eds.), *Agroforestry and small farm development.* Boca Raton, FL: CRC Press. pp. 195–210.
5. J. B. Alcorn, C. Hernández V. (1983). Plants of the Huastecan region of Mexico with an analysis of their Huastec names. *Journal of Mayan Linguistics, 4,* 11–118.
6. A. Ariel de Vidas. (2004). *Thunder doesn't live here anymore: The culture of marginality among the Teeneks of Tantoyuca.* Boulder: University Press of Colorado.
7. A. Barrera, A. Gomez-Pampa, C. Vazquez-Yanes. (1977). *El manejo de las selvas por los Mayas: Sus implicaciones silvicolas y agricolas.* Mexico: Instituto Nacional de Antropologia e Historia.
8. H. E. M. Braakhuis. (1990). The bitter flour: Birth-scenes of the tonsured Maize God. In: R. A. M. van Zantwijk, R. de Ridder, H. E. M. Braakhuis, (Eds.), *Mesoamerican dualism.* Utrecht RUU-ISOR. pp. 125–147.
9. F. D. Duran. (1965 [1581]). *Historia de las Indias de Nueva España y islas de tierra firme.* Mexico: Editorial Nacional.
10. B. W. Edmonson. (1988). A descriptive grammar of Huastec (Potosino Dialect). Ph.D. Dissertation. New Orleans: Department of Anthropology, Tulane University.
11. B. W. Edmonson. (2004). Investigación linguística del huasteco. In: J. Ruvalcaba M., J. M. Perez Z., O. Herrera, (Eds.), *La Huasteca: Un recorrido por su diversidad.* Centro de Investigaciones y Estudios Superiores en Antropología Social. Tamaulipas, Mexico: El Colegio de San Luís, y El Colegio de Tamaulipas. pp. 295–318.
12. M. S. Edmonson. (1971). *The book of counsel: The* Popol Vuh *of the Quiche Maya of Guatemala.* Middle American Research Institute Publication 35. New Orleans: Tulane University.
13. J. G. Fought. (1972). *Chorti (Mayan) Texts I.* Philadelphia: University of Pennsylvania Press.
14. E. Hernández Xolocotzi. (1985). *Obras de Efraím Hernández Xolocotzi. Xolocotzi Tomo 1.* Mexico City, D. F.: Revista de Geografia Agricola, Universidad Autónoma de México.
15. E. Hunt. (1977). *The transformation of the hummingbird.* Ithaca, NY: Cornell University Press.
16. A. Ichon. (1973). *La religión des Totonaques de la Sierra.* Paris: Eds. du CNRS. Mexico: Spanish Ed. Instituto Nacional Indígenista.
17. P. Kirchhoff. (1943). Mesoamerica: Sus limites geográficas, composición étnica y carácteres culturales. *Acta Americana 1,* 92–107.
18. R. M. Laughlin. (1969). The Huastec. In: R. Wauchope, E. Vogt, (Eds.), *Handbook of Middle American Indians.* Ethnology, Part I. Austin: University of Texas Press. pp. 298–311.
19. H. W. Law. (1957). Tamakasti: A gulf Nahuat text. *Tlalocan, 3,* 344–360.
20. T. F. Lynch. (1973). Harvest timing, transhumance and the process of domestication. *American Anthropologist, 75,* 1254–1259.
21. R. S. MacNeish. (1958). Preliminary archaeological investigations in the Sierra de Tamaulipas, Mexico. *Transactions American Philosophical Society, 48,* part 6, 1–210.
22. R. S. MacNeish. (1964). The food-gathering and incipient agricultural stage of prehistoric Middle America. In: R. Wauchope and R. C. West (Eds.), *Handbook of Middle American Indians I.* Austin: University of Texas Press. pp. 413–426.
23. J. Meade. (1942). *La Huasteca, epoca antigua.* México: Editorial Cossio México.
24. S. W. Miles. (1960). Mam residence and the maize myth. In: S. Diamond, (Ed.), *Culture in history.* New York: Columbia University Press.
25. J. V. Murra. (1960). Rite and crop in the Inca state. In: S. Diamond, (Ed.), *Culture in history.* New York: Columbia University Press.
26. C. M. Peters. (1983). Observations on Maya subsistence and the ecology of a tropical tree. *American Antiquity, 48,* 610–615.
27. D. E. Puleston, O. S. Puleston. (1971). An ecological approach to the origins of Maya civilization. *Archaeology, 24,* 330–337.
28. R. Redfield, A. Villa-Rojas. (1934). *Chan Kom.* Washington, D.C.: Carnegie Institute of Washington.
29. R. E. Reina, R. M. Hill. (1980). Lowland Maya subsistence: Notes from ethnohistory and ethnography. *American Antiquity, 45,* 74–79.
30. J. Rivard. (1967). Maize in the Maya codices. *The Chesopiean, 5,* 27–50.
31. R. L. Roys. (1931). *The ethnobotany of the Maya.* Middle American Research Series No. 2. New Orleans: Tulane University.
32. W. T. Sanders. (1978). *The lowland Huasteca archaeological survey and excavation 1957 field season.* Columbia: Museum of Anthropology, University of Missouri.
33. A. R. Sandstrom. (1981). *Traditional curing and crop fertility rituals among Otomi Indians of the Sierra de Puebla, Mexico.* The Lopez Manuscripts. Bloomington: Indiana University Museum.
34. A. R. Sandstrom. (1992). Ethnic identity and the persistence of traditional religion in a contemporary Nahua village. *Journal of Latin American Lore, 18,* 37–52.
35. A. R. Sandstrom, A. Gómez M. (2004). Petición a Chicomexochitl. Un canto al espíritu del maíz por la chamana nahua Silveria Hernández Hernández. In: J. Ruvalcaba M., J. M. Perez Z., O. Herrera, (Eds.), *La Huasteca, un recorrido por su diversidad.* Centro de Investigaciones y Estudios Superiores en Antropología Social. Tamaulipas, Mexico: El Colegio de San Luís, A.C., El Colegio de Tamaulipas. pp. 343–367.
36. G. Stresser-Pean. (1967). Problemes agraires de la Huasteca ou region de Tampico (Mexique). In: *Les problemes agraires des ameriques latines.* Paris: Colloques Internationaux de Centre National de la Recherche Scientifique. pp. 201–214.
37. G. Stresser-Pean. (1971). Ancient sources on the Huasteca. In: R. Wauchope, G. F. Eckholm, I. Bernal, (Eds.), *Archaeology of Northern Mesoamerica, Part 2. Handbook of Middle American Indian*s. Austin: University of Texas Press. pp. 582–602.
38. T. D. Sullivan. (1982). Tlazoteotl-Ixcuina: The great spinner and weaver. In: E. Boone, (Ed.), *Art and iconography of the Late Post Classic Period in Central Mexico.* Washington, D.C.: A conference at Dumbarton Oaks, October 22–23, 1977.
39. N. Stiles. (1985). The creation of the Coxtecame, the discovery of the corn, the rabbit and the moon and other Nahuatl folk narratives. *Latin American Indigenous Literatures Journal, 1,* 97–121.
40. J. E. S. Thompson. (1970). Maya history and religion. Norman: University of Oklahoma Press.
41. V. M. Toledo, B. Ortiz-Espejel, L. Cortes, P. Moguel, M. Ordónez. (2003). The multiple use of tropical forests by indigenous peoples of Mexico: A case of adaptive management. *Ecology and Society, 7,* 3. Available online at: http://www.consecol.org/vol7/isse/art9/
42. B. L. Turner, II. (1976). Prehistoric intensive agriculture in the Maya lowlands. *Science, 185,* 118–124.
43. C. Wagley. (1941). Economics of a Guatemalan village. *American Anthropological Society Memoir, 58.*
44. S. J. K. Wilkerson. Ethnogenesis of the Huastec and Totonac. Ph.D. Dissertation. New Orleans: Department of Anthropology, Tulane University.

CHAPTER

17

Maize in Word and Image in Southeastern Mesoamerica

BRIAN STROSS
Department of Anthropology, The University of Texas, Austin, Texas

Glossary

Atole Boiled maize dough in water, often combined with other additives to make a ceremonial drink.

Calendar The Mayans and the Mixe–Zoqueans both employed two main calendars, a sacred almanac of 260 days—composed of a number cycle from 1–13 running concurrently with a 20-day name cycle—and an agricultural calendar composed of eighteen 20-day months plus 5 nameless days for a total of 365 days.

Cycad A member of a small and ancient plant family going back at least 200 million years, which has several unique features and that looks rather like a tree fern.

Hearth of creation A Quiché Maya constellation in Orion that is currently believed to go back to Classic Maya times as a three stoned hearth with a central "fire" and to represent the celestial hearth from which the universe was created.

Homxïk A dwarf maize deity of the Sierra Popoluca, hatched from an egg and subject of a cycle of myths in which he undergoes a series of trials, eventually vanquishing Hurricane and making him promise to supply water for the maize crop.

Maize tassel As used here it is the spike or tassel in the flowering stage of maize development.

Maya glyphs The individual components of the Maya script are called glyphs, and these individually or in combination comprise generally squared blocks with rounded corners, usually read left to right two at a time from column tops to bottoms. Glyphs generally represent an individual syllable or a word, or both.

Mayan language family A family of some 30 distinct and separate languages, the Cholan branch of which is primarily implicated in developing the Classic Maya Script that was employed by members of the Yucatecan branch during Post-Classic and early Colonial times.

Mesoamerica The geographic area as employed here that includes Guatemala, Belize, and Mexico as far north as the Panuco River delta and Lerma-Santiago River.

Mixe–Zoquean language family A family that includes approximately 12 separate and distinct languages located immediately adjacent to and on both sides of the Isthmus of Tehuantepec in Mexico.

Popol Vuh The sacred book of the Quiché Maya, originally written down in the sixteenth century and recounting the origins, traditions, and history of the Quiché people, it is analogous to the western Bible.

Pozole Maize dough that can be added to water, creating a nutritious and delicious drink something like thick milk.

Sipakna A "crocodilian" antagonist of the "hero twins" who is destroyed by them in the first half of the Quiché Maya sacred book, the *Popol Vuh*.

Thipaak The Huastec culture hero who first cultivated maize is referred to also as the **ehatal** (soul) of maize, the **tz'itziin** (spirit) of maize, and the **ichiich** (heart, embryo) of maize.

Time depth The number of years since currently related languages were the same language, having diversified over time because of inexorable language change that affects all languages.

Tzeltal A Mayan language of highland Chiapas, having some 300,000 speakers at the turn of the second millennium.

From Preclassic times to the present, indigenous people of southeastern **Mesoamerica** told stories of maize, recited prayers for maize, and spoke directly to the maize. They depicted it in images, stylized it in glyphs, and created words to speak of its many forms, parts, and stages, as well as its transformation from sown seed to food prepared and eaten. The words, glyphs, images, and narratives concerning maize, which are found in the many Mayan and Mixe–Zoquean languages, create a complex picture of a staple that ordered the daily lives of those who subsisted on it. Several linguistic, iconographic, and narrative components of this picture are isolated here to reveal something of the effect of maize cultivation on the cultivators, of the symbolic continuities through time, and of the rich linguistic reflections of maize in their lives. Lexical relationships, creation stories, the iconography of power, food preparation, and some symbolic relations between maize and certain other plants are considered in this attempt to give a coherent smaller and necessarily partial view of the much larger picture of maize in discourse, symbolism and iconography of Mesoamericans.

INTRODUCTION

Anyone observing maize planting societies in Mesoamerica will notice the importance of maize to the people who plant it. It figures into every aspect of their lives, from the widespread tradition that humans were created from maize to the fact that they spend a large part of their waking time speaking of maize and the seasons, weather, and the plants and animals that affect its growth. One may even suspect that maize in one form or another colors almost every waking thought of those whose ancestors domesticated it and whose lives depend on its continued availability. One ethnographer has suggested that among the Yucatec Maya a man's identity is defined by his *milpa* (cornfield) and that the focus of all male and most female conversation is the *milpa* [30].[1]

This is no less true for the Mixe–Zoqueans and the Mayans of Southeastern Mesoamerica than for any other indigenous people of Mexico, Guatemala, or Belize, and the intent here is to present a small part of the linguistic and iconographic evidence supporting this contention by means of the following eight overlapping rubrics pertaining to symbolic thought and activity: 1) vocabulary, 2) narratives, 3) sayings, 4) rituals, 5) numbers, 6) glyphs, 7) images, 8) calendar. A ninth category treats some plants relating to maize, as evidenced through words and images.[2]

In accord with a conception of southeastern Mesoamerica comprising Guatemala, Belize, and Mexico east of the Isthmus of Tehuantepec and including the region occupied by Mixe speakers just west of the Isthmus, societies treated here are members of the Mayan and Mixe–Zoquean linguistic families. With a **time depth** of at least 4000 years, the Mayan family today includes 29 different Mayan languages spoken in numerous communities in Mexico, Guatemala, and Belize, and a 30th has become extinct since conquest. In Mexico most Mayan communities are in Chiapas, Quintana Roo, Campeche, Tabasco, and Yucatán, but Huastec Maya speakers, separated by 400 miles from the rest of the Maya, live far to the north in San Luis Potosí and northern Veracruz. Ancestors of at least one branch of the Maya family tree, the Cholans, created Classic Maya civilization (Figures 17-1 and 17-2) [22].

Approximately 12 Mixe–Zoquean languages can be distinguished, one of which disappeared after conquest. They are found mostly in western Chiapas, southwestern Oaxaca, and southeastern Veracruz (see Figure 17-1). Zoqueans are primarily on the Chiapas side of the Isthmus of Tehuantepec and at the northern end of it (Sierra Popoluca), whereas Mixeans are mostly on the Oaxaca side of the Isthmus. The Olmecs were probably Mixe–Zoquean speakers, which coincides well with the approximately 3500-year time depth ascribed to the language family [10].

Regardless of whether Mayan and Mixe–Zoquean languages are ultimately recognized as related genetically, contact and borrowing occurred among and between members of these families; in spoken vocabulary and in script [6, 10–12, 63]. The Olmec civilization, presumably Mixe–Zoqueans, from Teopantecuanitlan, Guerrero, eastwards to La Venta, Tabasco, in the Olmec heartland, flourished from around 1200 BC to about 400 BC. They may have had a glyphic script, but we have little direct evidence of it, unless they were responsible for a script attributed to Zapotecs at Monte Alban that began around 500 BC or a somewhat later script found at Kaminaljuyu, Guatemala, or both. Between approximately 300 BC and AD 200 Mixe–Zoquean speakers used a script around the Isthmus of Tehuantepec, recently discovered and partly deciphered, called the Isthmian script [21].

Florescence of Classic Maya culture, accompanied by an elaborate hieroglyphic script, began around AD 300 and

[1]Thanks to John Staller for insightful comments and editorial assistance in the preparation of this chapter, and also to an anonymous reviewer.

[2]Orthography here employed for Mayan languages is for each language referenced by a subset of the standardized set of letters approved by Academia de las Lenguas Mayas de Guatemala in 1986 for transcribing [13]. Phonemes are represented by **p t tz ch k, q' b' t' tz' ch' k'q' s x j h ' m n nh l r w y i e o a u ii ee oo aa uu ï ë ö ä ü** (note that **x** replaces the IPA **<s>**, **nh** replaces IPA's **<n>**, and the apostrophe represents a glottal stop). Mixe–Zoquean orthography is a normalized version of Wichmann's usage, with a couple of modifications, initial glottal stops are not indicated, and **ï** replaces barred-**i** (which is generally pronounced as an unrounded high or midcentral vowel). Indigenous words are bolded rather than italicized for easier recognition. Other conventions observed in this paper include an asterisk (*) preceding a reconstructed word, single quotes around word meanings, brackets around native words spelled as in the source, and proper names capitalized in indigenous words.

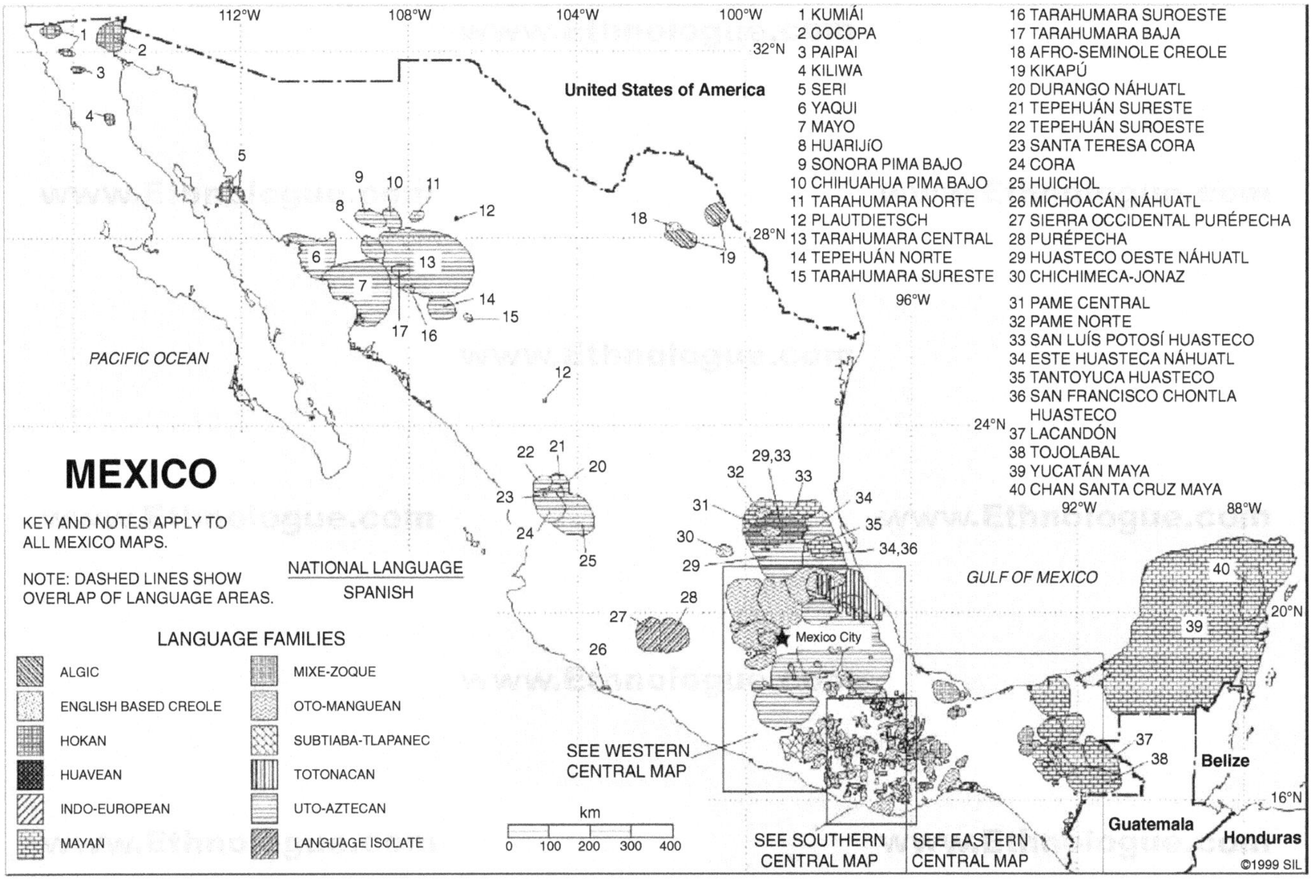

FIGURE 17-1 Map of Mexico with locations of Mayans and Mixe–Zoqueans. Of the latter, Mixeans are west of the Isthmus, Zoqueans to the east. (Courtesy Summer Institute Linguistics, www.ethnologue.com)

ended around AD 900 with the collapse of lowland Maya civilization that had been brewing for a hundred years. Although Olmec, Isthmian, and Maya civilizations were not contemporaneous, they were neighbors, and we can see Mixe–Zoquean contributions to Maya culture in the calendar and the script as well as in words spoken by Maya today [22, 40, 41, 44]. For example, the Maya script has a glyph depicting the earth, with a sound value of *na,* which is unlike any Maya word for "earth," but rather close to Mixe–Zoquean words for earth (*nax* or *nas*), and because the Isthmian script has a glyph formally and structurally similar to the Maya glyph for earth, borrowing from the earlier by the later script must be inferred [44, pp. 48–51, 45]. Maya borrowing of Mixe–Zoquean words occurred as well, exemplified by the Kekchían subgroup of Mayan languages having borrowed the word *po* (and *poh*) for "moon" from a Mixean language. Proto-Mayan **iik'* (moon) is different from Proto-Mixe–Zoquean **poy'a* (moon) [23, p. 135; 61, p. 438].

It is worth remembering that the Maya borrowed from the Mixe–Zoqueans, both in word and script, as we turn our attention toward words related to maize. Because much more has been written on Mayans than on Mixe–Zoqueans, attention to these families here may seem somewhat lopsided, a situation that will hopefully change with the passage of time before too many more words will have been lost forever.

VOCABULARY

The vocabulary of a people contains much information about their environment, their habits, and their thoughts. Elaboration of some domains of meaning implies the importance of those domains for the people, and we can infer that domains having few words to classify concepts are rather less important, whereas concepts without words are probably not recognized at all. Maize words of course tell us much about the people whose words they are and can also suggest historical contacts with other people.

Basic Maize Words

Maize words of the Maya reflect a set of words different from those of Mixe–Zoqueans although occasional tantalizing signs of contact exist even in such an important word

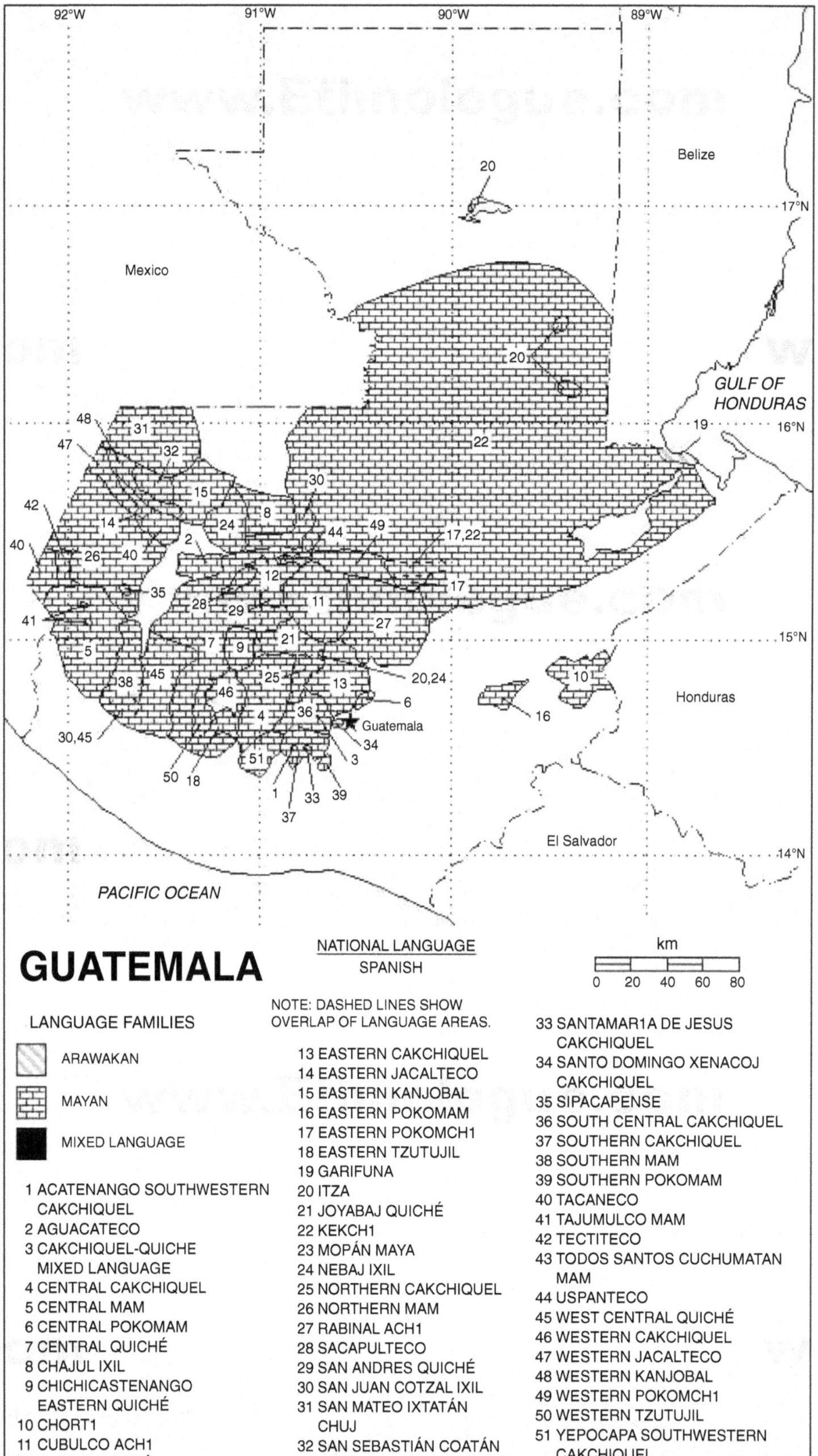

FIGURE 17-2 Map of Guatemala with approximate locations of Mayans. (Courtesy Summer Institute Linguistics, www.ethnologue.com)

domain. Beginning with the basic word for maize, reconstructed Proto-Mixe–Zoquean **ïksi* (maize [kernels]), which comes from Proto-Mixe–Zoquean **ïks* (to shell corn), and may be derived from Proto-Mixean **ïk* (to toast), only slightly resembles Proto-Mayan **ixi'm* (maize [kernels]) [23, 61]. As with Mixe–Zoquean, the Mayan word for corn kernels appears to come from the verb for "to shell corn," Proto-Mayan **ix* and comparing this with Proto-Mixean **ïkx* yields a more notable resemblance [23]. Subtracting the verb from **ixi'm* (maize kernels) leaves *i'm,* which is similar to Huastec Maya *eem* (maize plant), implying the possibility of a Proto-Mayan **i'm* (maize plant). The word *ixim* in many of today's Mayan languages means "maize, maize kernels." To demonstrate the importance of maize, these reconstructions have limited value. More instructive is the fact that Zinacantan Tzotzil Maya has at least 42 names for different kinds of indigenous maize [4].

Proto-Mixe–Zoquean *mo:k* (maize [generally], maize ear [specifically]) is comparable in meaning if not form to Proto-Mayan **nhal* (maize ear) [23]. Notably, Yucatec Maya retains *nal* (maize ear), whereas **Tzeltal** Maya *nal* (semen, seed) illustrates informative semantic change, as does Chol *na'al* (deity of abundance of plants and animals). Also of note, Huastec Maya may have borrowed *way'* (maize ear, *mazorca de maiz*) and a plausible donor might be reflected in Proto-Mixe–Zoquean **way* (to grind [maize]) [61]. While the *mazorca* (mature maize ear) provides the kernels for grinding, the *elote* (roasting ear) is freshly picked and not yet dried. Proto-Mixe–Zoquean **yaw-mok (elote)* is derived from the word for *mazorca,* whereas Proto-Mayan **ajn (elote)* is basic.

Maya words for "cornfield, *milpa*" were recruited from various sources, as seen in Chol *chol* (milpa), related to Tzeltal *chol* (to clear off) or to Tzotzil *chol* (to form rows or lines). Jacaltec *awal* (cornfield), by contrast, comes from "to plant seeds," and Tzeltal *k'al* likely came from *k'al* (clear a way, make an opening) (although one Tzeltal consultant suggested it came from Tzeltal *k'ahk'al* (sun, hot) recalling Chontal Maya *k'intuni* (plowed cornfield) from the word for sun). Tzotzil *chob* (formerly *chab*) (planted cornfield) is cognate with Yucatec *kab* (earth, land, *pueblo*) and Yucatec *kol* (*milpa*, to clear land) is cognate with Chol *chol* (cornfield), suggesting the latter's derivation from a word for "clearing land" rather than "forming rows."

Proto-Mixe–Zoquean **kama (milpa)* (cornfield), unlike the Mayan situation, was delivered with meaning intact to most of the daughter languages. A borrowed reflection may be found in Ixil Maya *ko'm* (cornfield) [61].

So many concepts involve maize, its planting, growth stages, harvesting, storage, preparation, and consumption, that most cannot even be enumerated here, let alone saying more about their expression in Mayan or Mixe–Zoquean languages, a fact that is itself informative. Named concepts involving maize are truly staggering in number. Therefore, I propose to address growth stages, briefly mention selected reconstructed words and their meanings, and then introduce interesting notes on the maize deity.

Maize Growth Stages

Named stages in the growth of maize are rarely encountered in the ethnographic literature. Notable exceptions are for Tzeltal Maya, Huastec Maya, Mochó Maya, and Tzutujil Maya [1, 2, 9, 29]. The following 13 stages were given by Tzeltal consultants (Table 17-1).. The theme of maize growth has significance to the calendar.

Among the stages of maize, the term for **maize tassel** *(flor de milpa)* is given as *tz'utoh* in Tzeltal. Stability in this word is attested by the Proto-Mayan reconstruction **tz'utuuj* [23]. The Proto-Mixe–Zoquean term for maize tassel is **mok-pïhy*, compounded from words for maize and flower [61]. The husk or *doblador* that protects the maize ear from intruders is reconstructed as **jomoch'* for Proto-Cholan [23]. Significantly, Proto-Cholan represents the subgroup of Mayan languages most closely related to the language of the Classic Maya. Several eastern Mayan languages call the husk *jo'q* and in Tzeltal is it *hohoch.* For similar reasons of variation the maize husk cannot be reconstructed for Proto-Mixe–Zoquean, valuable as this item is for Mesoamericans; but in Proto-Oaxaca Mixean, it is **ahktz* [61]. Once stripped of its kernels, the corn cob or *olote* remains. This is reconstructed for Proto-Mayan as **b'aqal* and is based on the word for "bone" [23]. Proto-Mixe–Zoquean **hï:pak* (corn cob) is similarly based on the word for "bone." Interchange of human body part terms with maize part nomenclature is particularly notable for Mayan languages, and this is another indication of the close connection between people and maize in Maya thought.

Huastecs recognize at least 12 named stages of maize development and tie the plant's development to the aging of the "spirit of the maize," seen as a little boy who grows up as the maize does, suggesting an equivalence between humans and maize that will be explored later [1, pp. 354–361]. Like the Tzeltal, Huastecs (or Teenek) also make a large number of named food items with maize, and they are particularly well known for making a large meat filled tamale called a *bolim* for special occasions, and an even larger tamale called a *sakawil* that feeds many and may take two people to carry it.

Food Preparation

Foods are prepared from maize at different stages of development, and remarkably, Tenejapa Tzeltal speakers prepare at least 32 named food dishes from maize and bean products alone. Representative maize dishes include boiled *hi* (immature maize ear [taking one ear early enlarges the one remaining]) and *ahan* (roasting ear, *elote*) roasted with

TABLE 17-1 Tzeltal Stages in the Life Cycle of Maize

Stage	Tzeltal name	Description	Days
1	lok'ok y-isim	Roots appear	1–20
2	luhpuk yuninal sni'	Its first leaf appears	21–40
3	lach'etal te yabenale	Leaves become compressed	41–60
4	chexlahetal te yabenal	Leaves become leaflike	61–80
5	pech'ok ta yakane	Stem shows sheath covering	81–100
6	tz'al haben te yabenale	Leaves appear as levels	101–120
7	tek'ni' te ste'el sok ya xlihk lok'el spakumal a te shiale	Staminate inflorescence of stem appears and first immature corn ear appears	121–140
8	tohkel stz'utohil	Staminate inflorescence born	141–160
9	lok'shial	Young immature corn ear emerges	161–180
10	me'hiiltikix	Mature jilote with immature styles (**hi** stage)	181–200
11	chamholtikix	Mature roasting ear, elote (literally "dead head" (**ahan** stage)	201–220
12	xk'anub yixmal k'alk'altik	The maize ripens	221–240
13	xtakeh yixmal k'alk'altik	The maize dries (**ixim** stage)	241–260

If we assume that each of the 13 named stages stands for a 20-day period—as Tzeltal consultants imply—then a 260-day period is completed at the end of the named stages. This theme of maize growth cycles has significance to the calendar.

the husk intact. Mature *ixim* kernels boiled with lime are occasionally eaten as hominy, and called *paynil,* but usually ground on a *metate*. Thrice ground maize known as *matz'* is placed in water and stirred, making a cold maize gruel also known as *matz'*, or made into tortillas by flattening a ball of *matz'* to be heated on a *samet* (griddle) and then called *wah* (tortilla), this same word being used also for "food in general." Tortillas are usually eaten with beans, chiles, and salt. Ground maize dough, in addition to being used to make tortillas or *matz'* (cold maize gruel), locally known in Spanish as *pozole,* can be fashioned into several types of tamales, called *patz',* which are then steamed in an urn. A drink, commonly consumed on festal occasions is made of boiled corn dough, water, and brown sugar and named *ul,* the glyphic spelling of which has been found on vessels, once filled with *ul* in Classic Maya tombs.

Proto-Mayan **q'oor* has been reconstructed for "maize dough, *masa,*" but this term was not passed consistently to the daughter languages [23]. "Maize dough" has been reconstructed for Proto-Mixean as **hïtzi* and for Proto-Zoquean as **ki'ti* [61]. Maize dough in cold water makes a maize drink that is reconstructed for Proto-Mayan as **maatz',* but the daughter languages have several different words for it; for example, Yucatec Maya **k'oyem* (from Proto-Mayan **q'oor [masa]*), Chol *sa',* Choltí *sipak,* Chuj *pichi',* and Ixil *jul.* In many Mixe–Zoquean languages the term for *pozole* stems from the Proto-Mixe–Zoquean words noted earlier for maize dough.

Maize dough boiled in water with other additives makes a ceremonial maize drink known in local Spanish as **atole,** and the Proto-Mayan word referencing this drink is **uul* [23]. Chamula Tzotzil *ul* references both "maize gruel" and "semen" in "men's speech" [16, p. 229]. *Atole* has many varieties, some sweetened, some soured, and the varieties have more names than there are additives. Ground parched maize is generally known as *pinole* in local Spanish; in Proto-Mayan it is **k'aj* [23]. It becomes a tasty drink with the same name when stirred in water. Proto-Mixean and Proto-Zoquean branches diverge with names for *pinole,* the former reconstructed as **mo:hk-way* (literally "ground maize") and the latter as **po'te* [61].

After maize has been harvested and shelled, the kernels are boiled with lime (calcium hydroxide), softening the kernels and creating soft wet hominy known in Spanish as *nixtamal.* Mayan languages have various names for this hominy, including some from the Proto-Mayan word for maize dough. Huastec has an especially interesting name for it, *pitzi',* which means both "hominy" and a "species of bird." It is said that the *pitzi'* bird was once a woman too lazy to make her husband tortillas, so she served him the more easily prepared hominy *(pitzi')* instead. Because of her laziness she was transformed into the lazy bird [1, p. 89]. Huastec must have borrowed the word for hominy from a Mixe–Zoquean language, for *nixtamal* (hominy) reconstructs to Proto-Mixe–Zoquean **pitzi* [61].

Ritual Names and Maize Deities

In the figurative and formal vocabulary of prayers and other ritual speech, many plants and animals have special names. Not surprisingly, in these discourse forms maize has special terms as well; Yucatec Maya *Grasya* (young maize plant) uses a word derived from Spanish *gracia* (grace, favor, charm), whereas *Santo grasya* means "maize ear." Chenalhó Tzotzil employs *Anhel* similarly, although this word can also mean "earthlord, mountain, lightning, or wild tobacco" (*Nicotiana rustica* or *N. glauca*). Quiché ritual speech uses *Qanan* (corn, literally "our mother") or *Mariya* (corn, literally "Mary" referencing the Virgin Mary), and Cakchiquel uses *Jolooma* (dried maize kernels, literally "little skulls"). Chamula Tzotzil Maya use the ritual term *Xohobal* "radiance, sun's halo" referencing maize foods. Chenalhó Tzotzil Maya employs *Ob* "maize (when addressed)" for ritual occasions, and Zinacantan Tzotzil Maya ritual speech includes both *Xohobal* (radiance), and *Xohobal ryos* (corn; sunbeam [literally analyzed "needle of God"]). It is notable that the ritual terminology not infrequently comes from Spanish.

Freidel, Schele, and Parker [15] identify the Classic Maya maize god as the "First Father" whose glyphic name was *Hun Nal Ye,* which they gloss as "One Maize Revealed" [p. 55]. J. Eric Thompson identifies from the Yucatec Maya Chilam Balam of Chumayel, a name for maize, *Ah Mun,* suggesting it to be a name of the Yucatec maize deity of the colonial period and interpreting the root *mun* as reference to an unripe crop or youth [36, p. 48]. Although Lacandón Maya uses *mu'un näl* to mean "young maize," the terms *munach* (slave) and Quiché *mun* (captive) in closely related Yucatec suggest additional dimensions of meaning that could inform our interpretation.

Contemporary Lacandón Maya identify a maize deity, lord of the *milpa,* whom they call *Äk'inchob* (*dios del maiz* [literally "cross-eyed sun lord"]) [7, p. 131, 26, p. 62] and whose wife is *Ixchel,* the goddess of pregnancy and childbirth. The Mochó Maya are also said to identify a maize deity, *Ham-iil paach-iil,* the meaning of whose name can be inferred from *ham* (aperture, crack, opening) and *paach* (home in the middle of the milpa, twinned ear of maize).

Chortí Maya identify *Ih p'en* as the (spirit of maize) or (soul of maize), another name for whom is *Ah k'anan* (he of the yellow maize) according to Charles Wisdom [62]. *Ih p'en* also names the (earth deity and protector of *milpas,* houses, and property) [62]. As the spirit of maize he is the male consort of the spirit of beans, *Ix k'anan,* "female of the yellow maize" by Wisdom's [62] analysis. Lacking a Maya root *p'en,* one can note simply that Mixe *pen* (to plant, sow) is appropriate in meaning, and suggests possible contact in the past (Mixe–Zoquean languages lack glottalization). Chortí *Ah yum nar* names the "four pellets of copal placed in the four corners within the storehouse to protect the maize (literally 'lord of maiz')," whereas *u mein e nar* (spirit of maize) comes from *mein* (spirit [of plant or animal], shadow). Whereas Chortí has no words from which to analyze *matulin* (a guardian spirit of the maize storehouse), it is noteworthy that Tzeltal Maya *matul* is a medicinal hallucinogen of the genus *Datura.* Although primarily male for the Chortí and Huastec Maya, the maize spirit is seen as female by Tzeltal and Tzotzil Maya speakers. The data are too scant at this point to draw conclusions about gender concepts in this regard.

The Tzeltal *X'ob* (spirit of maize) appears as a tiny young maiden and often is seen "to have a bloody nose." It is significant that Mitla Zapotec *xob* means "maize," specified in colonial Zapotec with the spelling [xooba] as "maize kernels." The Tzeltal word, because there are few traces in Mayan languages, could have historical connections with the Zapotec term [22]. Tzotzil X'ob (soul of maize [and beans]), is the maiden daughter of *Anhel* (the earthlord), and wife of *Yusumprun.* She once took an ear of maize from each corner of his *milpa,* and it miraculously multiplied, recalling images of the Olmec maize deity–ruler surrounded by an ear of maize in each corner of the image [52, p. 46]. Nevertheless he was an abusive husband and slapped her more than once, which explains the blood seen on her nose [17, p. 40; 218; 291]. Blood associated with the soul of maize, who, though married is seen as a virgin, reminds one of "Blood Woman" in the sacred Quiché Maya book, the ***Popol Vuh.*** Daughter of one of the lords of the Underworld *(Xibalba)* and still a virgin, she is impregnated by saliva from a skull on a forbidden tree and subsequently forced to flee the Underworld. Arriving in this world she picks a maize ear from the *milpa* of her brothers-in-law and makes it miraculously multiply, convincing their mother that she is indeed their daughter-in-law. The evidence suggests Blood Woman to be the "soul of maize," a point heretofore unrecognized.

Tzotzil *Ob,* a ritual term addressing maize in Chenalhó Tzotzil, would be *X'ob* if referential and is thus clearly related to the Tzotzil and Tzeltal *X'ob* (soul [or spirit] of maize). Less obvious is a relationship to Chol Maya *x'ob (Hamelia patens, coralillo,* scarlet bush), which is a plant that grows up to 9 feet tall (like maize), has long green leaves emerging in three from red nodes on the stalk, and long bright red flowers, recalling the Tzeltal and Tzotzil maize spirit's bloody nose. The plant has several medicinal properties and is referred to by Mopan Maya speakers as the "guardian (of the forest)" *X-kanan,* and by Yucatec Maya as *X-k'anan,* derived from a root for "yellow" [5, 35].

NARRATIVES

Although words when explained or put in context certainly lead to stories, narratives are themselves stories, and stories are sometimes seen as the basis for culture as well as a primary means of cultural continuity. A few interpreted snippets from narratives will suffice to illustrate some of the importance of maize to Mayans and Mixe–Zoqueans. For the Mayan languages, the illustrations will come primarily from the *Popol Vuh,* an early colonial period Quiché written narrative, from the *Chilam Balam of Chumayel,* a nineteenth-century Yucatec manuscript with passages copied from early colonial texts and from narratives collected by anthropologists in the twentieth century from various Mayan languages. All Mixe–Zoquean narrative material derives from twentieth-century collections by anthropologists and other visitors to indigenous communities.

The *Popol Vuh* was an oral narrative, probably written down in the middle of the sixteenth century, by a Quiché Maya speaker, recounting the origins, traditions, and history of the Quiché Maya people. According to this tradition, following the unsuccessful creation of humans first from mud, and then wood, in the third creation the first eight ancestors (four men and four women) were formed of yellow and white maize dough, by *Xmukane,* grandmother of the "hero twins" (themselves strongly identified with maize), who had been told about a mountain filled with maize, discovered by the fox, coyote, crow, and parrot. To keep humans from being too wise, the gods blew mist in their eyes, recalling the Huastec Maya creation narrative specifying maize pollen as the material blown into their eyes [1, p. 62]. The Cakchiquel neighbors of the Quiché record humans as created by the gods from maize dough mixed with the blood of the tapir and the serpent. Notably, "Great White Tapir" was one of *Xmukane's* epithets in the *Popol Vuh.*

Before the creation of humans, the *Popol Vuh* recounts adventures of the "hero twins." These twins are *Hun-Aj-Pu,* whose name is given as "One blowgunner" by Dennis Tedlock and *X-Balan-Ke* "Little Jaguar Deer" or "Little Jaguar Sun" (meaning "moon" according to Tedlock) [54]. They ultimately sacrifice themselves in a fire and rise into the heavens becoming the sun and the moon. Because *nal* is "maize ear" in the Cholan language of the glyphs, and *aj* is "fresh maize ear" in Quiché, I suspect that *Hun-Nal-Ye,* the Classic Maya maize deity, is the counterpart of the Quiché *Popol Vuh's Hun-Aj-Pu,* name for one of the hero twins, which—with an extra *hun* (one) prefixed—is also the name of their father. A number of scholars have proposed on other grounds that the hero twins are dual aspects of a maize deity or that they represent maize itself [54]. The twins' adventures in the upper and underworlds can be read both as a creation myth narrative and as an allegory of the cosmic cycle with a focus on the adventures of maize as it faces many trials underground before sprouting above ground. The hero twins appear to represent anthropomorphic maize and perhaps a single twin ear of maize. When they leave their grandmother to visit the underworld, for example, they each "plant" an ear in the center of the house, telling her that when the maize dries up it will be a sign of their death [54, p. 42]. She cries when the maize ears dry out, signaling the death of her grandsons, and rejoices when later the ears sprout, meaning that they have returned to life.

More obscure references to maize reside in the *Popol Vuh* as well. One, I believe, revolves around **Sipakna,** a crocodilian being—son of luciferian Seven Macaw who probably represents the sun of the previous era. The name Sipakna shares much with Aztec *Sipaktli* (first day of the 20-day month cycle, meaning crocodile and/or shark). Before leaving for the underworld, the hero twins set out to kill Sipakna who had himself just finished killing "the 400 boys" identified by Tedlock as the gods of alcoholic drinks [54, p. 36]. Earlier Sipakna had feigned death when the 400 boys tried to crush him with a huge log pole after having tricked him into digging a deep hole, as if he were maize being planted [cf. 54, p. 265]. Sipakna had hidden in a side pocket at the bottom, and when they planted the pole in the hole, he then had ants bring his hair and fingernails to the surface. The 400 boys saw these as tokens of his death and celebrated for 3 days by getting drunk on what I interpret as the maize fingernails and hair of *Sipakna,* based on an analogy with the Aztec maize deity *Sinteotl.* Grandson of the Aztec counterpart to the *Popol Vuh*'s *Xmukane, Sinteotl* was buried in a cavern in a mountain so that his body parts could emerge as plant foods to later feed humanity. His nails, for example, became a long variety of maize. Once the 400 boys were drunk, *Sipakna* crushed their house with them inside. Later they ascended to the skies, becoming the Pleiades, a constellation whose appearance alerts many peoples, including Mayans, to planting and harvesting activities.

Identification of *Sipakna* with some aspect of a maize deity is suggested by more than his name's similarity to the Aztec day name meaning "crocodile" and goes beyond the earlier interpretation of a *Popol Vuh* story in maize planting allegorical terms. **Thipaak** is a culture hero of Huastec Maya tradition who saved the people from having to sacrifice their children and who first cultivated maize and stored it in the huge pillar supporting the sky, long before it was made available to humans [1, p. 62, 90]. Long ago a bird drew attention to maize kernels being brought by leaf cutter ants from the depths of the pillar supporting the sky, *T'ithach,* currently the name of a 9000-foot peak in the Huastec region. The eastern rain deity opened up the pillar with a blast of lightning, making maize available to the people [1].

Noting that Huastec *th* corresponds to *s* in other Mayan languages, and that *-na* of *Sipakna* and the *-tli* of Aztec *Sipaktli* are removable suffixes, permits recognizing Thipaak as related to *Sipak,* the root form of the Aztec day name meaning "crocodile." *Sipak* is likely etymologically related to the term *ceiba* applied to the kapok tree *(Ceiba pentandra).* Referred to as the tree of breasts, the *ceiba* tree is said to sustain babies who die in childbirth. Janis Alcorn [1] notes that *Thipaak*'s maize storehouse was close by *Tampaxal* (literally "place of *Paxal*"), a name similar to the maize mountain, or "sustenance mountain" of the Guatemalan narratives of the *Popol Vuh* and *Annals of Cachiquels,* which is named variously *Paxal, Paxil,* or *Paxel* [1]. Remarkable traditional memory is evidenced, or perhaps some later contact, as the Huastec left their Guatemala homeland more than 3000 years ago. Just as remarkable, a *tz'ok* bird impregnated *Thipaak*'s virgin mother (a *ta'tam* bird). This occurred when she went to the creek to wash the *nixtamal* made of the maize precursor *ohox ("Brosimum alicastrum")* and the *tz'ok* whistled at her or defecated in her mouth, depending on who tells the story. The similarity to the *Popol Vuh* narrative involving Blood Woman and her virgin birthing of the hero twins is again striking.

Thipaak is called by Huastecs the *ehatal* (soul) of maize, the *tz'itziin* (spirit) of maize, and the *ichiich* (heart, embryo) of maize, a maize identification yet more certain in that local Nahuatl speakers refer to *Thipaak* as *Sintektli* (maize lord) [1, pp. 68, 204].

Hundreds of miles to the south, at the Mexico–Guatemala border, late Pre-Classic Stela 25 at Izapa depicts a large bird, perhaps representing Seven Macaw of the *Popol Vuh.* With no lower jaw, it holds a person's detached arm, whereas a man below, lacking an arm, seems to represent a precursor to *Hunahpu.* They face an upended crocodile whose tail is a leafy tree on which a smaller bird perches, inviting a conclusion that the crocodile represents *Sipakna,* while the bird perched on him, represents his brother *Kabrakan* (two legs), both being offspring of Seven Macaw. The similarity to the hero twins narrative of the *Popol Vuh* again puts the crocodile into a close relationship with maize.

In the Yucatec Maya *Chilam Balam of Chumayel,* the story of maize is couched in ritual language from which it is difficult to extract a complete outline, but it is clear that before heaven and earth there was a maize spirit, referred to as the "three cornered jade stone," who remained within the maize, later emerging with long locks of hair. Apparently maize was hidden within a stone pillar in this rendition, but the story is obscure here [36, p. 107, 55, p. 351]. The three-cornered stone refers to the maize ear, represented during the Maya Classic period as sometimes pointed pendant jadeite celts, occasionally alone, but typically in groups of three [52, p. 42]. Because the *Chilam Balam* refers to creation here, it is of note that the cosmic **hearth of creation,** the "First-Three-Stone-Place" identified by Freidel, Schele, and Parker [15, p. 69, 79] from Palenque inscriptions, consists of three stars forming an equilateral triangle hanging from Orion's belt in the constellation seen as the giant Orion. This hearth of creation is identified as a Quiché hearth by contemporary Quiché speakers (Figure 17-3).

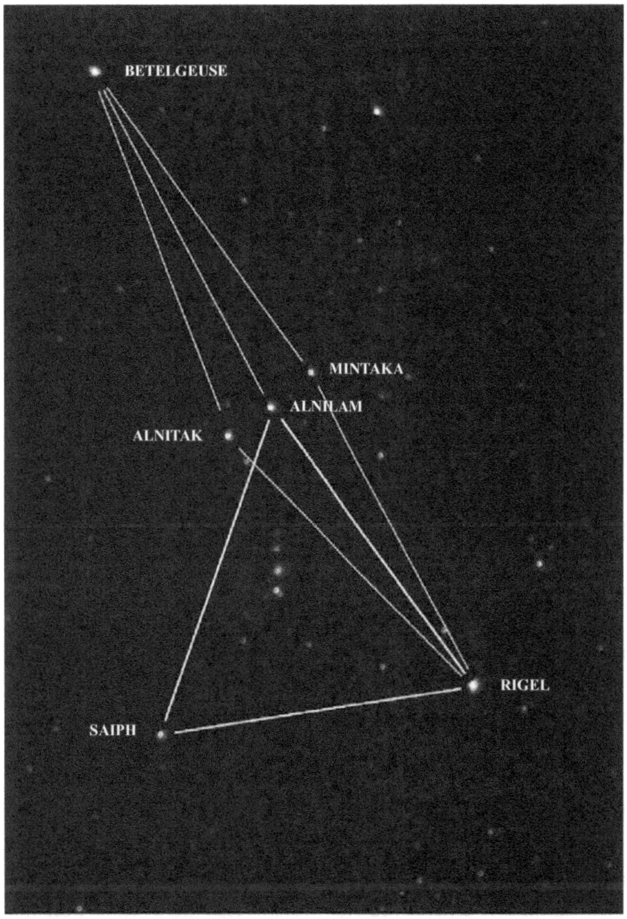

FIGURE 17-3 A photograph of the night sky taken in Ecuador showing Orion's belt and the associated stars making up the "hearth of creation," with labeled stars showing the hearth of creation and "child's hammock" (Photo taken by Natalia Biani).

> Today these three stars [Alnitak, Saiph, and Rigel in Orion] are said to be the three hearthstones of the typical Quiché kitchen fireplace, arranged to form a triangle, and the cloudy area they enclose (Great Nebula M42) is said to be the smoke from a fire. [54, p. 261]

I suggest that the Classic Maya three stone hearth hanging from Orion's belt as jade celt(s) used Alnilam, the central star of the belt, rather than Alnitak, because it is brighter, because it approximates a more equilateral triangle, and because Mixe neighbors of the Maya see Alnilam as an important star in their "hammock" constellation. If indeed this is the hearth of creation from which the maize deity emerged (see Figure 17-3), the three stars of Orions belt, which for the Maya was a constellation known as the Turtle, can be seen as a roasting ear of maize, being cooked on the hearth of the heavens [15, pp. 80–81]. From macrocosm to microcosm we can see the Maya place of creation in at least three arenas. In the heavens it is a triangle of stars equidistant from the Orion nebula M42. On earth in the middle of the house floor it is the three hearthstones with a central fire, and on the body it is the sacrum with its appended coccyx [47]. In this latter connection, Chamula Tzotzil Maya tradition holds that maize came from a piece of the sun's groin and included a part of his pubic hair, which is cornsilk. The moon gave beans (her necklace), and potatoes came from her breast milk [16, p. 40].

Most Mixe–Zoquean narratives differ considerably from Mayan ones, and although in Mixe lore twin brother and sister become the sun and the moon, twins do not appear in other narrative contexts [27, p. 218]. Elements of the Mixe narrative in which the twins become the sun and moon will be recognized by readers of the *Popol Vuh.* A maiden is impregnated by a bird suckling at her breast, and her twin children cause injury to another bird's beak—not a macaw in this case, but a vulture. The twins, who are hunters, live with their grandmother. Leaving the house they overcome a series of obstacles and eventually become the sun and the moon. The details of the narrative, however, differ significantly from those in the *Popol Vuh.*

In the Zoquean branch, Sierra Popoluca tradition regards a dwarf with golden hair named **Homxïk** as the "spirit of maize," patron of the *milpa* [37, 38]. Homxïk was hatched from an egg and claims to be "he who sprouts at the knee joints," and "he who is shelled," and "the one who is eaten." From the beginning he announced his mission to feed humankind.

Homxïk was discovered while still an egg by an old man and woman who brought him home with them. He hatched in 7 days, and in 7 more he could walk and talk. When his adoptive parents tried to eat him, he had a bat cut his father's throat; later his mother died by fire. Afterward, while drumming at the seashore, Homxïk was summoned to the land of Hurricane across the sea. After crossing the sea on the back of a tortoise he was put through a series of trials, much like those the Quiché Maya hero twins endured in the Maya underworld of *Xibalba.* One may note that the hurricane, like the dust devil and the whirlwind, is associated by Mesoamericans with the underworld, and that Zoque *supupi* (whirlwind) recalls *Xibalba* in form and sounds. Having passed the trials, *Homxïk* tricks the inhabitants of Hurricane's land into dying in a hammock swung over the sea, which falls into it. Hurricane, the lone survivor, albeit with an injured leg, capitulates, promising from then on to water *Homxïk*'s head when he is dry [14]. One can imagine *Homxïk* as the maize deity being swung far out over the sea in that hammock to reach his own land after having defeated Hurricane.

Mixe tell of a celestial hammock, a constellation named *Ungazaay-maza'* (Child's Hammock), extending from Betelgeuse to Rigel, with Orion's belt in the middle, that they petition for rain [25, p. 112]. The middle star of Orion's belt, Alnilam, is a special star to Mixe descendants of the Olmecs and neighboring predecessors of the Classic Maya, and they call it *Maza'nïï* (Star of the Sea). The central part of the Child's Hammock constellation, the three stars of Orion's Belt would seem an excellent locus for heavenly creation, when we note that a cradle—the hammock of cooler climes—is a metaphor for "origin point" as in "the cradle of civilization," and indeed an Olmec "cosmogram" on a celt appears to depict three circlets at the base, above which is a stepped mountain that is surmounted by a maize tree *axis mundi* under a symbol for the sky [52, p. 46]. The three circlets clearly are the foundation on which the rest is constructed. In short, Mixe celestial interpretations fit nicely as macrocosmic setting and partial analog for the Sierra Popoluca maize narrative, and they also represent a plausible analogue to the Maya locus of creation, also in Orion, but with a slightly different focus, whether the three creation stones in a row seen by Olmecs in the heavens were Betelgeuse, Alnilam, and Rigel together, or Orion's Belt itself.

SAYINGS, METAPHORS, AND BELIEFS

In addition to narratives relating events with reference to a chronological order, more pithy expressions of belief, instruction, and/or classification are frequently found in Mayan and Mixe–Zoquean discourse. These can be termed sayings, and they frequently employ metaphors and embody traditional beliefs, making them particularly useful to record and interpret. In the context of discourse, invocation of these sayings often has an educational objective, and many of them involve maize. Sayings and verbalized restrictions concern other foods as well, but not nearly so many as involve maize. A few are exemplified here, from various sources, mostly Mayan (see Table 17-2).

Rituals

Many Mayan and Mixe–Zoquean rituals involve maize, whether as offerings, as instruments of divination, or as the ritual's objective. Rituals attending childbirth, planting, harvest, rain petitions, curing, new house construction, prayers for relatives, community renewals, death, and so on, are often called *costumbre* in Spanish. These rituals are invariably accompanied by formulaic discourse that we might label prayers, invocations, and incantations, and are frequently performed by specialists. Maize in the form of tortillas, tamales, or cornmeal is a frequent sacrificial offering on altars, in caves, on bodies of water, or at the foot of trailside crosses or shrines, for these are the places that can become portals to the other world where those to whom the petitions are made reside.

An interesting divinatory ritual symbolically equates blood and maize, a widespread equation in Mesoamerica, and one delivered in many forms. One seventeenth-century

TABLE 17-2 Sayings about Maize

Language	Saying	Source
Mixe	Do not eat twin maize ears (**ko:ngmok** [king maize]). They mean a good harvest. Put them on the house altar and later hang them in the kitchen or granary for good luck and good harvest.	[25]
Mixe	Maize should be planted between the first crescent and the last quarter of the moon.	[25]
Mixe	Maize kernels spilled on the ground must be retrieved to avoid offending the Lightning deity and risking the loss of the next harvest.	[25, 27]
Tzotzil	It is bad to eat pieces of burnt tortilla straight off the griddle because people will then slander you.	[39]
Tzotzil	You should not eat the first tortilla that comes off the griddle or you will become talkative, saying bad things about other people.	[39]
Quiché	Do not eat the first tortilla that comes from the *comal* or you will become forever unsatisfied. Same if you eat another tortilla before finishing the first.	[34]
Tzeltal	To dream of selling maize means one's child will die (of soul loss).	field notes
Yucatec	To dream of a full granary foretells good health for the dreamer	field notes
Lacandón	To dream of roasting ears is a bad omen, foretelling vomiting, abdominal pains, and toothache.	[7]
Mochó	The atole jar must be kept clean. Atole sticking to the edges means that your children will have continually runny noses.	[29]
Mochó	A menstruating woman who weeds in the *milpa* will cause the maize kernels on the ears to be sparse and widely separated.	[29]
Mochó	Do not blow on atole (boiled maize gruel) to cool it or your teeth will fall out.	[29]
Mochó	A pregnant woman cannot eat a twin elotes or she will give birth to twins.	[29]
Quiché	Maize resents coffee, so tortillas should not be dunked in coffee.	[34]
Tzeltal	Maize is a protector of children. If one leaves an ear of yellow maize with a child that is left alone, this will prevent the child's soul from being stolen.	field notes
Cakchiquel	Maize is a protector of children. If one leaves an ear of yellow maize with a child that is left alone, this will prevent the child's soul from being stolen.	[34]
Tzotzil	Unless people eat tortillas they are not fully socialized and cannot learn to speak Tzotzil.	[58]
Quiché	If one sees kernels of maize strewn on the flower one must pick them up immediately, because it is our own flesh and blood that suffers. If this is not done, one could die of hunger.	[34]
Quiché	When shelling maize, one must do the whole load, or the kernels on the next harvest's ears will be uneven.	[34]
Quiché	Never sell or give away the first fruits of the maize harvest.	[34]
Quiché	It is bad to steal maize, to spit it out, or in any other way to demonstrate lack of respect for maize.	[34]
Cakchiquel	To steal elotes (roasting ears) is bad, but worse is to steal odd numbers of them. It is also bad to give away an odd number of elotes, and a woman who does so will never in her life find a mate.	[34]

writer noted that Pokomam Maya during childbirth united the newborn with maize by severing its umbilical cord over a multicolored ear of maize with a new obsidian knife, accompanied by prayers for its well being. The bloodied kernels of the maize ear were sown in the child's name, and that harvest was then sown, with the yield used to feed the child until old enough to plant his own *milpa* [55, p. 283]. Tzotzil Maya have a similar ritual, in which blood from the umbilicus is dripped on a maize ear, the kernels then being sown by the father in a little milpa called "the child's blood." The progress of this crop is said to predict the child's future, and the harvest is shared by members of the family as a kind of communion to introduce the new family member [17].

The Zinacanteco Tzotzil Maya Fiesta of San Sebastian has a ritual in which a "jousting target" called the *k'oltixyo* (probably from Spanish *colgadizo* [pendant]) is an important component that on specified days of the fiesta is struck by specified ritual actors. Pointed at the top, sprouting multicolored ribbons from the bottom, about a foot long, having a red background decorated with yellow, black, and silver rings, the *k'oltixyo* greatly resembles a maize ear fetish, and indeed must have at one time been precisely that [50]. Today, some say it represents the heart of San Sebastian [59, pp. 207–211, cf. 52, p. 72]. Sporting a cross at the top, it has five nodes, and the internodes are color striated with three sections each. Here we have a ritual in which a onetime maize symbol is ritually speared, almost as if it were a deer.

Quiché Maya divination as practiced in the *Popol Vuh* involved both maize kernels and the seeds of the *tz'ite'* (coral bean tree), *Erythrina* spp. and now in Chichicastenango, the Quiché curer diagnoses illness by maize divination as well [8, 54]. Tzotzil Maya curers of Larrainzar, to divine the outcome of an illness, take a specified magical number of maize kernels from a maize ear, blow on them, pray, and throw them into a gourd container with a little water. When the kernels reach the bottom the diviner studies the pattern that they make as an indication of the patient's outcome [18]. The Mam and Kanjobal Maya employed similar maize divination practices [24, 28].

Mixe–Zoqueans too use maize for more than sacrificial offerings. Maize divination is employed to discover the location of sorcery objects. If the cause of an illness is indeed sorcery, then the maize kernels will "jump" toward the culpable object, which will then be discovered and removed [25]. In a village protection ritual four kernels of maize are placed equidistant from one another and nine more are dropped one at a time into the center, and the resulting patterns are interpreted [25]. At planting time birds are used to divine the harvest of maize and beans. The Lightning deity *Ko:ng Ene* is entreated to let the supplicant's harvest be bountiful, and the heads of a newly killed male and of a female turkey are sacrificed over bags of seed maize and seed beans, allowing the blood to flow onto the seeds. Then the heads of the sacrificed fowl are dropped into the bags. Each seed sticking to the bloody part of the head represents a *carga* (43 kilos) that will be harvested in the fall. To begin the harvest, fowl are sacrificed over a pile of maize kernels, and an offering of tortillas is placed in a nearby hole, followed the next morning by an offering of five maize ears, but no divination is attempted. However, once the harvest is in, double ears of maize augur a good harvest next year (see Table 17-2), and they protect this year's other harvested ears [25].

Numbers

Numbers have had important symbolic significance for humans in many if not all societies, and the Mayans and Mixe–Zoqueans are no exception. To the Maya, for example, the number 9 symbolized the underworld with its 9 levels; 13 symbolized the sky world, with its 13 different levels; 7 was the number of layers of the earth; whereas 20 symbolized completion. Whereas the Aztecs associated the number seven with maize, the Maya associated maize with the number eight. Information from Mixe–Zoqueans is insufficient to ascertain reliably what numbers were associated with maize.

For the Classic Maya, eight is the number principally associated with the maize deity and with maize itself, and some reflection of that association remains in contemporary Mayan languages. For example, Chortí women eat only maize foods for 8 days following birth, when menses begins a Chorti girl will begin an 8-day maize fast, a newly married Chortí couple will fast on maize for 8 days, and relatives of deceased Chortí fast on maize for 8 days [62, pp. 289–301].

In the Classic Maya script each number from 1 to 20 has a bar and dot variant and a "head variant." The head variants are heads with headdresses of the deities associated with those numbers. The normal variant of eight is a bar and three dots. The head variant for eight is the young maize god's head with a "maize curls" forehead ornament, and a maize ear or maize foliage, or both, in his headdress. Additionally, each of the 20 named days is ruled by a deity that occasionally appears in the glyphs. The ruler of the day *K'an* is the maize deity, and *K'an* is the eighth day of the 20-day "month" when we start with *Kaban* (earth), which is ruled by the goddess of number one [56]. The word *k'an* in lowland Mayan languages means "yellow, ripe," and the glyph for the day *K'an* depicts either a maize kernel as some infer, or a maize tamale as currently preferred [51]. Closer to the present, the *Popol Vuh* says that the first people were made from maize; four men and four women, totaling eight. Thompson adds that the connection between eight and maize is sealed on Copan Stela 1, where one finds a bar and three dots within a cartouche, surmounted by affix T86, which depicts a maize ear (Figure 17-7C) [54, p. 289].

The number five is the planting number for some Maya groups, like the Chortí and Tzeltal, both because they often make an effort to put around five (or at least four) kernels of seed corn into each hole made by the digging stick, and because ideally each hole is surrounded by five equidistant planting holes at about the length of the digging stick, which is used for the measurement [42]. As the planting number, five creates a quincunx pattern that would seem to represent a model of the cosmos and also a model of the *milpa* where offerings are made to the center and to the four corners (see (Figure 17-4A). It is also a model of the altar on which offerings are often done with gourd bowls forming the quincunx. Karl Taube [52] has explicitly related the quincunx model to the Olmec "bar and four dots motif," which consists of four maize ears around a central element (Figure 17-4B) [52, p. 46, Figure F].

I maintain that the number three, independently associated with bloodletting, is also closely associated with maize, perhaps in part because three dots over a bar make the number eight [43]. Three kernels are frequently depicted in Mayan, Isthmian, and Olmec maize ears [43, 45]. This is well illustrated by the shark tail composed of a twinned maize ear on Isthmian La Mojarra Stela 1 (Figure 42-5). Three jade celts representing maize ears are worn on waist or chest of Maya and Olmec rulers, and when only a single celt, it may have three holes in it. Whether through its equivalence with blood or independently, three has a relationship with maize.

The number two has a definite but not exclusive association with maize. Twin ears of maize have already been mentioned as auguring luck, fertility, and a bountiful

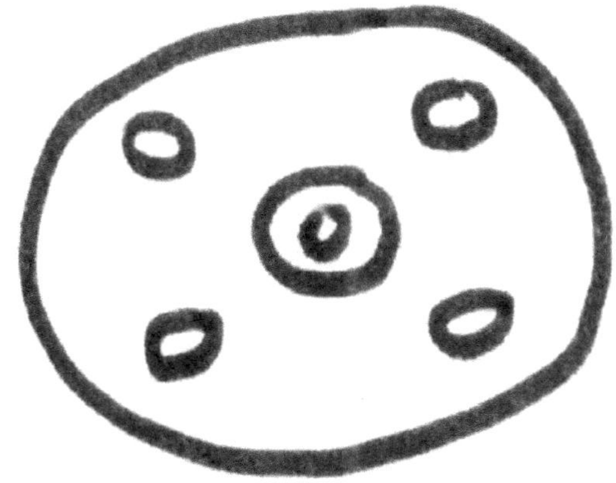

FIGURE 17-4A A quincunx glyph, showing the characteristic format. (Drawing by Brian Stross.)

harvest, for example, surely today and likely in times past. Double ears of maize are seen as a manifestation of the maize deity and the spirit of the maize kernel. Perla Petrich [29] notes that among the Mochó Maya of Chiapas, Mexico, when someone finds a double ear of maize in the *milpa* he cuts down the whole plant and brings it home. Placing it in the center of the patio, he surrounds it with the harvested maize and then places a candle at each of the four corners. Two more candles are burned in front of the mound of maize ears next to a brazier with smoking copal [p. 156]. Despite the ritual attending the finding of double maize ears, it must be said that any twinned fruit has similar implications of abundance and fertility, but maize is the most important food of all and is the subject of more sayings, more rituals, and more narratives.

Glyphs

A book could be written on maize represented in Maya hieroglyphic writing, and indeed should be. Here a few examples will be provided to supply the flavor of such an endeavor. The Classic Maya script has a glyph of the Maize Deity proper, generally identified as cognate with *Hun Hunahpu,* father of the hero twins of the previously mentioned *Popol Vuh.* We do not yet know how his name was pronounced in the Classic period, but we recognize him and know that his glyph is the head variant for 8. The head of his son *Hun Ahaw* is one glyphic variant of the 20th day name *Ahaw* (lord, master) (with a dot prefix and a large dot on the cheek), and the portrait head of his other son *Yax Balam* (with its *yax* (green) forehead ornament and spotted hairy cheek), serves as the head variant of the number nine. I suspect that the Classic cognate of *Hun Hunahpu* represents the mature roasting ear of maize (Tzeltal *ajan [elote]*), wheres his sons represent other stages of maize, *Hun Ahaw* being the immature ear (Tzeltal *ji [jilote]*) and *Yax Balam* being the older dried maize ear (Tzeltal *ach ixim* [new

FIGURE 17-4B Olmec celt with maize ears arranged around a maize deity/ruler to form a quincunx. (Drawing by F. Kent Reilly III.)

maize]) as indicated by his *yax* [*new*, green, fresh] forehead ornament (Figure 42-6).

Taube [52] recognized several glyphs relating to maize, identifying a "corn curl" glyphic element as having the sound value of *wa,* and as depicting the maize tamale, noting that it could substitute for T130, for example, already known to represent *wa*. He also identified the corn curl on the head variants of God N, *Pawatun,* and on head variants of the numbers 6 and 8 as a phonetic indicator or complement *(wa),* pointing out its presence in glyph T86 (a maize ear,

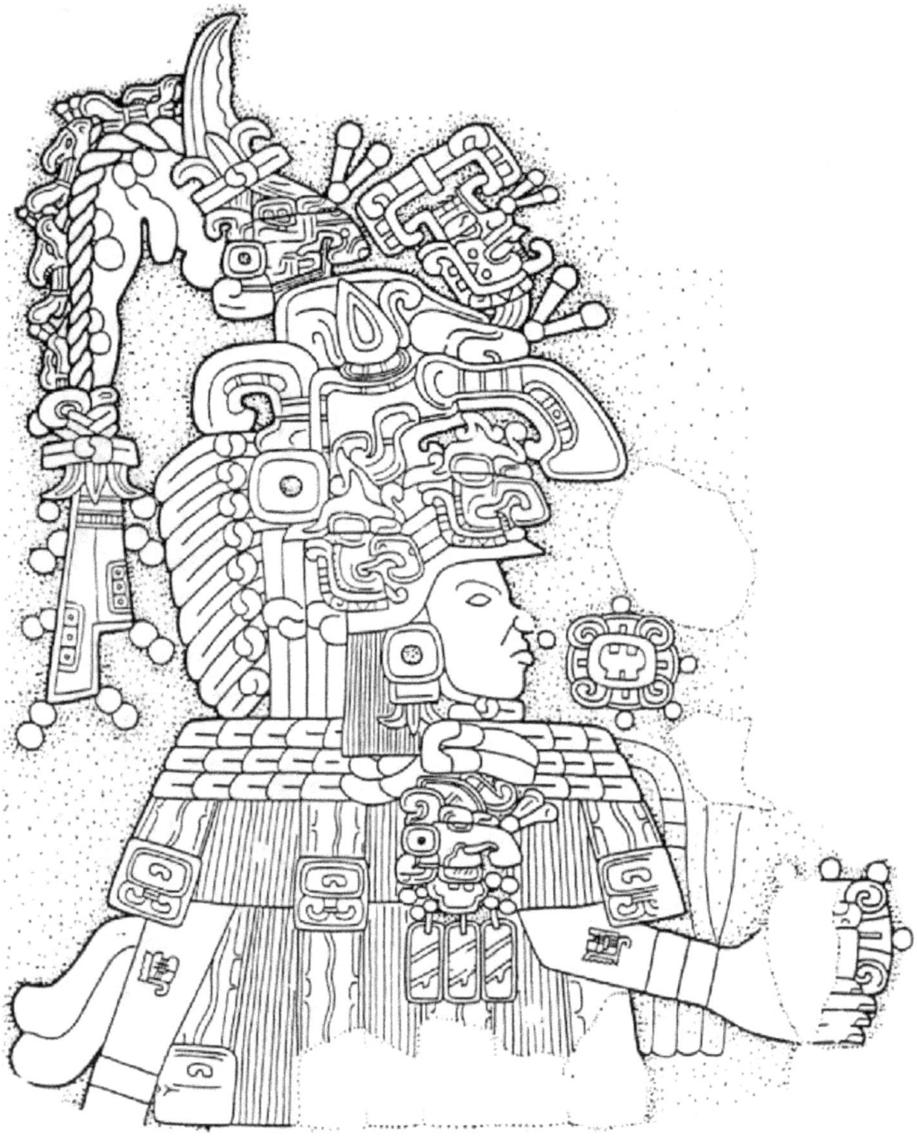

FIGURE 17-5 La Mojarra Stela 1, of late Preclassic Isthmian Civilization, in which the shark's tail behind the ruler's head is a twin ear of maize, and from the face on his pectoral dangle three celts, likely jade, representing three maize ears [52]. (Drawing by George Stuart, courtesy of the Center for Maya Research.)

see Figure 17-7C), and in the glyph for the day *K'an* [52]. He identified the glyph for *K'an,* T506, as a tamale, and suggested a T574 "shell variant" of the word for "day" to have the sound value *wa,* a suggestion amplified in later work [48].

The ruler of *Imix,* first of the 20 named days, also rules the previously mentioned number five. Imix has no other meaning in Mayan languages, sounds like a reversed *ixim* (maize), and apparently glyphically depicts a water lily leaf, fittingly, for Yucatec *ixim ha'* (water lily) is literally "maize (of the) water." *Imix* symbolizes earth and abundance and is equivalent to the Aztec day *Sipactli* (crocodile) [40, 41]. Mam diviners see the day as favorable to maize, and a colonial Yucatec source gives maize dough *(iximil wah)* as this day's "symbol."

In yet other examples of maize in script, a glyph at Yaxchilan shows the maize deity's head on a plate, his beheading constituting a major component of the creation narrative, and on the rim of some pottery buried with rulers, the ritual food *ul (atole)* is spelled out glyphically. The earlier examples illustrate maize as important in yet another Maya domain and much remains to be identified in the future. Our corpus of Olmec glyphs is insufficient to demonstrate anything about maize, but their probable descendants who used the Isthmian script also depicted maize, with more yet to be identified therein as well [21].

FIGURE 17-6 The Creation Plate, K1892, showing the maize deity emerging from a crack in the tortoise shell, flanked by Maya Classic period versions of the hero twins of the *Popol Vuh*. (From a color photograph © by Justin Kerr, K1892, reproduced with permission.)

Images

Many references to maize are known in images left us by the Olmecs and the Classic and Post-Classic Maya, and many more remain unidentified as such. Images identifiable as maize appear in the form of "world trees," objects held or worn by presumed rulers, and deities, sometimes as contextualizing representations, and likely sometimes as generalized ornamentation.

In contemporary times we are more apt to find visual representations of maize in the form of maize itself, including in ritual enactments (maize ears, tortillas, kernels, cornmeal), or buried in the symbolism of objects associated with ritual, such as the jousting target mentioned earlier, likely once a maize fetish. The trailside crosses in Maya communities, too, often painted green, or adorned with pine boughs, have been interpreted as maize stalks representing the *axis mundi* and its celestial representation, the cosmic tree visible at certain times in the heavens.

The image of the Classic Maya maize deity is discussed in depth by Taube [50], who identifies a youthful male with stylized maize atop his head depicted in Early Classic Maya art as a maize deity, developing in the Late Classic period into two different forms; one he calls the Tonsured Maize God representing ripe maize ears, and the other, he names the Foliated Maize God, representing tender growing maize. Contemporary Maya narratives sometimes help us to interpret images from the past, as for example the maize deity shown on Bonampak Stela 1, emerging from what must be "sustenance mountain" at creation (Figure 17-8). The Olmec maize and maize deity representations are well analyzed, if not yet fully, by several scholars [19, 20, 31, 32, 49, 52].

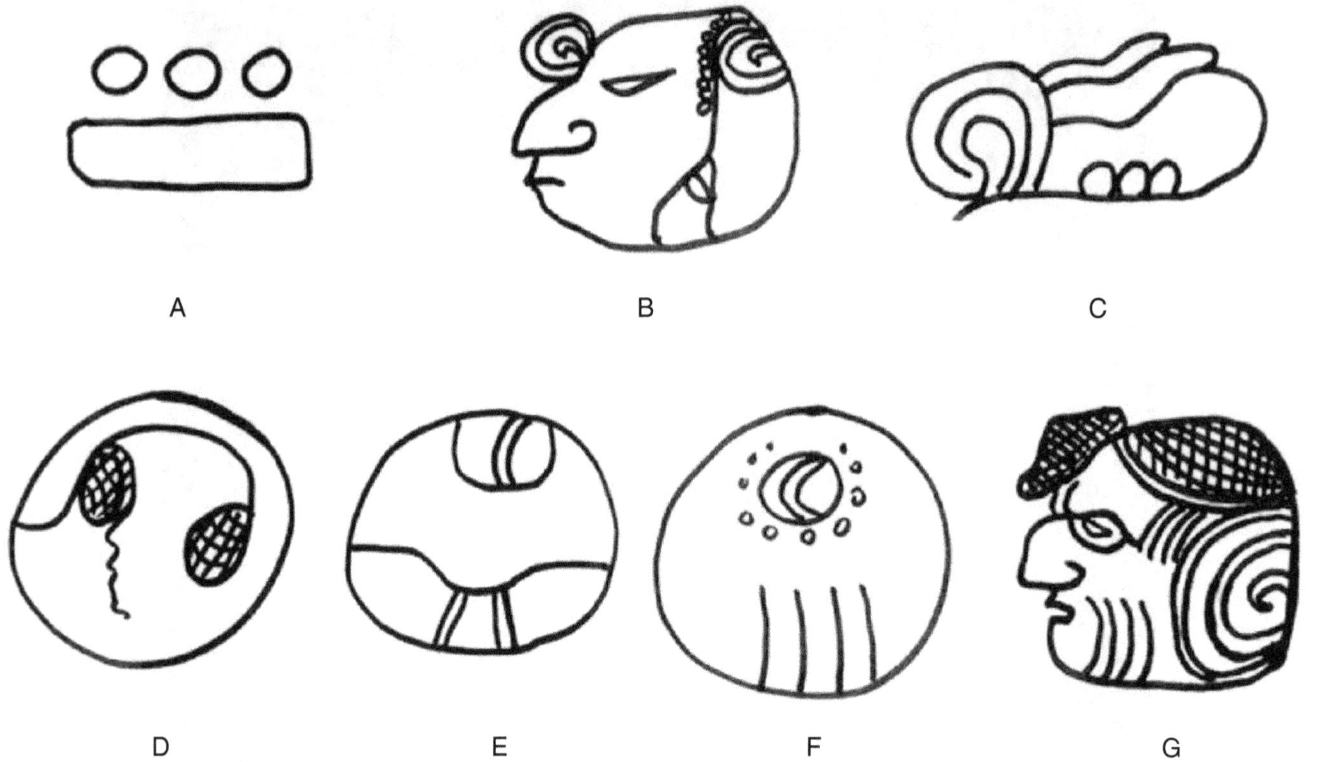

FIGURE 17-7 Various Maya Hieroglyphs related to maize (drawings by Brian Stross). (**A**) Normal variant of eight, a bar and three circlets (dots). (**B**) Head variant of eight and face of the ruler of number eight. (**C**) T86 a 'maize' affix in the glyphic script, a depiction of a maize ear. (**D**) Kaban 17th day of the almanac, ruled by the moon goddess who also rules number 1 and is a goddess of crop and human fertility. (**E**) K'an 4th day of the almanac, ruled by the maize deity who also rules the number eight. (**F**) Imix 1st day of the almanac, ruled by Pawatun, who also rules the number five. (**G**) Pawatun head variant, also known as God N, who may be an aged manifestation of a maize deity.

FIGURE 17-8 Bonampak Stela 1. Depiction of the Maya maize deity emerging from what has been called "sustenance mountain" where maize was stored. (Drawing by Linda Schele, © David Schele, courtesy Foundation for the Advancement of Mesoamerican Studies, Inc., www.famsi.org)

FIGURE 17-9A The Tablet of the Foliated Cross. Note the personified maize ears. (Drawing by Linda Schele, © David Schele, courtesy Foundation for the Advancement of Mesoamerican Studies, Inc., www.famsi.org)

The maize plant and its ears are depicted in many places and in several ways in Maya images accompanying the glyphic script. Particularly notable is a maize plant as the central image on Palenque's Tablet of the Foliated Cross, in which the ripe ears are personified in depiction as human heads (Figure 17-9A). A notable Olmec image also depicting a maize plant is on the Dallas Plaque mentioned earlier.

I have identified in the ruler's headdress on the Isthmian (Mixe–Zoquean) La Mojarra Stela 1, a shark with a tail simultaneously representing a twinned ear of maize [46]. Many other rulers are shown with representations of maize on their heads, on their chests, or in their hands, from most periods and most places in Mesoamerica, though not all have been so identified yet, and I have used these representations to illustrate some of the iconography of power in Mesoamerica [45–47]. One obvious illustration of this is from the early Maya Classic is on the Hauberg Stela, to be further analyzed later (Figure 17-10),

FIGURE 17-9B The Dallas Plaque. A cosmogram with an Olmec "quincunx" of maize (Drawing by Brian Stross.)

Calendar

The eighteen 20-day months plus 5 "nameless" days comprising the 365 day agricultural calendar still maintained by Chiapas Maya peoples has several months whose names reflect maize and maize cultivation. These named months vary widely from language to language within both

FIGURE 17-10A. The Hauberg Stela, from Tikal dating to around AD 200. (Drawing by Linda Schele, © David Schele, courtesy Foundation for the Advancement of Mesoamerican Studies, Inc., www.famsi.org)

(Continued)

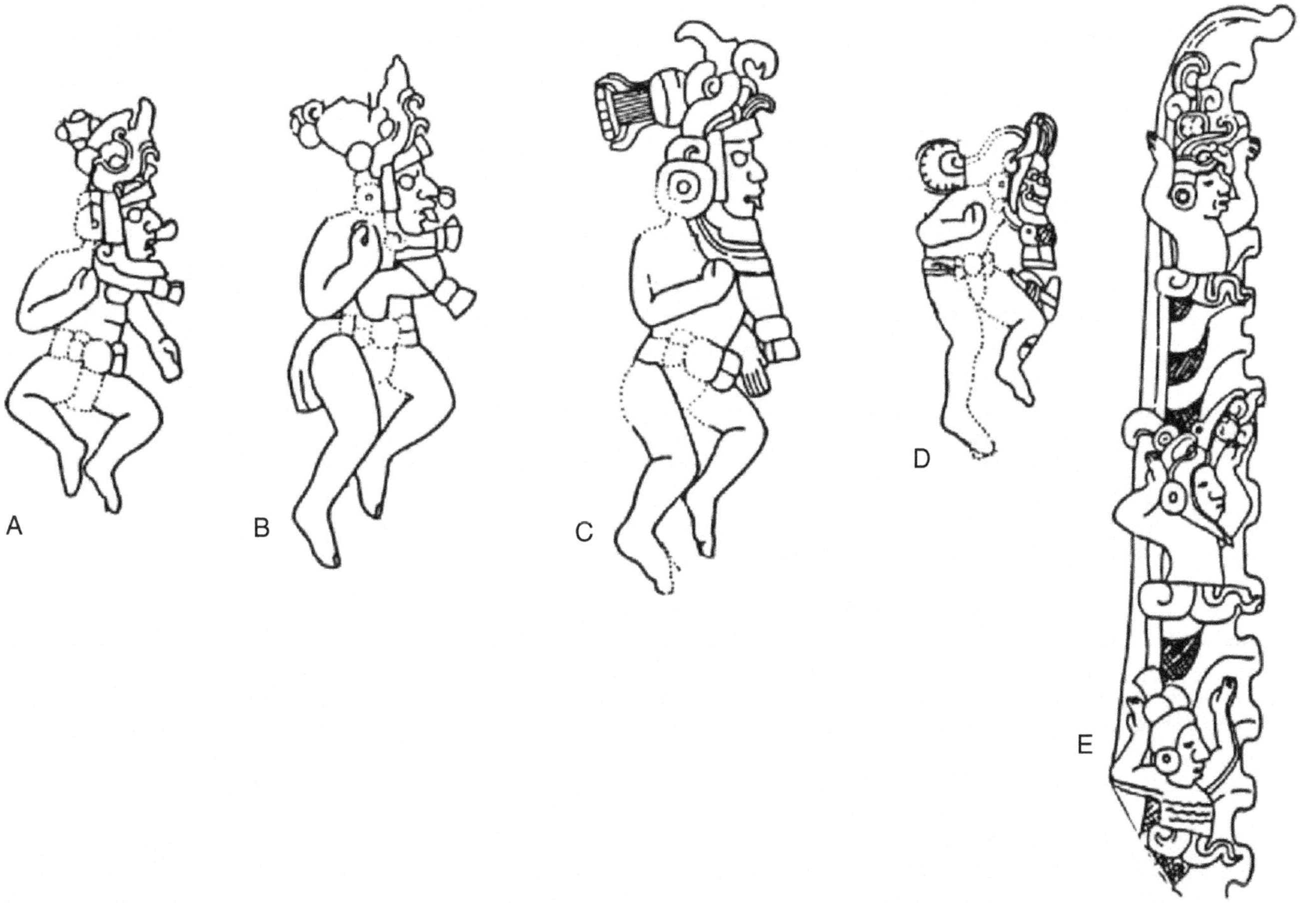

FIGURE 17-10B The "climbers" and "fallers" from the Hauberg Stela. (Drawing by Linda Schele, © David Schele, courtesy Foundation for the Advancement of Mesoamerican Studies, Inc., www.famsi.org)

Mayan and Mixe–Zoquean families today [2, 25, 27]. Less variable are the 20 day names of the 260-day sacred almanac still used by Guatemala Mayan peoples. These days each have meanings interpreted by diviners, and two of them, *K'an* and *Imix* are shown by lowland Colonial period documents to have been particularly favorable for maize. In the highlands as exemplified by the Quiché, the lowland day *Lamat* has a name *Q'anil* (yellow) that is interpreted like the near homophonous lowland day *K'an* [53, p. 114].

The 260-day cycle resulting from two concurrent cycles, permuting 13 numbers and 20 day names without repetition, has no obvious analog in celestial cycles, but this number is close to the nine months of human gestation, and also to the nine-month growing period of mountain maize, as noted by Tedlock [53, p. 190], I have maintained elsewhere that the names of the days likely indicate a deliberate connection to the life cycle of maize by attempting to connect the meanings of the first 13 day names in order, starting with the one ruled by the moon goddess, to named stages in the life cycle of maize at 20 day intervals [47, p. 29] (Table 17-3).

Whereas Table 17-3 suggests 13 named stages in the maize life cycle, I interpret the iconography of Classic period Maya monuments to yield 5 particularly important stages in the growth of the maize deity. These can be easily seen on the Hauberg Stela (see Figure 17-8), and have been interpreted by others as representing constellations of the zodiac [15].

Four of these stages in the maize deity's growth are the "climbers," clinging to the serpent whose head looks down on the ruler from the sky above; two on each side of the ruler. The climber closest to the top has a seed in his headdress. The top climber on the right side has what appears to be a maize spike or an early ear developing. Below him the deity wears a skull face mask and a mature ear on his headdress on the bottom right, suggesting the Tzeltal stage named "dead head." The ruler himself has the largest ear at the perfect stage of development. Schematically this appears to reproduce a more anthropomorphized refinement of an earlier representation on an Olmec celt (see Figure 17-4B). If we add up the four climbers, the three "fallers" on the blade of the knife, and the ruler with his roasting ear, we find that

TABLE 17-3 vidence That the 260-Day Calendar is Based on the Maize Life Cycle

Day name	Meaning and comment	Name and meaning shared concepts
Kaban	"*Earth*" is day 1 of the 260-day period, when one prepares the *earth* for the maize.	Earth
Ik'	"Wind, *breath, spirit, life,*" the sixth day name following **Kaban,** days 100–120 of the cycle, when the *breath* of *life* is in the maize; the *spirit* of maize has entered the kernel.	Spirit, life
Ak'bal	"*Night*" ("*house*" in the Aztec almanac), the seventh day following **Kaban,** days 120–140 of the cycle, when the kernels are inside a *house* or husk, and just waking up. The developing ear is now called **pak** in Tzeltal as it faces down toward the "*dark house* of *night*" (i.e., the underworld). One can connect this with the **Uk Ek' K'an** "seven-black-yellow" emblem that on occasion flanks the maize deity whose number eight follows.	Dark, house
K'an	"Yellow, *ripe,*" the eighth day name, days 140–160, when this stage is completed, the small edible ear, the jilote, is ready to be plucked for food.	Ripe
Chikchan	"*Snake* species," ninth day name, day 160–180 of the cycle. The more mature jilote (Tzeltal **me-ji** "mother jilote") in early days. Mixe **po:p xï'ïx** is both "*jilote*" and "white *viper,*" equating a snake to a maize stage. By 180th day it becomes a roasting ear *elote*. The **Bolon Mayel** "nine twenty" emblem (better as **Bolon K'alel**) flanks the young maize deity, and is *ophidian.*	Snake
Kimi	"*Death,*" the tenth named day following **Kaban,** Tzeltal **lah** ('10) also means "finished, *dead,*" days 180–200 in the cycle. The silks have "died," darkened, and dried up. Maize growth stops with maturity of the fruit. The maize ear may be plucked from the plant (equivalent to being beheaded). Tzeltal **cham hol** (mature roasting ear) is literally "*dead head.*"	Death, completion

A selection of 6 day names reflecting concern with the stages of maize, as evidence the 260-day calendar may have origins in the life cycle of maize. Italicized words indicate shared concepts between name meaning and comment. A slight mismatch suggests that the maize stages assigned to the thirteen 20-day periods would likely have originated and developed in a lowland region where maturation would be somewhat sooner.

the total stages add up to eight, the number associated with the maize deity. The three fallers would seem to be less important stages, having been cut in half (Figure 42-10B).

Plants

Other plants appear in stories about maize as well as in maize related words, demonstrating concerns with growth, harvest, and storage of maize, as well as deep interest in understanding why plants and animals are the way they are, and how they relate to one another. We still do not know to what extent root crops have figured in the diets of Mesoamericans before and in early stages of maize domestication, nor do we know the relationship to their diets of various other grains, such as *Setaria geniculata* (a relative of grain millets) amaranth, and even *teosinte (Zea mays* ssp. *parviglumis),* but we have hints in discourse about what some Mayans think preceded their introduction to maize. For example the Huastecs say that before they encountered maize they made their tortillas of ground *ramón* nuts *(Brosimum alicastrum),* the *iximte'*, literally "maize tree," of the Tzeltal, who use the seeds as famine food. Lacandón Maya know that where the *ramón* grows, their *milpa* will prosper. The Mam Maya of Colotenango tell of times before maize when they ate the large roots of a single stalked plant called *"txetxina"* (mother maize) [57, p. 239–241, 55, p. 352]. The word could be a corruption of the Nahuatl *teosintli, "teocinte,"* or even of the borrowed Spanish term for it.

In Honduras near where Chortí Maya live, the giant **cycad** *Dioon mejiae* is called *teucinte* by the local inhabitants, and its seeds are ground and eaten [3]. The name is a Pipil Nahuatl term for the grass with edible seeds from which maize was domesticated. Similarly Huastecs of San Luis Potosí call the local *Zamia* species of cycad *tzalaam thipaak* (shade of Thipaak [who brought maize to humans]), or just *thipaak.* Huastecs have other interesting terms for these cycads, like *tzakam way'* (little maize ear), or *tzakam thipaak* (little *Thipaak*), *tz'een thipaak* (sierra *Thipaak*), or *ahaatik a eem* (maize lord) [1]. Zamias are also called *teocintle* by more acculturated Huastecs [3]. Interestingly, these cycads, thought to be the "maize" of the ancestors, are allowed to grow alongside maize in the Huastec *milpa* because they are seen as maize "shepherds," and when they produce their cones, which look rather like shucked maize ears, it portends a good maize harvest [1].

Sap from several different incense producing trees, called *copal,* is believed to be a protector of maize in storage. The name *Ah Kohk Nar* "guardian of maize" is given by Chortí Maya to four maize ears modeled in copal *(Bursera bipinnata)* and put in the granary, one in each corner, to watch over the newly harvested maize. Three have Spanish names, and the fourth is called *Kumix* (little one) [62, p. 403]. When beginning prayers for blessing their maize, the head of a Mam household beheads a chicken and puts it on a heap of copal to mix with the blood. Then a little of the bloodied incense is burned on a little fire lit in front of the maize seed

[60, p. 25]. Many more plants have associations with maize, both positive and negative, but this exemplifies the nature of some of those associations.

CONCLUSION

The linguistic and iconographic evidence for the importance of maize to the indigenous inhabitants of southeastern Mesoamerica and their predecessors is almost overwhelming. Everything points to what one might expect from a staple whose nourishing blessing is an absolute requisite for the lives of those who plant it: reverence and respect for the food and its deities, ubiquity of maize in ritual activities, keen observation of its morphology and growth, and great variety in how it is prepared, utilized, and named, including special terminology for ritual occasions. Much ordinary conversation deals directly with maize, its cultivation, harvest, and consumption, and one could hardly find any aspect of Mayan or Mixe–Zoquean life that does not relate at least indirectly to maize.

Through words, images, script, metaphor, and other conventional forms of symbolic representation, indigenous societies of Mesoamerica have through the centuries indicated their dependence on maize, their reverence for maize, and the full integration of maize into their intellectual and artistic lives. The selection of examples presented here would seem to bear this out with little room for doubt in the minds of the societies' participants or of the observers. One observer says that Quiché Maya do not plant maize to live, but rather they live to plant maize [33]. Although this is a matter of perspective, it is a perspective shared by many regarding Mesoamerican subsistence maize farmers.

References Cited

1. J. B. Alcorn. (1984). *Huastec Mayan ethnobotany*. Austin: University of Texas Press.
2. B. Berlin, D. E. Breedlove, P. H. Raven. (1974). *Principles of Tzeltal plant Classification: An introduction to the botanical ethnography of a Mayan-speaking people of highland Chiapas*. New York: Academic Press.
3. M. Bonta, R. Osborne. (In press.). Cycads in the vernacular—a compendium of local names, *Proceedings of Cycad 2005, the 7th International Conference of Cycad Biology*, Xalapa, Veracruz, Mexico, January 2005.
4. D. Breedlove, R. M. Laughlin. (2000). *The flowering of man: A Tzotzil botany of Zinacantán*. Washington, D.C.: Smithsonian Institution.
5. V. R. Bricker, E. Po'ot Yah, O. Dzul de Po'ot. (1998). *A dictionary of the Maya language as spoken in Hocabá, Yucatán*. Salt Lake City: The University of Utah Press.
6. C. Brown, S. Witkowski. (1979). Aspects of the phonological history of Mayan-Zoquean. *International Journal of American Linguistics, 45*, 34–47.
7. R. D. Bruce. (1968). *Gramática del Lacandón*. Departamento de Investigaciones Antropológicas. I.N.A.H. Pub. No. 21. México, D.F.: I.N.A.H.
8. R. Bunzel. (1952). *Chichicastenango*. New York: American Ethnological Society.
9. J. H. Butler, D. E. Arnold. (1977). Tzutujil maize classification in San Pedro la Laguna. In: H. L. Neuenswander, D. E. Arnold, (Eds.), *Cognitive studies of southern Mesoamerica*. Summer Institute of Linguistics Museum of Anthropology Publication 3. Dallas, TX: Summer Institute of Linguistics Museum of Anthropology. pp. 185–205.
10. L. Campbell, T. Kaufman. (1976). A linguistic look at the Olmecs. *American Antiquity, 41*, 80–89.
11. L. Campbell. (1980). On Mesoamerican linguistics. *American Anthropologist, 82*, 850–857.
12. L. Campbell. (1983). Mesoamerican historical linguistics and distant genetic relationship: Getting it straight. *American Anthropologist, 85*, 362–372.
13. N. England. (1992). *Autonomía de los idiomas Mayas: Historia e identidad*. Guatemala: Editorial Cholsamaj.
14. G. Foster. (1945). *Sierra Popoluca folklore and beliefs*. American Archaeology and Ethnology XLII 2. Berkeley: University of California Press.
15. D. Freidel, L. Schele, J. Parker. (1995). *Maya cosmos: Three thousand years on the shaman's path*. New York: William Morrow and Co.
16. G. Gossen. (1974). *Chamulas in the world of the Sun*. Cambridge, MA: Harvard University Press.
17. C. Guiteras-Holmes. (1961). *Perils of the soul: The world view of a Tzotzil Indian*. New York: The Free Press of Glencoe.
18. W. R. Holland. (1963). *Medicina Maya en los altos de Chiapas*. México: Instituto Nacional Indigenista.
19. P. D. Joralemon. (1971). *A study of Olmec iconography*. Dumbarton Oaks Studies in Pre-Columbian Art and Archaeology, No. 7. Dumbarton Oaks, Washington D.C.
20. P. D. Joralemon. (1976). The Olmec dragon: A study in precolumbian iconography. In: H. B. Nicholson, (Ed.), *The origins of religious art and iconography in Preclassic Mesoamerica*. University of California at Los Angeles Latin American Center. pp. 27–72.
21. J. S. Justeson, T. Kaufman. (1993). A decipherment of epi-Olmec hieroglyphic writing. *Science, 271*, 1703–1711.
22. J. S. Justeson, W. M. Norman, L. Campbell, T. Kaufman. (1985). *The foreign impact on lowland Mayan language and script*. New Orleans: Middle American Research Institute, Tulane University.
23. T. S. Kaufman, W. M. Norman. (1985). An outline of Proto-Cholan phonology and vocabulary. In: J. S. Justeson, L. Campbell, (Eds.), *Phoneticism in Maya hieroglyphic writing*. Institute for Mesoamerican Studies S.U.N.Y Publication #9. Albany: SUNY Press. pp. 77–166.
24. O. LaFarge. (1947). *Santa Eulalia*. Chicago: University of Chicago Press.
25. F. Lipp. (1991). *The Mixe of Oaxaca: Religion, ritual, and healing*. Austin: University of Texas Press.
26. R. J. McGee. (1990). *Life, ritual, and religion among the Lacandon Maya*. New York: Wadsworth.
27. W. Miller. (1956). *Cuentos Mixes*. Mexcico: Instituto Nacional Indigenista, Biblioteca de folklore indigena.
28. M. Oakes. (1951). *The two crosses of todos santos*. New York: Bollingen Foundation.
29. P. Petrich. (1985). *La alimentación Mochó*. San Cristobal de las Casas, Mexico: Universidad Autónoma de Chiapas.
30. I. Press. (1975). *Tradition and adaptation: Life in a modern Yucatan Maya village*. Westport, CT: Greenwood Press.
31. F. K. Reilly. (1994). *Visions to another world: Art, shamanism, and political power in middle formative Mesoamerica*. Ph.D. Dissertation. Austin: University of Texas.
32. F. K. Reilly. (1987). *The ecological origins of Olmec symbols of rulership*. Masters Thesis. Austin: University of Texas.
33. N. F. Rodas, C. O. Rodas, L. F. Hawkins. (1940). *Chichicastenango; the Kiche Indians; Their history and culture; Sacred symbols of their dress and textiles*. Guatemala: Union Tipografica.

34. F. Rojas Lima. (1988). *La culura del maiz en Guatemala.* Guatemala: Ministerio de Cultura y Deportes.
35. R. L. Roys. (1931). *The ethnobotany of the Maya.* Tulane University M.A.R.I. Publication 2. New Orleans.
36. R. L. Roys. (1967). *The book of Chilam Balam of Chumayel.* Norman: University of Oklahoma Press.
37. K. Sammons. (1992). Translating poetic features in the Sierra Popoluca story of *Homshik.* In: B. Swann, (Ed.), *On the translation of Native American literatures.* Washington, D.C.: Smithsonian Institution. pp. 368–386.
38. K. Sammons. (1996). *Rhetorical functions of parallelism in Sierra Popoluca discourse performance.* Ph.D. Dissertation. Austin: University of Texas.
39. M. Sántiz Gomez. (1998). *Creencias de nuestros antepasados.* Mexico: CIESAS.
40. B. Stross. (1982). Maya hieroglyphic writing and Mixe-Zoquean. *Anthropological Linguistics, 24,* 73–134.
41. B. Stross. (1983). Oppositional pairing in Mesoamerican divinatory day names. *Anthropological Linguistics, 25,* 211–273.
42. B. Stross. (1986). Some observations on T585 (quincunx) of the Maya script. *Anthropological Linguistics, 28,* 283–311.
43. B. Stross. (1989). Maya bloodletting and the number three. *Anthropological Linguistics, 31,* 209–226.
44. B. Stross. (1990). Mesoamerican writing at the crossroads: The Late Formative. *Visible Language, 29,* 38–62.
45. B. Stross. (1992). Maize and blood: Mesoamerican symbolism on an Olmec vase and a Maya plate. *Res: Anthropology and Aesthetics, 22,* 82–107.
46. B. Stross. (1994). Maize and fish: The iconography of power in late formative Mesoamerica. *Res: Anthropology and Aesthetics, 25,* 10–35.
47. B. Stross. (1996). The Mesoamerican cosmic portal: A Zapotec example. *Res: Anthropology and Aesthetics, 299/30,* 82–101.
48. B. Stross. (1997). A note on three kin variants in distance numbers. *U-Mut Maya, 6,* 15–20.
49. C. Tate. (1999). Patrons of shamanic power: La Venta's supernatural entities in light of Mixe beliefs. *Ancient Mesoamerica, 18,* 169–198.
50. K. Taube. (1985). The classic Maya maize god: A reappraisal. In: M. G. Robinson *Fifth Palenque Round Table, 1983.* San Francisco, CA: Pre-Columbian Art Research Institute. pp. 171–181.
51. K. Taube. (1989). The maize tamale in Classic Maya diet, epigraphy and art. *American Antiquity, 54,* 31–51.
52. K. Taube. (1996). The Olmec maize god: The face of corn in Formative Mesoamerica. *Res: Anthropology and Aesthetics, 29/30,* 39–81.
53. B. Tedlock. (1992). *Time and the highland Maya.* Albuquerque: University of New Mexico Press.
54. D. Tedlock. (1985). *Popol Vuh.* New York: Simon and Schuster.
55. J. E. S. Thompson. (1970). *Maya history and religion.* Norman: University of Oklahoma Press.
56. J. E. S. Thompson. (1971). *Maya hieroglyphic writing: An Introduction.* Norman: University of Oklahoma Press.
57. L. A. Valladares. (1957). *El hombre y el maiz.* B. Costa-Amic. Mexico, D.F.
58. E. Z. Vogt. (1976). *Tortillas for the gods: A symbolic analysis of Zinacantceo rituals.* Cambridge, MA: Harvard University Press.
59. E. Z. Vogt, V. R. Bricker. (1996). The Zinacanteco fiesta of San Sebastian: An essay in ethnographic interpretation. *Res: Anthropology and Aesthetics, 29/30,* 203–222.
60. C. Wagley. (1957). *Santiago Chimaltenango.* Guatemala: Seminario de Integración Social Guatemalteca.
61. S. Wichmann. (1995). *The relationship among the Mixe-Zoquean languages of Mexico.* Salt Lake City: University of Utah Press.
62. C. Wisdom. (1940). *The Chorti Indians of Guatemala.* Chicago: University of Chicago Press.
63. S. Witkowski, C. Brown. (1981). Mesoamerican historical linguistics and distant genetic relationship. *American Anthropologist, 83,* 905–911.

CHAPTER

18

The Place of Maize in Indigenous Mesoamerican Folk Taxonomies

NICHOLAS A. HOPKINS

Department of Modern Languages and Linguistics, Florida State University, Tallahassee, Florida

Glossary

Ethnobotany The study of plant knowledge among non-Western peoples. Modern anthropological ethnobotany places primary emphasis on the native categorization of plants and seeks to understand the origin and development of such cognitive systems.

Ethnoscience The anthropological study of organized knowledge and belief among non-Western peoples, including ethnobotany, ethnozoology, and other systematic studies of cognitive and conceptual systems, for which an extensive methodology has been developed.

Folk science A derivative of ethnoscience, folk science (including folk botany, folk zoology, etc.) is content-oriented and less concerned with methodology.

Folk taxonomy Named and unnamed categories that organize folk knowledge in a system that features hierarchical inclusion; in ethnobotany, the non-Western equivalents of Linnaean classification.

Life form Major partitions of the animal and plant kingdoms (e.g., the common Mesoamerican ethnobotanical plant categories "tree," "vine," "grass," and "herb").

Mesoamerica An indigenous New World culture area that includes the regions occupied by major Pre-Columbian civilizations; roughly, from the latitude of Tampico, Mexico, south through Guatemala and along the Pacific Coast of Central America to the Gulf of Nicoya.

The investigation of indigenous Mesoamerican folk biology is still in its infancy, with few systematic exhaustive treatments of native categorization of plants and animals. Nevertheless, preliminary studies suggest major patterns in Mesoamerican ethnozoology and **ethnobotany.** Four major categories of animals are based on locomotion: walkers, crawlers, fliers and swimmers (mammals, reptiles and amphibians, birds and bats, and fish). Four major categories of plants correspond to the functionally distinct **life forms** "trees, vines, grasses, and herbs" (as described for the Tzeltal Maya by Berlin, Breedlove and Raven [8]. In Mesoamerican folk ethnobotanies, most plants fall into the four major life forms. However, culturally significant cultigens, including maize, often occupy categories parallel to the major life forms (in the Tzeltal study, these are referred to as "unaffiliated generic taxa"). This chapter reports the results of preliminary investigation and comparison of folk taxonomies representing two of the four major language families of **Mesoamerica** (Mayan and Otomanguean). Each gives differential treatment to the cultigens, placing some in the major life forms and others in unaffiliated taxa. Maize, for instance, may be found within the life form "grasses" or it may occur as an independent class of plants. Similar differences in treatment are found for beans, squashes, tomatoes, and peppers, the most significant Mesoamerican cultigens. No firm conclusions can be drawn from the meager sample of systems reported, but the data suggest a complex cultural evolutionary explanation (the precise nature of which is beyond the scope of the present chapter). However, general patterns are outlined, and questions are raised that may stimulate future research leading toward a comprehensive theory of the evolution of Mesoamerican ethnobotanical knowledge.

INTRODUCTION

In Mesoamerica, the homeland of maize, the systematic investigation of native plant categorization among the indigenous societies began less than 50 years ago, and there are few extensive studies. Available data suggest that one of the principal differences between Mesoamerican folk taxonomies is the treatment given to maize and the other cultigens with respect to the common life forms: tree, vine, grass, and herb. For instance, one Mesoamerican **folk taxonomy** places all the cultigens in life forms with their morphologically similar congeners (e.g., maize is categorized as a kind of grass); another places all cultigens apart from wild plants, in a set of matching life form categories. Other taxonomies are intermediate between these extremes. The place of maize in these taxonomies varies from group to group; the opposing principles of morphology and utility appear to be at the core of the differences in categorization. Further investigation of Mesoamerican folk botanies may cast light on important processes of cultural evolution and development. For instance, data from the Maya area show that plant cultivation predates the terminology for life forms, and that the creation of the latter is a product of first millennium (AD) Classic Period science. Supporting evidence comes from Classic Maya hieroglyphic inscriptions. Botany can now be added to the inventory of Maya (and general Mesoamerican) sciences.

The anthropological study of the folk botany of the indigenous people of Mesoamerica was dominated in the twentieth century by two paradigms. The earlier paradigm is illustrated by the work of Isabel Kelly and Angel Palerm [32] on the Totonac of Tajín (Veracruz, Mexico). Ample ethnobotanical data were presented, but the principles of organization were foreign to the Totonacan culture. Entries were ordered by common name in Spanish, followed by the Totonac plant name and its translation, taxonomic determination, habit and habitat, period of introduction into the region, utilization, informants' names, and commentary (on variant names, synonyms, etc.). The "taxonomic determination" was a Latin binomial rather than a Totonac classification. Throughout, emphasis was on the inventory of plants in the region and their utilization rather than on the nature of the botanical knowledge of the Totonacs themselves, other than pragmatic knowledge related to plant utilization. Informant categorizations of plants that conflicted with botanical classifications were considered to be errors, and no consideration was given to the possibility that a native categorization scheme might exist.

Around the middle of the twentieth century the earlier paradigm was challenged by a new approach to ethnobotany that privileges native knowledge and belief. "**Ethnoscience**" and its derivative "**folk science**" seek to discover the natives' categorization of the botanical world (and other areas of organized knowledge); an extensive methodology has been developed to elicit folk knowledge and beliefs [6]. Primary emphasis is put on how the local population classifies plants, which often relates directly to their patterns of utilization. In the Totonac case, in contrast, an informant who placed distinct species in the same category by grouping together three plants with a common use in the production of cane alcohol—a perfectly logical grouping—was considered to be in "error." In ethnoscientific research, information is organized in native cognitive terms, and botanical identifications serve only to relate this knowledge to the empirical world and facilitate comparisons between distinct native ethnobotanies. This approach has raised questions that were impossible to formulate in the early period. The investigation of folk botanical knowledge has stimulated speculations about human cognitive development in general and the capabilities and constraints that regulate its growth.

Maize is the quintessential cultigen among the people of Mesoamerica, and Mesoamerica is the homeland of maize. Nevertheless, within Mesoamerica the place of maize in folk taxonomies differs from society to society, as does the treatment of other cultigens. Why this should be the case is not yet understood; indeed, the question has barely been raised. In this chapter I will argue that comparative research into Mesoamerican folk taxonomies, especially with regard to the treatment of cultigens, has the potential to shed light on important questions of human cultural development. Although knowing the place of maize and other cultigens in a folk taxonomy may tell us little about the plant species themselves, it may tell us quite a bit about the relationship of local people to the plants, and the history of ethnobotanical knowledge in developing agricultural societies.

THE ETHNOBOTANY OF THE AMUZGO

The Amuzgo are a small indigenous ethnic group settled in the southwestern corner of the state of Oaxaca, Mexico, and adjacent parts of the state of Guerrero [19, 36]. The Amuzgo language, which has two major dialects and some 15,000 speakers, constitutes one branch of the Otomanguean language family [31, 37]. Otomanguean languages are found throughout Oaxaca (most notably, Mixtec and Zapotec), Puebla (Cuicatec, Mazahua), Central Mexico (Otomi and Mazahua), and the arid regions to the north (Pame and Chichimeca). Otomanguean is the language family most closely associated with the earliest known maize agriculturalists in Mesoamerica, judging from the evidence of the early cultures in the Tehuacan Valley, Puebla, and adjacent areas that are within the range of distribution of the proto-Otomanguean language [2, 17, 29, 33].

Currently, the principal population center of the Amuzgo is San Pedro Amuzgos, Oaxaca, located in "hot country" at the foot of the piedmont overlooking the Pacific coastal plain; data reported here are from San Pedro. Because the Amuzgo area is within the suspected homeland of maize and supports populations of wild teosinte, the knowledge of plants that native populations have developed over millennia of exploitation of local botanical resources is of potential interest. The organization of the plant world reflected in the folk taxonomy of the Amuzgo takes on special significance when this system is compared with other Mesoamerican ethnobotanies. Of particular interest is the treatment of the cultigens: Mesoamerican ethnobotanies feature two diametrically opposed treatments of cultigens, with intermediate systems that suggest a pattern of cultural evolution. Specifically, some folk taxonomies place maize and other cultigens in their morphologically appropriate life forms (maize is a grass for the Amuzgo). In the opposing model, no cultigens fall into the life forms (maize and all other cultigens are "unaffiliated generic taxa" for the Tzeltal). Intermediate ethnobotanical systems show some other pattern of treatment (e.g., maize and other grains constitute a life form themselves for the Chuj Maya, but other cultigens are classified under morphologically based life form taxa).

The ethnobotany of the Amuzgo came under study in the 1970s as part of a research project directed by the author (Proyecto II–2, Etnociencia y Etnobotánica [Ethnoscience and Ethnobotany]) in the Programa de Lingüística of the Centro de Estudios Superiores del Instituto Nacional de Antropología e Historia (CIS-INAH), an agency of the federal government of Mexico under the Secretaría de Educación. (At the end of 1980, the name of this institution was changed to Centro de Investigaciones y Estudios Superiores en Antropología Social, CIESAS.) In general, this project applied the research methodology pioneered by Berlin, Breedlove, and Raven [8] in their landmark study of Tzeltal ethnobotany. Amuzgo members of the research team were Fermín Tapia García, the principal informant [38–40], Modesta Cruz Hernández, research assistant [18], and Josefa Martínez, draftsman. Andrea Caso Aguilar and Rodrigo de la Torre, undergraduate student assistants, and Nicholas A. Hopkins, project director, were the other members of the Amuzgo ethnobotany research team. In the course of the project, Amuzgo plant names were registered and the native categorization of plant taxa was investigated, along with the criteria of classification. However, only a few field collections were made, and the project ended before adequate botanical identifications could be obtained. Most data are thus reported with local Spanish common names or translations of the Amuzgo terms: glosses for many Amuzgo terms are likewise not available.

AMUZGO ETHNOBOTANY AND FOLK TAXONOMY

The categorization of plants by the Amuzgo is similar to that of other Mesoamerican peoples in placing most plants under four major life forms, glossed tree, vine, grass, and herb, and in placing morphologically distinct plants in unaffiliated generic taxa (to use the language of the Tzeltal Maya ethnobotany of Berlin, Breedlove, and Raven [8], where 25% of the generic taxa fall outside the four life forms). In the case of Amuzgo, major categories other than the four principal life forms are maguey (including the subclass tree maguey, bromeliads), fern, wide-leafed plant (a category that includes Xanthosoma species, "elephant ears" and similar plants) and mushroom. The numbers of generic taxa included in these upper-level categories are similar to those of other reported Mesoamerican indigenous ethnobotanies: trees include some 150 generic taxa, vines include some 40 generics; there are about 15 grasses, 35 herbs, 10 wide-leafed plants, 10 magueys, five ferns, and five mushrooms [38, 39]. The corresponding numbers for the four major life forms from the much more extensive Tzeltal study are 178 trees, 24 vines, 35 grasses, and 119 herbs. Amuzgo generic taxa include from 2 to 10 specific taxa, and about 1000 specific taxa have been reported. A few varietal taxa occur among the cultivated plants, especially maize (with white, yellow, and purple varieties of some of the seven folk specifics).

Amuzgo ethnobotanical terminology lends itself to taxonomic and historical research. The names of superior taxonomic categories are commonly incorporated in the citation forms of generic and specific plant names, which are multimorphemic nominal constructions. The Amuzgo word "pole beans," for instance, is *tzö-ndë-n'on,* literally "vine-bean-pole," containing the life form term *tzö* (vine), the generic *ndë* (bean), and the specific designator *n'on* (pole). Tree names begin with *tz'on,* vine names with *tzö* (or, in free variation, *tz'ö*), grasses with *jndë,* herbs with *tzkojndë,* and so on. In context, the life form term may be dropped: *(tzö) ndë-n'on,* (vine) bean-pole. The great majority of names for specific taxa are formed from a generic term plus a modifier, often a reference to a color, size, or some other distinctive feature of the plant.

The criteria used in categorizing plants vary across plant types, but usually include habit, leaf and stem morphology, fruit and seed characteristics, edibility, and other utilization. Among the vines *(tzö)*, for instance (Table 44-1, reporting work by Tapia and Caso [21]), the beans *(ndë)* have in common a growth pattern (they are all "climbers" as opposed to "extenders"). They also share the relative length of stems ("medium," as opposed to "longer" and "longest"), branching characteristics (relatively few branchings, the least of four possibilities), leaf shape ("heart-shaped" leaves as opposed to "round, forked, saw-edged, long saw-edged, elongated, and rounded saw-edged" leaves), fruit type (pods), and edibility (all are edible).

The squashes *(tzken)* and 'pumpkins' *(të-nta')* tend to have in common a growth pattern (with one exception, they are extenders rather than climbers), relative length (with one exception, they are short), branching characteristics (relatively many branches), leaf shape (round or "fork-shaped" leaves), and seed type. Squashes and pumpkins form a

TABLE 18-1 Amuzgo tz'ö: Vine Names and Their Classification

Amuzgo name	Morpheme gloss	Local Spanish common name
1. Beans *(ndë)*		
All are edible.		
tzö-ndë-kichi'	vine-bean-white	*frijol blanco*
tzö-ndë-nd'oa	vine-bean-string	*frijol correa*
tzö-ndë-ndy'non'	vine-bean-slow	*frijol criollo*
tzö-ndë-n'on	vine-bean-pole	*frijol de vara*
tzö-ndë-kahbe	vine-bean-coffee	*frijol nescafé*
2a. Squashes *(tzkën)*		
All are edible.		
tzö-tzkën-be	vine-squash-red	*calabaza colorada*
tzö-tzkën-n'ian	vine-squash-*chompa*	*calabaza chompa*
tzö-tzkën-kindi	vine-squash-odor	*sandía*
tzö-tzkën-melon	vine-squash-melon	*melón*
tzö-tzkën-tzohndya	vine-squash-Spanish	*melón chico*
2b. Pumpkins *(tzö-të)*		
None are edible.		
tzö-të-nta'	vine-fruit-pumpkin	*chilacayote* (used for beverage)
tzö-të-nta'-hndë	vine-fruit-pumpkin-small	*chilacayotito* (used for soap)
3. Wild vines *(hn'on)*		
All are used for tying.		
tzö-hn'on-be	vine-*molato*-red	*bejuco molato*
tzö-hn'on-kichi'	vine-*molato*-white	*bejuco casero*
tzö-hn'on-chjen	vine-*molato*-fiber	*bejuco ratón blanco*
tzö-hn'on-ndotzian	vine-*molato-tlachicón*	*bejuco tlachicón*
tzö-nion-xkin-kityiu	vine-thorn-head-shrimp	*bejuco camarón*
4. Ungrouped vines		
Utility varies; one is edible.		
tzö-të-tzja'	vine-fruit-fiber	*estropajo* (used as a rasp)
tzö-tzko-xoa-kitye	vine-leaf-throat-lizard	*muchachito*
tzö-tzjan'-chi	vine-tuber-self sweetening	*guía camote morado* (edible)
tzö-të-tzondyo'	vine-fruit-acid taste	*uva*
tzö-tzjan-t'oa	vine-tuber-twist/sprain	*camote curativo* (medicinal)

covert category within the vines, and are distinguished from one another by the criterion of edibility: only the former have edible fruit (although some of the latter have edible seeds). Wild vines *(hn'on)* have in common their utility; they are all useful for tying. Otherwise, they differ internally in growth patterns, length, branchings, and leaf form; none are edible. Unclassified vines also vary internally and have no common features other than those that define them as vines (twining habit, nonerect stems).

In sum, the Amuzgo recognize two classes of vines with edible fruit or seeds (beans and the squashes and pumpkins), a class of vines useful for tying (the wild vines), and an ungrouped set of vines with no common uses or morphology.

Of greater interest for purposes of this chapter, there is an extensive class of grasses, which includes both maize and teosinte (Table 44-2). One of the defining criteria for membership in this class seems to be a long, thin leaf shape with parallel rather than branching veins, justifying the inclusion of onions and garlic. The major informant remarked that it is easiest to tell maize is a grass when it is just sprouting and most closely resembles other grasses.

In this taxonomic set, maize is treated as just another grass, parallel to canes, reeds, and the small class of grasses that includes teosinte (tall grasses). Four categories of smaller grasses and the introduced categories of rice, garlic, and onion complete the inventory, along with a half-dozen

TABLE 18-2 \muzgo jndë: Grass Names and Their Classification

Amuzgo name	Morpheme gloss	Local Spanish common name
1. (Covert) Tall grasses with erect stems		
1a. *jndë ndye* (tall grass)		
jndë ndye	grass-tall	zacate *alto*
jndë nkwjan		*maicillo silvestre* [teosinte]
jndë ntyunda		*zacate de falda*
1b. *ntjo* (maize)		
nna 'ndya		*maíz de zapalote* (has varietals)
nna kiobechi'		*cuarenteaño* (has varietals)
nna ñ'on		*maíz tepesentle* (has varietals)
nna su	maize-blue	*maizón azul*
nna t'oan		*maizón* (has varietals)
nna tyosu	maize-rabbit	*maíz de conejo* (has varietals)
nna xtyein		*maiz mixteco*
1c. *tzjo* (cane)		
tzjo be	cane-red	*caña alta morada*
tzjo bi		*caña alta blanca*
tzjo ki		*caña alta dura*
tzjo nthi		*caña alta rayada*
tzjo tzja'		*caña alta ceniza*
tzjo tzojndya		*caña pequeña morada*
1d. *tzoan* (reed)		
tzoan kitza'san		*carrizo silvestre*
tzoan kizoan	reed-wild-purple	*carrizo morado silvestre*
tzoan tz'oun		*carrizo cultivado*
2. *jndë ntkein* (rice) Grains		
jndë ntkein 'ndya		*zacate de arroz zapalote*
jndë ntkein jndë		*zacate de arroz silvestre*
jndë ntkein jndë tyo'		*zacate de arroz trigo*
jndë ntkein kind'o		*zacate de arroz boludo*
jndë ntkein ñ'on		*zacate de arroz tepesentle*
3. (Covert) Introduced grass-like plants		
3a. *tëxjo* (garlic)		
tëxjo be	fruit-metal-red	*ajo morado*
tëxjo kichi'	fruit-metal-white	*ajo blanco*
3b. *tëxjoin* (onion)		
tëxoin be	onion-red	*cebolla colorada*
tëxoin kichi'	onion-white	*cebolla blanca*
4. *jndë kotzama'* (spreading grass)		
4a. *jndë ton* (node grass)		
jndë ton		*bramillo*
jndë ton kintku		*zacate bramillo*
4b. *jndë tonko* (long grass)		
jndë jnon		*zacate pastura*
jndë jñinkio		*zacate aguate*
jndë nd'iki tzonde		*zacate cuita de vaca*
jndë ñoan		*zacate menudito*

(Continued)

TABLE 18-2 *(continued)*

Amuzgo name	Morpheme gloss	Local Spanish common name
5. (Covert)		
5a. *jndë kijndë* (thin grass)		
jndë b'a	grass-house	*zacate casero*
jndë nd'i		*zacate delgado*
5b. *jndë të* (rough grass)		
jndë limon	grass-lime	*zacate de limón*
jndë ndyue'		*zacatón de lomas*
jndë tzan	grass-hard	*zacate duro*
jndë tzja		*zacate rasposo*
6. (Covert)		
6a. (Covert)		
jndë nn'en		*zacate zopilote*
jndë snda'		*zacate mexicano*
jndë tyosu	grass-rabbit	*zacate de conejo*
6b. (Covert)		
jndë kichilo	grass-knife	*zacate filoso*
jndë ndiu		*zacate agujilla*
7. Ungrouped "grasses"		
jndë chein		*zacate bramillo blanco*
jndë chjein		*zacate guineo*
jndë jñinkio		*zacate velloso aguatoso*
jndë ndy'ia		*zacate delgado largo*
jndë noan		*carricillo acuático*
jndë ton'		*zacate de loma*

ungrouped grasses. The generic taxon 'maize' has at least the seven specific taxa listed previously, and this is one of the few plant classes that include varietal taxa. Thus, maize is at once the most extensive class of grasses, with more members than any other generic taxon, and the most elaborated, with a varietal level of taxa unmatched elsewhere in the botanical taxonomy.

The remaining major Amuzgo life forms are tree and herb: these have been detailed by Tapia García [38, 39].

COMPARATIVE MESOAMERICAN PLANT CATEGORIZATION

Tzeltal (Mayan) Ethnobotanical Categories

The prototypical ethnoscientific study of a Mesoamerican ethnobotany is the work of Berlin, Breedlove, and Raven [5–8]. Based on several years of research into the folk knowledge of the Tzeltal Maya of Tenejapa, Chiapas (Mexico), this project opened the eyes of anthropologists and botanists alike to the extraordinary amount of knowledge of the plant world shared by nonspecialists and to the impressive native organization of this knowledge. The ordinary Tzeltal controls a vocabulary of several hundred plant names, organized into a multilevel taxonomic hierarchy.

The botanical domain of the Tzeltal corresponds roughly to that of Western botanists, except that fungi, algae, and some other minor plant classes are not included. There is no Tzeltal term for "plant" in general, although the existence of the cognitive category can be demonstrated by empirical tests. Tzeltal and other Mayan languages require the use of a "numeral classifier" when anything is counted; the classifier follows the numeral and precedes the quantified noun (e.g., Chol *cha' tikil winik* [two men)], two-PERSON-man). In Tenejapa Tzeltal, members of the plant class require the use of a specific numeral classifier when they are counted, and this classifier, *tehk,* is only used with plants (e.g., *cha' tehk te'* [two trees], two-PLANT-tree). Thus, although the category is a "covert category" (a cognitive category that is not lexically labeled), its existence can be demonstrated through the application of ethnoscientific methodology, including sorting tasks [6].

Within the domain of plants, Berlin and colleagues [5–8] discovered an extensive taxonomic organization. Immediately included in the covert "unique beginner" (plants in general) are a number of "life forms," which in turn include "generic, specific and varietal types." Between the levels of life forms and generics are a few "intermediate types," covert categories that Berlin [4] takes to be an adaptation to the introduction of foreign species. In line with later discoveries in other folk botanies, Tzeltal has between 5 and 10 life forms, some 500 generics, around 1000 specific names, and only a few varietals (mostly cultivated plants or plants of special cultural interest).

Four major life forms include 75% of the generic types: *te'* (tree) includes 178 generic taxa; *wamal* (herb) includes 119 generic taxa, *'ak* (grass) includes 35, and *'ak'* (vine) includes 24. The majority of generic taxa are monotypic; only 15% are subdivided into specific taxa. The generic taxa not found in life forms are unaffiliated generic taxa. These taxa include morphologically aberrant plants, on the one hand, and all of the cultigens, on the other hand. Standing outside the four major life forms are the ungrouped Tzeltal categories of cultigens maize, beans, squash, tomato, chile, and so on, as well as the "aberrant" plants epiphites, cacti, agaves, ferns, mosses, bamboos, and so forth.

The Tzeltal system thus stands in stark contrast to the Amuzgo system in its treatment of cultigens. In the Amuzgo system, *all* cultigens fall into the major life forms; maize is a grass; tomatoes and beans are vines; chiles are herbs, and so on. The opposite is true of the Tzeltal system, where *no* cultigens fall into the four major life forms. In addition to these two ethnobotanies there are a number of other reported Mesoamerican folk taxonomies that treat the cultigens in some other fashion.

Chuj (Mayan) Ethnobotanical Categories

A case in point is the categorization of plants among the Chuj Maya of Guatemala, where these categories have become grammaticalized [34]. In the Chuj language [20] there is a set of "specifiers" related to the essential nature of the entities of the world and manifesting important aspects of Chuj worldview. Some 15 specifiers segregate the world into its essential natures: stone, earth, liquid, animal, male and female human, tree, grain, vine, other plants, cloth, sheet nylon (a recent innovation), salt, rains, and illnesses. Nouns representing empirical entities fall into one or more of these classes, depending on the basic substance of the entities. In their use as articles, the specifiers precede the noun or noun phrase whose nature they identify. The word for "bowl" is *jukup'*. A wooden bowl is *te' jukup'*, a metal (or stone) bowl is *k'en jukup'*, a ceramic (earthen) bowl is *lum jukup'*, a bowl made from animal hide would be *nok' jukup'*. Specifiers in these constructions are unstressed, similar to the Spanish gender articles *el* and *la* in *el niño, la niña* (the boy, the girl). In their use as pronouns, the specifiers may replace the noun or noun phrase; in context, a wooden bowl could be referred to simply as *te'* (the "wooden" thing we are talking about), a rabbit *(nok' chich)* could be referred to simply as *nok'* (the animal). This usage is parallel to the Spanish gender pronouns *él* (he) and *ella* (she). Significantly, animals that behave like humans in folktales are classified as humans: *winh chich* ([male human] rabbit) rather than *nok' chich* ([animal] rabbit).

Four of the specifiers cover the botanical world: *te'* (trees, wood, and wood products), *ch'anh* (vines and vine products), *'ixim* (grains, including maize and maize products), and *'anh* (grasses, herbs, and other plants and their products) [11–13]. Although the data are linguistic and not ethnoscientific, it is apparent that Chuj ethnobotany features these four major life forms, and between them they include all plants. In comparison with the Tzeltal four major life forms, it could be argued that in Chuj, maize and the introduced grains (wheat) have in a sense driven other grasses out of their life form and assigned them to the herbs (a default class, in any case). The category *'ixim* (maize–grains) includes 11 varieties of *Zea mays,* as well as Triticum aestivum [13].

There is a distinct utilitarian bias in the Chuj plant categorization. Of the 11 vines *(ch'anh),* uses are recorded for only 6: all 6 are used for tying (or for clothesline), including *Passiflora membranacea, Smilax subpubescens,* and *Clematis dioica/C. grossa.* Closely related species that are not suitable for tying are not classed as *ch'anh,* but as *'anh: Smilax subpubescens (kul),* used for tying, is *ch'anh,* but another species of *Smilax (k'ix tan),* which has edible growing tips, is classed as *'anh.* Likewise, *Passiflora membranacea (chuluma'ax),* used for tying, is *ch'anh,* but the edible *P. ligularis (karanáduh,* a loan from Spanish *granadilla),* is classed as *'anh te',* a subdivision of *'anh.* The botanical genus *Solanum* is divided into four folk generics, three of which *(pay te', kakaw k'i'ix,* and *túnhku')* are classed as *te'* (trees), but the fourth *(muh)* is placed in *'anh* [11–13]. Although much more research would be needed to test the utilitarian hypothesis, a first impression of the Chuj categorization is that it includes two small life forms with specialized uses (11 kinds of maize–grains and a similar number of vines–rope substitutes), and two large, less well-defined life forms (some 178 trees/firewood? and 192 herbs, about half of which are edible). Herb is perhaps not a suitable gloss for the latter category because it includes not only small herbaceous plants (including beans, tomatoes, squashes, peppers, greens, and tobacco), but also mushrooms, bananas, agaves, cacao, coconuts, and pineapples [11–13].

Like the flora, the animal kingdom is also divided into four major life forms, but the grammaticalization of these categories takes a distinct form. Animal names are associated with the *nok'* specifier regardless of their life form. But the nomenclature for male and female animals divides vertebrates into four life forms (the categorization of insects and other invertebrates is less clear). The gender of animals is designated by applying to the species name one of the "sex markers" [22]. Male animals may be designated *'ajtzo'* (rooster), *mam* (father), *xip'* ([male] dog), *winak* (man), or *ch'a'ak* (horn), as in *ch'a'ak k'ultakil chej* (male deer), *'ajtzo' chay* (male fish). Female animals may be designated by applying *'ix* (woman), *nun* (mother), or *chichim* (chick), as in *'ix k'ultakil chej* (female deer), *nun chay* (female fish). The intersection of the distinct male and female sex markers defines the four life forms of (vertebrate) animals. Reptiles and amphibians (and snails) take *'ajtzo'* (male) and *'ix* (female); fish take *'ajtzo'* (male) and *nun* (female); birds (and bats) take *'ajtzo'* (male) and *chichim* (female). Mammals generally take *xip'* (male) and *'ix* (female), but the horned animals deer and goats (an introduced species) mark males with *ch'a'ak* (horn), and the tapir, uniquely, takes *mam* (male) and *nun* (female).

There is a notable correlation between these classes of animals and their modes of locomotion. Reptiles and amphibians (and snails) are "crawlers," fish are "swimmers," birds (and bats) are "fliers," and mammals are "walkers." This four-way division of vertebrates has been noted elsewhere in Mesoamerica [22, 25].

Itzá (Mayan) Ethnobotanical Categories

The ethnobotanical classification reported for the Itzá Maya offers yet another scheme of botanical categorization [1]. Most Itzá plant taxa, with the notable exception of the cultigens, fall into four life forms, the expected tree, vine, grass, and herb categories; together, these constitute a covert category, the "wild flora." In direct contrast to the wild flora, the Itzá cultigens fall under a named taxon that includes all cultivated plants: *päk'aal,* from the verb *päk'* (to plant), literally, "planted (things)." The internal organization of cultivated plants parallels that of the wild plants (including the life forms trees, vines, etc.). In effect, intermediate categories have been created *above* the life forms to form a binary contrast between cultivated and wild plants; the life forms are replicated within each of these opposing plant classes. The treatment of cultigens is quite unlike that of Tzeltal, where cultigens constitute a number of unaffiliated generic taxa and do not (apparently) form even a covert category of "cultivated plants"; these intermediate categories are also distinct from those of Tzeltal, where the "intermediate taxa" occur *below* the level of life form.

Popolocan (Otomanguean) Ethnobotanical Categories

The Popolocan languages (Mazatec, Ixcatec, Chocho, and Popoloca) are members of the Otomanguean family spoken in the northern mountains of the state of Oaxaca (Mexico) and adjacent parts of the state of Veracruz. Like Amuzgo (also an Otomanguean language), these languages have sets of prefixes that attach to nouns and identify the semantic class of the noun's referent(s). A comparison of these prefixes and the nouns they are ultimately drawn from, by Annette Veerman-Leichsenring [41], shows that the noun categories can be reconstructed to the stage of Proto-Popolocan, the antecedent language to the Popolocan subfamily, spoken before about 500 BC.

Nine categories treat humans and group people by criteria of sex, age, kinship, and social status. Twelve classifiers treat the nonhuman world, including the general categories of "animal," "tree–wood," "liquid," "spherical" (applied to fruits), and "flower" and "plant/leaf" (applied to plant parts, not plant classes). The remaining six classifiers mark the classes maize, bean, squash, and chile, as well as "pulque" (an alcoholic drink made from agave) and a category that includes the Zapotaceae, early domesticates in this region [2]. These categories are quite stable in the Popolocan languages, and most of the languages studied have matching sets of cognate terms. The maize classifier prefix and its antecedent noun, for instance, can be reconstructed as ***niu* in Proto-Popolocan from Popoloca *nu-* ~ *ñu-, niu*; Chocho *ñu-, ñu;* Ixcatec *ñu-, niñu;* and Proto-Mazatec (in turn reconstructed from several varieties) **ñu-*, **niñu* (vowel tone and nasalization omitted) [41].

The inference that can be drawn from this linguistic evidence is that in the ethnobotanical vocabulary of the speakers of Proto-Popolocan in the first millennium BC, there was one term at the level of life form (tree), and folk generics for the major cultigens. The latter resemble the unaffiliated fork generic taxa of the Tzeltal system reported earlier. This system is especially interesting in light of the evolutionary process proposed by Cecil Brown [15] discussed later in the chapter. Because the prefixation of nominal classifiers occurs not only in Popolocan languages, but in other Otomanguean languages (cf. Amuzgo, preceding), this marking system is either a retention from Proto-Otomanguean (spoken before about 4000 BC), a retention from some early development from Proto-Otomanguean, or the result of widespread diffusion at some later (but still ancient) time. In either case, the systematic comparison of nominal classification across Otomanguean languages should be rewarding.

THE EMERGENCE OF MESOAMERICAN LIFE FORMS

The inspection of Mesoamerican folk botanical systems suggests that there is a wide variety of systems. Most share the four major life forms, but each system is likely to give different treatment to the cultigens. This situation suggests that a developmental or cultural evolutionary process is at work, and that the distinct systems represent different alternatives in a common process of change. The two most distinct systems seem to be the Amuzgo taxonomy, which places all plants within a restricted set of life forms, be they cultigens or not, and the Itzá taxonomy, which has two matching sets of life forms under the taxa "wild plants" (covert) and "cultivated plants" *(päk'aal)*. All other reported systems can be seen as intermediate between these two extremes in their treatment of the cultigens. Chuj has maize (grains) as a life form, but other cultigens are found in the life form herbs. The Popolocan languages have tree as a life form, and treat the cultigens like Tzeltal, which has all the cultigens outside the major life forms, but not organized into a class of cultivated plants.

The variants suggest a developmental process whereby either 1) cultigens are being removed from life forms, perhaps as they become aberrant morphologically and acquire more cultural significance, or 2) the life forms have come into existence since the domestication of plants, and the cultigens have not yet been placed within them, perhaps due to their aberrant morphology or cultural significance.

The second possibility is suggested by the work of Cecil Brown [15]. Brown has proposed an evolutionary model based on an examination of botanical life forms around the world. Applying a model and a methodology developed by Brent Berlin and Paul Kay [10] in their study of the developmental sequence of basic color terms (categories analogous to botanical life forms), Brown argued that terms for life form categories evolved late in human history, and that they have tended to appear in cognitive systems and languages in a particular order. The first life form term to emerge is tree, which contrasts with the unlabeled nontrees. The contrast is essentially "plant larger than a human" versus "plant smaller than a human." In the next stage of development, the nontrees are labeled ("grerb," a term Brown coins from grass and herb). In later stages, the life forms vine, grass, and herb partition grerb, but in no particular order.

With respect to Mesoamerica, Brown [16] also carried out a historical comparative study of the terms for life form plant classes in the languages of the Mayan family, and the data support his argument that life form terms appeared rather late in history. The identification of cognates, words in related languages that are developments of the same word in the common ancestral language, allows for the hypothetical reconstruction of the vocabularies of languages spoken thousands of years ago. Because culturally significant vocabulary is more likely to survive over time to become cognate sets [9], we would expect to be able to find cognate sets in the domain of plant classes, if such terms existed in the proto-language. The only life form term that reconstructs to the stage of Proto-Mayan (the ancestor of all the languages of the Mayan family, spoken before 2200 BC) is **te'* (tree), as Brown's evolutionary model would predict. (The Popolocan data discussed earlier, not available when Brown formulated his hypothesis, also support this model.) Not until about one thousand years ago (at the end of the Classic Period), did Mayan life form vocabulary begin to increase, and the other life forms emerged. Having entered the vocabularies after the diversification of the protolanguage into independent languages, these life form terms are not cognate across the family, although they are cognate within some subbranches.

Additional support for this model comes from Maya hieroglyphic writing and its associated iconography, where semantic signs mark illustrated objects as being of the substances of categories similar to those represented by the Chuj specifiers: wood, stone, water, and so on [24, 28]. The only plant class that has been securely identified so far is wood. On the other hand, there are other signs that may well refer to botanicals, including something that may be either vine or herb or, for that matter, grerb.

CLASSIC MAYA SCIENCE

If Mayan life form terms can be shown to have existed around AD 1000, at the end of the Classic Period, then the assignment of most botanical generic taxa to four life forms is most likely an accomplishment of Classic Maya science. This would add botany to the inventory of sophisticated sciences among the Classic Maya, along with fields of inquiry such as astronomy, where Maya expertise is well documented. It is notable that principal Maya partitions of the world tend to have four opposing divisions. The "four directions" are quadrants defined on the east and west by the rising and setting ranges of the sun, with the north and south quadrants poorly represented in modern Maya vocabulary but named in Classic Period hieroglyphic texts [25]. The directions are matched by four colors (east, red; north, white; west, black; south, yellow), with a fifth color (blue–green) associated with the center.

In the animal kingdom, vertebrates are divided into four classes based on locomotion: walkers (mammals), crawlers (reptiles and amphibians), flyers (birds and bats), and swimmers (fish) [22]. These classes are manifest in the Post-Classic codices, where offerings typically feature one animal from each class (e.g., Dresden 30B–31B), where deer, turkey, iguana, and fish are given in successive offerings [14, 42]. It is notable that in this text there are directional associations with these offerings: the East (sunrise) offering is deer, often associated with the sun in Mesoamerica; the North (zenith, sky) offering is turkey, a bird. The West (sunset) offering is iguana, a reptile (reptiles, especially snakes, are associated with openings to the underworld), and the South (nadir, underworld) offering is fish; the Maya often depicted the underworld as watery.

Significantly, the remains of just this set of life forms are often found together in archaeological deposits associated with other offerings (William Fash, personal communication). The evidence thus suggests that Classic Maya science tended to categorize things in terms of four opposing categories, and the creation of four botanical life forms would be consonant with other Classic Maya partitions of the world.

As to the origin of the Mayan botanical life form terms, Brown [16] argues that terms for "rope" came to be used for vine, and terms for "roofing grass" or "thatch" have become grass; terms for the default class grerb variously derive from "weed," "medicinal plant," "underbrush" or "green." That is, in Mayan languages at least, the life form terms likely came from the names of common materials derived from the plants in each life form. Thus, *te'* (tree) would have originally referred to wood, *ch'anh* (vine) was originally rope, grass was originally thatch, and so on. Although agreeing that the life form categories have a striking relationship to their utilitarian uses, I would argue that the transfer of meaning could easily have gone the other way. In the case of vine and rope, there is every reason to believe that Mayas used vines for tying long before they had rope, and twisted rope is still a rare item in rural Maya communities, where vines are routinely used in the construction of houses, fences, and other artifacts. Brown's proposal was further elaborated by Stanley Witkowski, Cecil Brown, and Paul Chase [44], but has been criticized on theoretical grounds by Scott Atran [3]; see also the comments of Eugene Hunn [26, 27] and Anna Wierzbicka [43].

Whatever the origin of the life form terms, there is a clear pragmatic component to the classification of plants into life forms. As Duane Metzger and Gerald Williams [35] have shown, the categories tree *(te')* and firewood *(si')* are coterminous in Tzeltal; any plant with an erect woody stem is a *te'* and is potential firewood, although there are clearly better and worse woods for making fires. Likewise, nothing that is not a *te'* is suitable for firewood. In Amuzgo, the class of wild vines includes only those vines that are useful for tying: utility is clearly a criterion of classification. It is not too far-fetched to imagine that the basis for creating the life forms was a pragmatic one: trees are those plants used for firewood and construction, vines for tying, grasses for thatching, and herbs include all those other plants. Herb continues to be a default class in modern Mesoamerican ethnobotanical systems—not easy to define except as opposed to the other three classes, which can be characterized by morphology as well as by patterns of usage. If utility were the original intent of the classification, then it would be reasonable *not* to place into the life forms plants that were used for a unique purpose: food.

If Brown's model and its chronology are correct, the differences between the Mesoamerican systems discussed earlier are because of cultigens not having been incorporated into the emerging life form categories, rather than having been removed from them, as I have argued earlier [23]. Agriculture among the Maya clearly predates the existence of all the life form terms other than tree. Reconstructed Proto-Mayan vocabulary (ca. 2200 BC) includes only the life form tree, but includes generic terms for maize, beans, squash, peppers, tubers, and a host of other cultigens, along with names for principal plant parts (ear of maize, etc.) and products derived from domesticated plants [30]. Archaeological data support a much earlier date for Maya agriculture, and this is consistent with the already well-developed agricultural vocabulary of Proto-Mayan.

CONCLUDING REMARKS

Despite the fact that one of the prototypical studies of folk botanical taxonomies was carried out in Mesoamerica, Mesoamerican ethnobotanies in general are scarcely known [8]. The preliminary observations offered here suggest that their study would be rewarding. Mesoamerica offers the chance to do comparative work on the languages and cultures of societies that have developed each in its unique environs but all in direct or indirect contact within a single culture area, sharing basic concepts of the world as well as technologies and patterns of social organization. Mesoamerican native populations speak the languages of five language families (Mayan, Mixe–Zoquean, Otomanguean, Uto–Aztecan, and Totonacan), one large language isolate (Tarascan), and a number of minor languages of unknown affiliation (Xinca, Chontal de Oaxaca, etc.). Although there is considerable linguistic diversity, there is also considerable cultural similarity: all of these various linguistic groups share the "Mesoamerican triad," the agricultural complex of maize, beans, and squash, along with other cultigens.

Newly domesticated plants spread quickly across the region, and Mesoamerican societies thus faced similar problems in the conceptual treatment of cultigens with respect to wild plants. The variety of responses—to create life forms, to incorporate cultigens in life forms or to exclude them, to create intermediate categories encompassing the life forms, and so on—must respond both to local conditions and previous history and to whatever cognitive tendencies and universals guide the behavior of humans in these pursuits.

The Mesoamerican culture area constitutes a virtual laboratory for the study of the evolution of ethnobotanical knowledge and belief. Scores of languages encode common conceptual categories in distinct orders and distinct fashions. Historical comparison and reconstruction of the botanical vocabulary of languages within each family would shed light on developmental sequences, and maize is clearly a major player in this development.

The questions raised by the little we know about Mesoamerican folk taxonomies cannot be resolved without much more extensive research into the botanical knowledge of many more Mesoamerican indigenous groups. What is clear from this preliminary examination, however, is that it is inappropriate to take the epoch making pioneer study of Tzeltal ethnobotany as a model for all of Mesoamerica, or even, for that matter, as a model of the ethnobotanies of the Mayan family.

References Cited

1. N. L. Adamson, C. A. Hofling, F. F. Tesucún. (1997). Flora taxonomy; taxonomía de flora. In: C. A. Hofling, (Ed.), *Itzaj Maya–Spanish–English Dictionary*. Salt Lake City: The University of Utah Press. pp. 61–71.
2. M. A. Hernández, P. Casasa García. (1979). Un análisis cultural de juegos léxicos reconstruídos del proto-otomangue. In: N. A. Hopkins, J. K. Josserand, (Eds.), *Estudios lingüísticos en lenguas otomangues*. Colección Científica, 68 (Lingüística). México, D. F., Instituto Nacional de Antropología e Historia. pp. 13–19.
3. S. Atran. (1990). *Cognitive foundations of natural history: Towards an anthropology of science*. Cambridge, MA: Cambridge University Press.
4. B. Berlin. (1972). Speculations on the growth of ethnobotanical nomenclature. *Language in Society, 1,* 63–98.
5. B. Berlin, D. E. Breedlove, P. Raven. (1966). Folk taxonomies and biological classification. *Science, 154,* 273–275.
6. B.Berlin, D. E. Breedlove, P. Raven. (1968). Covert categories and folk taxonomies. *American Anthropologist, 70,* 290–299.
7. B.Berlin, D. E. Breedlove, P. Raven. (1973). General principles of classification and nomenclature in folk biology. *American Anthropologist, 75,* 214–242.
8. B. Berlin, D. E. Breedlove, P. Raven. (1974). *Principles of Tzeltal plant classification: An introduction to the botanical ethnography of a Mayan-speaking people of highland Chiapas*. New York: Academic Press.
9. B. Berlin, D. E. Breedlove, R. M. Laughlin, P. Raven. (1973). Cultural significance and lexical retention in Tzeltal-Tzotzil ethnobotany. In: Munro S. Edmonson, (Eds.), *Meaning in Mayan languages: Ethnolinguistic studies*. Janua Linguarum, series practica, 158. The Hague: Mouton. pp. 143–164.
10. B. Berlin, P. Kay. (1969). *Basic color terms: Their universality and evolution*. Berkeley: University of California Press.
11. D. E. Breedlove, N. A. Hopkins. (1970). A study of Chuj (Mayan) plants, with notes on their uses, Parts I–III. *Wasmann Journal of Biology, 28,* 275–298.
12. D. E. Breedlove, N. A. Hopkins. (1971). A study of Chuj (Mayan) plants, with notes on their uses, Parts I–III. *Wasmann Journal of Biology, 29,* 107–128.
13. D. E. Breedlove, N. A. Hopkins. (1971). A study of Chuj (Mayan) plants, with notes on their uses, Parts 1–III. *Wasmann Journal of Biology, 29,* 189–205.
14. V. R. Bricker. (1997). The structure of almanacs in the Madrid Codex. In: V. R. Bricker, G. Vail, (Eds.), *Papers on the Madrid Codex*. Middle American Research Institute, Tulane University, Publication 64. New Orleans: Tulane University. pp. 1–25.
15. C. H. Brown. (1977). Folk botanical life forms: their universality and growth. *American Anthropologist, 79,* 317–342.
16. C. H. Brown. (1979). Growth and development of folk botanical life-forms in the Mayan language family. *American Ethnologist, 6,* 366–385.
17. D. S. Byers, (Ed.). (1967). *The prehistory of the Tehuacan Valley*. Vol. 1, Environment and Subsistence. Austin: University of Texas Press.
18. M. Cruz Hernández. (1993). *N'oan nan kobijnd'ue n'an tzjon noan: Los usos de la madera entre los amuzgos*. México, D. F.: Centro de Investigaciones Superiores en Antropología Social.
19. V. M. Franco Pellotier. (1995). Amuzgo. In: J. W. Dow, R. Van Kemper, (Eds.), *Encyclopedia of world cultures*. Volume VIII: Middle America and the Caribbean. Boston: G. K. Hall. pp. 3–7.
20. N. A. Hopkins. (1967)/ The Chuj language. Ph.D. dissertation. Chicago: University of Chicago.
21. N. A. Hopkins. (1978). Amuzgo and general Mesoamerican ethnobotanical nomenclature. Paper presented to the American Anthropological Association, Annual Meeting, November 14–19, Los Angeles, California.
22. N. A. Hopkins. (1980). Chuj animal names and their classification. *Journal of Mayan Linguistics, 2,* 13–39.
23. N. A. Hopkins. (1987). Etnobotánica y evolución: un comentario sobre Mesoamérica. In: S. Glanz, (Ed.), La heterodoxia recuperada; en Torno a ángel Palerm. México, D. F.: Fondo de Cultura Económica. pp. 203–225.
24. N. A. Hopkins. (1994). Days, kings, and other semantic classes marked in Maya hieroglyphic writing. Paper presented to the American Anthropological Association, Annual Meeting, November 30–December 1, Atlanta, Georgia.
25. N. A. Hopkins, J. K. Josserand. (2001). Directions and partitions in Maya world view. Paper presented to the 19th Maya Weekend, University of Pennsylvania, Philadelphia, March 24, 2001.
26. E. Hunn. (1982). The utilitarian factor in folk biological classification. *American Anthropologist, 84,* 830–847.
27. E. Hunn. (1987). Science and common sense: A reply to Atran. *American Anthropologist, 89,* 146–149.
28. J. K. Josserand, N. A. Hopkins. (2002). Classic Maya social interaction and linguistic practice: Evidence from hieroglyphic inscriptions and Mayan languages. In: V. T. Blos, R. Cobos, M. G. Robertson, (Eds.), *Tercera mesa redonda de palenque* [new series]. México, D. F.: Instituto Nacional de Antropología e Historia. pp. 355–372.
29. J. K. Josserand, M. C. Winter, N. A. Hopkins, (Eds.), *Essays on Otomanguean culture history*. Vanderbilt University Publications in Anthropology 31. Nashville, TN: Vanderbilt University.
30. T. S. Kaufman. (1964). Materiales lingüísticos para el estudio de las relaciones internas y externas de la familia de idiomas mayanos. In: E. Z. Vogt, A. Ruz L., (Eds.), *Desarrollo cultural de los Mayas*. México, D. F.: Universidad Nacional Autónoma de México. pp. 81–136.
31. T. S. Kaufman. (1978). *Meso-American Indian languages*. Encyclopedia Britannica, 15th edition, Vol. 11. Chicago: Encyclopedia Britannica. pp. 956–963.
32. I. Kelley, A. Palerm. (1952). *The Tajin Totonac*. Part 1: History, subsistence, shelter and technology. Smithsonian Institution, Institute of Social Anthropology, Publication No. 134, Washington, D.C.: Smithsonian Institution Press.
33. R. S. MacNeish, A. Nelken-Terner, I. W. Johnson. (1967). *The prehistory of the Tehuacan Valley*. Vol. 2. The non-ceramic artifacts. Austin: University of Texas Press.
34. J. M. Maxwell. (1995). Chuj. In: J. W. Dow, (Ed.), *Encyclopedia of world cultures*. Vol. VIII: Middle America and the Caribbean. Boston: G. K. Hall. pp. 70–74.
35. D. G. Metzger, G. E. Williams. (1966). Some procedures and results in the study of native categories: Tzeltal 'firewood.' *American Anthropologist, 68,* 389–407.
36. R. Ravicz, A. K. Romney. (1969). The Amuzgo. In: E. Z. Vogt, (Ed.), *The handbook of middle American Indians*. Vol. 7: Ethnology, Part One. Austin: University of Texas Press. pp. 417–433.
37. C. R. Rensch. (1966). Comparative Otomanguean phonology. Ph.D. dissertation. Philadelphia: University of Pennsylvania.
38. F. Tapia García. (1978). *Etnobotánica de los amuzgos*. Parte 1: Los árboles, Cuadernos de la Casa Chata 14. Tlalpan, México, D. F.: Centro

de Investigaciones Superiores del Instituto Nacional de Antropología e Historia.
39. F. Tapia García. (1980). *Etnobotánica de los amuzgos*. Parte 2: Los bejucos, zacates, yerbas y otras plantas. Cuadernos de la Casa Chata, 28. Tlalpan, México. D. F.: Centro de Investigaciones Superiores del Instituto Nacional de Antropología e Historia.
40. F. Tapia García. (1985). *Las plantas curativas y su conocimiento entre los amuzgos; árboles grandes y arbustos*. México. D. F.: Centro de Investigaciones Superiores en Antropología Social.
41. A. Veerman-Leichsenring. (2004). Popolocan noun classifiers: A reconstruction. *International Journal of American Linguistics, 70,* 416–451.
42. J. A. Villacorta C., C. A. Villacorta. (1976). *Codices mayas,* 2nd ed. Guatemala: Tipografía Nacional.
43. A. Wierzbecka. (1992). What is a life form? Conceptual issues in ethnobiology. *Journal of Linguistic Anthropology, 2,* 3–29.
44. S. R. Witkowski, C. H. Brown, P. K. Chase. (1981). Where do tree terms come from? *Man, 16,* 1–14.

C H A P T E R

19

The Historical Linguistics of Maize Cultivation in Mesoamerica and North America

JANE H. HILL

Department of Anthropology, University of Arizona, Tucson, Arizona

Glossary

Cognate set A set of words that resemble one another in form and meaning across several members of a language family and that can be related by regular sound correspondences, showing that they probably descended from a common ancestral word.

Gloss A preanalytic meaning or rough translation assigned to a word, whether attested or reconstructed.

Glottochronological date Glottochronology assumes a constant rate of replacement of vocabulary, and from the amount of core vocabulary shared between two languages known to be related, calculates the date when they diverged from their common ancestor. The method has been vigorously challenged, especially as an absolute dating method, but is used here informally as a guess at the relative ages of the major Mesoamerican language families.

Linguistic isolate A language that cannot be demonstrated to share a common ancestor with any other language.

Loan word A word in one language that resembles in form and meaning a word in another language, because it was borrowed from that language through the mediation of bilingual speakers.

Maize complex All the vocabulary labeling the maize plant, its parts, its stages of maturation, processes of cultivation, processes of preparation and the resulting foods, and special implements used in cultivation and cuisine.

Proto-language The ancestral language common to all the languages of a language family, reconstructed using the Comparative Method of historical linguistics.

Reconstruction, reconstructed word A word in a protolanguage that exhibits a shape derived from comparison of regular sound correspondences among words in the daughter languages, and a meaning also derived from comparison of meanings of words in the daughter languages, keeping in mind known principles of meaning change.

Historical linguistic evidence in the form of reconstructed vocabularies for maize and its cultivation and processing reflects the antiquity of cultivation in Mesoamerica, the early diffusion of maize to the U.S. Southwest, and the late arrival of maize in the eastern United States. Maize vocabularies can be reconstructed for the **proto-languages** of all the major Mesoamerican families, and languages there are linked by **loan words** for maize at several historical stages. Similar **reconstructions** are shown for the languages of the U.S. Southwest. However, maize vocabularies cannot be reconstructed for the proto-languages of families in the eastern United States.

INTRODUCTION

Historical–linguistic evidence has long been taken to support the proposal that the communities ancestral to three of the major Mesoamerican language families—Otomanguean, Mixe–Zoquean, and Mayan—included maize cultivators. Recent results suggest that a fourth major

FIGURE 19-1 Major North American language families with maize cultivators.

Mesoamerican language family, Uto–Aztecan, also descends from a community of cultivators. Although we cannot definitively determine from the linguistic evidence which of the four major linguistic groups descend from the first domesticators, or, indeed, if any of them does, loan words among the Mesoamerican languages suggest that communities speaking an ancient stage of Otomanguean may have been the earliest donors of loan vocabulary into the system.

Linguistic evidence attests to the diffusion of maize vocabulary, probably alongside maize cultivation itself, from the core Mesoamerican language families in two cases: from Northern Uto–Aztecan into Kiowa–Tanoan on the Colorado Plateau about 3000 years ago, and from Mayan languages into Xincan-speaking communities of Guatemala, at an undetermined date. However, other language families outside Mesoamerica reveal no linguistic trace of the diffusion that must have occurred. In the Southwest, the Yuman languages and two **linguistic isolates,** Keresan and Zuni, show no sign of linguistic influence from Uto–Aztecan or from any of the other Mesoamerican groups. In the eastern United States, maize vocabulary can neither be reconstructed for the language families of the region, nor can any maize vocabulary of the eastern languages be linked to Mesoamerican or Southwestern sources. (See Figure 19-1 for the approximate locations of the major language families discussed in this chapter.)

HISTORICAL–LINGUISTIC METHODS

Historical linguists use several methods to study prehistory. The Comparative Method tests sets of languages for descent from a single common ancestral language. The members of a language family—all the languages descended from a single common ancestor—will share words that resemble one another in both form and meaning. The formal resemblance is crucial: the sounds of such resemblant words must be relatable by regular sound correspondence. From the inventory of this "cognate" vocabulary the Comparative Method reconstructs the form and meaning of lexical items in the ancestral language, the proto-language.

The Comparative Method distinguishes true cognates, words that resemble each other because they descend from a common ancestral word, from words that resemble each other because they have been borrowed from one language to another. Like cognates, loan words resemble each other in form and meaning, but the resemblance in form cannot be accounted for by the same sound laws that relate true cognates. The study of the sources, trajectories, and meanings of loan words yields clues to the history of the diffusion of both material and symbolic culture.

Words can also resemble each other by chance. Distinguishing between chance resemblances and loan words requires attention to a variety of factors. For instance, resemblance among several syllables of multisyllabic words is less likely to be because of chance than is resemblance among monosyllabic words. The presence of several resemblant items in a single semantic field is less likely to be the product of chance than is resemblance in an isolated pair of words. Proposals for isolated monosyllabic loan words thus must be regarded as speculative, unless strong nonlinguistic evidence supports the hypothesis of a loan.

When the donor language of a loan word can be identified, linguists assume that the word and its meaning were linked by the borrowing group to the cultural practices of the donor group. But the presence of a loan word does not necessarily mean that the item labeled was entirely new to the culture of the borrowing community or that it originated in the donor community. Furthermore, borrowed cultural items and ideas are often labeled with indigenous linguistic materials, and not with a label from the donor language. It is not always obvious which language is the donor and which language is the receiver of a loan word. Many lines of reasoning are used to determine the direction of the loan. For instance, if a complex word has a morphological analysis in one language but not in another, the first language will be taken to be the donor. Thus, determining loan word trajectories requires considerable specialist expertise in the grammar of the languages involved.

Wörter-und-Sachen or "Word and Thing" methods, like the study of loan words, depend on the proto-language reconstructions of the Comparative Method. A word reconstructed in a proto-language signals the presence of a referent known to its speakers. If we can determine the meaning of the word, we can establish what this referent was. For instance, if we can assign the meaning "corn cob" to a reconstructed form, because most of the daughter languages share this or a closely related meaning, the *Wörter-und-Sachen* method permits us to suggest that corn cobs were known to speakers of a proto-language.

Many of the historical–linguistic results discussed in this chapter continue to be debated. Detailed evaluation of the evidence for every claim made is beyond the scope of this chapter, but where they have been published, I provide references to opposing views.

RECONSTRUCTED MAIZE COMPLEX VOCABULARIES IN MESOAMERICAN LANGUAGES

Mesoamerica, the site of the earliest domestication of *Zea mays,* is among the world's most linguistically diverse regions. Jorge Suárez [69] inventories 14 language families and isolates. Even the largest families are not large by world standards. They are 1) Uto–Aztecan, with about 30 known daughter languages and perhaps 20 spoken by cultivators; 2) Oto–Manguean, with about 23 major groups (of which one, Zapotecan, includes up to 55 languages); 3) Mixe–Zoquean, with 12 languages; and 4) Mayan, with about 30 languages. Several much smaller families are 5) Totonac–Tepehua of Mexico, 6) Xincan of Guatemala, 7) Lencan of Honduras, and 8) Misumalpan of Nicaragua. Four linguistic isolates are found in Mexico: 1) Tarascan, 2) Cuitlatec, 3) Tequistlatec–Jicaque, 4) Huave.

Glottochronological dates are of uncertain status and should be invoked with great caution. However, those that have been proposed suggest that the initial radiations of the large Mesoamerican language families occurred well after maize first appears in the archaeological record. The dates suggested for the families discussed in this chapter are shown in Table 19-1.

TABLE 19-1 Glottochronological Dates Proposed for Major Mesoamerican Language Families

Family	Date	Source
Otomanguean	4600 BC	Hopkins [35, p. 31]
	4400 BC	Campbell [7, p. 159]
	4000 BC	Kaufman [38, p. 94]
Uto–Aztecan	4000 BC	Miller [47]
	3100 BC	Suárez [69, p. 28]
	2000 BC	Hale [27, p. 107]
Mayan	2200 BC	Kaufman [38, p. 104]
Mixe–Zoquean	1500 BC	Campbell and Kaufman [9, p. 80]

These dates are "separation dates;" the date of the deepest split in the language family.

Application of *Wörter-und-Sachen* techniques to the reconstructed proto-languages of the language families shown in Table 46-1 suggests that their ancestral speech communities practiced cultivation. Words for a **maize complex** can be reconstructed for all of them (cf. for Uto–Aztecan [28]; for Otomanguean [1, 35, 39]; for Mayan [38]; for Mixe–Zoquean [9, 74, 78, 79]). This reconstructed maize complex vocabulary is given for each of these major families in Tables 46-2, 46-3, and 46-4.[1]

My claim, that Proto-Uto–Aztecan (PUA) was spoken in a community of cultivators originating in the northwest quadrant of Mesoamerica, is controversial. Until recently I shared the view, still held by many Americanist linguists, that the Uto–Aztecans descend from a proto-community of hunters and gatherers of the U.S. Southwest and northwestern Mexico and were latecomers into Mesoamerica [6, 7, 38, 40]. Important evidence for this perspective is the fact that only a few groups in the large Northern Uto–Aztecan subfamily, including the Hopi and a few bands of the Southern and Central Numic of the Colorado Plateau and Great Basin, were cultivators at the time of Spanish contact, and these were believed to have acquired this practice by diffusion from southern neighbors who had, in turn, acquired cultivation from further south in Mesoamerica. Although I identified Hopi and Numic cognates in the maize complex, the

[1]Asterisks before linguistic items indicate that these are forms reconstructed by the comparative method. Two asterisks mean that the form reconstructs to the deepest level of the protolanguage. In Uto–Aztecan (Table 46-2), single asterisks are used for two forms that can be reconstructed to Proto-Uto–Aztecan, but where meanings within the maize complex are not attested in Northern Uto–Aztecan.

For citation of linguistic materials, including reconstructed forms, I have adopted the orthography for Mesoamerican languages developed many years ago by the Proyecto Lingüístico Francisco Marroquín and used by Kaufman [41], because it avoids problems with font reproducibility. For a few symbols that may not be transparent: /7/ is glottal stop, contrasting with /'/ representing glottalization of a consonant. /x/ is "sh", /j/ is phonetic [x], nh is "ng" (pronounced as in final position in English), /tz/ is "ts" (pronounced as in final position in English). Vowels have their Spanish values. Central vowels usually written as "barred i" or "schwa" are here written /i/; no language cited distinguishes the two vowels so they need not be differentiated here. The midhigh front vowel of English "it" is represented as /ɪ/. Long vowels are written with double vowels, and superscript "n" following a vowel means that the vowel is nasalized. The transcription will be a bit jarring for linguists who are used to their own local orthographic conventions, but consistency throughout the manuscript was my primary consideration. Original sources should be consulted for the orthographies preferred by their authors.

Data sources not directly referenced in the text include the following: Atakapa [24]; Cocopa [11]; Huave [66]; Kiliwa [50]; Maricopa [23]; Mojave [51]; Tequistlatec (Chontal of Oaxaca) [71]; Totonac [52]; Walapai [74]; Zuni [20, 53].

TABLE 19-2 The Uto–Aztecan Maize Complex [30, 32]

1. **sunu* (corn ear, maize)
2. ***sono* (maize byproducts such as cobs, leaves, cane)
3. ***pa7tz-i/a* (green corn ear ("elote"), corn kernel, seed)
4. **saki* (popcorn, parched corn, to make parched corn or popcorn)
5. ***wïra* (to shell corn)
6. ***o7ri/a* (ear of corn, cob)
7. ***sura* (embryo of corn, tender ear of green maize)
8. ***tïma* (tortilla, tamale)
9. ***komal* (cooking surface for maize cakes)
10. ***kuumi/u* (to nibble small pieces of food, especially corn on the cob or popcorn)
11. ***wika* (digging stick)
12. ***ïca* (to plant)

possibility cannot be dismissed that these reflect early diffusion of loan words from the southern languages rather than descent from Proto-Uto–Aztecan [30, 32]. Second, the meanings of some words in the Uto–Aztecan maize complex in some of the daughter languages are not associated with maize. For instance, in several of the southern languages the cognate of item 3 in Table 46-2 means "seed in general," and not the maize kernel. Such shifts of meaning, from "seed" to "green corn ear, corn kernel," might have occurred after the breakup of the proto-language community. The southern Uto–Aztecan languages, especially Aztecan, do not share many features considered to be typical of the Mesoamerican areal complex [10]. Finally, there is no question that Uto–Aztecan is marginal linguistically within the larger complex of cultivation. Unique among the major Mesoamerican protolanguages, Proto-Uto–Aztecan includes no words for squash or beans [30]. Lyle Campbell [8] has raised these and other questions. I have replied to Campbell's arguments and present in this chapter additional evidence for cultivation in the community ancestral to Northern Uto–Aztecan, in the form of a set of loan words exchanged with Proto-Kiowa–Tanoan [32, 33]. If this evidence holds, it means that Proto-Northern Uto–Aztecan was a community of cultivators, and those groups that were not cultivators at contact do not represent the ancestral adaptation, but instead must have undergone "agricultural regression," abandoning maize cultivation as they moved beyond the limits of the appropriate climatic regimes [2].

The Otomanguean reconstructions given in Table 19-3 include many segments of uncertain status, and meanings in the daughter languages for each item are so diverse that Calvin Rensch [57] does not venture to reconstruct **glosses** for his proto-forms. The glosses in the table summarize the diverse meanings for the cognate words in the daughter languages. However, although we cannot confidently reconstruct the meanings of individual reconstructions, the totality of the set makes us confident that Proto-Otomanguean speakers were maize cultivators—perhaps the first maize cultivators. Otomanguean is the oldest of the major Mesoamerican language families. The current distribution of the Otomanguean languages includes all of the areas where the earliest Mesoamerican maize is attested archaeologically, including the Balsas River Basin where the initial domestication from teosinte may have occurred, as well as at Guilá Nahquitz, Oaxaca, and the Tehuacan Valley of Puebla where the earliest preserved cobs and kernels are found. The initial spread of the family is probably reflected in the formation of the Tehuacan Tradition at 5000–2300 BC [35, 39].

TABLE 19-3 The Otomanguean Maize Complex (The numbers in parentheses with # are cognate set numbers in Rensch [57]; forms with (K) are from Kaufman [39].)

1. ***7te(n)*[3] (corn gruel, atole) (#23); ***7tye* (atole) (Oto–Chinanteco only)
2. ***ta(h)* (tortilla, cooked corn) (#48) ; ***ta* (tortilla, hominy)
3. ***(7)ta(n)*[1/2] (clay griddle, clay pot, pottery) (#56) (possibly related to (#48) and to "stone"); ***ta* (clay griddle) (K)
4. ***(H)tu(h)(n)*[3] (to plant, to sow, to dress oneself, cornfield, corn crop, jungle growth, vegetation) (#91)
5. ***(n) (7)tu (h) (n)*[1] (moon, month, trough, manger, corn measure (almost anything hollowed out or concave), clay griddle, tortilla, spherical, land, oven) (#93)
6. ***ki* (to grind, to twist, twisted, to be ground) (# 100; perhaps related to 366)
7. ***k^we* (dust, powder, ground dried corn) (#154) (may be related to #283 "dust, powder, sand")
8. ***(n)(h) k^we(7)(n)*[3] (dough, maize, corn dough, corn, roasting ear, corn gruel, dry, shelled corn, kernels of corn) (#177) (may be related to 285) see green; ***k^wau* (K) (maize); ***hme* (K) (maize) (that is, Kaufman splits Rensch's set into descendants from two different ancestral forms)
9. ***(n)(7)k^wa(h)(n)*[2] (twist, grind) (#224)
10. ***(Y)(n)si* (clay griddle) (#248), (may be related to words meaning "round"); ***ci* (K) (clay griddle, metate)
11. ***(Y)(n)(H)se(h)(n)* (roasting ear, corncob, cooked corn, ground dry corn) (#285) (may be related to #177); ***s(a7)ai(n)* (K) (roasting ear)

(Table 3 continued on page 249.)

The reconstructions in Table 46-3 are shaky as to both form and meaning, primarily because Otomanguean languages present severe challenges to the historical linguist. The radiation of the family is an old one, making connections among the daughter groups—and even affiliation to the family—relatively obscure. A special problem is that some derivation, of uncertain function, was apparently accomplished in the proto-language by prefixes that have been absorbed into the beginning of stems, yielding mutations of

TABLE 19-4 The Mayan Maize Complex (Forms cited are from [41][a] unless otherwise noted.)

1. **7e7m* (corn the plant) and **7ixi7m* from *7ix = e7m* (corn, as substance)
2. **q'al* (corn field, milpa)
3. **7ix.i* (to shell corn)
4. **q'ut* (to grind on metate, mash, shell corn)
5. **kee7.e* (to grind)
6. **nhal* (ear of corn, mazorca)
7. **pa7at* (ear of corn with several ears joined)
8. **tz'utuj* (corn tassel)
9. **7ajn* (roasting ear, elote)
10. **qas* (to eat only tortillas [without meat, drink, etc.])
11. **k'aj* (ground parched corn, pinole, flour)
12. **maatz'* (corn gruel, atole)
13. **7uul* (corn gruel, atole) (from Proto-Gulf Zoquean *7uunu7* [41, p. 1186; Zoquean reconstruction from [78], Wichmann's 7U #033; Kaufman cites this form as "Sokean")
14. **q'oor* (corn dough/masa, hominy/pozole, atole)
16. **wah* (tortilla) [18]; Kaufman [41] reconstructs this form only to Eastern Mayan. This is the phonetic reading for the glyph sign for "offering of tamal."
17. **baqal(aq)* (corn cob, olote) [18]; Kaufman [41] lists Lowland and Western Mayan *b'aqal* and Eastern Mayan *b'aqlaq*, but does not reconstruct to Proto-Mayan.

[a]Kaufman [41] does not give reconstructed glosses for all forms. I have given summaries from glosses in the daughter languages.

stem–initial consonants. Thus different derivations of the same stem begin with different consonants, such as /t/~/y/~/n/ and /k/~/w/~/n/. The result is that cognate forms in the daughter languages may be descended from different consonantal grades. The indeterminacy introduced by the consonant alternations makes the reconstruction of tone especially crucial [71]. Suárez [71, p. 51] suggests that Rensch's [57] reconstructions of tone—where he attempts such at all—can be rated as no better than "provisional." The Rensch reconstructions have also been criticized by Terrence Kaufman [39], who proposes many new reconstructions, although he does not reconstruct tone.

With the exception of Huastec, spoken on the Gulf Coast of Mexico in Vera Cruz, the Mayan languages are found primarily in Yucatan and Chiapas in extreme southern Mexico, and in Guatemala. Thus the distribution of Mayan is relatively southerly, and Mayan languages may have been involved in the diffusion of maize vocabulary and perhaps maize cultivation into the language groups on the southern fringes of Mesoamerica.

Kaufman's [41] *Preliminary Etymological Dictionary* of Mayan, from which Table 46-4 is largely derived, includes **cognate sets,** proposals of loan word connections, and other

TABLE 19-5 The Mixe–Zoquean Maize Complex [78]

1. **mook* (corn)
2. **hïïpak* (olote) (note Maya "olote" no. 4)
3. **pah* (kernel)
4. **7iïks* (to shell corn)
5. **pitz-i* (leached corn, nixtamal)
6. **7an-e* (tortilla)
7. **hoo7s* (moler masa)
8. **kïï7t* (moler pinole)
9. **way* (to grind)
10. **7akwa7n* (comal, fogon)
11. **nïïp'* (to plant)
12. **kama, omook-kama* (cornfield)
13. **tze7s* (granary)

data and comments. Dienhart [18] is a much shorter online etymological dictionary that often disagrees with Kaufman on the details of reconstruction. In Table 46-3, I give two forms from Dienhart to discuss them in connection with the loan word complexes that follow.

The Mixe–Zoquean languages comprise the smallest, and probably the youngest, major Mesoamerican family, with only 12 languages, all spoken in southern Mexico. The languages are of special interest because of claims that early forms of Mesoamerican writing may have been in early Zoquean, and because of their geographical overlap with the earliest Mesoamerican high culture sites in southern Vera Cruz and Tabasco [37]. Campbell and Kaufman [9] associate the Mixe–Zoquean radiation with the rise of the Olmec tradition. Søren Wichmann's [78] magnum opus is the most useful single source for the historical linguistics of this group, and his reconstructions confirm the view that Mixe–Zoquean languages descend from an ancestral language spoken by cultivators.

Because well-documented Mesoamerican languages often exhibit at least 100 lexical items associated with maize, the short lists of **reconstructed words** given in Tables 46-2, 46-3, 46-4, and 46-5 must be a pale reflection of the rich vocabularies that actually existed in the proto-languages of these four families. One important point with reference to the earlier lists must be made. The preparation of corn considered today to be especially typical of Mesoamerican cuisine, the flat *tortilla* patted out from a dough made of finely ground maize kernels that have been leached with lime, cooked on a pottery griddle (called *comal* in Nahuatl), appears only during the first millennium BC [21]. Kaufman [41, p. 192] states that, "tortillas were not made in Southern Mesoamerican before 700–800 CE." Patricia Fournier [21, p. 24] confirms that on the Gulf Coast and in the Maya region, the *comal* appears only in the Post-

Classic period. Leaching of corn and the *comal* are not attested in the U.S. Southwest before 1300 AD [65].[2] All four proto-language communities are thus much older than the tortilla and the *comal,* and words in Tables 19-2, 19-3 19-4, and 19-5 glossed as tortilla or clay griddle (the *comal*) must have meant something different in the proto-languages. The kinds of meaning these words might have had are suggested by related words. For instance, in Uto–Aztecan the word for *comal* is related to words meaning "flat," whereas in Otomanguean it is related to words for "round." One of the Otomanguean words for tortilla also means "round, concave objects" including the moon, bowls, and troughs. In Mayan languages the word for tortilla is generally the same as the word for *"tamal"* or for food in general.[3]

The words glossed today as tortilla or *comal* are, of course, not the only forms that have changed their meaning. The range of meanings shown for the Otomanguean maize complex in Table 19-3 make clear the degree of mutability that is involved. The meanings of words shift around within the maize complex. For instance, a form may be preserved as the verb "to grind" in one language, but in another only as a derived noun meaning, "flour." In one language a word may mean "to shell corn," and in another "corn kernel." The Uto–Aztecan language Hopi reversed the meanings of the Uto–Aztecan etyma for "corn cake" and "griddle."[4] In the Otomanguean languages the same reversal appears between Popolocan and the other branches [39].

Shifts in meaning occur not only within the maize complex but also between it and other semantic domains. This raises the question as to when it is appropriate to reconstruct meanings within the maize complex for the proto-language. For instance, words for maize are often related to words for grass or seed (see Chapter 18). \ well-attested type of semantic change is "marking reversal," and the shift of meaning from grass to maize may be of this type, where maize becomes the prototypical grass and then the only grass that can be thus labeled. The word for grass comes to mean maize, and other grasses, the former prototypes, are left with labels like "wild maize." Indeed, the Nahuatl word *teosintli* (teosinte), which literally means "god-maize," looks like the product of such a marking reversal process involving *sintli* (ear of corn, mazorca). If an identical change from a meaning outside the maize complex to one within it is attested for the same word across several subfamilies in a language family, it is more likely that the change took place within the proto-language, rather than independently within several already differentiated daughter groups. However, because relationships like the grass–maize and seed–maize shifts are so widely attested, they may represent natural cognitive associations that could have developed independently within descendant languages.

Because of the mutability of meaning, historical linguists require the largest possible lists of lexical items within the maize complex and related domains for each of the daughter languages to identify cognates among resemblant items that may be rather different in meaning. Given the paucity of forms reconstructed for the proto-languages, the likelihood that we will be able to identify historical relationships between them is greatly reduced.

MAIZE COMPLEX LOAN WORDS IN MESOAMERICAN LANGUAGES

Are there any connections among the maize vocabularies of the four major Mesoamerican proto-languages, either because of descent from some language ancestral to one or more of the large families, or because of borrowing between the proto-languages? I will not discuss here the highly controversial possibility of common inheritance from some remote proto-language ancestral to two or more of the Mesoamerican families.[5] However, we can identify loan words of an early date. There are at least three levels of such loan vocabulary. These are: 1) Words from early stages of Otomanguean and, in one case, Uto–Aztecan, that appear in other language families, which may attest to the earliest diffusion of maize cultivation out of the communities of its primary domesticators; 2) words associated with the Olmec tradition and the earliest stages of Mesoamerican high culture, which probably originate in Mixe–Zoquean, and appear widely in other language families; 3) small sets of loan words that link only one or two languages or language subgroups and are thus probably of more recent date. The following critical discussion attempts some sorting out of the three levels.

Campbell and Kaufman [9] proposed that an important vocabulary complex diffused from Mixe–Zoquean into a

[2]Jaenicke and Smith (Chapter 6) may have a genetic explanation for this late appearance of tortillas, relating to the *sul* allele for starch, which is not present in cobs dated to before 1500 BC. The presence of sherds from the comal or flat clay griddle is the main source of archaeological evidence for the preparation of maize as tortillas. This preparation requires "nixtamalization," soaking maize kernels in water mixed with lime to soften the hulls and release nutrients. Data from residue analysis that would identify the presence of lime in vessels used for maize preparation were not available at the time that Fournier [21] prepared her review on this question. In contrast, Snow [65] depends heavily on residue analysis data to date the appearance of nixtamalization in the U.S. Southwest.

[3]I thank an anonymous referee for this information, and for the note on the "tamal" glyph in item (16) Table 46-4.

[4]Maize preparations with lime do not appear in the Hopi region until after AD 1300 [65], so these Hopi terms cannot have referred to piki bread or the piki stone as these are known ethnographically.

[5]Wichmann [77] has proposed that Uto–Aztecan and Mixe–Zoquean may ultimately share a common ancestor. Several scholars have proposed a "Macro-Mayan," linking Mayan, Mixe–Zoquean, Totonacan, and perhaps Huave. Campbell [7] critiques these proposals. Greenberg's "Central Amerind," which includes most of the Mesoamerican languages in a single subphylum of his Amerind, is not accepted by specialists.

variety of other Mesoamerican languages during the "Olmec" phase, 1200–300 BC [38]. Such loans are late in relation to the original domestication of maize, so shed no light on the ethnic affiliation of the earliest cultivators. However, there is evidence that some of Campbell and Kaufman's examples may, in fact, not be Olmec period loans, but belong to the first, earlier level and are thus relevant to this question. Some others, though, should be demoted to the third level. I treat here only the maize vocabulary. However, I note that Karen Dakin [15] and Dakin and Wichmann [16] argue that some of Campbell and Kaufman's most important loan items involving Olmec prestige goods, such as the words for *cacao* and chocolate, probably originated in Uto–Aztecan, not Mixe–Zoquean.

Wichmann [78] was unable to confirm several of Campbell and Kaufman's [9] Mixe–Zoquean reconstructions. He argued that others could not be associated with Olmec civilization, because they dated only to Proto-Zoquean, to Proto-Mixean, or to even shallower branches of the family, not to Proto-Mixe–Zoquean itself. An example is seen in item 13 in Table 19-4, where Kaufman [41, p. 1186] proposed that Proto-Mayan ***7uul* (corn gruel, atole) came from Zoquean **unu* of the same meaning. However, Wichmann reconstructs the form only to a lower level, Proto-Gulf Zoquean. Because Proto-Mayan is considerably older, either the resemblance is because of chance, or the loan is in the other direction, or both forms descend from some more ancient loan.

A second problem involves resemblances based on little phonological evidence. Thus Campbell and Kaufman [9, p. 85] relate Proto-Mixe–Zoquean ***mook* (see item 1 in Table 46-5) to Proto-Mayan ***7e7m* (corn plant). Other loans that they believe reflect the same Mixe–Zoquean origin include Proto-Mixtecan **yam* and Proto-Chiapanec–Mangue **-ma* (both of these are Otomanguean groups); Xinca *ayma* and Lenca *ima, ema, ama;* Cacaopera and Matagalpa *ayma* and Sumu *ama* (all Misumalpan), and Tarascan *ema.* Although Campbell and Kaufman suggest *ex hypothesi* that the diffusion of the item is because of Mixe–Zoquean Olmec influence, we cannot be confident that the Mixe–Zoquean word is related to the other words by borrowing rather than by chance resemblance because the Mixe–Zoque form shares with them only the phoneme /m/! Campbell [5] has made an excellent case for Mayan languages as the proximal source of a number of loans in Xinca, including this item, and it seems most likely that the Mayan form is the source for the diffusion into the other Central American languages. Thus, the presence of a resemblant form in Tarascan is the main evidence for an Olmec origin of this widespread cluster. However, I judge the Mixe–Zoquean source proposed by Campbell and Kaufman to be dubious.

Campbell and Kaufman's [9] second candidate for a widespread loanword with an "Olmec" source is item 9 in Table 19-5, Proto-Mixe–Zoquean ***way* (to grind corn) [p. 85]. Their resemblant set includes Totonac *wa7t* (tamale), Proto-Mayan ***wah* (tortilla) (see item 16 in Table 19-4), two Otomanguean forms—Proto-Chinantecan **wih(n)* (tortilla) and Proto-Chipanec–Mangue **wih7* (tortilla), Xinca *iwa* (to make tortillas), and Jicaque *we* (nixtamal, tamale). Several problems with this account have arisen. Kaufman [38] no longer believes that ***wah* is a Proto-Mayan form: he reconstructs it only to Eastern Mayan, although somewhat similar forms appear in the other branches in related meanings, such as Lowland and Western Mayan **waaj* (tortilla, to feed someone). That such a loan probably occurred after the breakup of Proto-Mayan, however, actually improves the likelihood of an Olmec source of the form because the radiation of Mayan probably began by 2000 BC, before the beginning of the Olmec horizon. A more serious problem with this proposal suggests that this complex should be elevated to level 1, loan words ultimately of Otomanguean origin. Rensch [57] reconstructs the source word, not just in Chinantecan and Chiapanec–Mangue, but in Oto–Pamean as well, and thus to Proto-Otomanguean as ***(H)wi(h)(n)* (tortilla) (#366; see item 14 in Table 19-3), Rensch suggests that this form in turn may be related to Proto-Otomanguean ***ki,* "to grind" (#100; see item 6 in Table 19-3), based on a productive consonant alternation /k/~/w/. Thus, if this is indeed a loan item (the cross-linguistic match is only in the initial /w/), its origin is probably Otomanguean and not Mixe–Zoquean, because Proto-Otomanguean is surely considerably older than Proto-Mixe–Zoquean. Suárez [70] makes a related point. In the Otomanguean language Tlapanec, Suárez identified an unpossessed form derived from ***-nwin.* However, the possessed form comes from ***-wiHn,* without the initial nasal. Notes Suárez, "It seems unlikely that such an irregular word could be a borrowing. On the contrary, the way it fits in the complicated structure of Proto-Otomanguean almost ensures that this word, like the Chinantecan and Chiapanec Manguen [sic] ones, reflects an old Proto-Otomanguean root" [70, p. 575].

Campbell and Kaufman [9] identify Proto-Mixe–Zoquean ***pitz-i* (leached corn, nixtamal) (see item 5 in Table 19-5) as the source of an Olmec period loan appearing as Totonac *pa7s(a)* (to shell corn), Nahua *patza* (to grind, mash), Totontepec–Mixe *pootza* (tamal), Tequistlatec *xpatz'a7i* (tamal de elote), Huave *peatx* (tortilla), Tzeltal and Tzotzil (Mayan) *patz* (tamal), and Xinca *paatz7i* (to grind, corn dough). This loan complex, however, should perhaps also be elevated to an earlier period, because Proto-Uto–Aztecan presents a competing source, ***pa7tz-i/a* (corn ear, corn kernel, seed) (see item 3 in Table 19-2), The Uto–Aztecan daughter-language reflexes thus far identified are Hopi *paatzama* (hominy); Taracahitan **patzi* (semilla, elote); Cora *hatzí* (elote), Nahuatl *aach-tli* (seed) (usually not a corn seed, but note Nahuatl *xin-(?)aachi-wa* [to plant seeds], used in a passage in Sahagún's [61] *General History* *of the Things of New Spain* that refers to corn seeds taken from images of the maize gods). The Uto–Aztecan range of meanings is a good semantic match to the rest of the items in the complex. Furthermore, the Uto–Aztecan source item matches across the languages in several phonological elements, the initial /p/, the vowel /a/, and the spirant /ts/ or /s/. The glottal stop in the Proto-Uto–Aztecan root matches the glottal stop in Totonac *pa7s(a)* and derived Totonac forms like *pa7sma* (corn cob, olote) and *tapa7sni7* (maize as kernels). Furthermore, the *pa-* element may reflect a Uto–Aztecan root for seeds (especially the edible seeds of chenopods and amaranths) and eggs [16]. Nahua *patza* (to grind, mash) is not directly derived from this set (the true Nahua cognate is *aach-tli*), and is probably a late loan, perhaps from Totonacan [cf. 40]. The Nahua languages regularly lost Proto-Uto–Aztecan *p in word-initial position, so Campbell and Kaufman [9] correctly identify any Nahua item beginning in /p/ as a likely loan word.

Campbell and Kaufman [9] suggest Proto-Mixe–Zoquean **hïtz* (to grind) as the source for Proto-Mayan ***xuch'* (to grind) and Xinca *huts'i* (nixtamal, corn dough). But Wichmann [78] states that * *hïtz* dates only to Proto-Mixean. Kaufman [41] now reconstructs the Mayan form only for Lowland and Western Mayan. A better Mixe–Zoquean source for the Mayan and Xinca items might be item 7 in Table 19-5, Proto-Mixe–Zoquean **hoo7s* (moler masa). However, this set perhaps should be demoted to the third, more local level.

The last maize-related form for which Campbell and Kaufman [9] suggest an Olmec provenience is Proto-Mixe–Zoquean **po7t* (to grind corn) as the source for Nahua *potonki (harina muy molida)*. Wichmann [78] argues that **po7t* dates only to Proto-Zoquean. This may not be too late to be Olmec, but on the model that Campbell and Kaufman share, Nahua languages should not have been in Mesoamerica until well after the Olmec period. Thus some intermediate source is required as a later donor into Nahua.

Three other items in the tables invite research. Proto-Mixe–Zoquean ***hïïpak* (olote) (see item 2 in Table 19-5 Wichmann's PA#024) resembles Western Mayan **b'aqal* and Eastern Mayan **b'aqlaq* (corn cob, olote) (see item 17 in Table 19-4). This might be an Olmec period loan.

A second is a series of forms with initial /k/, sometimes labialized, reconstructible in Uto–Aztecan, Otomanguean, Mixe–Zoquean, and Tequistlatec–Jicaque, that may reflect an ancient loan out of Otomanguean. The series includes Proto-Uto–Aztecan ***kuumi/u* (to nibble, especially corn) (the cognate in the northern Uto–Aztecan languages means simply "corn") and **wi~mi* (seeds, grain) [13]; Otomanguean ***ki* (to grind, twist) (see item 6 in Table 19-3)., its consonant-alternation mate ***wi* (tortilla) (#366; see item 14 in Table 19-3), and **k^we (maize, corn dough, etc.) (see item 9 in Table 19-3. I have stripped these reconstructions of Rensch's [57] indeterminate elements); Proto-Mixe–Zoquean *kïï7t* (moler pinole), and Jicaque–Tequistlatecan ***k7V* (roasting ear) (#108 in Oltrogge [54]; the vowel in Jicaque is /u/). An Otomanguean root with a /k/~/w/ alternation could be the source for this resemblant set. Of special interest here is the Uto–Aztecan alternant form proposed by Dakin [13], ***wi~mi* (seeds, grain). No such alternation is productive in Uto–Aztecan. However, Otomanguean languages exhibit an alternation ***wi* (tortilla) and, in Poplocan languages **-mh<ec>* (tortilla) (prefixed with **na-* or **nu-* [corn]), ***kwe* (maize, etc.) and again, in Popolocan, ***mh<e>*. Kaufman [39] reconstructs ***hme* as a separate cognate set in each of these items. The alternations are strikingly parallel to the Uto–Aztecan alternation, and almost certainly originate in Otomanguean, well-known for productive alternations of initial consonants.

The third is Kaufman's [39] Proto-Otomanguean ***na* (corncob) (see item 17 in Table 19-4). Two Uto–Aztecan forms shown in Table 46-2—items 5, ** *wïra* (to shell corn) and 6, ***o'ra* (corn, corncob)—are difficult and irregular. They involve a formative, apparently a verb stem ***-ra*, with instrumental prefixes, ***wï* (long object) in 5 and ***o-* (small round object) in 6. Item 7, ***sura* (embryo of maize, tender ear of green maize) may be part of the same series, and if so has ***-ra* with ***su-* (green). I have preferred to reconstruct Proto-Uto–Aztecan **r in these words. However, Wick Miller [62] preferred to reconstruct **n. If he is correct, the Uto–Aztecan proto-form is not ***-ra* but ***-na*, the form still attested in Northern Uto–Aztecan languages and may represent a loan blend with Otomanguean ***na* combined with Uto–Aztecan prefixes.[6]

Abstract semiotic principles such as metaphors and metonyms, like words, can be borrowed between languages. The Mesoamerican conceptualization of maize involved an intricate web of symbolism (see Chapter 17). We know it mainly from iconography, beginning with the Olmec period at about 1200 BC. However, the roots of the system must be more ancient, and historical linguistic techniques permit us to explore these.

Appearing in iconography in the first millennium BC, a set of symbolic connections links maize to two of the most important Mesoamerican prestige goods, greenstone (jade, jadeite, or serpentine, often carved into "celts" that represent maize ears and leaves) and iridescent green quetzal feathers [72]. An obvious metaphoric link between maize and these two precious materials is the color green.

[6] If Miller is in fact correct that Proto-Uto–Aztecan **n, not **r, should be reconstructed—and this possible Otomanguean loan is evidence in favor of his view—then the change **n > *r would constitute an argument for the unity of Southern Uto–Aztecan, as it occurs in all branches in this family (albeit with dialect complications surveyed by Dakin [14]).

Linguistic derivations between words for maize and words for green can be suggested at least for Uto–Aztecan and Otomanguean languages. This linguistic association may represent a precursor of the Olmec metaphorical system.

The Uto–Aztecan items 1 and 2 in Table 19-2 *sunu (ear of corn) and **sono, (corn waste), are phonologically and semantically similar, and may reflect early variants of a single word. Both of these items have cognates in some of the daughter languages that refer, not to maize, but to other grasses. The Hopi reflex of *sunu, sonho, refers to the large reed *Calamovilfa gigantea,* with important ritual uses for its stalks and tassels. Cognates of **sono refer to diverse grasses in the Numic languages, and the **so- element in **sono shows up in many words for grass across the Uto–Aztecan family. Cognates of the first syllable in *sunu (which may also appear in item 7 in Table 19-2, **sura (embryo of corn kernel, tender ear of green maize)) appears in words for green (e.g., Classical Nahuatl xiwi- [grass, green stone, turquoise, year] and Tepiman *suu-dagi [water, literally green-ness]). Nahuatl has another set of words for green that begin with xo-, including *xoo-pan* (spring, the green time of the year), *xoxoktik* (something made green, green, unripe), *xoxoowik* (something green, unripe). Alexis Manaster Ramer [44] derives these from PUA **siHV, where *H is "some kind of glide," with other cognates including Tarahumara siyo- "ser verde o azul" and Mayo sia-li "green." Dakin [12] derives these from Proto-Uto–Aztecan **sï- (with various second elements). In summary, there are puzzling irregularities, but likely relationships, in a set of Uto–Aztecan words for corn, grass, green. Such irregularities are what we might expect in a set of loan words.

Both Rensch's [57] and Kaufman's [39] cognate sets for Otomanguean also suggest relationships between green, grass, and words in the maize complex. We begin with Rensch's Proto-Otomanguean **(h)k^{w}e(7)(n) (green, blue, raw) (#174). Two of Rensch's sets in the maize complex show the **k^{w}e core seen in this word: **k^{w}e (dust, powder, ground dried corn) (#154; see item 7 in Table 19-3). and **(n)(h)k^{w}e(7)(n)^3 with meanings ranging from "corn dough" to "roasting ear" (#177 see item 8 in Table 19-3. Kaufman's **k^{w}au). This latter set is identical to the green set in every detail except for the tone and the optional pre-nasalization represented by the initial (n). Rensch himself suggests additional conections: that **(Y)(n)(H)se(h)(n) (roasting ear, corncob, cooked corn, ground dry corn) (#285; see item 12 in Table 46-3) is a consonant alternation derivation of **(n)(h)k^{w}e(7)(n)^3 (corn dough, etc.) (#177). This set can also be related as a consonant alternation to Rensch's #272 **(H)se(n)$^{1/2}$ (grass, hay) (#272). Finally, Rensch relates **(H)wi(h)(n) (tortilla) (#366; see item 14 in Table 19-3). to his #177, **(n)(h)k^{w}e(7)(n)^3, observing that, with regard to the word for tortilla, "it might be better to reconstruct this set with the vowel **e. This set may also be related to POM 100" [57, p. 302]: his #100 is **ki (to grind) (see item 6 in Table 19-3). Thus, an intricate series of relationships involving consonant alternations known to be productive in this proto-language connects several of the most important items in the Proto-Otomanguean maize complex with the word for green, blue, and raw and with a word for grass.

Otomanguean provides another provocative resemblance involving a second word for green. Suárez [71, p. 59] reconstructs Proto-Otomanguean **Ysa (green, raw) (Rensch's [57] #303 **sa(n), Kaufman's [39] **(n)sa), yielding Subtiaba of Nicaragua—*xa* (green) [54, 68]. In Huave the word for green ear of corn, elote is reconstructed by Suárez [66, p. 60] as *ása. Huave, generally considered to be an isolate, is unlikely to be an Otomanguean language, so this is a probable loan from Otomanguean, and it supports the green–maize connection (although we should keep in mind that the usage for green corn, *elote* may reflect the raw, unripe dimension of meaning rather than the color term) [7].

The Uto–Aztecan and Otomanguean green–maize connections may represent an ancient loan not simply of a metaphor, but of the phonological forms of the relevant lexical items, from Otomanguean into Uto–Aztecan. This is suggested by their phonological similarity. In Proto-Uto–Aztecan the relevant words began with **su, **so, or **sï/**si. These Uto–Aztecan relationships are irregular and are not found in other derivations. The Otomanguean sets show **k^{w}e alternating with **wi/e and **se or **sa. Suárez [67] observes that Rensch [57] reconstructs an alternating set **k ~ **w ~(**n), with "some evidence that **s and **y" also participated in the set [67, p. 50]. That is, the Otomanguean consonant alternations are well-established as productive. The indeterminacy of the Uto–Aztecan vowels may reflect several independent borrowings of the Otomanguean words into dialects of Uto–Aztecan at an early stage of its radiation, and the initial consonant **s may reflect an origin in the Proto-Otomanguean **s series. It is unlikely that the derivation was in the other direction, given the Otomanguean forms with initial **k and **w, an established alternation in that family. Thus, Otomanguean may be the ultimate source, not only for the metaphor, but for the Uto–Aztecan words for green, maize.

As a final set of connections in the green–grass–maize series, Papantla Totonac *saqat* (grass) resembles proto-Mixean *sokot and proto-Zoquean *so7k (grass) [36, p. 27]. Dakin [15] derives this set from proto-Uto–Aztecan *sa-kaa (dry leaves for forage and other uses). However, I noted earlier that proto-Otomanguean exhibits **sa(n)) (green). This must be considered as a possible ultimate source for the first syllable of these resemblances.

One element of these connections has been noted, although, I believe, at the wrong level, by Campbell [6]. He has several times suggested that Uto–Aztecan *sunu (corn), which he believed was attested only in the southern Uto–Aztecan languages, is "almost certainly a loan from Otomanguean" [6, p. 949]. However, Campbell's position cannot be sustained if, as I have argued earlier, the Uto–Aztecan word reconstructs to the protolanguage (as item 1 in Table 19-2). Instead, this item, along with the *sa-* set, is an excellent candidate, not for a late loan from some Otomanguean language into southern Uto–Aztecan, but for a loan from the earliest stages of the spread of cultivation out of Otomanguean-speaking communities into Proto-Uto–Aztecan, perhaps spreading along with an ideational complex constructing maize as the prototypical green item.

I have not been able to identify green–maize connections in the lexicon in either Mayan or Mixe–Zoquean. However, in Tequistlatec (Oaxacan Chontal), we find a root of the shape /x(a)k/ in some maize words, as well as the word for green, (e.g., *uxkua* [green], *-uxak* [olote], *-xkuh* [tortilla]) [74]. These are not, as far as I know, regular derivations in the contemporary language, but the resemblances invite research on possible fossilized derivational alternations.

MAIZE VOCABULARIES IN THE SOUTHWESTERN UNITED STATES

Maize arrived in the U.S. Southwest by about 4000 years ago. The language families found in the region, in addition to Uto–Aztecan, are 1) Yuman; 2) Kiowa–Tanoan; 3) Keresan; and 4) Zuni. Both geography (see Figure 46-1) and historical–linguistic evidence points to speakers of Uto–Aztecan languages as the most probable agents of the early spread of maize into the southwestern United States. I have recently identified important new evidence for this role, in the form of a set of loan words between Proto-Northern Uto–Aztecan and Proto-Kiowa–Tanoan. Table 46-6 shows seven Proto-Northern Uto–Aztecan words borrowed into Kiowa–Tanoan. Table 19-7 shows eight Proto-Kiowa–Tanoan words that were borrowed into Northern Uto–Aztecan (this list adds new items to those noted in my earlier work [31]).

Several properties of the loan sets shown in Table 19-6 and 19-7 strongly suggest that speakers of Proto-Northern Uto–Aztecan and Proto-Kiowa–Tanoan were in intimate contact. The two sets are large: compare these with the scattered evidence for the more ancient loans from Otomanguean into Proto-Uto–Aztecan reviewed previously. The loan process exhibits several phonological regularities, suggesting that the entire set was exchanged during a fairly short period of time in a community where bilingualism was not uncommon. The loans from Kiowa–Tanoan in Table 19-7 are all unique to Northern Uto–Aztecan and are not found in other Uto–Aztecan languages. The Kiowa–Tanoan words reconstruct to Proto-Kiowa–Tanoan, suggesting an early contact.[7] The semantics of the sets is consistent with a contact between Northern Uto–Aztecan cultivators and Kiowa–Tanoan hunter–gatherers, with the loans from Northern Uto–Aztecan into Proto-Kiowa–Tanoan in Table 19-6 including several forms in the maize complex, and loan words from Proto-Kiowa–Tanoan into Northern Uto–Aztecan in Table 19-7 including major noncultivated economic plants and game animals that are common on the Colorado Plateau but rare or absent in the deserts to the south. Three words in Table 19-7, "Yamp" (item 10), "Sego Lily" (item 11), and "Chrysothamnus" (item 13) are certainly of Kiowa–Tanoan origin, because they include morphological elements that are highly productive in those languages, but are meaningless in Uto–Aztecan. For yamp the first element is the Kiowa–Tanoan word for "earth" and the second is the word for a fruit or tuber. For "Sego Lily" the *gobe* element is clearly from Kiowa–Tanoan root for "plant." For "Chrysothamnus," *si-p'o is literally "fall-water" (which in Kiowa means "rain") plus an element *p'e* (stick) or *p'u*^{n}, attested in Tewa words for Chrysothamnus ("Rabbit Brush" in English).

This set of loan words is probably the linguistic face of a contact situation identified by R. G. Matson [45] between the Western Basketmaker II, an intrusive group of cultivators, and the Eastern Basketmaker II, who shared a continuous tradition with the Middle Archaic of the region attested in lithics, basketry, and house construction. The cultivators arrived on the Colorado Plateau in the Four Corners region by 3000 BP, and by 2500 BP the Eastern Basketmaker had adopted cultivation [45]. Of special interest in connecting the loan set with this archaeological context is item 12

TABLE 19-6 Proto-Northern Uto–Aztecan Loans into Proto-Kiowa–Tanoan

1. PNUA *ïya (to plant)	KT *7ia '(corn) [17, p. 373]
2. PNUA **kuma (corn)	KT *k^hu^n, *k^hë (corn, seed) [76, p. 621; 16, p. 370]
3. PUA **pa7tzi (corn, corn kernel, seed)	KT *p'ëa (fresh corn) [17, p. 369]
4. PUA **wika (planting stick, dibble)	Taos *xwia-d-* (hoe) [74, p. 622), Tewa *xwèè*^{n}*-toh* (to plow) *-toh* (to work) [19, p. 142]
5. PUA **k^wïs C (to carry, take)	Taos x^w*ia* (to harvest) [76, p. 623]
6. PNUA *tehoi- (male person)	KT * tz^hïo (person) [17, p. 36]
7. PNUA *wïnï (stand)	KT *g^winV- ~ *k^winV- (stand) [76, p. 620; 28, p. 117; 17, p. 59)

[7]Those that do not reconstruct are simply not attested in more than one language. Our lexical resources for Kiowa–Tanoan languages are restricted.

TABLE 19-7 Proto-Kiowa–Tanoan Loan Words into Proto-Northern Uto–Aztecan

8. PNUA **tïpat* (piñon, pine nut)	<	KT **t'ou* (pine nut) [76, p. 622; 17, p. 370]
9. PNUA **k*^{w}*inha-*~*k*^{w}*iya* (Oak spp.)	<	KT **k*^{w}*e*^{n} (oak) [17, p. 371]
10. PNUA **yampa* (Indian potato)	<	Tewa *nam-p'u* 9Earth-potato) [59]
11. PNUA **siko* (Sego lily)	<	Tewa *sæ*^{n}*gobe* (a white-flowered plant with tubers similar to potatoes) [59, p. 73]
12. PNumic **wi-* (string, hemp)		KT **wi-* (fiber, string, rope)
13. PNUA **sibu* (*Chrysothamnus* spp.) Hopi *sivàapi*	<	KT ***si-p'o-p'u*^{n}*/p'e* (Rain Tübatulabal *siba-pul, Chrysothamnus) (Chrysothamnus bigelovii)* [59]
14. PNumic **kuhtzuN* (buffalo, cow)	<	KT **kon* (buffalo) [76, p. 620; 17, p. 43]
15. PNUA **tïhï~tïnï* (deer, antelope)	<	KT **t'o* (deer, antelope) [76, p. 622]

in Table 19-7, the word for "fiber, string." The Eastern Basketmaker II usually used *Apocynum* or Indian Hemp as their main fiber plant, whereas the Western Basketmaker II used *Yucca* [25]. Thus we might predict that the earliest Northern Uto–Aztecans would borrow the word for *Apocynum*, and indeed they seem to have done so.

Published glottochronological dates are consistent with the hypothesis of contact between Proto-Northern Uto–Aztecan and Proto-Kiowa–Tanoan. Kenneth Hale and David Harris [29] date Proto-Kiowa–Tanoan dates to 1000–500 BC. Dates of separation for Northern Uto–Aztecan languages range from 3206 years ago (between Cahuilla and Northern Paiute) to 2229 (between Tübatulabal and Numic) [27]. Catherine Fowler [22, p. 239] estimates "3000 plus years ago" as the period of formation of the PNUA speech community. These dates are a good match to Matson's [45] dates for the Western–Eastern Basketmaker II contact.

Uniquely among the peoples of the Southwest and the Great Basin, members of Uto–Aztecan groups exhibit Albumin Mexico, an albumin allele also found in Mesoamerica [64]. Interestingly, though, studies of mitochondrial DNA show that Uto–Aztecan groups do not share a single mtDNA haplogroup, but instead resemble non-Uto–Aztecan people in the same region [43]. This suggests that Uto–Aztecan groups have incorporated women who do not speak Uto–Aztecan over a long period, a practice that would be consistent with the presence in Northern Uto–Aztecan of Kiowa–Tanoan words for major economic plants that would have been the province of women in Kiowa–Tanoan hunter–gatherer communities (in Uto–Aztecan communities, men are the maize cultivators). William Haas [25] has also suggested that the entire Basketmaker II complex looks like a single community in terms of marriage exchange, as indexed by styles of cordage.

No loans from Uto–Aztecan appear in the maize complex vocabulary of the other Southwestern language groups. Among the Yuman languages, three of the four subgroups (Kiliwa, River Yuman, and Upland Yuman or Pai) share a word for corn that is transparently formed as the plural of the word for seed, making it likely that the derivation descends from a Proto-Yuman way of speaking. Cognate forms include Kiliwa *ta7yit* (from *t-yit* "seed"), Walapai *dhiyách*, Yavapai *teyách* (*–yaách* "seed"), and River Yuman (Mojave, Maricopa) *tadiich* (*–diich* "seed"). Cocopa in the Delta–California branch has a different word, *jtzaas* (corn). This word also appears in the expression for "hail," called "sky corn," suggesting a possible marking-reversal source for the Cocopa corn word. Cocopa *yas* (seed), is cognate with the forms in the other branches, but is not used to refer to maize. The languages share a word for tortilla, cognate throughout the group: Kiliwa *myal* (tortilla), Cocopa *muyály* (tortilla, dough), *myály* (to roll out dough), Mojave *madiily, modiily* (to bake bread), Walapai *miyál* (bread) (the word for tortilla is a compound). Throughout the Yuman languages, other items in the maize complex, such as cobs, ears, husks, and the like, are binomials that incorporate the word for corn plus a modifying element. This binomial tendency means that we have only two maize complex roots, the corn word (which is derived from seed) and the tortilla word, and neither resemble any maize words in Uto–Aztecan or in other southwestern language families.

Keresan and Zuni both have fairly elaborate maize complex vocabularies, unrelated to any Uto–Aztecan, Kiowa–Tanoan, Yuman, or Mesoamerican forms. For Keresan Miller and Irvine Davis [48] reconstruct the following maize vocabulary: **ya:chínɪ* (corn), **spíníní* (dwarf corn), **yábáxɪ* (corn silk), **y'úskúm'á* (corn cob), **díískámí* (corn husk). These long forms break down into smaller derivational elements: for instance, *-ni* ~ *nɪ* and *-xɪ* are suffixes. The *yaa* element in the word for corn is a frozen formative meaning "lots of little" [46, p. 164]. Thus it is clear that Keresan, like Yuman, formed its maize complex vocabulary from indigenous elements.

Zuni has a particularly interesting maize vocabulary that is also entirely its own, with no connection to any other languages. *Aatowa* (corn) literally means "ancient things, traditional things." The forms for the various colored corns are modified color words that gloss as "blue things," "yellow things," "white things," and so on. The remaining words in the complex are short and are clearly nonderived forms: They are *mi* (corn ear), *xe* (corn husk), *chu* (corn kernel), *mu* literally "a piece of solid substance" but usually glossed as "bread" (presumably, originally a cake of corn dough or tamale). A rich vocabulary for maize cuisine is also recorded for Zuni, much of it clearly derived from the basic words listed earlier plus various modifiers for preparation procedures.

MAIZE VOCABULARY IN THE EASTERN UNITED STATES

In the eastern United States, unlike in Mesoamerica and the U.S. Southwest, we cannot reconstruct maize complex vocabulary for the proto-languages of any of the major language families. This historical–linguistic picture is consistent with the archaeological evidence for the late arrival of maize in the region, certainly no earlier than 200 BC, well after the initial radiations of the major language families [58]. Patrick Munson's [52] essay remains a useful survey of this question. However, he did not look at the maize complex as I have defined it, but surveyed only words glossed as "corn, ear of corn," almost certainly missing many connections. But since there is a strong tendency for the eastern languages to form maize complex words by compounding or binomial constructions on a small set of roots, a more extensive lexical survey might not alter his basic findings.

The major language families that included descendant communities of cultivators in the eastern United States are 1) Muskogean, 2) Caddoan, 3) Siouan, 4) Iroquoian, and 5) Algonquian. Algonquian, at the northern and eastern margins of the spread of maize cultivation, is not relevant to my concerns here, and I do not consider it further. Several language isolates of the southeastern United States are also spoken by cultivators, including Atakapa, Chitimacha, Natchez, Timucua, Tunica, and Yuchi. A "Coahuiltecan" group of dubious affinity includes Cotoname, Comecrudo, and Coahuilteco. Tonkawa and Karankawa are other language isolates of the Texas Gulf coast. Speakers of some of these languages were primarily hunter–gatherers, but because they are located on a possible route of diffusion from Mesoamerica, their languages should be examined.

Rankin (see Chapter 41) points out that no terms in the maize complex can be reconstructed even for the major daughter branches of Siouan. Many Siouan maize words are transparently derived from premaize vocabulary, and others are obvious loan words from neighboring languages in the Algonquian, Caddoan, and Muskogean families.

Maize vocabulary does not reconstruct to the proto-language for Caddoan, although the major words are resemblant. Attested forms include Wichita *tées* (corn), *kaasás'a* (parched corn), and *kufaakhir7a* (corn cob) [60]; Arikara *nichíx-u7* (corn); Caddo *kisi7* (corn) [73]; Pawnee *rikiis-u* (corn) [55]. For the long-extinct language Adai, Alan Taylor [73] records "ocasuck," probably a representation of something like the Caddo or Pawnee words.

Aaron Broadwell [4] suggests that **(t/tz)achi* (corn) can be reconstructed for Proto-Muskogean, from Creek and Seminole *ahtzi,* Hitchiti–Mikasuki *asp-i,* Choctaw *ta*^{n}*chi,* and Chickasaw *tanchi7.* Alabama–Koasati has *chassi,* which Broadwell suggests is not cognate. Broadwell also holds that words for to shell corn and corn riddle (a type of basket) can be reconstructed to Proto-Muskogean, but gives no cognate sets. He observes of the set for corn that "this is not a perfect set, by any means" [4, p. 26]. Furthermore, it "constitutes a genuine conflict between the linguistic and archaeological data," because glottochronology dates Proto-Muskogean to 950 BC (±380 years) [4, p. 28]. No one has argued that maize appears in the southeastern United States before 200 BC at the earliest. Given the problems with sound correspondences in the set, it is probably the result of borrowing among the geographically contiguous Muskogean languages at a date well after the breakup of Proto-Muskogean.

Marianne Mithun [49] reports that no agricultural words reconstruct to Proto-Iroquoian. Two words for corn, **-neeh-* and **-neehst-,* trace to Proto-Northern Iroquoian. Mithun observes that "There is a good chance that both of these words originally referred to seeds in general" [49, p. 272]. She does not reconstruct a word for "cornhusk," but gives the following resemblant forms: Mohawk *onóora7,* Oneida *–nol-otxu* (to husk), Onondaga *onóoya7,* Cayuga *onóonya7,* Seneca *–nowi-,* Tuscarora *utùureh,* and Cherokee *nuulaki.*

Although the remaining languages of the U.S. Southeast have from time to time been linked to one another or to one or another of the major families, they are today usually treated as individual isolates. Maize vocabulary (indeed, all vocabulary) is poorly attested for these languages. Maize words attested by Munson [52] include Tunica *háhka* (corn), Natchez *hokko* (probably loan words, but directionality may be undeterminable), Timucua *hulupa,* Tawasa *chesapà,* Chitimacha *kaseman,* all meaning corn. The Tunica word serves as the stem for many binomial derivations of other maize vocabulary [26]. For Atakapa, a language isolate of southern Louisiana, Albert S. Gatschet and John R. Swanton [24] collected several words in the maize complex, including *nexooum* (corn), *tzó-otz* (corn) (where *–otz* (grain, seed)), *tzí* (corn cob). Other words in the Atakapa complex are binomials incorporating *tzó-otz.* Yuchi *tzot7o7* (flint corn) might be a loan from Atakapa.

Swanton [71] assembled a few maize words for the languages of the Texas Gulf Coast. These include Karankawa *kwiám* (corn) and Comecrudo *tawеló* (corn, ear of corn), the

TABLE 19-8 A Possible Loan Set in the U.S. Southeast (Modified from Munson [52, p. 127])

tées	maize	Wichita	Caddoan
chassi	maize	Alabama	Muskogean
totosi	blue flint corn	Biloxi	Siouan
tzo7ots	maize		Atakapa
tzot7o7	flint corn		Yuchi

base for several binomials, and *pakenhá* (corn cob, corn kernels). Tonkawa has *tol7ajan* (maize) [52, p. 127; 31a].

Munson [48] proposed several loan complexes of limited scope, including Tunica *háhka* and Natchez *hokko,* noted earlier, and Alabama *chassi* (corn) and Chitimacha *chaxa* (seed). In addition, he suggests that the items in Table 46-8 may be linked.

Given the minimal phonetic similarities among most of these words, the set must be regarded as merely speculative unless additional resemblant items in the same semantic domain can be recovered. Munson [52] thought that he had recovered such a set for squash and gourd. However, the resemblances among words for cucurbits that Munson proposes are equally vague and scattered. Furthermore, gourds and squash are premaize domesticates in the southeastern United States, so it is unlikely that labels for them would share a history with words for maize [63].

In summary, the linguistic evidence suggests a number of independent developments of maize vocabulary in the eastern United States. Loan words there surely are, but these exhibit no consistent geographical trajectory that would support any hypothesis about routes of diffusion (for instance, from Florida or from the U.S. Southwest), and the sources of loans appear to all be within the region, unconnected to donor languages outside it [58].

SUMMARY AND CONCLUSION

The application of historical linguistic techniques to the history of maize holds most promise for the languages of Mesoamerica and the Southwest. Among these languages, it is likely that the adoption of cultivation played a role in stimulating the initial radiation of the families, consistent with the hypothesis of Peter Bellwood [3], and the reconstructibility of maize complex vocabularies supports such a hypothesis. The study of loan vocabulary, initiated for the Mesoamerican languages by Campbell and Kaufman [9], is a highly promising line of investigation, permitting us to study not only the diffusion of maize, as from Mayan into Xincan and from Northern Uto–Aztecan into Kiowa–Tanoan, but the diffusion of ideas about it. The cultivation vocabularies for the languages of the eastern United States remain enigmatic. Attention from specialists will be required to explore the relationships between maize complex lexicons beyond the single-word approach taken by Munson [52], and the frustrating gaps in documentation for languages that became extinct early may mean that linguistic research will not be able to contribute much to our understanding of maize in the region. However, our inability to reconstruct maize vocabularies for eastern proto-languages, where we know that maize arrives fairly late, compared with our ability to reconstruct extensive maize vocabularies for the several protolanguages of Mesoamerica and the Southwest, where maize is ancient, is a reassuring confirmation of the basic validity of historical linguistic methods.

References Cited

1. M. Amador Hernández, P. Casasa García. (1979). Un análisis cultural de juegos léxicos reconstruidos del proto-otomangue. In: N. A. Hopkins, J. K. Josserand, (Eds.), Estudios lingüísticos en lenguas Otomangues. Collección Científica Lingüística 68. México, D.F.: SEP-INAH. pp. 13–24.
2. W. Balée. (1994). Footprints of the forest. New York: Columbia University Press.
3. P. Bellwood. (1997). Prehistoric cultural explanations for widespread linguistic families. In: P. McConvell, N. Evans, (Eds.), Archaeology and linguistics: Aboriginal Australia in global perspective. Melbourne: Oxford University Press., pp. 123–134.
4. G. A. Broadwell. (2005). Reconstructing proto-Muskogean language and prehistory: preliminary results. Available at: http://www.albany.edu/anthro/fac/broadwell/flora.pdf. Accessed March 15, 2005.
5. L. Campbell. (1972). Mayan loan words in Xinca. *International Journal of American Linguistics, 38,* 187–190.
6. L. Campbell. (1979). Middle American languages. In: L. Campbell, M. Mithun, (Eds.), *The languages of Native America: Historical and comparative assessment.* Austin: University of Texas Press. pp. 902–1000.
7. L. Campbell. (1997). *American Indian languages: The historical linguistics of Native America.* New York: Oxford University Press.
8. L. Campbell. (2003). What drives linguistic diversification and language spread? In: P. Bellwood, C. Renfrew, (Eds.), *Examining the farming/language dispersal hypothesis.* Cambridge, MA: McDonald Institute for Archaeological Research. pp. 49–64.
9. L. Campbell, T. Kaufman. (1976). A linguistic look at the Olmecs. *American Antiquity, 41,* 80–89.
10. L. Campbell, T. Kaufman, T. Smith-Stark. (1986). Mesoamerica as a linguistic area. *Language, 62,* 530–570.
11. J. M. Crawford. (1984). *Cocopa dictionary.* University of California Publications in Linguistics Vol. 114. Berkeley: University of California Press.
12. K. Dakin. (1994). El sinna7aβe, koyo:tl y xolo:tl. Paper presented to the III Encuentro de Lingüística en el Noroeste. Hermosillo, Sonora, November 16–18, 1994.
13. K. Dakin. (2001). Animals and vegetables, Uto-Aztecan noun derivation, semantic classification and culture history. In: *Historical Linguistics 1999: Selected papers from the 14th international conference on historical linguistics, Vancouver, BC 9–13 August 1999.* Philadelphia: John Benjamins. pp. 105–117.
14. K. Dakin. (2001). Isoglosas e innovaciones yutoaztecas. In: J. L. Moctezuma Z., J. H. Hill, (Eds.), *Avances y balances de lenguas yutoaztecas.* Serie Lingüística, Instituto Nacional de Antropología e Historia. México: INAH. pp. 313–344.

15. K. Dakin. (2004). Nahuatl *-ka* words: Evidence for a proto-Uto-Aztecan derivational pattern. *STUF-Sprachtypologie und Universalienforschung, 57,* 6–22.
16. K. Dakin, S. Wichmann. (2000). Cacao and chocolate: A Uto-Aztecan perspective. *Ancient Mesoamerica, 11,* 55–75.
17. I. Davis. (1989). A new look at Aztec-Tanoan. In: M. R. Key, H. M. Hoenigswald, (Eds.), *General and Amerindian ethnolinguistics: In remembrance of Stanley Newman.* Berlin: Mouton de Gruyter. pp. 365–390.
18. J. M. Dienhart. (1997). *The Mayan languages: A comparative vocabulary.* Odense, Denmark: Odense University. Available at: http://maya.hum.sdu.dk. Accessed on November 19, 2004.
19. E. P. Dozier, H. Hoijer. (1949). The phonemes of Tewa, Santa Clara dialect. *International Journal of American Linguistics, 15,* 139–144.
20. W. Eriacho, Sr., R. Gonzales. (1998). Bena:we Dana:we/*Word Categories.* Zuni Public School District No. 89. Zuni, NM: Public School District.
21. P. Fournier. (1998). El complejo nixtamal/comal/tortilla en Mesoamerica. Boletín de Antropología Americana, Julio 1998. Instituto Panamericana de Geografía e Historia.
22. C. S. Fowler. (1983), Some lexical clues to Uto-Aztecan prehistory. *International Journal of American Linguistics, 49,* 224–257.
23. J. A. Frisch. (1968). Maricopa foods. *International Journal of American Linguistics, 34,* 16–20.
24. A. S. Gatschet, J. R. Swanton. (1932). *A dictionary of the Atakapa language: Accompanied by text material.* Smithsonian Institution Bureau of American Ethnology Bulletin 108. Washington, D.C.: Government Printing Office.
25. W. R. Haas. (2003). The social implications of Basketmaker II cordage style distribution. Master's Thesis. Flagstaff, AZ: Northern Arizona University.
26. M. R. Haas. (1953). *Tunica dictionary.* University of California Publications in Linguistics, Vol. 6, No. 2. Berkeley: University of California Press. pp. 175–332.
27. K. L. Hale. (1958). Internal diversity in Uto-Aztecan: I. *International Journal of American Linguistics, 24,* 101–107.
28. K. L. Hale. (1967). Toward a reconstruction of Kiowa-Tanoan phonology. *International Journal of American Linguistics, 33,* 112–120.
29. K. L. Hale, D. Harris. (1979). Historical linguistics and archaelogy. In: A. Ortiz, (Ed.), *Handbook of North American Indians.* Vol. 9: Southwest. Washington, D.C.: Smithsonian Institution. pp. 170–177.
30. J. H. Hill. (2001). Proto-Uto-Aztecan: A community of cultivators in central Mexico? *American Anthropologist, 103,* 913–934.
31. J. H. Hill. (2002). Toward a linguistic prehistory of the Southwest: "Azteco-Tanoan" and the arrival of maize cultivation. *Journal of Anthropological Research, 58,* 457–476.
32. J. H. Hill. (2003). Proto-Uto-Aztecan and the northern devolution. In: P. Bellwood, C. Renfrew, (Eds.), *Examining the farming/language dispersal hypothesis.* Cambridge, MA: McDonald Institute for Archaeological Research. pp. 331–440.
33. J. H. Hill. (2004). The non-agricultural origin of the proto-Uto-Aztecan vocabulary of maize use. In: Z. Estrada F., (Ed.), *VIII Encuentro sobre la lingüística en el noroeste.* Hermosillo: Universidad de Sonora, Hermosillo.
34. H. Hoijer. (1949). An analytical dictionary of the Tonkawa language. University of California Publications in Linguistics, Vol. 5. Berkeley: University of California Press.
35. N. A. Hopkins. (1984). Otomanguean linguistic prehistory. In: J. K. Joserand, M. Winter, N. A. Hopkins, (Eds.), *Essays in Otomanguean culture history.* Vanderbilt University Publications in Anthropology No. 31. Nashville, TN: Vanderbilt University. pp. 25–64.
36. J. S. Justeson, W. M. Norman, L. Campbell, T. Kaufman. (1985). *The foreign impact on lowland Mayan language and script.* Middle American Research Institute 53. New Orleans, LA: Tulane University.
37. J. S. Justeson, T. Kaufman. (1993). A decipherment of epi-Olmec hieroglyphic writing. *Science, 259,* 1703–1711.
38. T. Kaufman. (1976). Archaeological and linguistic correlations in Mayaland and associated areas of Meso-America. *World Archaeology, 8,* 101–118.
39. T. Kaufman. (1990). Early Otomanguean homelands and cultures: Some premature hypotheses. *University of Pittsburgh Working Papers in Linguistics, 1,* 1–90.
40. T. Kaufman. (2001). The history of the Nawa language group from the earliest times to the sixteenth century: Some initial results. Available at: http://www.albany.edu/anthro/maldp/Nawa.pdf. Accessed November 19, 2004.
41. T. Kaufman with the assistance of J. Justeson. (2003). A preliminary Maya etymological dictionary. Available at: http://www.famsi.org/reports/01051/pmed.pdf. Accessed January 20, 2005.
42. S. LeBlanc. (2004). The case for a farmer migration into the greater American Southwest. Peabody Museum of Harvard University manuscript.
43. R. S. Malhi, H. M. Mortenson, J. A. Eshleman, B. M. Kemp, J. G. Lorenz, F. A. Kaestle, J. R. Johnson, C. Gorodezky, D. G. Smith. (2003). Native America mtDNA prehistory in the American Southwest. *American Journal of Physical Anthropology, 120,* 108–124.
44. A. Manaster-Ramer. (1996). /ih/-, /ah/-: The joys of Nahuatl historical phonology. Manuscript in possession of author.
45. R. G. Matson. (1991). *The origins of Southwestern agriculture.* Tucson: University of Arizona Press.
46. W. R. Miller. (1965). *Acoma grammar and texts.* University of California Publications in Linguistics No. 40. Berkeley: University of California Press.
47. W. R. Miller. (1983). Uto-Aztecan languages. In: A. Ortiz, (Ed.), *Handbook of North American Indians.* Vol. 10: Southwest. Washington, D.C.: Smithsonian Institution. pp. 113–124.
48. W. R. Miller, I. Davis. (1963). Proto-Keresan phonology. *International Journal of American Linguistics, 29,* 310–330.
49. M. Mithun. (1984). The Proto-Iroquoians: Cultural reconstruction from lexical materials. In: M. K. Foster, J. Campesi, M. Mithun, (Eds.), *Extending the rafters: Interdisciplinary approaches to Iroquioan studies.* Albany, NY: SUNY Press. pp. 259–281.
50. M. J. Mixco. (1996). *Kiliwa del Arroyo León.* Baja California, México, D.F.: El Colegio de México.
51. P. Munro, N. Brown, J. G. Crawford. (1992). *A Mojave dictionary.* UCLA Occasional Papers in Linguistics, No. 10. Los Angeles: Department of Linguistics, University of California.
52. P. J. Munson. (1973). The origins and antiquity of maize-beans squash agriculture in eastern North America: Some linguistic implications. In: D. W. Lathrap, J. Douglas, (Eds.), *Variation in anthropology: Essays in honor of John C. McGregor.* Urbana: Illinois Archaeological Survey. pp. 107–136.
53. S. Newman. (1958). Zuni dictionary. *International Journal of American Linguistics 24:* 1, Part II. Bloomington, IN: Publication Six of the Indiana University Research Center in Anthropology, Folklore, and Linguistics.
54. D. Oltrogge. (1977). Proto Jicaque-Subtiaba-Tequistlateco: A comparative reconstruction. In: D. Oltrogge, C. Rensch, (Eds.), *Two studies in Middle American comparative linguistics.* Summer Institute of Linguistics Publications in Linguistics, Publication Number 55. Dallas: Summer Institute of Linguistics-University of Texas at Arlington. pp. 3–52.
55. D. R. Parks. (1976). A grammar of Pawnee. New York: Garland Publishing.
56. A. A. Reid, R. G. Bishop. (1974). *Diccionario Totonaco de Xicotepec de Juárez.* Vocabularios Indígenas 17. México, D.F.: Instituto Lingüístico del Verano y Secretaría de Educación Pública.
57. C. R. Rensch. (1976). Comparative Otomanguean phonology. Language Science Monographs, Vol. 14. Bloomington: Indiana University.

58. T. J. Riley, R. Edging, J. Rossen. (1990). Cultigens in prehistoric eastern North America: Changing paradigms. *Current Anthropology, 31,* 525–541.
59. W. W. Robbins, J. P. Harrington, B. Freire-Marreco. (1916). *Ethnobotany of the Tewa Indians*. Smithsonian Institution Bureau of American Ethnology Bulletin No. 55. Washington, D.C.: Government Printing Office.
60. D. Rood. (1976). *Wichita grammar.* New York: Garland Publishing.
61. B. de Sahagún. (1981). General history of the things of New Spain. Book 2—The Ceremonies. Trans. by A. J. O. Anderson, C. E. Dibble. Santa Fe, NM, and Salt Lake City: The School of American Research and The University of Utah.
62. S. Silver, W. R. Miller. (1997). *American Indian languages: Cultural and social contexts.* Tucson: University of Arizona Press.
63. B. D. Smith. (1992). *Rivers of change: Essays on early agriculture in Eastern North America.* Washington, D.C.: Smithsonian Institution Press.
64. D. G. Smith, J. G. Lorenz, B. K. Rolfs, R. L. Bettinger, B. Green, J. A. Eshleman, B. Schultz, R. A. Malhi. (2000). Implications of the distribution of Albumin Naskapi Mexico for New World Prehistory. *American Journal of Physical Anthropology, 11,* 557–573.
65. D. Snow. (1990). "Tener comal y metate": Protohistoric Rio Grande maize use and diet. In: P. E. Winnis, C. L. Redman, (Eds.), *Perspectives on Southwestern prehistory.* Boulder, CO: Westview Press. pp. 289–300.
66. J. A. Suárez. (1975). *Estudios huaves*. Collecci6n Científica, Serie Lingüística 22. México, D.F.: SEP-INAH.
67. J. A. Suárez. (1980). Review of comparative Otomanguean phonology (Rensch). *International Journal of American Linguistics, 46,* 49–54.
68. J. A. Suárez. (1980). Review of two studies in Middle American Comparative Linguistics (Oltrogge and Rensch). *International Journal of American Linguistics, 46,* 56–60.
69. J. A. Suárez. (1983). *The Mesoamerican Indian languages*. Cambridge, MA: Cambridge University Press.
70. J. A. Suárez. (1985). Loan etymologies in historical method. *International Journal of American Linguistics, 51,* 574–576.
71. J. R. Swanton. (1940). *Material from the tribes of southern Texas and northeastern Mexico.* Smithsonian Institution Bureau of American Ethnology Bulletin 127. Washington, D.C.: Government Printing Office.
72. K. Taube. (2000). Lightning celts and corn fetishes: The formative Olmec and the development of maize symbolism in Mesoamerica and the American Southwest. In: J. Clark, M. Pye, (Eds.), *Olmec art and archaeology.* Washington, D.C., and New Haven, CT: National Gallery of Art and Yale University Press. pp. 297–337.
73. A. R. Taylor. (1963). Comparative Caddoan. *International Journal of American Linguistics, 29,* 113–131.
74. P. Turner, S. Turner. (1971). *Chontal to Spanish–English/Spanish to Chontal Dictionary.* Tucson: University of Arizona Press.
75. L. Watahomigie, J. Bender, A. Yamamoto. (1982). *Hualapai reference grammar.* Los Angeles: UCLA American Indian Studies Center.
76. B. L. Whorf, G. L. Trager. (1937). The relationship of Uto-Aztecan and Tanoan. *American Anthropologist, 39,* 609–624.
77. S. Wichmann. (1994). On the relationship Between Mixe-Zoquean and Uto-Aztecan. *Kansas Working Papers in Linguistics, 24,* 101–113.
78. S. Wichmann. (1995). The relationship among the Mixe-Zoquean languages of Mexico. Salt Lake City: University of Utah Press.

(Table 3 continued from page 238)

TABLE 19-3 The Otomanguean Maize Complex (The numbers in parentheses with # are cognate set numbers in Rensch [57]; forms with (K) are from Kaufman [39].)

1. **7te(n)^{3} (corn gruel, atole) (#23); **7tye (atole) (Oto–Chinanteco only)
2. **ta(h) (tortilla, cooked corn) (#48) ; **ta (tortilla, hominy)
3. **(7)ta(n)$^{1/2}$ (clay griddle, clay pot, pottery) (#56) (possibly related to (#48) and to "stone"); **ta (clay griddle) (K)
4. **(H)tu(h)(n)^{3} (to plant, to sow, to dress oneself, cornfield, corn crop, jungle growth, vegetation) (#91)
5. **(n) (7)tu (h) (n)^{1} (moon, month, trough, manger, corn measure (almost anything hollowed out or concave), clay griddle, tortilla, spherical, land, oven) (#93)
6. **ki (to grind, to twist, twisted, to be ground) (# 100; perhaps related to 366)
7. **k^{w}e (dust, powder, ground dried corn) (#154) (may be related to #283 "dust, powder, sand")
8. **(n)(h) k^{w}e(7)(n)^{3} (dough, maize, corn dough, corn, roasting ear, corn gruel, dry, shelled corn, kernels of corn) (#177) (may be related to 285) see green; **k^{w}au (K) (maize); **hme (K) (maize) (that is, Kaufman splits Rensch's set into descendants from two different ancestral forms)
9. **(n)(7)k^{w}a(h)(n)^{2} (twist, grind) (#224)
10. **(Y)(n)si (clay griddle) (#248), (may be related to words meaning "round"); **ci (K) (clay griddle, metate)
11. **(Y)(n)(H)se(h)(n) (roasting ear, corncob, cooked corn, ground dry corn) (#285) (may be related to #177); **s(a7)ai(n) (K) (roasting ear)
12. **(n)(h)san "tamale, tortilla, boiled corn" (#306) (see green, raw); **sau (+wV) (K) "hominy, tamale, tortilla"
13. **(H)wi(h)(n) tortilla (#366) (may be related to #177)
14. **7yu3 "corn fodder, dried corn-stalks, bamboo, sugar cane" (#424) (probably related to #80)
15. **yo (K) "clay griddle"
16. **nu (K) "cornfield"
17. **na (K) "corncob"
18. **ren (K) "corncob"

CHAPTER

20

Glottochronology and the Chronology of Maize in the Americas

CECIL H. BROWN

Department of Anthropology, Northern Illinois University, DeKalb, Illinois

Glossary

Ancestral language Language of the past from which modern languages have developed (e.g., French, Spanish, and Italian developed from Latin, their common ancestral language); also referred to as a parent language or proto-language.

Chi square A formula that measures the statistical significance of an association or correlation.

Cognates Words of two or more genetically related languages, typically similar in sound and meaning, all of which originate from a single word (proto-word) of an ancestral language (e.g., English deer and German *tier* [animal]).

Gamma A statistic that measures the strength of an association or correlation. Its value ranges from 0.00 (indicating no correlation whatsoever) to 1.00 (indicating a perfect correlation). Conventionally, gamma scores ranging from 0.50 to 1.00 are considered indicative of strong associations.

Genetically related languages Languages that have developed from the same ancestral language. For example, French, Spanish, and Italian are genetically related languages, all of which developed from Latin.

Glottochronology Controversial method of linguistics used to determine the minimum number of centuries since genetically related languages diverged from their common ancestral language.

Historical linguistics The study of language change and how languages are related to one another.

Language genetic group A group of languages that have all developed from the same ancestral language (e.g., the Siouan family of languages).

Lexical replacement A common process of language change wherein a word for a particular referent in a language is replaced by a different word for that referent (e.g., Anglo Saxon *hund* (dog) is replaced by dog in later English).

Maize term evaluations Refers to judgments made by the investigator with regard to the occurrence of a proto-term for maize in an ancestral language. Evaluations are limited to three results: A (evidence indicates a *probable* proto-word for maize), B (evidence indicates a *possible* proto-word for maize), and C (there is *no* evidence of a proto-word for maize).

Offspring language A language that has developed (or descended) from a parent language. For example, French is an offspring language of Latin.

Proto-word Word of an ancestral language from which words of two or more descendant or offspring languages have originated, also referred to as a proto-form.

Reflex A feature of an offspring language that has developed from a particular feature of the language's parent language.

Similarity sets Refers to maize terms in languages of a genetic group or subgroup that are phonologically similar to one another. Any genetic group or subgroup may have one or more such sets. In addition, some similarity sets may have only one member.

Statistical significance Indication of whether or not a particular correlation or association is because of chance. If a correlation is found to be statistically significant, it is highly unlikely that it is coincidental. Conventionally, if an association is calculated to occur by chance only five times or less in a hundred ($p < .05$), it is deemed to be statistically significant.

Archaeological dates relating to the early New World occurrence of *Zea mays* L. are compared with glottochronological dates for ancient Amerindian languages in an effort to evaluate glottochronology's potential as a tool for investigating language and culture prehistory. **Glottochronology** is a linguistic method for determining the minimum number of centuries since **genetically related languages** diverged from a common ancestor, a dating approach believed by many historical linguists to produce flawed results. Analysis of 626 words for maize in 591 Amerindian languages and dialects affiliated with 51 different genetic groups (e.g., the Siouan language family of North America) yields evaluations concerning whether or not a word for maize pertained to parent languages of each of the 51 genetic groups and also to parent languages of certain subgroups of genetic groups (e.g., Mississippi Valley Siouan, a subgroup of Siouan). Maize terms posited for parent languages suggest the great cultural importance of maize for their speakers. Parent languages, for which maize terms are posited, strongly tend to have glottochronological dates indicating that they were spoken around or after 3000 years BP. This is the period of time shown by stable carbon isotope studies of human bone collagen to have been the era in which consumption of maize developed significantly for most Amerindian groups. On the other hand, where there is no evidence of a maize term for an **ancestral language,** this suggests a minor or even no role for maize in the lives of its speakers. Such parent languages strongly tend to have glottochronological dates indicating that they were spoken in years before 3000 BP when maize consumption in most areas of the New World is shown by stable carbon isotopes studies not to have been significant. These findings indicate that glottochronology can prove useful in the study of prehistory. Finally, results suggest that the accuracy of glottochronology can be improved by subtracting 800 years from glottochronological dates of ancestral languages older than 2500 years BP.

INTRODUCTION

This study evaluates glottochronology's potential as a tool for investigating language and culture prehistory. It does so by comparing archaeological dates relating to the early New World occurrence of *Zea mays* L. with glottochronological dates for ancient Amerindian languages. Results indicate that although glottochronology probably does not yield dates having the accuracy of those obtained through archaeological dating methods, it can nonetheless provide a chronological perspective of an approximate nature that can be useful in sorting out prehistory.[1]

Glottochronology was devised by Morris Swadesh in the mid-twentieth century as a method for determining the number of centuries since genetically related languages diverged from a common ancestor.[2] This involves comparing the core vocabulary[3] of two languages to determine the degree to which words are similar in the languages. Lesser lexical similarity indicates greater chronological depth for a split and more lexical similarity indicates less chronological depth. Swadesh, working with the assumption that **lexical replacement** of core vocabulary occurs at a constant rate, developed a formula for determining the number of centuries since a language divergence occurred. Applying this formula to the number of similar words in the core vocabulary list found for two languages, the number of centuries that have passed since the two split from a common ancestor can be computed.

Fifty-one **language genetic groups** distributed throughout the Western Hemisphere—including families such as Mayan and Uto–Aztecan—are used in this study. Words for maize are assembled from as many member languages of each of the 51 groups as possible. These words are analyzed with respect to whether or not they constitute evidence that a term for maize was found in the lexicon of a genetic group's parent or ancestral language. For example, maize term evidence indicates that Proto-Mayan probably had a word for maize, whereas such evidence is not suggestive of a Proto-Uto–Aztecan maize term.

A glottochronological date is determined for each genetic group. For example, Proto-Mayan has a date of 4200 years BP and Proto-Uto–Aztecan has a date of 4800 BP. Given such information, conclusions can be drawn regarding the dates at which *Zea mays* was known to prehistoric Amerindian groups. For example, because Proto-Mayan probably had a maize term, maize was likely to have been a part of the cultural inventory of Proto-Mayan speakers of more than 4000 years ago. Conversely, because maize term evidence does not attest to a Proto-Uto–Aztecan word for maize, its speakers of 4800 years ago may have been unfamiliar with *Zea mays*.

Independent evidence from archaeology can be used to evaluate such conclusions. For example, if ancient maize cobs from archaeological sites in areas in which Mayan lan-

[1]Other recent studies [e.g., 1, 13, 48] report positive correlations between glottochronological dates and archaeological dates and, thus, lend additional support to the current study's favorable evaluation of glottochronology.

[2]For a description of this method see Swadesh [43], translated from Spanish to English by Joel Sherzer [44, pp. 284–271].

[3]Typically this is a list of 100 items including, for example, such common things as seed, blood, and water, and such ordinary activities as eat, sleep, and hear.

guages are spoken were shown through dating techniques, such as accelerator mass spectrometry (AMS), to be older than 4200 years BP, this would dovetail with the conclusion that Proto-Mayan, for which a maize term is posited, was spoken around 4200 years ago. Similarly, if archaeological sites from regions in which Uto–Aztecan languages are spoken were to produce no evidence of maize before 4200 BP, this would accord with the conclusion that Proto-Uto–Aztecan, for which no maize term is posited, was spoken 4800 years BP.[4] If such isomorphic associations were found for most if not all of the 51 genetic groups used in this study, this would constitute some important counter-evidence to the widely held belief among linguists that glottochronological dates are, at best, grossly inaccurate and, at worst, useless.

GLOTTOCHRONOLOGY

Although glottochronological dates are frequently cited in literature dealing with language prehistory, many linguists are critical of the method and its results. Campbell [11] provides a comprehensive consideration of the methodological and analytical inadequacies. He finds shortcomings and flaws in all basic propositions underlying the approach,[5] concluding [11, pp. 186]: "[G]lottochronology is not accurate; all its basic assumptions have been severely criticised. It should not be accepted; it should be rejected."

Apparently, glottochronological dates are not so easy to reject, even for Campbell [10]. For example, there are numerous instances in his book, *American Indian Languages: The Historical Linguistics of Native America,* in which he cites glottochronological dates (e.g., pp. 112, 123, 132, 133, 138, 139, 142, 159, 165, 167, 172, 174, 175). When doing so, he usually adds a disclaimer such as that found parenthesized in the following quote, "Glottochronological calculations (rejected by most linguists) indicate that Proto-Apachean split from the Northern Athabaskan languages at about A.D. 1000" [10, pp. 112]. Other reputable scholars use glottochronological dates in their work as well, but like Campbell, tend to alert readers to their deficiencies.

If glottochronology is so problematic, why should scholars continue to use it at all? **Historical linguistics** is by definition diachronic, and, thus, lends itself to a chronological framework provided at this time only by glottochronology. No other method has yet been devised that links findings of historical linguistics to a real time scale.[6] In addition, glottochronology is the only existing method allowing the possibility of relating results of historical linguistics to findings of prehistoric archaeology.

Among the numerous criticisms directed toward glottochronology, the most serious are those raising questions about the presumed constancy of the rate of lexical replacement across all languages. This presumption is the bedrock on which glottochronology is built. If lexical replacement does not take place at a predictable rate, it would seem impossible to develop an accurate formula for calculating time of language divergence.[7] On the other hand, if languages do replace lexical items at a predictable rate, then glottochronology can and will work despite other problems with the approach that have been raised.

Evidence to the effect that lexicons of different languages have been replaced at different rates is available.[8] For example, as Campbell [11] notes, we know from surviving early written texts that Icelandic and English show greatly differing percentages of lexical retention, respectively, 97.3% and 67.8%. Others have assumed in linguistic theories that language change in general is far from constant over time. Dixon [12] presents a model of language change entailing "punctuated equilibrium," in which languages are described as changing at a steady rate over time, but undergoing periods of rapid change caused by events external to language phenomena such as natural disasters, material innovations, development of aggressive tendencies, and so on, from time to time. Indeed, the Icelandic–English difference mentioned earlier may relate to such a dynamic, wherein the lower retention rate of English is because, at least in part, of a punctuating event in British history (i.e., the Norman invasion of England in AD 1066 and the subsequent considerable influence of French on the English language).

Lexical replacement, then, almost certainly does not take place at a constant rate at all points in time. At any given moment in time a particular language may be in a state of equilibrium in which replacement occurs at a steady rate or in a state of punctuation in which replacement is accelerated. Over a long period of time, say greater than 2500 years, any single language will randomly experience a number of episodes of both equilibrium and punctuation. The total

[4] The archaeological dates mentioned in this paragraph are for illustrative purposes only and have no basis in reality.

[5] For these and a detailed description of the method, readers are invited to consult Campbell [11, pp. 177–186]. Also, the pros and cons of glottochronology are discussed at considerable length in various papers of a recent book edited by Renfrew, McMahon, and Trask [37].

[6] While there have been some recent efforts to retool features of glottochronology to improve its accuracy, no totally new, radically different approach to language dating has been advanced since Swadesh's invention of the lexical dating technique [14, 22, 26, 29, 41].

[7] For example, Starostin [41, p. 229] writes, "[I]t is obvious that in the absence of a constant rate the whole procedure of glottochronology becomes senseless."

[8] Among a number of works in which such evidence has been discussed, Blust [6] is especially noteworthy.

lexical replacement that may occur during this long period is an aggregate of steady replacement that took place during some parts of the long period and rapid replacement that occurred at other times. Most languages over long periods of time are subject *by chance* to roughly the same numbers of periods of equilibrium and punctuation. Therefore, over a long period of time, for most languages the accruing lexical replacement would be roughly the same.[9]

I suggest that lexical replacement for any two particular languages tends to approach parity as the period of time in which replacement occurred is augmented. Thus, for example, if language A and language B are genetically related, but have been separated from one another for a long period of time, then the degree of lexical replacement that has taken place in each language during that period of time will tend to be more or less the same. On the other hand, if languages A and B have been split for only a short period of time, say, less than 2500 years, the amount of lexical replacement that has taken place in each during that period may be substantially different, although not necessarily so.

The implication of this for glottochronology is that if we could determine the rate of lexical replacement that ordinarily occurs over a long period of time, the accuracy of glottochronology could be significantly enhanced, but possibly only for language families of considerable chronological depth (e.g., those older than 2500 years[10]).

Glottochronological dates for genetic groups cited in this study are from several sources. Most dates are from Swadesh [42] and Kaufman [23–25]. Others are from Broadwell, [7] Campbell, [10] Fowler, [15] Hill, [19] Holman, [20] and Wichmann [47]. Dates for some genetic groups have been calculated with the help of Pamela Brown (personal communication). These include Northern Athapaskan, Apachean (of Athapaskan), Eastern Algonquian, Mississippi Valley Siouan, Ofo–Biloxi (of Siouan), Missouri River Siouan, Northern Iroquoian, Yuman, and Tacanan (of Pano–Tacanan).

I have calculated and include here other glottochronological dates, all for South American genetic groups, which should be regarded as highly tentative. These calculations did not entail Swadesh's core vocabulary of 100 items, but rather short vocabularies for languages provided by Loukotka [27] in his *Classification of South American Indian Languages*.[11] Cognate percentages determined by use of these short lists were converted into minimum centuries through use of Swadesh's formula. The genetic groups for which these tentative dates were calculated include Tupi–Guaraní, Tuparí Branch (of Tupian), Ramarama Branch (of Tupian), Muran, Western Tucanoan, Eastern Tucanoan, Zaparo Group (of Zaparoan), Iquito–Cahuarano Dialects (of Zaparoan), Eastern Dialects (of Nambiquaran), Central Dialects (of Nambiquaran), and Jabutian.

METHODOLOGY

For this study, 626 words for *Zea mays* in 591 Amerindian languages and dialects were collected from dictionaries and other sources.[12,13,14] These languages and dialects are affiliated with 51 different genetic groups. Genetic groups recognized here include only those identified by Campbell [10] as being bona fide genetic units such as Ge, Panoan, and Cariban, three South American language families. Campbell considers many controversial classifications such as Greenberg's [17] large Ge–Pano–Carib grouping, which unites the latter three groups, as invalid. Although I attempted to collect words for maize found in all languages of each of the 51 genetic groups, this was not always possible because of data limitations. Consequently, maize term inventories for some genetic groups are not as close to being complete as those for others.

[9]A key word in this and in the immediately preceding sentence is *most*. A few languages (at the extreme ends of normal distribution) over a long span of time may undergo more periods of equilibrium and fewer periods of punctuation than is typically the case, or more periods of punctuation and fewer periods of equilibrium than is usual. However, for most languages, over long time periods, episodes of equilibrium and punctuation should by chance work out to roughly the same. Similar evaluations have been recently expressed by Ehert [13, p. 373], by Jaxontov [22, p. 52], and by Lohr [26, p. 219].

[10]The constant rate of lexical replacement that Swadesh uses in his formula for calculating glottochronological dates is based on the lexical change observed when comparing early texts of historical languages with their modern versions. Thus, Swadesh's constant is founded on lexical replacement occurring over time periods of no more than 2000 years or so. This constant may not be the same as the typical rate of lexical change that occurs in languages over a period of more than 2500 years.

[11]Dictionaries for most South American Indian languages do not exist or are not accessible. Consequently, 100-item core vocabularies cannot be assembled for them. In many cases, Loukotka's short vocabularies are the only lexical materials forthcoming for a language. These vocabularies range in size from around 10 to 15 words. While some words of these short lexicons refer to things on Swadesh's list such as head, ear, water, and fire, they also usually include some "cultural" items as well such as axe and maize.

[12]Some languages have more than one term for maize.

[13]Many of the dictionaries, vocabularies, and word lists from which Amerindian-language maize terms have been extracted are listed in Appendix B of Brown [8, pp. 184–220]. The major source of maize words in languages of South America is Loukotka [27], supplemented by Huber and Reed [21, p. 162] for terms for maize in languages of Colombia. Some words for maize in languages of Eastern North America are from Munson [30]. Hill [19] has been the source for maize terms of some Uto-Aztecan languages.

[14]Amerindian-language terms for *Zea mays* found in lexical sources are often glossed by single words such as English maize or Spanish *maíz*. In some instances, such glosses may be oversimplifications because words may actually refer more specifically to parts of the maize plant, for example, to maize kernels or to the corncob rather than to the whole plant. In other instances, words may be polysemous, indexing two or more dimensions of maize such as is the case with American English corn which refers both to the plant that produces maize kernels and to the kernels themselves.

Once terms were collected for each group, they were evaluated with regard to whether or not they indicate the occurrence of a word for maize in the genetic group's parent language. This process involved (1) looking for phonological similarities among a genetic group's maize terms and (2) attending to how phonologically similar terms are distributed across major subgroups of a genetic unit.

The task of determining term similarity was complicated by the many different orthographies employed in lexical sources. My experience dealing with different orthographies for recording Amerindian languages helped in fleshing out similarities obscured by different ways of symbolically representing the same sound [8]. Another complication relates to the fact that I control knowledge of the regular sound correspondences of a limited number of Amerindian language families. Because of regular sound shifts, some maize terms found in two or more languages of a genetic group can appear phonologically dissimilar when they are in fact **reflexes** of the same ancestral maize term. Although I may have failed to recognize some instances of term cognation, a reasonably thorough familiarity with processes of phonological change helped keep such problems to a minimum.

The manner in which a group of similar maize terms are distributed across subgroups of a language family is indicative of whether these terms are reflexes of a word for maize, which may have pertained to the family's ancestral language. For example, if similar maize terms occur in languages of most major subgroups, this is reasonably strong evidence that the words are cognate and, hence, trace to a maize term in the proto-language pertaining to the genetic group.[15]

An example of such a distribution is that found for Mayan, a Mesoamerican family whose member languages are spoken in Mexico and northern Central America. In Table 20-1, all Mayan languages for which a maize term has been found are listed and organized in the list by family subgroup affiliation.[16] There are 30 such languages. I have identified maize terms in 29 of these languages as being phonologically similar. These 29 words are presented in Table 20-1, in bold type (e.g., Yucatec ***ixi'im***).

Yucatec *ixi'im* and the 28 other maize words similar to it are found in languages of all five major branches of the family—Huastecan, Yucatecan, Greater Tzeltalan, Eastern Mayan, and Greater Qanjobalan—robustly suggesting that a term similar to *ixi'im* designated maize in the lexicon of Mayan's ancestral language. The Mayan evidence (Table 20-1). then, attests to a *probable* **proto-word** for maize.

A distribution not indicative of a proto-term for maize is one in which no two languages belonging to different major subgroups of a family show phonologically similar terms. Maize terms in languages of the Uto–Aztecan family (American Southwest and Middle America) illustrate such a distribution. In Table 20-2 phonologically similar maize terms in Uto–Aztecan languages are identified by use of a parenthesized number suffixed to terms. Maize words with the same suffixed number are identified as being similar. There are 10 groups of similar terms or **similarity sets** found in Uto–Aztecan languages (some similarity sets of which have only one member). The largest group of these 10 consists of maize terms suffixed with (7). All 10 of the latter terms are restricted to languages of one major subgroup, Southern Uto–Aztecan.

With one exception, similar maize terms do not occur in both the Northern and Southern branches of Uto–Aztecan. The exception involves four similar words suffixed with (2) (i.e., maize words found in Panamint, Cahuilla, and Cupeño of Northern Uto–Aztecan and in Mountain Pima of Southern Uto–Aztecan). All four words are unambiguous loans from Spanish *(maíz)*. Consequently, this distribution does not constitute evidence for a Proto-Uto–Aztecan maize term. Because no other similar terms overlap Northern and Southern Uto–Aztecan, Table 20-2 offers no evidence supporting the proposal of a Proto-Uto–Aztecan maize term.

Other distributions of terms across languages of genetic groups are suggestive of a proto-word for maize, but not strongly so, such as is the maize term distribution found for the Tucanoan family of South America (Brazil, Colombia, Ecuador, and Peru) (see Table 20-3).

The Tucanoan family has three main branches, Western, Central, and Eastern Tucanoan. Only terms suffixed with (2) in Table 47-3 are found in languages of more than one of the main branches. Terms suffixed with (2) occur in languages of Western and Central branches, but not in those of the Eastern branch. This distribution suggests a Proto-Tucanoan word for maize, but not as strongly as would a distribution involving all three branches of the family. Maize words suffixed with (2) occur in eight Western Tucanoan languages and in Cubeo, the only member of the Central Tucanoan branch. Conceivably, Cubeo's maize term could be a loan from some Western Tucanoan language. If so, Tucanoan maize term distribution would not constitute evidence for a Proto-Tucanoan term for maize. Given such considerations, I interpret the Tucanoan data as indicating a *possible* proto-word for maize.

Using approaches described earlier, I have evaluated each of the 51 genetic groups and, in many instances, subgroups

[15]To state with strong confidence that similar terms trace to a single proto-word, it must be established that these words demonstrate regular sound correspondences.

[16]The subgrouping of Mayan presented in Table 47-1 follows that used by Brown and Wichmann [9, pp. 129–130]. In this study, all analyses, except that for Mayan, Nambiquaran, and Southern Uto-Aztecan, are based on subgrouping of genetic units as given in Campbell [10]. Nambiquaran is based on Loukotka's [27] classification and Southern Uto-Aztecan on Kaufman's [24] classification. Subgroupings for Mayan, Nambiquaran, and Southern Uto-Aztecan used here are believed by the author to be more up-to-date and/or complete than those found in Campbell [10] for the same three language groups.

TABLE 20-1 Terms for Maize in Mayan Languages Sorted by Classification

Mayan			
Huastecan			
	Huastec: ***e:m***, *ithith*		
	Chicomuceltec: *isis*		
Yucatecan			
	Yucatec: ***ixi'im***		
	Mopán: ***ixi'im***		
	Itzaj: ***ixi'im***		
	Lacondón: ***ix-im***		
Greater Tzeltalan			
	Cholan		
		Chol: ***ixim***	
		Chontal: ***ixim***	
		Ch'orti': ***ixim***	
		Cholti': ***ixim***	
	Tzeltalan		
		Tzeltal: ***ixim***	
		Tzotzil: ***ixim***	
Eastern Mayan			
	Mamean		
		Teko: ***ixi'm***	
		Awakateko: ***isi'n***	
		Ixhil: ***ixi'm***	
		Mam: ***ixi'ŋ***	
	K'ichee'an		
		Kaqchikel: ***ixin***, *jel*	
		Tz'utujiil: ***ixiim***	
		K'iche': ***ixim***	
		Sakapulteko: ***ixiim***	
		Uspanteko: ***ixiim***	
		Poqomam: ***ixim***	
		Poqomchi': ***ixim***	
		Q'ekchi': ***ixim***	
Greater Q'anjob'alan			
	Chujean		
		Chuj: ***ixim***	
		Tojolabal: ***ixim***	
	Q'anjob'alan		
		Q'anjob'al-Jakalteko	
			Akateko: ***ixim***
			Jakalteko: ***ixim***
			Kanjobal: ***ixim***
		Mocho	
			Mocho: ***ix(i')m***

TABLE 20-2 Terms for Maize in Uto-Aztecan Languages Sorted by Classification

Uto-Aztecan			
Northern Uto-Aztecan			
	Hopi: *qaa'ö* (1)		
	Numic		
		Central	
			Panamint: *maisi* (2)
			Comanche: *haníibi̲* (3)
			Shoshone: *how-eep* (4)
		Southern	
			Southern Paiute: *hauvivgh* (4)
			Southern Ute: *kumúy* (5)
			Chemehuevi: *ha'wiv(i)* (4)
	Takic		
		Cahuilla-Cupeño	
			Cahuilla: *máys* (2)
			Cupeño: *m̲a̲ayis* (2)
Southern Uto-Aztecan			
	Aztecan		
		Pochutla: *teyu'l* (6)	
		Nahuatl	
			Classical Nahuatl: *tlaulli* (6), *cintli* (7)
			Nahuatl (Tetelcingo): *tlauli* (6)
			Nahuatl (Xalita): *tlayohli* (6)
			Nahuatl (Zacapoaxtla): *tago̲l* (6)
			Nahuatl (Huazalinguillo): *cintli* (7)
			Nahuatl (Mecayapan): *sinti* (7)
			Nahuatl (Pajapan): *tayo:l* (6)
			Pipil: *sin-ti* (7)
	Sonoran		
		Cahita	
			Yaqui-Mayo: *batchi* (8)
		Opata-Eudeve	
			Eudeve: *sunú-t* (7)
		Corachol	
			Cora: *yuuri* (9)
			Huichol: *'icú* (10)
		Tarahumara-Guarijío	
			Tarahumara: *sunú* (7)
			Guarijío: *sunú* (7)
		Tepiman	
			O'odham: *huhni* (7)
			Pima Bajo: *hunu* (7)
			Mountain Pima: *huun* (7), *mai* (2)

TABLE 20-3 Terms for Maize in Tucanoan Languages Sorted by Classification

Tucanoan		
Western Tucanoan		
	Tanimuca: *o'aka* (1)	
	Koreguaje Dialects	
	Tama: *keá* (2)	
	Koreguaje: *'Bea* (2)	
	Cóto (Orejon): *bea* (2)	
	Pioje-Siona	
	Pioje: *wéa* (2)	
	Siona: *we'a* (2)	
	Amaguaje: *bea* (2)	
	Secoya: *'we'a* (2)	
	Yahuna: *oáka* (1)	
Central Tucanoan		
	Cubeo: *we'á* (2)	
Eastern Tucanoan		
	Macuna Dialects	
		Omöa: *ohólika* (3)
		Sära: *ohólika* (3)
		Buhágana: *oholika* (3)
		Macuna: *oho-rika* (3), *áre* (4)
	Yupua-Durina	
		Yupua: *óo* (5)
	Coretu: *mitólikere* (6)	
	Desano-Siriano	
		Desano: *ohodi'ká* (3)
		Siriano: *o'hódika* (3)
	Bará-Tuyuka	
		Bará: *'ódika* (3)
		Waimaja: *'ódika* (3)
		Tuyuca: *'hórika* (3)
		Tsolá: *ohólika* (3)
	Tucano-Carapano	
		Yuruti: *'órika* (3)
		Tucano: *ohóka* (3)
		Uasona: *olíka* (3)
		Urubu-Tapuya: *oriká*(3)
		Tatuyo: *'ó-'ríka* (3)
		Barasana: *oh'ó-riká* (3)
		Carapana: *'ó-ríka* (3)
	Wanano: *jo* (5)	
	Piratapuyo: *'joo* (5)	

within groups as well, to ascertain whether or not a word for maize pertains to a group's or subgroup's parent language. My evaluations are limited to three results: A (evidence indicates a *probable* proto-word for maize), B (evidence indicates a *possible* proto-word for maize), and C (there is *no* evidence of a proto-word for maize).

THEORETICAL CONSIDERATIONS CONCERNING TERMS FOR MAIZE IN ANCESTRAL LANGUAGES

Earlier in the chapter I assert that evidence gathered for Uto–Aztecan languages (see Table 20-2) is not indicative of a Proto-Uto–Aztecan (PUA) word for maize. I am not implying that PUA lacked such a word, only that the assembled evidence (Table 20-2) does not attest to one. Additional evidence from Uto–Aztecan languages may well indicate that a maize term did pertain to the PUA lexicon. Indeed, such evidence recently has been brought together by Jane Hill [19].

Hill [19] assembles a number of plausible cognate sets consisting of words that may be reflexes of PUA terms that referred to things and activities connected with maize. According to Hill, words of all sets show expected sound correspondences, except for a few which she deals with in detailed discussions. For each set she reconstructs a proto-language word and its referent or referents.

An example is Set 1 [19, p. 918], which includes the following proposed **cognates:** Hopi *so–owï* (sand grass), Proto-Tepiman **hu:nu-i* (maize), Eudeve *sunú-t* (maize), Guarijío *sunú* (maize), Tarahumara *sunú* (maize), Mayo *súnnu* (maizefield), and Proto-Nahuatl **sin-tli/*sen-tli* (maize). For this set Hill reconstructs **sunu* (maize ear, maize) as a proto-form.

The Hopi reflex of **sunu* is the only word in Set 1 that does not refer to a maize-related item. It is also the only word in the set from a language belonging to Northern Uto–Aztecan (all other languages of the set are Southern-Uto–Aztecan) (see Table 20-2). If **sunu* (maize ear, maize) pertained to PUA, then the Hopi reflex, *so–owï,* has lost its original meaning, replacing it with the nonmaize-related, but nonetheless botanical referent "sand grass." In addition, one other reflex, Mayo *súnnu,* has shifted in meaning from (maize ear, maize) to (maizefield).

Assuming that **sunu* (maize ear, maize) was a PUA word, then a relatively large number of Uto–Aztecan languages have replaced it with new words for maize: 17 of the 27 Uto–Aztecan languages of Table 20-2, or 63%, have done so. At least two of these 17, Hopi and Mayo, have retained the term but have shifted its referent from maize to a new one.

This plausible development in Uto–Aztecan raises a question: Why should languages of some genetic groupings, such as Mayan (see Table 20-1), , retain a maize term of an ancestral language, whereas others, such as Uto–Aztecan languages, replace such a word with new terms? An answer to this may relate to the possibility that maize terms in some ancestral languages have been more salient for their speakers than maize words in other ancestral languages.[17] The salience of a word for maize relates to the degree to which a people speaking a language have embraced maize products and agriculture or, in other words, to maize's cultural significance. The salience of a maize term will be low when the cultural significance of maize is low and high when the cultural significance of maize is high.

There is evidence that words for plants of high salience tend to be retained by **offspring languages,** whereas those for plants of low salience tend to be replaced. Some years ago, Berlin and colleagues [5] published the results of an ethnobotanical study showing an extremely robust positive correlation between the lexical retention of plant names and the cultural significance of the plants to which they refer. Plant names were collected from speakers of Tzeltal and Tzotzil, two closely related Mayan languages of Chiapas, Mexico, both of which are immediate daughter languages of Proto-Tzeltalan. Plant names in each of the two languages that refer to at least one identical species were compared for lexical similarity and possible cognation. Plants designated by these terms were grouped into four categories delimiting their cultural significance (from high to low): 1) cultivated plants, 2) protected plants, 3) wild-useful plants, and 4) wild-insignificant plants.

A total of 257 plant species have both Tzeltal and Tzotzil names [5, p. 161]. Of these, paired terms (both naming the same species) for 111 species were found cognate attesting to a plant term's pertinence to Proto-Tzeltalan's lexicon, and paired terms for 146 species were found noncognate. Of terms pertaining to cultivated plants, 14 are cognate and 2 noncognate; of those pertaining to protected plants, 29 are cognate and 7 noncognate; of those pertaining to wild-useful plants, 52 are cognate and 63 noncognate; and of those pertaining to wild-insignificant plants, 16 are cognate and 74 noncognate. The correlation between cognation and cultural significance is extremely strong and highly statistically significant: **gamma** = 0.97 (on a scale from 0.00 to 1.00, where 0.00 indicates no association whatsoever and 1.00 is a perfect correlation), **chi square** = 41.15753, $p < .001$.[18] In other words, plant names of Proto-Tzeltalan strongly tend to be retained by its offspring languages (Tzeltal and Tzotzil) when the plants these terms designate are high in cultural importance and strongly tend to be replaced when designated plants are low in cultural importance.

I assume earlier that when most major branches of a genetic group show similar terms for maize, these words are probably reflexes of a term for maize that occurred in the group's parent language. Given the exceptionally robust results of the Berlin and colleagues [5] study, I also assume that retention of the term in most major branches is at least in part because of the high salience of the word and the great cultural importance of maize for speakers of the parent language. In short, if similar maize terms appear in most major branches of a genetic group, I assume that speakers of the ancestral language not only cultivated maize but that it constituted a dietary staple. Consequently, the postulation of a probable proto-term for maize for a genetic group's parent language is considered here indicative of the high cultural significance of *Zea mays* for speakers of that ancestral language.

In instances like Uto–Aztecan, where there is evidence that a maize term pertained to a parent language, yet where reflexes of the latter having the designation maize do not occur in most major branches of a group, I assume that the proto-word for maize was not especially salient for parent language speakers. This would be attributable to the low cultural significance of maize, perhaps involving situations in which people are familiar with maize, but do not cultivate it themselves or in which people grow maize, but only as a supplement to major food resources.

CHRONOLOGY OF MAIZE IN THE AMERICAS

It is widely agreed by scholars that maize was first domesticated in Mesoamerica. The earliest evidence of Mesoamerican maize comes from the Río Balsas region of southwestern Mexico, where it was domesticated before 6000 BP [3, 4, 28]. During Pre-Columbian times, from this homeland it spread to other parts of Mexico and into Central and South America, and eventually north to the American Southwest and finally to Eastern North America.

In addition to well-known ancient cobs from Tehuacán, evidence of maize's early spread through Mexico comes from Oaxaca where maize cobs are found at Guilá Naquitz Cave dating to 5400 BP [34]. Central America also shows early evidence of maize. Maize phytoliths, pollen, and starch grains—but not cobs—have been found in Panama with an estimated date of at least 7000 years BP [33–36]. There are some early dates for *Zea,* which may have been *Zea mays,* from South America including *Zea* pollen and phytoliths from Ecuadorian Amazonia dated at 5300 BP [32]. Directly dated cob phytoliths or rondels from

[17]A contributing factor, in addition to the one described in this and following paragraphs, is that languages of older genetic groups have had more time to lose lexical items that pertained to their common parent language than have languages of younger genetic groups.

[18]These statistics are based on information given in Table 5 of the Berlin et al. [5, p. 161] paper. Gamma and chi square scores have been calculated by using a 2 × 2 table in which numbers of terms for cultivated and protected plants were combined into one category and numbers of terms for wild-useful and wild-insignificant plants were combined into another category.

Coastal Ecuador have generated AMS dates ranging between 4200 and 3900 BP and the area has yielded archaeological remains of maize dating between ca 4000 and 3500 BP [40, 45].

Maize probably reached the American Southwest by 4000 BP and macrobotanical evidence from the Holding site near St. Louis, Missouri, attests to maize in Eastern North America as early as 2150 BP [38, see Chapter 20].

As Smalley and Blake [39] point out, until recently the presence of maize in any form at an archaeological site—be it cobs, pollen, phytoliths, or starch grains—has been generally assumed demonstrative of the important role of maize in the diet of the ancient humans associated with it. However, paleoethnobotanical evidence of a quantitative nature suggests that early maize was typically of low dietary significance [18]. Smalley and Blake have assembled evidence indicating that in most parts of the Americas maize was not a dietary staple before 3000 BP. Around 3000 BP, according to them, "[T]hroughout the New World . . . there was a dramatic increase in maize use, and it eventually became the staple we know today" [39, p. 678].[19]

They bring together for analytical comparison a number of stable carbon isotope studies of human bone collagen samples from archaeological sites in Latin America [39]. The stable carbon isotope is robustly indicative of the extent of maize consumption in ancient times. They describe their approach as follows:

> We looked at nine published isotope studies spanning the period from the first appearance of maize to the Spanish conquest. These included 622 individuals from well-dated sites, giving a general overview of trends of maize consumption over 8,000 years. [39, p. 684]

In general, the assembled studies show that consumption of maize in Latin America begins to become significant for most groups beginning around 3000 BP. Smalley and Blake [39] describe their detailed findings as follows:

> In the Mesoamerican samples, . . . ratios . . . suggesting maize as a dietary staple . . . appear only in Middle Formative times (i.e., after about 3,000 B.P.). There are four exceptions: two individuals from early Tehuacán and two from the Soconusco region. . . . Samples from 16 individuals at Cerro Mangote in Pacific Panama also show . . . ratios indicating moderate maize consumption at an early date (between 7,000 and 4,500 B.P.). . . . In contrast, at most South American sites, there was a longer delay between the first appearance of maize and its increasing use as a major food crop, beginning around 3,000 B.P. in some areas and much later in others. [p. 684]

According to Coltrain and colleagues (see Chapter 20), reliance on maize was widespread in the American Southwest by 3000 years BP. Maize came to dominate the Amerindian diet in Eastern North America considerably later in time. Maize farming villages appear archaeologically in Eastern North America at around 1400 BP [16]. In the Southeast, maize developed as a dietary staple after around 1200 BP [46].

MAIZE CHRONOLOGY AND GLOTTOCHRONOLOGICAL DATES

Table 20-4 lists Amerindian genetic groups and subgroups with their respective glottochronological dates, **maize term evaluations**, and other information. It sorts groups into three geographic categories: those of North America (north of the U.S.–Mexico border), those of Middle America (Mexico and northern Central America), and those of South America (including southern Central America). Languages of only one genetic group, Uto–Aztecan, straddle two of these geographic areas. In Table 20-4, Uto–Aztecan is placed with genetic groups of Middle America rather than with those of North America, mainly because more Uto–Aztecan languages of my sample are spoken in Middle America than in North America. Within geographic areas, genetic groups are ranked according to magnitudes of associated glottochronological dates, from lowest to highest. In addition, given for each genetic group is my evaluation of whether or not lexical evidence indicates the pertinence of a proto-term for maize in an associated ancestral language (again, A indicates a probable term for maize, B indicates a possible term for maize, and C indicates that there is no evidence of a term for maize). Table 20-4 also gives for each genetic group the number of member languages consulted, the total number of maize terms found for consulted languages of a group, and the number of sets of similar maize terms (refer to Methodology section in this chapter). In addition, for genetic groups for which a possible or a probable proto-term for maize is posited, a sample reflex of the proto-term from a contemporary language is presented.

Table 20-4 shows that for the vast majority of genetic groups of North America (north of the U.S.–Mexico border) there is no evidence for positing a proto-language maize term. For example, for parent languages of 10 of the 11 genetic groups of North America of which most or all respective member languages are spoken in Eastern North America, there is no evidence for a maize term.[20] All but one of these 10 groups has a glottochronological date older than 1400 BP (Missouri River Siouan is dated at 540 BP). This

[19]Smalley and Blake [39] argue convincingly that maize was domesticated for its sugary pith or other edible parts rather than for its grain. They propose that the stalk sugar of maize was its principal attraction for groups until around 3000 BP after which kernels of maize surpassed stalk sugar in importance and *Zea mays* widely became a dietary staple.

[20]These 11 groups are Missouri River Siouan, Ofo-Biloxi, Mississippi Valley Siouan, Southwestern Muskogean, Southern Muskogean, Northern Iroquoian, Eastern Algonquian, Muskogean, Caddoan, Algonquian, and Iroquoian.

TABLE 20-4 Amerindian Language Genetic Groups and Subgroups within Respective Glottochronological Dates, Maize Term Evaluations, and Other Information

North American genetic groups and subgroups[1]	Number of consulted languages	Number of maize terms	Number of sets of similar terms	Glottochronological dates	Maize-term evaluation for ancestral language	Sample reflex of probable or possible proto-word for maize
Apachean (of Athapaskan)	4	5	2	460 BP[2]	A	Navajo: ***naadáá'***
Missouri River Siouan	2	2	2	540 BP[3]	C	
Ofo–Biloxi (of Siouan)	2	2	1	1300 BP[4]	A	Ofo: ***atce'ki***
Northern Athapaskan	6	6	5	1600 BP[5]	C	
Mississippi Valley Siouan	8	10	5	1600 BP[6]	C	
Southwestern Muskogean	4	5	3	1600 BP	C	
Yuman	8	9	4	1600 BP[7]	A	Mohave: ***tadiich***
Southern Muskogean	6	7	4	1900 BP	C	
Northern Iroquoian	9	9	2	2200 BP[8]	C	
Eastern Algonquian	11	12	8	2400 BP[9]	C	
Athapaskan	10	11	7	2500 BP	C	
Interior division (of Salishan)	3	4	3	2500 BP	C	
Miwokian (of Utian)	4	4	1	2500 BP	C	
Wintuan	2	2	2	2600 BP	C	
Siouan (core)	14	18	7	3000 BP	C	
Muskogean	7	8	5	3000 BP	C	
Yukian	2	2	2	3000 BP	C	
Kiowa–Tanoan	6	7	7	3000 BP	C	
Caddoan	4	5	5	3300 BP	C	
Algonquian	26	31	20	3500 BP	C	
Iroquoian	10	10	3	3500 BP	C	
Northern Uto–Aztecan	9	9	4	3900 BP	C	
Salishan	5	6	5	4500 BP	C	
Utian	5	5	1	4500 BP	C	
Middle American genetic groups and subgroups[10]						
Chinaantecan (of Oto–Pame–Chinantecan)	4	4	1	1500 BP	A	Chinateca (Lealoa): ***mil[l]cuii[3]***
Zapotecan (of Popolocan-Zapotecan	9	9	1	2400 BP	A	Zapotec (Mitla): ***xob***
Totonacan	4	5	2	2600 BP	A	Totonac (Papantla): ***cux'i***
Lencan	2	2	1	3000 BP	A	Honduran Lena: ***ama***
Popolocan–Zapotecan (of Eastern Oto–Manguean)	15	15	2	3500 BP	C	
Mixe–Zoque	10	10	1	3500 BP	A	Rayon Zoque: ***mojk***
Otopamean (of Oto–Pame–Chinantecan)	5	5	1	3600 BP	A	Otomi (Mezquital): ***detha̲***
Mixtecan (of Popolocan–Zapotecan	7	7	1	3700 BP	A	Mixtec (Xochapa): ***nuní***
Southern Uto–Aztecan	18	20	8	3900 BP	A	Tarahumara: ***sunú***
Oto–Pame–Chinantecan (of Western Oto–Manguean)	9	9	2	4000 BP	C	

(Continued)

TABLE 20-4 *(continued)*

Middle American genetic groups and subgroups[10]	Number of consulted languages	Number of maize terms	Number of sets of similar terms	Glottochronological dates	Maize-term evaluation for ancestral language	Sample reflex of probable or possible proto-word for maize
Tlapanec–Manguean (of Western Oto–Manguean)	3	3	2	4100 BP	C	
Mayan	30	32	3	4200 BP	A	Yucatec: ***ixi'im***
Western Oto–Manguean	12	12	4	4700 BP	C	
Eastern Oto–Manguean	23	23	3	4700 BP	C	
Uto–Aztecan	27	29	10	4800 BP	C	
Oto–Manguean	35	35	14	6000 BP	C	
South American genetic groups and subgroups[11]						
Jivaroan	4	4	1	140 BP[12]	A	Aguaruna: ***sháa***
Iquito-Cahuarano dialects (of Zaparoan)	2	2	1	300 BP*	A	Iquito: ***shakárok***
Chocoan	10	11	3	700 BP	A	Catio: ***be***
Central dialects of Nambiquaran	2	2	1	700 BP*	A	Kokozú: ***kayátsu***
Karaja	2	2	2	900 BP	C	
Huarpe	2	2	1	1100 BP	A	Allentiac: ***telag***
Eastern dialects of Nambiquaran	4	4	2	1300 BP*	C	
Timotean	4	4	1	1500 BP	A	Mocochi: ***chixsak***
Rarmarama branch (of Tupian)	4	4	2	1500 BP*	A	Urukú: ***náya***
Panoan (of Pano–Tacanan)	22	26	4	1600 BP	A	Cashinahua: ***shöki***
Matacoan	5	7	6	1700 BP	C	
Jirajaran	3	3	1	1900 BP	A	Jirajara: ***dos***
Salivan	3	3	1	2000 BP	A	Sáliva: ***yamo***
Chimuan	4	4	2	2100 BP	A	Eten: ***mang***
Nimbiquaran	7	7	4	2200 BP	C	
Guajiboan	6	6	1	2300 BP	A	Guajibo: ***hétsa***
Tiniguan	2	2	2	2300 BP	C	
Chipaya–Uru	2	2	1	2400 BP	A	Chipaya: ***tara***
Tupi–Guarani	26	26	1	2600 BP*	A	Chiriguano: ***avatí***
Zaparao Group (of Zaparoan)	4	4	1	2700 BP*	A	Záparo: ***sáuk***
Tacanan (of Pano–Tacanan)	9	9	2	2700 BP[13]	A	Tacana: ***d'ije***
Zamucoan	2	2	2	2800 BP	C	
Eastern Tucanoan	21	21	2	3100 BP*	A	Desano: ***ohodi'ká***
Kamakanan	5	5	2	3200 BP	A	Mangaló: ***kesho***
Purian	3	3	2	3200 BP	B	Puri: ***maky***
Barbacoan	6	6	1	3300 BP	A	Cayapa: ***pishu***
Muran	3	3	2	3300 BP*	B	Mura: ***chihuha***
Jubutian	3	3	2	3300 BP*	B	Yabutí: ***tsitsi***
Chapacuran	10	10	3	3700 BP	A	Abitana: ***mapa:k***
Cariban	41	41	23	3700 BP	A	Trio: ***anai***

(Continued)

TABLE 20-4 · *(continued)*

South American genetic groups and subgroups[11]	Number of consulted languages	Number of maize terms	Number of sets of similar terms	Glottochronological dates	Maize-term evaluation for ancestral language	Sample reflex of probable or possible proto-word for maize
Maxakalian	3	4	4	3800 BP	C	
Tupari branch (of Tupian)	5	5	2	3900 BP*	B	Macuráp: ***atiti***
Western Tucanoan	9	9	2	4000 BP*	A	Siona: ***we'a***
Zaparoan	6	6	2	4100 BP	C	
Guaykuruan	4	4	4	4100 BP	C	
Eastern Chibchan	14	14	6	4200 BP	A	Kuna: ***oba***
Misulmapan	6	6	1	4300 BP	A	Miskito: ***aya***
Maipurean	74	74	22	4500 BP	A	Palicur: ***mahiki***
Pano–Tacanan	31	35	6	4700 BP	C	
Ge	15	15	6	5400 BP	C	
Tupian	47	48	11	6000 BP	B	Oyampi: ***awasi***
Chibchan	29	34	17	6700 BP	C	
Tucanoan	31	35	6	4500+ BP	B	Cubeo: ***we'á***
Witotoan	8	8	5	5400 BP	C	
Puinavean (Maku)	2	2	2	57000 BP	C	

Note: Names of genetic groups are given in bold type and names of subgroups are given in normal type.
*Indicates highly tentative glottochronological date calculated on the basis of a short vocabulary list (see Glottochronology section in this chapter).
A = evidence of a probable proto-term in the genetic group's parent language.
B = evidence of a possible proto-term in the genetic group's parent language.
C = no evidence of a proto-term in the genetic group's parent language.
[1]Includes only genetic groups of North America north of the U.S.-Mexico border.
[2]Date calculated by the author with the help of Pamela Brown. The calculation is based on the lexical comparison of Navajo and Western Apache.
[3]Date calculated by the author with the help of Pamela Brown. The calculation is based on the lexical comparison of Crow and Hidatsa.
[4]Date calculated by the author with the help of Pamela Brown. The calculation is based on the lexical comparison of Ofo and Biloxi.
[5]Date calculated by the author with the help of Pamela Brown. The calculation is based on the lexical comparison of Ahtna and Western Carrier.
[6]Date calculated by the author with the help of Pamela Brown. The calculation is based on the lexical comparison of Osage and Lakota.
[7]Date calculated by the author with the help of Pamela Brown. The calculation is based on the lexical comparison of Yavapai and Kiliwa.
[8]Date calculated by the author with the help of Pamela Brown. The calculation is based on the lexical comparison of Tuscarora and Oneida.
[9]Date calculated by the author with the help of Pamela Brown. The calculation is based on the lexical comparison of Delaware and Western Abenaki.
[10]Includes genetic groups of Mexico and northern Central America.
[11]Includes genetic groups of South America and southern Central America.
[12]Date calculated by the author. The calculation is based on the lexical comparison of Aguaruna and Huambisa.
[13]Date calculated by the author with the help of Pamela Brown. The calculation is based on the lexical comparison of Ese-eja and Tacana.

accords with the fact that maize did not develop as a dietary stable among any Amerindian groups of Eastern North America before 1400 BP [16].

Probable maize terms are posited for only 3 of 24 North American groups, these being Apachean (of Athapaskan), Ofo–Biloxi (of Siouan), and Yuman. Glottochronological dates for these 3 groups fit closely with emergence dates for maize as a dietary staple in the region. The latter date for the American Southwest is around 3000 BP. Yuman languages and Apachean languages, spoken in the Southwest, show respective glottochronological dates of 1600 BP and 460 BP, both of which are, of course, considerably younger than 3000 BP. The Ofo and Biloxi languages of the U.S. Southeast have a parent language which is glottochronologically dated to 1300 BP. This date is just slightly older than the date at which maize first developed as a dietary staple in the area (i.e., around 1200 BP).

Table 20-5 presents mean (average) glottochronological dates for genetic groups and subgroups of Middle America and of South America based on dates given in Table 20-4 The mean glottochronological date, based on dates for groups of Latin America (Middle and South America dates combined) for which probable maize terms are posited (A languages), is 2607 BP, and the mean glottochronological date, based on dates for groups of Latin America for which there is no evidence of a maize term (C languages), is 3948 BP.

These averages dovetail with the known chronology of maize as a dietary staple in Latin America. Probable maize

TABLE 20-5 Mean Glottochronological Dates for Respective Ancestral Languages Having a Probable Maize Term (A) and for Which There is No Evidence of a Maize Term (C) Calculated for Ancestral Languages of Middle America, of South America, and of Middle and South America Combined

	Mean date of A languages	Mean date of C languages
Middle America	3156 BP	4543 BP
South America	2410 BP	3650 BP
Middle and South America	2607 BP	3948 BP

TABLE 20-6 Association of Glottochronological Dates and Maize Term Evaluations for Middle America and South America Genetic Groups and Subgroups

	A	C
3000 years or less BP	20	6
3000+ years BP	14	15

gamma = 0.56.
chi square = 4.76651.
$p < .05$.

terms for ancestral languages suggest the great cultural importance of maize for their speakers. The glottochronological mean of 2607 BP for Latin American groups having a probable maize term is not unexpected given that consumption of maize is known to have become significant for most Indian groups of Latin America around or after 3000 BP [39]. A mean glottochronological date older then 3000 BP would have been unexpected.

On the other hand, where there is no evidence of a term for maize in ancestral languages, this suggests a minor or even no role for maize in the lives of their speakers. The glottochronological mean of 3948 BP for Latin American groups for which there is no evidence of a maize term is not unexpected given that consumption of maize in most areas of Latin America is known not to have been significant before 3000 BP.

Mean glottochronological dates calculated separately for groups of Middle America and South America respectively show a similar pattern. However, the mean date for groups of Middle America having a probable maize term is slightly greater than 3000 BP (3156 BP). This mean date perhaps reflects the fact that maize was first domesticated in and diffused earliest through Mesoamerica, a culture region of Middle America.

Table 20-6 presents a two-by-two table cross-tabulating maize term evaluations (A versus C) with glottochronological dates (3000 years or less BP versus 3000+ years BP) for genetic groups and subgroups of Latin America, this information having been recovered from Table 20-4. Table 20-6 shows a strong correlation between dates and evaluations (gamma = 0.56) that is statistically significant ($p < .05$). Thus, genetic groups for which there is a probable proto-term for maize (A) strongly tend to have dates of 3000 years or less BP and genetic groups for which there is no evidence of a proto-term for maize (C) strongly tend to have glottochronological dates older than 3000 years BP. This robust result accords with the understanding that beginning around 3000 BP in Latin America, the consumption of maize as a dietary staple began to become important for many Amerindian groups.

TABLE 20-7 Association of Adjusted Glottochronological Dates and Maize Term Evaluations for Middle America and South America Genetic Groups and Subgroups

	A	C
3000 years or less BP	28	7
3000+ years BP	6*	14

gamma = 0.81.
chi square = 13.48039.
$p < .001$.
*These six genetic groups, with adjusted glottochronological dates, are: Southern Uto-Aztecan (3100 BP), Mayan (3400 BP), Western Tucanoan (3200 BP), Eastern Chibchan (3400 BP), Misulmapan (3500 BP), and Maipurean (3700 BP).

Nevertheless, maize chronology suggests that glottochronology may tend to somewhat inflate some genetic group dates, specifically dates pertaining to groups that have considerable chronological depth, say, more than 2500 years. By subtracting 800 years from each of the Latin American glottochronological dates older than 3300 BP found in Table 20-4. and by reducing all glottochronological dates from 2600 BP through 3200 BP to 2500 BP, the gamma score for the association of dates and evaluations significantly increases in strength and in **statistical significance.** Table 20-7. presents the calculation of a new gamma score using these adjusted glottochronological dates. The resulting gamma is an extremely robust 0.81 and is highly statistically significant ($p < .001$).

Under the interpretation of Table 20-7. only six Latin American genetic groups for which a probable maize term is posited show glottochronological dates older than 3000 BP. These groups are, with parenthesized adjusted dates,

Southern Uto–Aztecan (3100 BP), Mayan (3400 BP), Western Tucanoan (3200 BP), Eastern Chibchan (3400 BP), Misulmapan (3500 BP), and Maipurean (3700 BP).

Uto–Aztecan and Mayan are Mesoamerican groups. As Smalley and Blake [39] note, stable carbon isotope values for Mesoamerican samples indicate that maize typically appears as a dietary staple in the area after about 3000 BP. However, they call attention to some atypical examples. There are two individuals from the Soconusco region of Mexico and Guatemala, dating from 5000–4800 BP, for which stable carbon isotopes ratios show high maize consumption [39]. High grain consumption is also in evidence for Tehuacán individuals dating from 8800–7000 BP. Thus, glottochronological dates older than 3000 BP pertaining to Southern Uto–Aztecan and Mayan accord with some of the stable carbon isotope findings for Mesoamerica.[21]

Western Tucanoan, Eastern Chibchan, Misulmapan, and Maipurean are genetic groups of the South America category. The original glottochronological date for Western Tucanoan (see Table 20-4) is based on a highly truncated list of vocabulary items and is not to be uncritically accepted (see Glottochronology section in this chapter). Original dates for the other three groups are based on complete core vocabulary lists following Swadesh.

Languages of both Misulmapan and Eastern Chibchan are spoken in southern Central America. One of the stable carbon isotope results among those surveyed by Smalley and Blake [39] is for a site in southern Central America, i.e., Cerro Mangote in Pacific Panama. Isotope ratios for 16 individuals at Cerro Mangote indicate moderate maize consumption between 7000 and 4500 years BP [40]. The adjusted glottochronological dates of 3500 BP and 3400 BP, respectively for Misulmapan and Eastern Chibchan for which probable terms for maize are posited, although older than 3000 BP, are nonetheless in accordance with the Cerro Mangote archaeological results.

The revised date for Maipurean,[22] 3700 BP, is the earliest adjusted date for any genetic group of Table 20-4 for which a probable proto-term for maize is posited. Maipurean, the vast majority of whose member languages are spoken in South America, is the most successful genetic group in the New World with respect to size (number of member languages) and breadth of member language geographical distribution. At present, no stable carbon isotope results suggest high cultural significance for maize anywhere in South America before around 3000 BP. Consequently, the early glottochronological date for Maipurean is not anticipated.

ADJUSTING GLOTTOCHRONOLOGY

Glottochronological dates for genetic groups of North America (north of the U.S.–Mexico border) are in accord with the known chronology of maize for that region without need for any kind of adjustment. On the other hand, although glottochronological dates for groups in Latin America are strongly correlated with archaeological dates for maize in the area, reduction of older glottochronological dates by 800 years brings the vast majority of dates closely in line with our current understanding of when maize began to become a dietary staple in prehistoric Latin America.

Maize became culturally significant in most of North America considerably later than it did in most of Latin America. As a consequence, parent languages of genetic groups and subgroups of North America, if they developed terms for maize at all, acquired them considerably later than did Latin American groups. Thus, ancestral languages of Latin America possessing maize terms tend to be substantially older than ancestral languages of North America having words for maize.

I propose earlier that although there is almost certainly no constant rate of lexical change in the sense that at all points in time all languages show the same rate of lexical replacement, most languages during long periods of time, say, greater than 2500 years, by chance typically undergo about the same magnitude of lexical change. The accuracy of glottochronology could be considerably improved if we could determine the rate of lexical replacement that typically occurs over a period of time greater than 2500 years.

The constant rate of lexical replacement calculated by Swadesh [43, 45] is based on rates of lexical change determined by comparing early texts of historical languages with their modern versions. His constant, then, is based on lexical change occurring over time periods of no more than 2000 years or so. This constant may not be the same as the typical rate of lexical change occurring over time periods greater than 2500 years.

Maize chronology suggests that the constant used by Swadesh [43, 45] in glottochronological analysis and the constant relating to lexical change over a period of time greater than 2500 years are not equivalent. Swadesh's constant seems to work reasonably well for chronologically shallow genetic groups such as those of North America.[23] On the other hand, it does not work as successfully for chronologically deep genetic groups of Latin America. The current study suggests that this complication might be overcome by simply subtracting 800 years from any Swadesh glottochronological date older than 3300 years BP and by

[21]John Staller (personal communication, 2005) notes that there is strong reason to suggest that these Mesoamerican signatures result from intensive consumption of crustaceans and of fresh and marine fish rather than maize.

[22]The name Arawakan is sometimes used as a synonym for Maipurean.

[23]As Bellwood [2, p. 194] notes, most linguists who do not totally reject glottochronology regard the method as being most effective when dealing with the past 500–2500 years.

reducing glottochronological dates from 2600 BP through 3200 BP to 2500 BP.[24]

CONCLUSION

Most glottochronological dates calculated for parent languages of Amerindian genetic groups are closely in line with archaeological dates pertaining to the chronology of *Zea mays* in the Americas. From this, I conclude that glottochronological dates are, at worst, "in the ballpark" and, at best, fairly accurate. Thus, there is reason for believing that glottochronology can contribute productively to the study of prehistory.

For example, 3700 BP, the adjusted glottochronological date for Maipurean, a genetic group for which a probable maize term is posited, can be interpreted as indicating that maize was a dietary staple for some South American Indian groups much earlier than 3000 BP. This date anticipates that stable carbon isotope studies of human bone collagen samples for individuals from archaeological sites in South America that have yet to be excavated will some day reveal high to moderate consumption of maize at chronological depths hundreds of years earlier than 3000 BP.[25]

As noted above, Maipurean is the most successful genetic group of the New World with regard to size and breadth of geographic distribution. Perhaps it is no mere coincidence that this genetic group, which has the oldest glottochronological date of all Amerindian genetic groups for which a probable proto-term for maize is posited (see Table 20-4), has also been so developmentally productive.

Evidence assembled here indicates that, counter to Campbell's [11] call for its rejection, glottochronology should be retained as a tool for historical linguistics, albeit, one of somewhat more circumscribed use than was originally intended by its inventor. Campbell claims that the approach is not accurate. I would agree that glottochronology does not provide dates that approach the precision of those obtained through archaeological dating techniques such as AMS. However, it does provide at least an approximate chronological perspective that can help to augment knowledge of prehistoric languages and cultures.

[24]The reduction number, 800 years, is an educated guess. If the number were 1500 years, this would produce a perfect correlation between glottochronological and archaeological dates. However, such a large number seems unnecessary, especially in view of the fact that for Mesoamerica and southern Central America there are stable carbon isotope readings suggesting high or moderate consumption of maize for some individuals dating earlier than 4800 BP. A reduction number of 800 years leaves a few genetic groups for which maize terms are posited that are older than 3000 BP. These groups can be linked, as I have attempted to do in the immediately preceding section, to the few examples of stable carbon isotope ratios indicating significant consumption of maize earlier than 3000 BP. On the other hand, a reduction number lower than 800 years (e.g., 400 years) would not seem to go very far in reducing the number of genetic groups for which maize terms are posited that are older than 3000 BP. Hopefully, in future studies a more objective method for arriving at a reduction number can be followed.

[25]If so, this would constitute some strong evidence that glottochronology can contribute significantly to understanding human prehistory.

Acknowledgments

A number of individuals in various different ways contributed to this study. I would like to thank Peter Bellwood, Bruce F. Benz, Michael Blake, George Aaron Broadwell, Pamela Brown, Wallace Chafe, Elizabeth Chilton, Joan Brenner Coltrain, Bernard Comrie, Jared Diamond, Major Goodman, Diana M. Greenlee, Martin Haspelmath, Christine A. Hastorf, Jane H. Hill, Eric W. Holman, Anne Katzenberg, Robert Lusteck, Yoshi Matsuoka, Marianne Mithun, Bruce D. Smith, John E. Staller, Brian Stross, Yves Vigouroux, and Søren Wichmann.

This study is dedicated to the memory of my son, Rufus Stennis Brown, who died unexpectedly of natural causes at the age of 32 years while I was preparing this chapter.

References Cited

1. P. Bellwood. (2000). The time depth of major language families: An archaeologist's perspective. In: C. Renfrew, A. McMahon, L. Trask, (Eds.), *Time depth in historical linguistics,* Vol. 1. Cambridge, England: McDonald Institute for Archaeological Research. pp. 109–140.
2. P. Bellwood. (2005). *First farmers: The origins of agricultural societies.* Oxford: Blackwell, Publishing.
3. B. Benz. (1994). Reconstructing the racial phylogeny of Mexican maize: Where do we stand? In: S. Johannessen, C. A. Hastorf, (Eds.) *Corn and culture in the prehistoric New World.* Boulder, CO: Westview Press. pp. 157–180.
4. B. Benz. (1999). On the origin, evolution, and dispersal of maize. In: M. Blake, (Ed.) *Pacific Latin America in prehistory: The evolution of Archaic and Formative Cultures.* Pullman: Washington State University Press. pp. 25–38.
5. B. Berlin, D. E. Breedlove, R. M. Laughlin, P. H. Raven. (1973). Cultural significance and lexical retention in Tzeltal-Tzotzil ethnobotany. In: M. S. Edmonson, (Ed.) *Meaning in Mayan languages.* The Hague: Mouton. pp. 143–163.
6. R. Blust. (2000). Why lexcostatistics doesn't work: The "universal constant" hypothesis and the Austronesian languages. In: C. Renfrew, A. McMahon, L. Trask, (Eds.), *Time depth in historical linguistics.* Vol. 2. Cambridge, England: McDonald Institute for Archaeological Research. pp. 311–331.
7. G. A. Broadwell. (2005). Reconstructing Proto-Muskogean language and prehistory: Preliminary results. Available at: www.albany.edu/anthro/fac/broadwell/ flora.pdf+Muskogean+glottochronology&hl=en.
8. C. H. Brown. (1999). *Lexical acculturation in Native American languages.* New York: Oxford University Press.
9. C. H. Brown, S. Wichmann. (2004). Proto-Mayan syllable nuclei. *International Journal of American Linguistics, 70,* 128–186.
10. L. Campbell. (1997). *American Indian languages: The historical linguistics of Native America.* New York: Oxford University Press.
11. L. Campbell. (1998). *Historical linguistics: An introduction.* Cambridge, MA: The MIT Press.
12. R. M. W. Dixon. (1997). *The rise and fall of languages.* Cambridge, MA: Cambridge University Press.
13. C. Ehret. (2000). Testing the expectations of glottochronology against the correlations of language and archaeology in Africa. In: C. Renfrew,

A. McMahon, L. Trask, (Eds.), *Time depth in historical linguistics*. Vol. 2. Cambridge, England: McDonald Institute for Archaeological Research. pp. 373–399.
14. S. Embleton. (2000). Lexicostatistics/glottochronology: From Swadesh to Sankoff to Starostin to furture horizons. In: C. Renfrew, A. McMahon, L. Trask, (Eds.), *Time depth in historical linguistics*. Vol. 1. Cambridge, England: McDonald Institute for Archaeological Research. pp. 143–165.
15. C. S. Fowler. (1983). Some lexical clues to Uto-Aztecan prehistory. *International Journal of American Linguistics, 49,* 224–257.
16. G. J. Fritz. (1995). New dates and data on early agriculture: The legacy of complex hunter-gatherers. *Annals of the Missouri Botanical Garden, 82,* 3–15.
17. J. Greenberg. (1987). *Language in the Americas*. Stanford, CA: Stanford University Press.
18. C. A. Hastorf, S. Johannessen. (1994). Cultural meanings: Introduction to part four. In: S. Johannessen, C. A. Hastorf, (Eds.), *Corn and culture in the prehistoric New World*. Boulder, CO: Westview Press. pp. 395–398.
19. J. H. Hill. (2001). Proto-Uto-Aztecan: A community of cultivators in central Mexico? *American Anthropologist, 103,* 913–934.
20. E. W. Holman. (2004). Why are language families larger in some regions than in others? *Diachronica, 21,* 57–84.
21. R. Q. Huber, R. B. Reed. (1992). *Comparative vocabulary: Selected words in indigenous languages of Colombia*. Santafé de Bogotá, Colombia: Instituto Lingüístico de Verano.
22. S. Jaxontov. (1999). Glottochronology: Difficulties and perspectives. In: V. Shevoroshkin, P. J. Sidwell, (Eds), *Historical linguistics and lexicostatistics*. Melbourne, Australia: Association for the History of Language. pp. 51–59.
23. T. Kaufman. (1990). Language history in South America: What we know and how to know more. In: D. L. Payne, (Ed.) *Amazonian linguistics: Studies in lowland South American languages*. Austin: University of Texas Press. pp. 13–73.
24. T. Kaufman. (1994a). The native languages of Mesoamerica. In: C. Moseley, R. E. Asher, (Eds.), *Atlas of the world's languages*. London: Routledge. pp. 34–45.
25. T. Kaufman. (1994b). The native languages of South America. In: C. Moseley, R. E. Asher, (Eds.), *Atlas of the world's languages*. London: Routledge. pp. 46–76.
26. M. Lohr. (2000). New approaches to lexicostatistics and glottochronology. In: C. Renfrew, A. McMahon, L. Trask, (Eds.), *Time depth in historical linguistics*. Vol. 1. Cambridge, England: McDonald Institute for Archaeological Research. pp. 209–222.
27. Č. Loukotka. (1968). *Classification of South American Indian languages*. Los Angeles: Latin American Center, University of California.
28. Y. Matsuoka, Y. Vigouroux, M. M. Goodman, J. Sanchez, G. E. Buckler, J. Doebley. (2002). A single domestication for maize shown by multilocus microsatellite genotyping. *Proceedings of the National Academy of Sciences of the United States of America, 99,* 6080–6084.
29. A. Militarev. (2000). Towards the chronology of Afrasian (Afroasiatic) and its daughter languages. In: C. Renfrew, A. McMahon, L. Trask (Eds.), *Time depth in historical linguistics*. Vol. 1. Cambridge, England: McDonald Institute for Archaeological Research, pp. 267–307.
30. P. J. Munson. (1973). The origins and antiquity of maize-beans-squash agriculture in Eastern North America: Some linguistic implications. In: D. W. Lathrap, J. Douglas, (Eds.), *Variation in anthropology: Essays in honor of John C. McGregor*. Urbana: Illinois Archaeological Survey. pp. 107–135.
31. L. Norr. (1995). Interpreting dietary maize from bone stable isotopes in the American tropics: The state of the art. In: P. W. Stahl, (Ed.), *Archaeology in the lowland American tropics: Current analytical methods and recent applications*. Cambridge, MA: Cambridge University Press. pp. 198–223.
32. D. M. Pearsall, D. R. Piperno. (1990). Antiquity of maize cultivation in Ecuador: Summary and reevaluation of the evidence. *American Antiquity, 55,* 324–337.
33. D. R. Piperno. (1999). The origins and development of food production in Pacific Panama. In: M. Blake, (Ed.), *Pacific Latin America in prehistory: The evolution of Archaic and Formative cultures*. Pullman: Washington State University Press. pp. 123–134.
34. D. R. Piperno, K. V. Flannery. (2001). The earliest archaeological maize (*Zea mayz* L.) from Highland Mexico: New accelerator mass spectrometry dates and their implications. *Proceedings of the National Academy of Sciences of the United States of America, 98,* 2102–2103.
35. D. R. Piperno, D. M. Pearsall. (1998). *The origins of agriculture in the lowland Neotropics*. San Diego: Academic Press.
36. D. R. Piperno, A. J. Ranere, I. Holst, P. Hansell. (2000). Starch grains reveal early root crop horticulture in the Panamanian tropical forest. *Nature, 407,* 894–897.
37. C. Renfrew, A. McMahon, L. Trask, (Eds.), *Time depth in historical linguistics*. Vols. 1 and 2. Cambridge, England: The McDonald Institute for Archaeological Research.
38. T. J. Riley, G. R. Walz, C. J. Bareis, A. C. Fortier, K. E. Parker. (1994). Accelerator mass spectrometry (AMS) dates confirm early *Zea mays* in the Mississippi River Valley. *American Antiquity, 59,* 490–498.
39. J. Smalley, M. Blake. (2003). Sweet beginnings: Stalk sugar and the domestication of maize. *Current Anthropology, 44,* 675–703.
40. J. E. Staller, R. G. Thompson. (2002). A multidisciplinary approach to understanding the initial introduction of maize to Coastal Ecuador. *Journal of Archaeological Science, 29,* 33–50.
41. S. Starostin. (2000). Comparative-historical linguistics and lexicostatistics. In: C. Renfrew, A. McMahon, L. Trask, (Eds.), *Time depth in historical linguistics*. Vol. 1. Cambridge, England: McDonald Institute for Archaeological Research. pp. 223–265.
42. M. Swadesh. (1959). *Mapas de Clasificación Lingüística de México y las Américas*. Mexico, D.F.: Universidad Nacional Autonoma de Mexico.
43. M. Swadesh. (1960). *Estudios sobre Lengua y Cultura*. Mexico, D.F.: Instituto Nacional de Antropologia e Historia.
44. M. Swadesh. (1971). *The origin and diversification of language*. Chicago: Aldine Atherton. pp. 271–284.
45. R. H. Tykot, J. E. Staller. (2002). The importance of early maize agriculture in coastal Ecuador: New data from La Emerenciana. *Current Anthropology, 43,* 666–677.
46. P. J. Watson.(1985). The impact of early horticulture in the upland drainages of the Midwest and Midsouth. In: R. I. Ford, (Ed.), *Prehistoric food production in North America*. Ann Arbor: Museum of Anthropology, University of Michigan. pp. 99–148.
47. S. Wichmann. (n.d.) Neolithic linguistics. In: G. Barjamovic, I. Elmerot, A. Hyllested, B. Nielsen, and B. O. Skuarup. (Eds.), *Language and prehistory of the Indo-European peoples: A cross-disciplinary perspective*. Budapest: Archaeolingua.
48. S. Wichmann, D. Beliaev, A. Davletshin. (2005). Posibles correlaciones lingüísticas y arqueológicas involucrando a los olmecas. Paper presented at the proceedings of the Mesa Redonda Olmeca: Balance y Perspectivas, Museo Nacional de Antropología, México City, March 10–12, 2005.

The Antiquity, Biogeography and Culture History of Maize in Mesoamerica

BRUCE F. BENZ* AND JOHN E. STALLER

*Texas Wesleyan University
†The Field Museum

The Culture History of Maize in the Americas: Contextual Considerations
Antiquity: First appearances
Biogeography: Dispersal and racial diversification
Culture history – staple, variety and cultural acceptance
Dietary reliance
Maize in Language, Myth and Legend
Extinction and Cultural Impact

Glossary:

Domestication genetic change in population due to interaction with humans that leads to a dependence relation.
Agriculture the mutual dependence of crop plant and humans
Anagenesis the persistence of one or a suite of biological traits that over time leads to varietal divergence
Cladogenesis the development of evolutionary novelty through the extinction of preexisting forms
Dietary reliance subsistent dependence on a specific crop or trophic level.

THE CULTURE HISTORY OF MAIZE IN MESOAMERICA: CONTEXTUAL CONSIDERATIONS

Over fifty years ago, Richard S. MacNeish was engaged in the search for the origins of settled village life in and Kent V. Flannery was piecing together the transition from food collection to food production [7, 16; see Chapter 1]. Their legacy of published research and museum collections provides the intellectual foundation, geographic framework and methodological guidelines for research focused on plant and animal domestication and the origin of agriculture. Their efforts continue to reward those of us who seek to understand the beginnings of food production in the Americas. Our understanding of the processes underlying domestication in the Americas can not be adequately addressed without detailed data on when, where, how and why maize (and other cultigens) originated and adaptively radiated to produce the variety of geographically distinct races of maize that penetrated virtually all arable habitats of the New World. As the contributions to this volume and its predecessor attest, we continue to strive to understand these processes through integration of anthropological and biological research efforts. MacNeish and Flannery provided the theoretical and methodological foundations for addressing many of the hypotheses being tested by authors of this volume. The scholarship presented in the preceding chapters provides contributions, some with decidedly different answers to questions MacNeish, Flannery and many others had postulated or anticipated. We nevertheless acknowledge their path-breaking efforts and contributions.

Our intention here is to first sketch a rough outline of the contributors' ideas, hypotheses or conclusions, and second, to identify general consensus and disagreement where such exist. We approach this summary in an objective fashion even while we are fully cognizant that such objectivity is simply a rationalization of our own deceptions. Nevertheless this summary is grounded in empirical evidence provided by the volume's contributors allowing the reader to independently identify our own rationalizations. This summary uses the contributions to focus on three principal areas of interest: 1) a culture history of maize in Mesoamerica, 2) the biological patterns and processes of maize's evolution in Mesoamerica, and 3) a brief, discussion of the impact of extinction on maize populations and its impact on our understanding of how the prehistoric evidence can be reconciled with the extant populations of maize.

ANTIQUITY: FIRST APPEARANCES

Perhaps the most persistent questions posed by MacNeish [16] and recently answered in part by Flannery [7] surrounds the antiquity of maize cultivation in the New World. The contributors to this volume provide compelling evidence that document human experimentation with *Zea mays* (maize and spontaneous subspecies) throughout the Americas. The widely accepted dates on maize remains provide a framework of our current understanding and identify where additional information is needed and hypotheses concerning discrepancies or confusing relationships can be postulated. We ask the reader to recognize that the contributors have included dates that are based on a variety of chronology-building tools. Thus, absolute dates based on AMS analyses of radiocarbon in maize are combined with conventional dates of stratigraphically associated organic material, as well as relative dates based on ceramic seriation, architectural typology, as well as glottochronology based on cognates, loan words, and myth. Citation of the author's work serves to index the literature cited therein. For detailed information, the cited work should be consulted.

In the simplest and most general terms, the archaeological and paleoecological evidence specifically points to south central Mesoamerica as the site of maize's origin [30, see Chapter 4]. This is anticipated given the abundance of genetic evidence (Chapters 2 and 3) that point to Balsas teosinte in southern highland Mexico as the ancestor of maize. On this count, the linguistic evidence is corroborated by the archaeological and genetic evidence (Chapter 18, 19, and 20). The various lines of evidence however do not agree, on the antiquity, or geography of dispersal.

Macrobotanical archaeological evidence is unequivocal on the antiquity of maize. A reasonably facsimile of maize is present at Guilá Naquitz in the Valley of Oaxaca at 4290 cal B.C. (see Chapters 2, 3, and 4). Maize is common in assemblages from the Tehuacan Valley of Puebla, Mexico by 3550 cal B.C. and remains abundant throughout the sequence from its earliest presence to post-Columbian time (Chapter 5). It makes its first appearance outside of the Tehuacán – Oaxaca Valley area far to the north in the present-day state of Tamaulipas 1000 years later, 2400 cal B.C., and appears at the same time or at most 200 years later in the American Southwest (see Chapter 3). The movement southeastward into the Maya area, Central and South America is inferred to have occurred around 3030 cal. B.C. in Costa Rica and along coastal plain of Chiapas (Chapters 8, 9, and Chapter 14) within 2100 years, or by 1640 cal. B.C. If the macrobotanical evidence is an accurate measure of relative antiquity, the implication is that maize moved more quickly to the north, rather than south or east which is consistent with the genetic evidence (see Chapter 2).

Maize pollen provides a slightly different chronological history with regard to its dispersal (see Chapters 9, and Chapter 10). While the macrobotanical evidence is chronologically consistent, it may not be accurate about the time of maize's domestication from teosinte because genetic estimates of divergence, pollen and phytoliths all suggest domestication and dispersal occurred thousands of years prior to the earliest macrobotanical evidence. Maize pollen appears to be slightly earlier on the Mesoamerican Gulf Coast than any type of evidence anywhere except maize phytoliths in Central America (see below). Despite the morphological evidence which shows the Guilá Naquitz cobs have greater similarity to teosinte than maize (distichous, two of three have a single spikelet per node), *Zea* pollen is consistently earlier than macrobotanical evidence in nearly all areas of the Americas where both have been identified (Chapters 2, 4, and Chapters 8 and 9). Maize pollen precedes macrobotanical remains by perhaps as little as 1500 years in Oaxaca. Elsewhere, in the highlands of Mesoamerica, *Zea* pollen is coeval with teosinte fruitcases, but much earlier than maize (Chapter 7) while it is consistently earlier (100 to 1500 years) at lower, more tropical elevations, than the macrobotanical evidence recovered from mid-elevation sites.

These chronological relationships are apparent when we consider the presence of maize in North America north of Mexico, in Central America and South America. Where macrobotanical evidence is available nearby, it is coeval or later in Central America (see Chapters 7, 8, and 9) and much later in South America (Chapter 8) Costa Rica represents the exception, in that the earliest macrobotanical evidence is accompanied by pollen (Chapter 9). Phytoliths and *Zea* pollen in more recent contexts in Honduras and El Salvador consistently predate the macrobotanical evidence (see Chapter 4). A similar pattern is evident in the Andes, where the macrobotanical remains from the Lake Titicaca Basin are consistently later in time by almost millennia than maize phytoliths identified in food residues from ancient pottery (Chapter 4).

Several concerns need to be addressed in the future concerning the apparent chronological discrepancy between macrobotanical and pollen evidence: context of discovery, contextual perturbation and microfossil identification. With the exception of pollen from Guilá Naquitz (Chapter 8), all the earliest pollen evidence for *Zea* from the Americas comes from lacustrine or littoral contexts. In nearly all cases, the earliest pollen grains discovered are generally smaller in diameter than later grains (see Chapters 2, 4, 8, 9). The same contexts rarely ever yield pollen and macrobotanical evidence. Where they do, the macrobotanical remains are later. The lack of co-occurrence may be expected because one does not expect corncobs in lake-bottom deposits. Nevertheless, where both have been recovered, a chronological discrepancy occurs in every case.

Recent research has focused on clarifying this apparent disparity of maize phytoliths from decidedly early contexts far outside the current geographic range of maize's ancestor. Pollen from early stratigraphic estuarine contexts from the Gulf Coast has been convincingly identified as *Zea*, suggesting an either more widespread distribution of teosinte than currently exists or early incursions of maize cultivators into the lowland tropical forests of Mesoamerica. The inability to distinguish maize from teosinte on the basis of pollen grain size [11] leaves one to question as to which hypothesis the evidence might support. These data also pose a serious problem as to when and where maize first appears in the tropical lowlands of Mesoamerica, those regions where most of the pollen research on maize origins has been undertaken [chapters 8 and

9; see also 11). Moreover, the earliest evidence of alteration of the landscape around the lakebed near Tuxpan dates to 3000 BP [25:11881]. Similarly, paleoecological data from Central Balsas show signs of maize microfossils and herbaceous disturbance taxa indicative of agriculture dated solely to 4100 BP [*ibid.*]. Proponents of each hypothesis find supporting evidence in the literature. Remarkably, pollen from lakebeds and a swamp near Iguala, Guerrero confirms the presence of *Zea* 7.25 meters below the surface (dating by association to between 11200 – 9000 years cal. BP). Zea pollen associated with a significant stratigraphic (both sediments and pollen spectra change demonstrably) transition in lake sediments around 7000 year cal. BP is suspected to be maize though pollen grain size cannot demonstrate this to be the case, and phytoliths do not exist to test this hypothesis [25].

Excavation at the Xihuatoxtla rockshelter [26, 27] yielded starch from stone tools and phytoliths from stratigraphic deposits the authors claim represent maize was domesticated 8700 cal BP. Maize and teosinte inflorescences are distinguished using microfossils based on presence of certain morphologies of short-cell phytoliths (rondels and surface sculpturing) and based on the relative proportion of larger cross-shaped phytoliths. Maize and teosinte stalks are distinguished based on the occurrence of deeply notched bilobate phytoliths and some other undescribed phytoliths seen in maize but not in teosinte. Maize and teosinte grain starch granules are similar but permit infraspecific distinction based on size – maize are slightly larger that teosinte – and irregular shapes and facets – maize granules are irregularly shaped and present compression facets. The wealth of evidence documenting the use of teosinte grains is derived from an unspecified number of cob-type, and wavy and ruffle-top rondel phytoliths from Zones D and E and approximately 200 stratigraphically associated starch grains from ground and chipped stone tools. No macrobotanical evidence for teosinte nor maize was recovered [27]. This hampers our efforts to understand human motivations for domesticating teosinte and document prehistoric human intentionality. The absence of macrofossil evidence such as cobs, fruitcases, stems or other recognizable plant part at Xihuatoxtla is perplexing though documented in other rockshelters in Mesoamerica. We are left to contemplate whether the quality of evidence [10] is inadequate or conclude that rockshelters in western Mexico simply are not sites of harvesting, processing and seed selection.

Why do maize microfossils dated by association consistently predate the macrobotanical evidence? It may be due to the fact that pollen is more widely distributed, thus, more readily encountered in lacustrine and littoral contexts, while macrobotanical remains occur almost exclusively in archaeological contexts. Such archaeological deposits are generally local isolates, and thus, are rare, and difficult to locate even though they are more contextually secure. Furthermore, the higher probability of finding widely dispersed material is sufficiently great we should expect to find *Zea* pollen more readily than macrobotanical remains. On the other hand, even though pollen might be more widely dispersed than macrobotanical evidence when appropriate contexts are sampled, the opportunity for perturbation and its displacement through the profile is as likely as macrobotanical remains in a multicomponent site. Yet, few paleoethnobotanists deal with the problem of identifying the geomorphological processes involved in fluvial sedimentation. Moreover, aquatic vertebrates nest and benthic invertebrates are as prone to burrow in lacustrine or fluvial deposits, as terrestrial vertebrates are to nest and burrow in open sites and rockshelters. Horn (Chapter 9) and Dull (Chapter 8) both wisely express their concerns for systematic treatment and analysis of pollen for appropriate comparison of pollen grain size of *Zea* taxa, both among sites and laboratories. Moreover, the pollen of maize and Zea mays ssp parviglumis overlap in size [25]. Unfortunately, paleoethnobotanists working with phytoliths from archaeological soils have yet to express a similar methodological concern.

Phytolith analysis of maize origin and dispersal now comprises a significant and widely cited literature [see e.g., 20, 21, 23, 24]. As mentioned above, this literature posits quite a different scenario for the origin and initial dispersal of maize (Chapters 2 and 4). The earliest *Zea* phytoliths dated by association appear at archaeological sites in Guerrero, Mexico and Panama and Ecuador, and are coeval or earlier than domesticated teosinte from Guilá Naquitz and significantly earlier — 650 years and 650 to 2250 years respectively — than maize from the Tehuacán caves [19, 20, 21, 23]. As maize spreads out of Mesoamerica much later in time, directly dated phytoliths predate macrobotanical remains. Directly dated Zea phytoliths from food residues provide date ranges consistent to a much greater extent with the presence of early maize macrobotanical remains in different regions of the Americas [see e.g., Chapter 4). Laden (Chapter 5) expresses concern for accuracy in the identification of maize landraces using rondel phytoliths, because of the small ethnographic sample sizes of maize varieties. The assumed length of time any given ethnographic variety is believed to have existed and can be identified from archaeological material is complicated by natural and human selection pressure through time and the archaeological and historical transformations that phytolith assemblages might have been subject to prior to recovery and analysis (see below). Much would be accomplished by continued research with rondel phytoliths and phytoliths in general from food residue samples and sediments by currently active analysts, but especially by specialists with less published research at stake.

BIOGEOGRAPHY: DISPERSAL AND RACIAL DIVERSIFICATION

Most authors have elected to avoid taxonomic terminology in reference to maize (Chapter 3), though some apparently find no better analog for archaeological populations than extant geographical landraces. Justification of avoiding race names to refer to archaeological maize seems warranted, excepting perhaps the Northern Flints. Nevertheless, tracing morphological variation, and the evolutionary process it represents (anagenic versus cladogenetic) through time and across space is productive and can lead to a greater understanding of the dispersal of landraces and/or human migration and markets (see Chapter 7). Maize races appear to evolve through cultural mechanisms, i.e., human selection for certain traits or characteristics (see Chapters 3 and 7) as well as through biogeographic-based selection pressures. The importance of landraces in chapters dealing with certain areas apparently derives from the close ethnic association with certain places and to particular ethnic groups within those regions (Chapter 11). Yet, we are only now just beginning to appreciate and characterize morphological variation in a manner in which the relevance of geography to cultural significance can be tested and their consistency with the genetic evidence can be ascertained [8, 18]. The apparent reluctance of some scholars to consider the names given these various landraces is related in part to their origin and association with Mangelsdorf's tripartite model for maize origins, which is no longer tenable. On the other hand, sometimes race names take on a life of their own (Chapter 6).

The dispersal of maize from the Southwest into the Eastern US appears to have occurred only once. The most recent entry into the Eastern US and Northern Great Plains may well represent a classic case of allopatric speciation. Following dispersal north and eastward from the American Southwest, a period of climatic amelioration (ca. A.D. 500 to 1000 and A.D. 1200 to 1300). These paleoclimatic changes allowed maize to find capable and willing cultural husbands who attended, incorporated and developed the crop so that it fit into existing subsistence systems. The genetic evidence indicates it is likely that this second wave of maize dispersal into the eastern US, the so-called radiation of Northern Flints (Eastern Eight Row), was received by indigenous groups that had in many cases previously embraced maize cultivation though it did not appear to already have been developed as an important dietary component. The morphological distinctiveness of the Northern Flints from varieties indigenous to southern and southwestern US and its relatively late introduction were already evident to Hugh Cutler in the late 1960s. Archaeologists, geneticists and botanists working in South America have long sought to develop a similar framework and this is in progress [1, 3].

Characterization of the identity of maize landraces based on archaeological evidence is ongoing. Apart from the efforts to document Northern Flints (Eastern Eight Row, using archaeologically appropriate terminology), archaeologists working in the American Southwest are characterizing morphological change through time. Longer sequences from single sites will enable us to distinguish between anagenic and cladogenic change in maize lineages and address the question surrounding, for example, the origin of Eastern Eight Row (i.e., Northern Flint), the eight-rowed races of west and southeast Mexico and Central America, the pyramidal-shaped ears of races in the highlands of Mexico and Central America, and the races with grenade-shaped ears in the Andes, and the Coroicos of South America as well. Phytolith specialists posit that such taxonomic relationships can be discerned in the archaeological record assuming, of course, that appropriate methodological steps which insure adequate sampling of extant landraces has been accomplished (Chapter 5). And even while documenting similarity will always be a challenge to the analytical techniques, questions about evolutionary relationships of extant landraces and archaeological contexts will remain confounding variables until the genetic basis of racial variation is suitably documented and environmentally induced variation has been accounted for.

CULTURE HISTORY STAPLE, VARIETY AND CULTURAL ACCEPTANCE

Documenting the presence of maize in the archaeological record has improved greatly with the advent of pollen and phytolith analysis [19, 22, 23] however, documenting their relative antiquity as mentioned above, can pose problems. On the other hand, phytoliths from food residues can be directly dated and application of stable isotope geochemistry can provide a measure of their economic role and importance [4, 33, 34, 35, 36]. While phytoliths typically derived from cultural sediments have fanned the flames of controversy surrounding maize's domestication, the silica bodies common in the vegetative organs and chaff of maize have been extracted from ceramic food residues [34] suggesting perhaps that the early use of maize adoption in the Andes and other regions of Central and Mesoamerica included boiling ears or when such residues are found in ancient bottles, consumed as beer or other type of beverage [34, 35, see also 33, 34].

Donald Lathrap [13] was one of the first to challenge the highland-lowland dichotomy outlined by various archaeologists, which perceived complexity as directly associated with food production and agricultural intensification — particularly in relation to the early cultivation of maize in the Mesoamerican highlands. He maintained instead that the tropical forest, rather than the highlands, had chronological primacy in a number of technological and cultural innovations including pottery technology, social stratification, intensive agriculture and the introduction of maize [12]. This provocative revision countered then prevailing currents of thought, and as a consequence generated considerable debate on a number of important theoretical and historical issues of maize origins throughout the Americas. Despite differences of opinion, most scholars envisioned differences in environment as critical to understanding sociocultural development, and interpreted culture change related to maize agriculture as a result of diffusion, migration, or invasion [see 28, pp. 13-24, 39-49]. Nevertheless, it was the theoretical differences among these various scholars that formed a basis for research design and theoretical focus for most of the subsequent archaeological research on the biogeography and economic role of *Zea* to sociocultural development in the Americas [see e.g., 19, 21, 22, 24, 28, 31, 32].

The occurrence of maize remains in archaeological contexts has formed the basis for discussion of the importance of maize to New World archaeological societies. First, the mere presence of maize in cultural deposits or food residues was deemed necessary and sufficient for interpretations that maize formed a critical if not essential role to the prehistoric economy and cultural complexity. Ubiquity indices followed as we attempted to ascertain whether maize was relatively more or less abundant than any other plant or animal unearthed in archaeological sediments..

The rationale for the maize agriculture model proposed by Lathrap was initially based upon a series of premises that were essentially unsupported by the archaeological evidence at the time they were published. His repeated contention that Valdivia represented tropical forest culture, that domesticates such as manioc, and maize were originally cultivated east of the Andes, that such domestic cultigens arrived along the coast and into an already existing food producing economy [12, 14, pp. 13-14, 21] have as yet never been substantiated archaeologically. In order to bolster the single piece of macrobotanical evidence, Lathrap maintained that maize kernel impressions on Valdivia pottery was presumably made by large kernels of an eight-row variety of flint corn [14, Figures 2-3; 15, 33].

A series of lake core samples were taken from the tropical forest region of southeastern Ecuadorian Amazon [5, 6]. Significantly, vertical distributions of maize phytoliths in the lake cores correspond almost exactly with maize pollen distributions [23, p. 173]. Since phytoliths, unlike pollen, are sensitive to changes created by fire, the lake cores show indications of burning forest vegetation in the stratigraphic record, an activity generally associated with slash and burn agriculture [5, 6]. High amounts of particulate carbon were found in core levels dated as far back as 5300 years ago and were said to reflect the introduction of swidden agriculture [5, 22, p. 665; 23, p. 173]. Clear evidence of intensification of agriculture dates to between 2500 to 800 B.P. [5, pp. 303-304; 6]. Again, the associated dates in the core for the earliest microfossils are internally inconsistent with various lines of evidence presented in this volume.

The legacy of the thinking and writings of Donald Lathrap are still apparent in the literature, as various scholars continue to give primacy to the tropical forest and the *cieja de montaña* or Upper Amazon for the introduction of maize into the coast and are also apparent in the previously outlined research in the tropical lowlands of Mesoamerica. Nevertheless, dates from directly and associated dates emphasize a later initial introduction into South America via the Andean coast and a more recent introduction into the lowlands of Soconusco and the Veracruz coast in Mesoamerica.

The application of stable isotope geochemistry to human diets has provided a significant contribution to our understanding of when and how maize achieved importance in the diet of prehistoric people. Study of stable isotopes of carbon and nitrogen at one time suggested an early reliance on maize in Mesoamerica (see Chapters 4 and 14) that has been reassessed to point at either other C^4 plant resources or marine resources as critical dietary elements. We find a similar pattern in Andean South America. Stable isotope signatures of Formative Period (3500-300 B.C.) skeletons; indicate that the integration of maize to the agricultural economy was gradual and developed fully during the final portion of the Formative Period at around 1000 to 800 B.C. [36, pp. 29- 31].

The principal conclusions to be drawn from the phytolith and pollen evidence is that there was a gradual adaptive shift away from maritime resources, to a subsistence economy specialized upon an agricultural way of life in various lowland regions of the Americas. These data however do not reflect the extent to which maize was a catalyst to increased nucleation of prehistoric populations or to developments to more complex forms of social organization. The stable isotope evidence presented herein and elsewhere suggests that the importance of domesticated varieties of maize to the early agricultural economies of Mesoamerica and the Andes was minimal. A strong possibility exists that in its early stages, maize was primarily a ritual plant used to make fermented beer or *chicha* that was consumed in the context of gift giving, ritual feasts and religious ceremonies [34, 35, 36]. Burger and van der Merwe [4 p. 92] emphasize the ritual importance of maize to Andean sociocultural development with reference to the spread of early civilizations in highland Peru;

> "The link between maize cultivation, *chicha* production, and religious ritual may explain why maize was depicted on the bottles discovered at the temples of Kotosh and Chavín de Huántar. Though it cannot be proven at this time, it is possible that these pottery vessels were designed to serve the *chicha* used in ceremonies and feasts in these public centers. Perhaps the decorations on these bottles were not intended to commemorate the preeminence of maize in the daily diet, as Lanning believed, but rather reflected the specific role of maize within the ideology and ritual life of early highland Peruvian civilization."

Maize does not appear to have displaced native cultigens in the early periods because C_4 cultigens were not specifically adapted to these environments without an intensive labor investment in the development of hydraulic technology [4]. In addition, viable local alternatives of C_3 foods were better adapted ecologically and therefore were more highly productive [4 pp. 92-93]. This interpretation was later supported by isotope signatures from various regions of the Andes by Tykot and Staller [31], and are consistent with the isotopic data from Mesoamerica reported in this volume.

A thread which runs through the chapters on paleodiet is that maize appears to have attained the role of economic staple much later in time than had been suggested previously, and that its initial role may have had more to do with social and cultural factors than as a food source. Maize as a staple crop appears to have occurred in Mesoamerica at least by the early first millennium B.C., when it became a major ingredient in the subsistence diet in the American Southwest! The technique has also identified the Maya ruling elite as a maize dependent class within a society that focused on maize production though did not appear to be uniformly dependent on it (see Chapters 12 and 13).

Dietary dependence occurs significantly later - ca. < 2000 years ago - in the higher latitudes of North America and permits recognition of use versus dependence [2]. The relatively early arrival of Eastern Eight Row in the Mississippi Valley, Ohio and Ontario documents use but slow acceptance (ca. A.D. 500 – 900) and later reliance (A.D. 1000 – 1200). Strontium isotopic evidence document the movement of maize from various regions to the Great Houses and Kivas of Chaco Canyon. The long-distance movement and storage, and redistribution of maize reflect its dietary as well as religious and ritual importance. The Anasazi moved maize over one hundred kilometers to the Chaco Canyon in some cases, suggesting that the authority and power of the religious specialists was based at least in part on their ability to bring good harvests and draw increasing quantities and regional varieties of maize to store and redistribute to regional population during period of scarcity. Similar research has recently been suggested for Formative Period Mesoamerica. Like in Mesoamerica, these cultural patterns speak to the religious importance of maize to native populations of the American Southwest — patently obvious to anyone familiar with ethnographically documented sand paintings, myths, legends and symbolism from this region.

The evidence for dietary reliance on maize in Mesoamerica is similar to what has been identified from skeletons in other regions of the Americas. Depending on whether one accepts the early phytoliths evidence for the introduction of *Zea*, the integration of maize into the prehistoric diet is considerably (by well over a millennium) or only slightly later in time. The carbon isotope signatures from Mesoamerica and Central America clearly indicate that maize takes on an economic importance much later than had previously been assumed in the archaeological literature from those areas. This is true in almost every case.

Soconusco, a region in southern coastal Mexico is close geographically to the place of origin for *Zea*, yet stable isotope signatures suggest that maize is not incorporated into the local economy until several thousand years after its appearance in that region (see Chapter 14).

Similarly, in Andean South America, maize does not represent a staple crop, in the bone chemistry until almost two millennia later in time, than has been suggested by associated dates with microfossils. Like dates on early macrobotanical remains from Tehuacán and other early cave and rockshelter sites, directly dated macrobotanicals and residues from ancient pottery all support a much later introduction and a significantly different economic role than had been previously suggested in the literature.

These quantitative data contradict many of long held assumptions about the economic role of maize to the rise of civilization in the Americas, and reinforce the ethnographic evidence and ethnohistoric accounts, which repeatedly tell us that its role was critical to ethnic identification and ceremonial ritual. The ethnohistoric accounts and archaeological research from the Caribbean suggest that it never achieved economic importance as a food staple — although it may have been important to long-distance exchange and recognition of elite status (see Chapter 10). It may have been this importance to status and ritual exchange that explains its apparently rapid dispersal into different regions of the Americas.

Isotopic combined with micro and macrobotanical evidence provide complementary data sets that clearly establish that maize became a dietary staple relatively later in its history of use, than had been assumed previously [17]. By most counts, teosinte might have been used and/or manipulated longer than maize has been domesticated (Chapters 2 and 4), and the combined duration of use and domestication would far exceed the three and a half millennia of reliance on maize in one or more culture areas of the Americas.

MAIZE IN LANGUAGE, LEGEND AND MYTH

The chapters on maize linguistics are surprisingly consistent with what has been presented with other lines of evidence presented in this volume. For example, the relatively late dietary reliance is corroborated by historical linguistics, more specifically, maize glottochronology (Chapter 20), and by its importance in Mayan myth, language, ritual and iconography (see Chapters 16, 17, and 19), as well as by Otomanguean language family member societies (see Chapter 18) and the apparent independent development of maize cultivation vocabularies for all four major linguistic families in Mesoamerica: Uto-Aztecan, Otomanguean, Mixe-Zoque and Mayan (see Chapters 19 and 20).

The linguistic evidence from the three endemic language groups indicates a time depth of no more than 4000 years for maize terminology. Moreover, these data suggest that Proto-Otomanguean speakers were involved in the initial domestication of teosinte (see Chapter 18). This is attested to by a generic term for maize among the Amuzgo that recognizes it as grass, while Mayan speakers tend to separate maize out as distinct from other life form categories, suggesting maize was incorporated into the lexicon after taxonomic developments had established the major subdivisions of the plant kingdom. Generic level recognition by the Amuzgo would suggest they knew or recognized maize as a grass, while Mayan speakers added maize on to a preexisting taxonomic framework that already knew varietal taxa with specialized use. Jane Hill [Chapter 19] has demonstrated that Proto-Uto-Aztecan speakers, like other Mesoamerican families, included maize vocabularies, which is equally intriguing as it also suggests early adoption of maize and maize cultivation by members of all of the four aforementioned language families.

The imagery and hieroglyphic as well as ethnographic information generated by linguistic chapters on Mesoamerican speaking societies infer that maize played a central role in ethnic identity in those regions as well. Moreover, they indicate that its long history as a secondary food source among these societies and those of the Andes speaks to its importance in the manufacture of fermented intoxicants. Maize beer was central to fostering social alliances in the context of long-distance interaction, and Spanish chroniclers remark that the Inca redistributed *aqha* or *chicha* as a form of reciprocity in exchange for corveé labor to their subject populations. The Nahuatl rather than Arawakan roots of the term "*chicha*" implies that such terminology may have more ancient uses in the Andes before the Spaniards spread this word with reference to all kinds of fermented beverages (Stross, personal communication to Staller, 2006). The linguistic data provide another dimension to the puzzle surrounding its early uses and apparently rapid spread throughout different regions of the Americas. They also provide one additional line of evidence by which to understand the stable isotope research.

EXTINCTION

With all the sophistication these contributions provide for understanding the evolution of maize and concomitant cultural developments dependent on maize's preeminence, none contemplate the possibility that some of the gaps in our understanding might stem from a lost roll of the dice, failed attempts to plant when favorable weather conditions didn't present themselves, or the seed saved could not withstand the infestation of graniverous insects, or the onslaught of a exotic rhizophagous beetle or stray fungal spores produce an organism capable of consuming all products of agricultural activity. These problems occur among subsistence agriculturalists in Latin America today and certainly occurred in the past. Fortunately for many, high population densities provide a web of kin groups who can provide replacement seed or work options in exchange for maize to feed the family or community. Replacement seed may have been impossible in the past, in particular during the early stages of domestication when maize's dependence on humans was incipient yet human intentionality was inconstant or fleeting. The lack of replacement seed might also explain the paucity of macrofossil evidence in other preceramic localities in western Mexico. Presently we assume that once initiated, domestication led to agriculture. Local extinction should be an expected phenomenon in subsistence agriculture; with greater repercussions the earlier in time it happened. We might expect that extinction might explain some of the patterns we recognize in the archaeological record of maize's use and evolution throughout the Americas. For example, local extinction might explain the lag in initial appearances of maize in the pollen or phytoliths record (Chapters 4, 8, and 20) or the narrow genetic bottlenecks (Chapter 2). Similarly the apparent time lags – more than 2500 years between the earliest putative maize phytoliths and starch grains in Guerrero and the first indication of domesticated teosinte in Oaxaca, at least 700 years between domesticated teosinte appearing in Guilá Naquitz and the predominance of maize at San Marcos Cave in the Tehuacán Valley, the millennium or longer between maize's appearance in Tehuacán and in Tamaulipas (Chapter 3) is probably a result of repeated extinctions as groups involved in the down-the-line exchange of seed risked it to cultivate maize for the first time or as a result of abandoning it in favor of some less risky subsistence pursuits.

Much of the conjecture concerning the evolution of maize races is steeped in untested theory and misapplied terminology. Those who wish to find extant races in prehistoric contexts subscribe to anagenetic change as the mechanism by which maize races evolve. Persistence over time of races characterized by one or a suite of traits obliges one to conceive of varietal divergence through anagenesis. Stated another way, if one recognizes two distinct forms in one stratigraphic horizon that are preceded stratigraphically by one of the subsequent forms, human and natural selection are invoked as having given rise to one new variety while retaining the distinct predecessor. Punctuated equilibrium is the evolutionary model used to describe this phenomenon while species selection is the mechanism [9]. When we describe a race whose geographic distribution can be explained by invoking geographic phenomena, we have no problem acknowledging the persistence of the ancestral form within its own range. In this context for example, anagenesis explains the existence of two maize races where one existed before. In this scenario, extinction would not appear to be important. The alternative, cladogensis, however, envisions extinction as a major player.

In the cladogenic view, the development of evolutionary novelty, the autapomorphies of evolutionary biology, would lead to the extinction of the ancestor and preexisting form. Technological innovation or social reorganization might be expected to give rise to cladogenic speciation as a product of cultural selection pressure. Extinction plays an important role in this scenario. Endogenous origins to changes in agricultural practices or accommodating hegemonic demands imposed by external polities might be expected to produce cladogenesis that is archaeologically evident by the disappearance of one form and the appearance of another. Applying a maize race name defined from extant populations to archaeological populations is justified if evidence can be produced to justify anagenic change. Application of an extant maize race's taxonomic designation when cladogenesis is anticipated is done so without justification. While both scenarios might be expected to have occurred prehistorically, neither has been appropriately tested using archaeological evidence. This is unfortunate. Inappropriate designation of archaeological maize without explicit testing of morphological difference (cobs, rachis, phytoliths or even pollen) is poor science. We hope appropriate steps will be taken in the future so racial designations are applied appropriately.

The culture history of maize, including considerations of its domestication and antiquity, its dispersal and biogeography, as well as evidence documenting maize's importance to humans and vice versa, was catapulted into the domain of archaeology with the work of MacNeish and Flannery during the 1950's and 60s. Their legacy, formed in part by their various publications and plentiful collections of artifacts and ecofacts from numerous sites in many countries, fosters intellectual pursuits by a wealth of academic talents [e.g., 31, 32]. "Everyone is studying domestication and agriculture," one colleague remarked to Benz while he studied one of numerous collections of archaeological maize by the late Richard MacNeish. While we recognize that his initial, and in fact his final published discussions [17] continue to stimulate research, we are pleased to acknowledge that largely through his efforts and those of Flannery and others, we have material to study and ideas to contemplate. Even while many of the contributions offer alternative explanations to many of the early hypotheses, we recognize the importance of their efforts and applaud their tenacity. We hope the contributions offered here do service to their early and very important efforts.

References Cited

1. E. Anderson (1947). Popcorn. *Natural History* 56(5): 227-230.
2. M. Diehl, (2005). Morphological observations on recently recovered early agricultural period maize cob fragments from southern Arizona. *American Antiquity* 70: 361-375.
3. N.V. Federoff, Transposable Genetic Elements in Maize. *Scientific American* (June 1984) 84-98.
4. R.L. Burger, N.J. van der Merwe, (1990). Maize and the Origin of highland Chavín Civilization. *American Anthropologist* 92: 85-95.
5. M.B. Bush, D.R. Piperno, and P.A. Colinvaux (1989). A 6000 year history of Amazonian maize cultivation. *Nature* 340 303-305.
6. M.B. Bush, P.A. Colinvaux, M.C. Weimann, D.R. Piperno, K.B. Liu (1991). Pleistocene temperature depression and vegetation change in Ecuadorian Amazonia. *Quaternary Research* 34: 330-345
7. K.V. Flannery (1986). The research problem. In K. V. Flannery (Ed.) *Guilá Naquitz: Archaic Foraging and Early Agriculture in Oaxaca, Mexico.* pp. 3-18. Academic Press Orlando.
8. F.O. Freitas, G. Bandel, R.G. Allaby, T.A. Brown, DNA from primitive maize landraces and archaeological remains: implications for the domestication of maize and its expansion into South America, *Journal of Archaeological Science* 30 (2003) 901–908.
9. S.J. Gould (2002). *The Structure of Evolutionary Theory*. The Belknap Press of Harvard University Press, Cambridge, MA.
10. J.R. Harlan, J.M.J deWet (1973). On the quality of evidence for origin and dispersal of cultivated plants. *Current Anthropology* 14: 51-62.
11. I. Holst, J. Moreno, D. Piperno, (2007). Identification of teosinte, maize and *Tripsacum* in Mesoamerica by using pollen, starch grains and phytoliths. *Proceedings of the National Academy of Sciences USA* 104: 17608-17613.
12. D.W. Lathrap (1968). Aboriginal occupations and changes in the river channel on the central Ucayali, Peru. *American Antiquity* 33: 62-79.
13. D.W. Lathrap (1970). *Upper Amazon.* Thames and Hudson Publications
14. D.W. Lathrap, D. Collier, D. H. Chandra (1975). *Ancient Ecuador Culture, Clay, and Creativity 3000-300 BC.*, Field Museum of Natural History. Chicago.
15. D.W. Lathrap, J.G. Marcos, J.E. Zeidler (1977). Real Alto: An Ancient Ceremonial Center. *Archaeology* 30(1): 2-13.
16. R.S. MacNeish (1967). An Interdisciplinary Approach to an Archaeological Problem. In D. Byers (Ed.) *The Prehistory of the Tehuacan Valley Vol. 1: Environment and Subsistence*. The University of Texas Press, Austin.
17. R.S. MacNeish, M.W. Eubanks (2000). Comparative analysis of the Rio Balsas and Tehuacan models for the origin of maize. *Latin American Antiquity* 11(1): 3-20.
18. Y. Matsuoka, Y. Vigouroux, M. Goodman, J. Sanchez G., E. Buckler, J. Doebley (2002). A single domestication for maize shown by multilocus microsatellite genotyping. *Proceedings of the National Academy of Sciences USA* 99: 8060-8064.
19. D.M. Pearsall (1992). The origins of plant cultivation in South America. In C. Wesley Cowan and P.J. Watson (Eds.) *The Origins of Agriculture: An International Perspective*. with the assistance of N.L. Benco pp. 173-205. Smithsonian Institution Press, Washington D.C. and London.
20. Pearsall, D. M. Issues in the analysis and interpretation of archaeological maize in South America. In S.Johannesson, , and C.A. Hastorf (Eds.) (1994). *Corn and Culture in the Prehistoric New World*, Westview Press, Boulder., pp. 245-272.
21. D.M. Pearsall, D. M., and D. R. Piperno (1990). Antiquity of maize cultivation in Ecuador: Summary and reevaluation of the evidence. *American Antiquity* 55 324-337
22. D.R. Piperno (1990) Aboriginal agriculture and land usage in the Amazon Basin, Ecuador. *Journal of the Archaeological Science* 17: 665-677.
23. D.R. Piperno (1991) The status of phytolithic analysis in

the American tropics. *Journal of World Prehistory* 5 (2): 155-191.

24. D.R. Piperno (1994). On the Emergence of Agriculture in the New World. *Current Anthropology* 35: 637-643.
25. D.R. Piperno, J. Moreno, J. Iriarte, I. Holst, M. Lachniet, J.G. Jones, A.J. Ranere, R. Constanzo (2007). Late Pleistocene and Holocene environmental history of the Iguala Valley, Central Balsas Watershed of Mexico. *Proceedings of the National Academy of Sciences USA* 104: 11874-11881.
26. D.R. Piperno, A.J. Ranere, I. Holst, J. Iriarte, R. Dickau, (2009). Starch grain and phytolith evidence for early ninth millennium B.P. maize from the Central Balsas River Valley, Mexico. *Proceedings of the National Academy of Sciences USA* 106: 5019-5024.
27. A.J. Ranere, D.R. Piperno, I. Holst, R. Dickau, J. Iriarte (2009) The cultural and chronological context of early Holocene maize and squash domestication in the Central Balsas River Valley, Mexico. *Proceedings of the National Academy of Sciences USA* 106:5014-5018.
28. A.C. Roosevelt (1980). *Parmana. Prehistoric Maize and Manioc Subsistence along the Amazon and Orinoco*. Academic Press N.Y.
29. C.O. Sauer (1950). Cultivated Plants of South, and Central America. In J. H. Steward (Ed.) *Handbook of South American Indians Volume 6: Physical Anthropology, Linguistics, and Cultural Geography of South American Indians.* pp. 487-543. Smithsonian Institution, Bureau of American Ethnology, Bulletin 143. Washington D.C.
30. J. Smalley, M. Blake (2003). Sweet Beginnings. Stalk Sugar and the Domestication of Maize. *Current Anthropology* 44: 675-703.
31. B.D. Smith, (2001). Documenting plant domestication: the consilience of biological and archaeological approaches. *Proceedings of the National Academy of Sciences USA* 98: 1324-1326.
32. B.D. Smith (2005). Reassessing Coxcatlan Cave and the early history of domesticated plants in Mesoamerica. *Proceedings of the National Academy of Sciences USA* 102: 9438-9444.
33. J.E. Staller (2001). Reassessing the chronological and developmental relationships of the Formative of coastal Ecuador. *Journal of World Prehistory* 15: 193-256.
34. J.E. Staller, R.G. Thompson (2002). A multidisciplinary approach to understanding the initial introduction of maize into coastal Ecuador. *Journal of Archaeological Science* 29: 33-50.
35. R.H. Tykot, J.E. Staller (2002). On the importance of early maize agriculture in coastal Ecuador: New data from the Late Valdivia Phase site of La Emerenciana. *Current Anthropology* 43(4): 666-677.
36. N.J. van der Merwe, J.A. Lee-Thorp, J.S. Raymond (1993). Light, Stable Isotopes and the subsistence base of formative cultures at Valdivia, Ecuador. In J.B. Lambert, G. Grupe (Eds.) *Prehistoric Human Bone Archaeology at the Molecular Level*. Springer Verlag Ltd., pp. 63-97.

Index

For Product Safety Concerns and Information please contact our EU
representative GPSR@taylorandfrancis.com
Taylor & Francis Verlag GmbH, Kaufingerstraße 24, 80331 München, Germany

www.ingramcontent.com/pod-product-compliance
Lightning Source LLC
LaVergne TN
LVHW080648011026
849160LV00014B/1360

* 9 7 8 1 5 9 8 7 4 4 9 6 5 *